STATISTICAL METHODS FOR FINANCIAL ENGINEERING

STATISTICAL METHODS FOR FINANCIAL ENGINEERING

BRUNO RÉMILLARD

CRC Press is an imprint of the
Taylor & Francis Group, an **informa** business
A CHAPMAN & HALL BOOK

CRC Press
Taylor & Francis Group
6000 Broken Sound Parkway NW, Suite 300
Boca Raton, FL 33487-2742

First issued in paperback 2022

Version Date: 20130214

ISBN 13: 978-1-03-247749-7 (pbk)
ISBN 13: 978-1-4398-5694-9 (hbk)

DOI: 10.1201/b14285

Publisher's Note
The publisher has gone to great lengths to ensure the quality of this reprint but points out that some imperfections in the original copies may be apparent.

Library of Congress Cataloging-in-Publication Data

Remillard, Bruno.
Statistical methods for financial engineering / Bruno Remillard.
pages cm
Includes bibliographical references and index.
ISBN 978-1-4398-5694-9 (hardcover : alk. paper)
1. Financial engineering--Statistical methods. 2. Finance--Statistical methods. I. Title.

HG176.7.R46 2013
332.01'5195--dc23 2012050917

Visit the Taylor & Francis Web site at
http://www.taylorandfrancis.com

and the CRC Press Web site at
http://www.crcpress.com

Contents

Preface xxi

List of Figures xxv

List of Tables xxix

1 Black-Scholes Model 1

Summary 1
1.1 The Black-Scholes Model 1
1.2 Dynamic Model for an Asset 2
1.2.1 Stock Exchange Data 2
1.2.2 Continuous Time Models 2
1.2.3 Joint Distribution of Returns 4
1.2.4 Simulation of a Geometric Brownian Motion 4
1.2.5 Joint Law of Prices 5
1.3 Estimation of Parameters 5
1.4 Estimation Errors 6
1.4.1 Estimation of Parameters for Apple 7
1.5 Black-Scholes Formula 9
1.5.1 European Call Option 9
1.5.1.1 Put-Call Parity 10
1.5.1.2 Early Exercise of an American Call Option . 10
1.5.2 Partial Differential Equation for Option Values 11
1.5.3 Option Value as an Expectation 11
1.5.3.1 Equivalent Martingale Measures and Pricing of Options 12
1.5.4 Dividends 13
1.5.4.1 Continuously Paid Dividends 13
1.6 Greeks 14
1.6.1 Greeks for a European Call Option 15
1.6.2 Implied Distribution 16
1.6.3 Error on the Option Value 16
1.6.4 Implied Volatility 19
1.6.4.1 Problems with Implied Volatility 20
1.7 Estimation of Greeks using the Broadie-Glasserman Methodologies 20

1.7.1 Pathwise Method 21
1.7.2 Likelihood Ratio Method 23
1.7.3 Discussion 23
1.8 Suggested Reading 24
1.9 Exercises 24
1.10 Assignment Questions 27
1.A Justification of the Black-Scholes Equation 27
1.B Martingales 28
1.C Proof of the Results 29
1.C.1 Proof of Proposition 1.3.1 29
1.C.2 Proof of Proposition 1.4.1 30
1.C.3 Proof of Proposition 1.6.1 30
Bibliography 30

2 Multivariate Black-Scholes Model 33
Summary 33
2.1 Black-Scholes Model for Several Assets 33
2.1.1 Representation of a Multivariate Brownian Motion . . 34
2.1.2 Simulation of Correlated Geometric Brownian Motions 34
2.1.3 Volatility Vector 35
2.1.4 Joint Distribution of the Returns 35
2.2 Estimation of Parameters 36
2.2.1 Explicit Method 36
2.2.2 Numerical Method 37
2.3 Estimation Errors 37
2.3.1 Parametrization with the Correlation Matrix 38
2.3.2 Parametrization with the Volatility Vector 38
2.3.3 Estimation of Parameters for Apple and Microsoft . . 40
2.4 Evaluation of Options on Several Assets 41
2.4.1 Partial Differential Equation for Option Values 41
2.4.2 Option Value as an Expectation 42
2.4.2.1 Vanilla Options 43
2.4.3 Exchange Option 43
2.4.4 Quanto Options 44
2.5 Greeks 47
2.5.1 Error on the Option Value 47
2.5.2 Extension of the Broadie-Glasserman Methodologies for Options on Several Assets 48
2.6 Suggested Reading 50
2.7 Exercises 51
2.8 Assignment Questions 53
2.A Auxiliary Result 54
2.A.1 Evaluation of $E\left\{e^{aZ}\mathcal{N}(b+cZ)\right\}$ 54
2.B Proofs of the Results 54
2.B.1 Proof of Proposition 2.1.1 54

2.B.2 Proof of Proposition 2.2.1 55
2.B.3 Proof of Proposition 2.3.1 56
2.B.4 Proof of Proposition 2.3.2 56
2.B.5 Proof of Proposition 2.4.1 57
2.B.6 Proof of Proposition 2.4.2 59
2.B.7 Proof of Proposition 2.5.1 59
2.B.8 Proof of Proposition 2.5.3 59
Bibliography 61

3 Discussion of the Black-Scholes Model 63
Summary 63
3.1 Critiques of the Model 63
3.1.1 Independence 63
3.1.2 Distribution of Returns and Goodness-of-Fit Tests of Normality 66
3.1.3 Volatility Smile 68
3.1.4 Transaction Costs 68
3.2 Some Extensions of the Black-Scholes Model 69
3.2.1 Time-Dependent Coefficients 69
3.2.1.1 Extended Black-Scholes Formula 70
3.2.2 Diffusion Processes 70
3.3 Discrete Time Hedging 72
3.3.1 Discrete Delta Hedging 73
3.4 Optimal Quadratic Mean Hedging 74
3.4.1 Offline Computations 74
3.4.2 Optimal Solution of the Hedging Problem 75
3.4.3 Relationship with Martingales 76
3.4.3.1 Market Price vs Theoretical Price 76
3.4.4 Markovian Models 77
3.4.5 Application to Geometric Random Walks 77
3.4.5.1 Illustrations 79
3.4.6 Incomplete Markovian Models 83
3.4.7 Limiting Behavior 89
3.5 Suggested Reading 89
3.6 Exercises 90
3.7 Assignment Questions 92
3.A Tests of Serial Independence 93
3.B Goodness-of-Fit Tests 94
3.B.1 Cramér-von Mises Test 95
3.B.1.1 Algorithms for Approximating the P-Value . 95
3.B.2 Lilliefors Test 96
3.C Density Estimation 96
3.C.1 Examples of Kernels 97
3.D Limiting Behavior of the Delta Hedging Strategy 97
3.E Optimal Hedging for the Binomial Tree 98

3.F A Useful Result . . . 100
Bibliography . . . 100

4 Measures of Risk and Performance **103**
Summary . . . 103
4.1 Measures of Risk . . . 103
4.1.1 Portfolio Model . . . 103
4.1.2 VaR . . . 104
4.1.3 Expected Shortfall . . . 104
4.1.4 Coherent Measures of Risk . . . 105
4.1.4.1 Comments . . . 106
4.1.5 Coherent Measures of Risk with Respect to a Stochastic Order . . . 107
4.1.5.1 Simple Order . . . 107
4.1.5.2 Hazard Rate Order . . . 107
4.2 Estimation of Measures of Risk by Monte Carlo Methods . . . 108
4.2.1 Methodology . . . 109
4.2.2 Nonparametric Estimation of the Distribution Function 109
4.2.2.1 Precision of the Estimation of the Distribution Function . . . 109
4.2.3 Nonparametric Estimation of the VaR . . . 111
4.2.3.1 Uniform Estimation of Quantiles . . . 113
4.2.4 Estimation of Expected Shortfall . . . 114
4.2.5 Advantages and Disadvantages of the Monte Carlo Methodology . . . 116
4.3 Measures of Risk and the Delta-Gamma Approximation . . . 116
4.3.1 Delta-Gamma Approximation . . . 117
4.3.2 Delta-Gamma-Normal Approximation . . . 117
4.3.3 Moment Generating Function and Characteristic Function of Q . . . 118
4.3.4 Partial Monte Carlo Method . . . 119
4.3.4.1 Advantages and Disadvantages of the Methodology . . . 120
4.3.5 Edgeworth and Cornish-Fisher Expansions . . . 120
4.3.5.1 Edgeworth Expansion for the Distribution Function . . . 120
4.3.5.2 Advantages and Disadvantages of the Edgeworth Expansion . . . 121
4.3.5.3 Cornish-Fisher Expansion . . . 121
4.3.5.4 Advantages and Disadvantages of the Cornish-Fisher Expansion . . . 122
4.3.6 Saddlepoint Approximation . . . 122
4.3.6.1 Approximation of the Density . . . 123
4.3.6.2 Approximation of the Distribution Function 124

4.3.6.3 Advantages and Disadvantages of the Methodology . . . 124
4.3.7 Inversion of the Characteristic Function . . . 125
4.3.7.1 Davies Approximation . . . 125
4.3.7.2 Implementation . . . 125
4.4 Performance Measures . . . 126
4.4.1 Axiomatic Approach of Cherny-Madan . . . 126
4.4.2 The Sharpe Ratio . . . 127
4.4.3 The Sortino Ratio . . . 127
4.4.4 The Omega Ratio . . . 128
4.4.4.1 Relationship with Expectiles . . . 128
4.4.4.2 Gaussian Case and the Sharpe Ratio . . . 129
4.4.4.3 Relationship with Stochastic Dominance . . . 130
4.4.4.4 Estimation of Omega and $\bar{G}$ . . . 130
4.5 Suggested Reading . . . 131
4.6 Exercises . . . 131
4.7 Assignment Questions . . . 134
4.A Brownian Bridge . . . 134
4.B Quantiles . . . 135
4.C Mean Excess Function . . . 135
4.C.1 Estimation of the Mean Excess Function . . . 136
4.D Bootstrap Methodology . . . 136
4.D.1 Bootstrap Algorithm . . . 136
4.E Simulation of $\mathbb{Q}_{F,a,b}$ . . . 137
4.F Saddlepoint Approximation of a Continuous Distribution Function . . . 137
4.G Complex Numbers in MATLAB . . . 138
4.H Gil-Pelaez Formula . . . 139
4.I Proofs of the Results . . . 139
4.I.1 Proof of Proposition 4.1.1 . . . 139
4.I.2 Proof of Proposition 4.1.3 . . . 140
4.I.3 Proof of Proposition 4.2.1 . . . 141
4.I.4 Proof of Proposition 4.2.2 . . . 141
4.I.5 Proof of Proposition 4.3.1 . . . 142
4.I.6 Proof of Proposition 4.4.1 . . . 143
4.I.7 Proof of Proposition 4.4.2 . . . 143
4.I.8 Proof of Proposition 4.4.4 . . . 144
Bibliography . . . 144

5 Modeling Interest Rates **147**
Summary . . . 147
5.1 Introduction . . . 147
5.1.1 Vasicek Result . . . 147
5.2 Vasicek Model . . . 148
5.2.1 Ornstein-Uhlenbeck Processes . . . 149

5.2.2 Change of Measurement and Time Scales 149
5.2.3 Properties of Ornstein-Uhlenbeck Processes 150
5.2.3.1 Moments of the Ornstein-Uhlenbeck Process 150
5.2.3.2 Stationary Distribution of the Ornstein-Uhlenbeck Process 151
5.2.4 Value of Zero-Coupon Bonds under a Vasicek Model . 151
5.2.4.1 Vasicek Formula for the Value of a Bond . . 152
5.2.4.2 Annualized Bond Yields 152
5.2.5 Estimation of the Parameters of the Vasicek Model Using Zero-Coupon Bonds 153
5.2.5.1 Measurement and Time Scales 154
5.2.5.2 Duan Approach for the Estimation of Non Observable Data 154
5.2.5.3 Joint Conditional Density of the Implied Rates 155
5.2.5.4 Change of Variables Formula 156
5.2.5.5 Application of the Change of Variable Formula to the Vasicek Model 156
5.2.5.6 Precision of the Estimation 158
5.3 Cox-Ingersoll-Ross (CIR) Model 160
5.3.1 Representation of the Feller Process 160
5.3.1.1 Properties of the Feller Process 162
5.3.1.2 Measurement and Time Scales 163
5.3.2 Value of Zero-Coupon Bonds under a CIR Model . . . 163
5.3.2.1 Formula for the Value of a Zero-Coupon Bond under the CIR Model 164
5.3.2.2 Annualized Bond Yields 165
5.3.2.3 Value of a Call Option on a Zero-Coupon Bond 165
5.3.2.4 Put-Call Parity 166
5.3.3 Parameters Estimation of the CIR Model Using Zero-Coupon Bonds 166
5.3.3.1 Measurement and Time Scales 167
5.3.3.2 Joint Conditional Density of the Implied Rates 167
5.3.3.3 Application of the Change of Variable Formula for the CIR Model 168
5.3.3.4 Precision of the Estimation 169
5.4 Other Models for the Spot Rates 170
5.4.1 Affine Models 171
5.5 Suggested Reading 171
5.6 Exercises 172
5.7 Assignment Questions 175
5.A Interpretation of the Stochastic Integral 175
5.B Integral of a Gaussian Process 176
5.C Estimation Error for a Ornstein-Uhlenbeck Process 176
5.D Proofs of the Results 178
5.D.1 Proof of Proposition 5.2.1 178

5.D.2 Proof of Proposition 5.2.2 178
5.D.3 Proof of Proposition 5.3.1 179
5.D.4 Proof of Proposition 5.3.2 180
5.D.5 Proof of Proposition 5.3.3 180
Bibliography 180

6 Lévy Models 183
Summary 183
6.1 Complete Models 183
6.2 Stochastic Processes with Jumps 184
6.2.1 Simulation of a Poisson Process over a Fixed Time Interval 185
6.2.2 Jump-Diffusion Models 185
6.2.3 Merton Model 186
6.2.4 Kou Jump-Diffusion Model 187
6.2.5 Weighted-Symmetric Models for the Jumps 187
6.3 Lévy Processes 188
6.3.1 Random Walk Representation 188
6.3.2 Characteristics 189
6.3.3 Infinitely Divisible Distributions 190
6.3.4 Sample Path Properties 190
6.3.4.1 Number of Jumps of a Lévy Process 191
6.3.4.2 Finite Variation 191
6.4 Examples of Lévy Processes 192
6.4.1 Gamma Process 192
6.4.2 Inverse Gaussian Process 193
6.4.2.1 Simulation of $T_{\alpha,\beta}$ 193
6.4.3 Generalized Inverse Gaussian Distribution 194
6.4.4 Variance Gamma Process 194
6.4.5 Lévy Subordinators 195
6.5 Change of Distribution 197
6.5.1 Esscher Transforms 197
6.5.2 Examples of Application 198
6.5.2.1 Merton Model 198
6.5.2.2 Kou Model 199
6.5.2.3 Variance Gamma Process 199
6.5.2.4 Normal Inverse Gaussian Process 199
6.5.3 Application to Option Pricing 199
6.5.4 General Change of Measure 200
6.5.5 Incompleteness 201
6.6 Model Implementation and Estimation of Parameters 203
6.6.1 Distributional Properties 204
6.6.1.1 Serial Independence 204
6.6.1.2 Lévy Process vs Brownian Motion 204
6.6.2 Estimation Based on the Cumulants 205

6.6.2.1 Estimation of the Cumulants 206
6.6.2.2 Application 207
6.6.2.3 Discussion 209
6.6.3 Estimation Based on the Maximum Likelihood Method 209
6.7 Suggested Reading 215
6.8 Exercises 215
6.9 Assignment Questions 216
6.A Modified Bessel Functions of the Second Kind 217
6.B Asymptotic Behavior of the Cumulants 218
6.C Proofs of the Results 219
6.C.1 Proof of Lemma 6.5.1 219
6.C.2 Proof of Corollary 6.5.2 219
6.C.3 Proof of Proposition 6.6.1 220
6.C.4 Proof of Proposition 6.4.1 220
Bibliography 221

7 Stochastic Volatility Models **223**
Summary 223
7.1 GARCH Models 223
7.1.1 GARCH(1,1) 224
7.1.2 GARCH(p,q) 226
7.1.3 EGARCH 226
7.1.4 NGARCH 227
7.1.5 GJR-GARCH 227
7.1.6 Augmented GARCH 227
7.2 Estimation of Parameters 228
7.2.1 Application for GARCH(p,q) Models 229
7.2.2 Tests 230
7.2.3 Goodness-of-Fit and Pseudo-Observations 230
7.2.4 Estimation and Goodness-of-Fit When the Innovations Are Not Gaussian 232
7.3 Duan Methodology of Option Pricing 235
7.3.1 LRNVR Criterion 235
7.3.2 Continuous Time Limit 237
7.3.2.1 A New Parametrization 238
7.4 Stochastic Volatility Model of Hull-White 239
7.4.1 Market Price of Volatility Risk 239
7.4.2 Expectations vs Partial Differential Equations 240
7.4.3 Option Price as an Expectation 240
7.4.4 Approximation of Expectations 242
7.4.4.1 Monte Carlo Methods 242
7.4.4.2 Taylor Series Expansion 242
7.4.4.3 Edgeworth and Gram-Charlier Expansions . 243
7.4.4.4 Approximate Distribution 245
7.5 Stochastic Volatility Model of Heston 246

7.6 Suggested Reading 247
7.7 Exercises 247
7.8 Assignment Questions 249
7.A Khmaladze Transform 250
7.A.1 Implementation Issues 250
7.B Proofs of the Results 251
7.B.1 Proof of Proposition 7.1.1 251
7.B.2 Proof of Proposition 7.4.1 253
7.B.3 Proof of Proposition 7.4.2 254
Bibliography 254

8 Copulas and Applications **257**
Summary 257
8.1 Weak Replication of Hedge Funds 257
8.1.1 Computation of g 258
8.2 Default Risk 259
8.2.1 n-th to Default Swap 259
8.2.2 Simple Model for Default Time 260
8.2.3 Joint Dynamics of X_i and Y_i 261
8.2.4 Simultaneous Evolution of Several Markov Chains . . 262
8.2.4.1 CreditMetrics 262
8.2.5 Continuous Time Model 264
8.2.5.1 Modeling the Default Time of a Firm 266
8.2.6 Modeling Dependence Between Several Default Times 266
8.3 Modeling Dependence 266
8.3.1 An Image is Worth a Thousand Words 267
8.3.2 Joint Distribution, Margins and Copulas 269
8.3.3 Visualizing Dependence 269
8.4 Bivariate Copulas 271
8.4.1 Examples of Copulas 271
8.4.2 Sklar Theorem in the Bivariate Case 272
8.4.3 Applications for Simulation 274
8.4.4 Simulation of $(U_1, U_2) \sim C$ 274
8.4.5 Modeling Dependence with Copulas 275
8.4.6 Positive Quadrant Dependence (PQD) Order 276
8.5 Measures of Dependence 276
8.5.1 Estimation of a Bivariate Copula 278
8.5.1.1 Precision of the Estimation of the Empirical Copula 278
8.5.1.2 Tests of Independence Based on the Empirical Copula 278
8.5.2 Kendall Function 280
8.5.2.1 Estimation of Kendall Function 281
8.5.2.2 Precision of the Estimation of the Kendall Function 282

8.5.2.3 Tests of Independence Based on the Empirical Kendall Function . . . 282
8.5.3 Kendall Tau . . . 286
8.5.3.1 Estimation of Kendall Tau . . . 286
8.5.3.2 Precision of the Estimation of Kendall Tau . 287
8.5.4 Spearman Rho . . . 287
8.5.4.1 Estimation of Spearman Rho . . . 288
8.5.4.2 Precision of the Estimation of Spearman Rho 288
8.5.5 van der Waerden Rho . . . 289
8.5.5.1 Estimation of van der Waerden Rho . . . 290
8.5.5.2 Precision of the Estimation of van der Waerden Rho . . . 290
8.5.6 Other Measures of Dependence . . . 291
8.5.6.1 Estimation of $\rho^{(J)}$ . . . 291
8.5.6.2 Precision of the Estimation of $\rho^{(J)}$ . . . 292
8.5.7 Serial Dependence . . . 292
8.6 Multivariate Copulas . . . 293
8.6.1 Kendall Function . . . 294
8.6.2 Conditional Distributions . . . 294
8.6.2.1 Applications of Theorem 8.6.2 . . . 294
8.6.3 Stochastic Orders for Dependence . . . 295
8.6.3.1 Fréchet-Hoeffding Bounds . . . 295
8.6.3.2 Application . . . 296
8.6.3.3 Supermodular Order . . . 296
8.7 Families of Copulas . . . 297
8.7.1 Independence Copula . . . 297
8.7.2 Elliptical Copulas . . . 297
8.7.2.1 Estimation of ρ . . . 298
8.7.3 Gaussian Copula . . . 298
8.7.3.1 Simulation of Observations from a Gaussian Copula . . . 299
8.7.4 Student Copula . . . 299
8.7.4.1 Simulation of Observations from a Student Copula . . . 300
8.7.5 Other Elliptical Copulas . . . 300
8.7.6 Archimedean Copulas . . . 301
8.7.6.1 Financial Modeling . . . 301
8.7.6.2 Recursive Formulas . . . 301
8.7.6.3 Conjecture . . . 303
8.7.6.4 Kendall Tau for Archimedean Copulas . . . 303
8.7.6.5 Simulation of Observations from an Archimedean Copula . . . 304
8.7.7 Clayton Family . . . 304
8.7.7.1 Simulation of Observations from a Clayton Copula . . . 305

8.7.8 Gumbel Family . . . 305
8.7.8.1 Simulation of Observations from a Gumbel Copula . . . 306
8.7.9 Frank Family . . . 306
8.7.9.1 Simulation of Observations from a Frank Copula . . . 307
8.7.10 Ali-Mikhail-Haq Family . . . 308
8.7.10.1 Simulation of Observations from an Ali-Mikhail-Haq Copula . . . 308
8.7.11 PQD Order for Archimedean Copula Families . . . 309
8.7.12 Farlie-Gumbel-Morgenstern Family . . . 309
8.7.13 Plackett Family . . . 310
8.7.14 Other Copula Families . . . 310
8.8 Estimation of the Parameters of Copula Models . . . 311
8.8.1 Considering Serial Dependence . . . 311
8.8.2 Estimation of Parameters: The Parametric Approach . 312
8.8.2.1 Advantages and Disadvantages . . . 312
8.8.3 Estimation of Parameters: The Semiparametric Approach . . . 312
8.8.3.1 Advantages and Disadvantages . . . 313
8.8.4 Estimation of ρ for the Gaussian Copula . . . 313
8.8.5 Estimation of ρ and ν for the Student Copula . . . 313
8.8.6 Estimation for an Archimedean Copula Family . . . 314
8.8.7 Nonparametric Estimation of a Copula . . . 314
8.8.8 Nonparametric Estimation of Kendall Function . . . 315
8.9 Tests of Independence . . . 315
8.9.1 Test of Independence Based on the Copula . . . 316
8.10 Tests of Goodness-of-Fit . . . 316
8.10.1 Computation of P-Values . . . 317
8.10.2 Using the Rosenblatt Transform for Goodness-of-Fit Tests . . . 318
8.10.2.1 Computation of P-Values . . . 318
8.11 Example of Implementation of a Copula Model . . . 319
8.11.1 Change Point Tests . . . 320
8.11.2 Serial Independence . . . 320
8.11.3 Modeling Serial Dependence . . . 320
8.11.3.1 Change Point Tests for the Residuals . . . 320
8.11.3.2 Goodness-of-Fit for the Distribution of Innovations . . . 320
8.11.4 Modeling Dependence Between Innovations . . . 321
8.11.4.1 Test of Independence for the Innovations . . 321
8.11.4.2 Goodness-of-Fit for the Copula of the Innovations . . . 323
8.12 Suggested Reading . . . 325
8.13 Exercises . . . 326

8.14 Assignment Questions 330
8.A Continuous Time Markov Chains 331
8.B Tests of Independence 332
8.C Polynomials Related to the Gumbel Copula 333
8.D Polynomials Related to the Frank Copula 334
8.E Change Point Tests 334
8.E.1 Change Point Test for the Copula 335
8.F Auxiliary Results 336
8.G Proofs of the Results 336
8.G.1 Proof of Proposition 8.4.1 336
8.G.2 Proof of Proposition 8.4.2 337
8.G.3 Proof of Proposition 8.5.1 338
8.G.4 Proof of Theorem 8.7.1 338
Bibliography 339

9 Filtering **345**
Summary 345
9.1 Description of the Filtering Problem 345
9.2 Kalman Filter 346
9.2.1 Model 346
9.2.2 Filter Initialization 347
9.2.3 Estimation of Parameters 348
9.2.4 Implementation of the Kalman Filter 348
9.2.4.1 Solution 348
9.2.5 The Kalman Filter for General Linear Models 353
9.3 IMM Filter 354
9.3.1 IMM Algorithm 354
9.3.2 Implementation of the IMM Filter 356
9.4 General Filtering Problem 356
9.4.1 Kallianpur-Striebel Formula 356
9.4.2 Recursivity 357
9.4.3 Implementing the Recursive Zakai Equation 358
9.4.4 Solving the Filtering Problem 358
9.5 Computation of the Conditional Densities 358
9.5.1 Convolution Method 359
9.5.2 Kolmogorov Equation 360
9.6 Particle Filters 360
9.6.1 Implementation of a Particle Filter 360
9.6.2 Implementation of an Auxiliary Sampling/Importance Resampling (ASIR) Particle Filter 361
9.6.2.1 ASIR_0 363
9.6.2.2 ASIR_1 363
9.6.2.3 ASIR_2 364
9.6.3 Estimation of Parameters 365
9.6.3.1 Smoothed Likelihood 365

9.7 Suggested Reading 366
9.8 Exercises 367
9.9 Assignment Questions 368
9.A Schwartz Model 369
9.B Auxiliary Results 370
9.C Fourier Transform 371
9.D Proofs of the Results 371
9.D.1 Proof of Proposition 9.2.1 371
Bibliography 372

10 Applications of Filtering **375**
Summary 375
10.1 Estimation of ARMA Models 375
10.1.1 AR(p) Processes 375
10.1.1.1 MA(q) Processes 376
10.1.2 MA Representation 376
10.1.3 ARMA Processes and Filtering 377
10.1.3.1 Implementation of the Kalman Filter in the Gaussian Case 378
10.1.4 Estimation of Parameters of ARMA Models 379
10.2 Regime-Switching Markov Models 380
10.2.1 Serial Dependence 380
10.2.2 Prediction of the Regimes 381
10.2.3 Conditional Densities and Predictions 382
10.2.4 Estimation of the Parameters 383
10.2.4.1 Implementation of the E-step 383
10.2.5 M-step in the Gaussian Case 384
10.2.6 Tests of Goodness-of-Fit 385
10.2.7 Continuous Time Regime-Switching Markov Processes 388
10.3 Replication of Hedge Funds 389
10.3.0.1 Measurement of Errors 390
10.3.1 Replication by Regression 391
10.3.2 Replication by Kalman Filter 391
10.3.3 Example of Application 391
10.4 Suggested Reading 395
10.5 Exercises 396
10.6 Assignment Questions 397
10.A EM Algorithm 398
10.B Sampling Moments vs Theoretical Moments 401
10.C Rosenblatt Transform for the Regime-Switching Model . . . 401
10.D Proofs of the Results 403
10.D.1 Proof of Proposition 10.1.1 403
10.D.2 Proof of Proposition 10.1.2 404
Bibliography 404

A Probability Distributions **407**
Summary 407
A.1 Introduction 407
A.2 Discrete Distributions and Densities 408
A.2.1 Expected Value and Moments of Discrete Distributions 408
A.3 Absolutely Continuous Distributions and Densities 410
A.3.1 Expected Value and Moments of Absolutely Continuous Distributions 410
A.4 Characteristic Functions 412
A.4.1 Inversion Formula 413
A.5 Moments Generating Functions and Laplace Transform . . . 413
A.5.1 Cumulants 414
A.5.1.1 Extension 415
A.6 Families of Distributions 415
A.6.1 Bernoulli Distribution 415
A.6.2 Binomial Distribution 416
A.6.3 Poisson Distribution 416
A.6.4 Geometric Distribution 417
A.6.5 Negative Binomial Distribution 417
A.6.6 Uniform Distribution 417
A.6.7 Gaussian Distribution 418
A.6.8 Log-Normal Distribution 418
A.6.9 Exponential Distribution 419
A.6.10 Gamma Distribution 420
A.6.10.1 Properties of the Gamma Function 420
A.6.11 Chi-Square Distribution 421
A.6.12 Non-Central Chi-Square Distribution 421
A.6.12.1 Simulation of Non-Central Chi-Square Variables 421
A.6.13 Student Distribution 422
A.6.14 Johnson SU Type Distributions 423
A.6.15 Beta Distribution 423
A.6.16 Cauchy Distribution 424
A.6.17 Generalized Error Distribution 424
A.6.18 Multivariate Gaussian Distribution 425
A.6.18.1 Representation of a Random Gaussian Vector 425
A.6.19 Multivariate Student Distribution 426
A.6.20 Elliptical Distributions 426
A.6.21 Simulation of an Elliptic Distribution 429
A.7 Conditional Densities and Joint Distributions 429
A.7.1 Multiplication Formula 429
A.7.2 Conditional Distribution in the Markovian Case . . . 430
A.7.3 Rosenblatt Transform 430
A.8 Functions of Random Vectors 430
A.9 Exercises 433

Bibliography . . . 434

B Estimation of Parameters **435**

Summary . . . 435
B.1 Maximum Likelihood Principle . . . 435
B.2 Precision of Estimators . . . 437
B.2.1 Confidence Intervals and Confidence Regions . . . 437
B.2.2 Nonparametric Prediction Interval . . . 437
B.3 Properties of Estimators . . . 438
B.3.1 Almost Sure Convergence . . . 438
B.3.2 Convergence in Probability . . . 438
B.3.3 Convergence in Mean Square . . . 438
B.3.4 Convergence in Law . . . 439
B.3.4.1 Delta Method . . . 440
B.3.5 Bias and Consistency . . . 441
B.4 Central Limit Theorem for Independent Observations . . . 441
B.4.1 Consistency of the Empirical Mean . . . 442
B.4.2 Consistency of the Empirical Coefficients of Skewness and Kurtosis . . . 442
B.4.3 Confidence Intervals I . . . 445
B.4.4 Confidence Ellipsoids . . . 445
B.4.5 Confidence Intervals II . . . 445
B.5 Precision of Maximum Likelihood Estimator for Serially Independent Observations . . . 446
B.5.1 Estimation of Fisher Information Matrix . . . 446
B.6 Convergence in Probability and the Central Limit Theorem for Serially Dependent Observations . . . 448
B.7 Precision of Maximum Likelihood Estimator for Serially Dependent Observations . . . 448
B.8 Method of Moments . . . 450
B.9 Combining the Maximum Likelihood Method and the Method of Moments . . . 452
B.10 M-estimators . . . 453
B.11 Suggested Reading . . . 454
B.12 Exercises . . . 454
Bibliography . . . 454

Index **455**

Preface

The aim of this book is to guide existing and future practitioners through the implementation of the most useful stochastic models used in financial engineering. There is a plethora of books on financial engineering but the statistical aspect of the implementation of these models, where lie many of the challenges, is often overlooked or restricted to a few well-known cases like the Black-Scholes and GARCH models. So in addition to a basic presentation of the models, my objective in writing this book was also to include the relevant questions related to their implementation. For example, the chapter on the modeling of interest rates includes the estimation of the parameters of the proposed models, which is essential from an implementation point-of-view, but is usually ignored. Other such important topics, including the effect of estimation errors on the value of options, hedging in discrete time, dependence modeling through copulas and hedge fund replication, are also covered. Overall, I believe this book fills an important gap in the financial engineering literature since I faced many of these implementation issues in my own work as a part-time consultant in the financial industry. Another aspect covered here that is largely ignored in most textbooks pertains to the validation of the models. Throughout the chapters, in addition to showing how to estimate parameters efficiently, I also demonstrate, whenever possible, how one can test the validity of the proposed models. Many techniques I used in this book appeared in research papers (many I have authored myself), and while powerful, are too rarely applied in practice. A companion website also offers MATLAB® and R programs that are likely to help practitioners with the implementation of these tools in the context of real-life financial problems.

The content of this book has been developed in the last ten years for a graduate course on statistical methods for students in finance and financial engineering. Since the course can be taken by first-year graduate students, I try to avoid as much as possible any reference to stochastic calculus. The book is also self-contained in the sense that no financial background is required, although it would definitely help. Rigor is shown by proving most results. However, for the sake of readability, the proofs are presented in a series of appendices, together with more advanced topics, including two appendices on probability distributions and parameter estimation, which both provide theoretical support for the results presented in the book. In every chapter, I try to introduce the statistical tools required to implement the models taken from the cornerstone articles in financial engineering. The use of each tool is facilitated by examples of application using MATLAB programs that are available on my website at `www.brunoremillard.com`.

Starting with the pioneering contribution of Black & Scholes, properties of univariate and multivariate models for asset dynamics are studied in Chapters 1 and 2, together with estimation techniques which are valid for independent observations. The effect of parameter estimation on the value of options is also covered. Furthermore, using techniques developed by Broadie-Glasserman, I show how Monte Carlo simulations can be used to estimate option prices and sensitivity parameters known as "greeks."

In Chapter 3, the limits of the Black-Scholes model are discussed, statistical tests are introduced to verify some of its assumptions, and a section discusses the challenges of dynamic hedging in discrete time.

Next, in Chapter 4, the estimation of risk and performance measures is covered, starting with a discussion on the axioms for coherent risk measures. The main tools used in this chapter are Monte Carlo methods, and other statistical tools such as nonparametric estimation of distribution and quantile functions, Edgeworth and Cornish-Fisher expansions, saddlepoint approximations, and the inversion of characteristic functions.

In Chapter 5, I present the foundations of the spot interest rate modeling literature using the article of Vasicek, and I show especially how to estimate parameters of the so-called Vasicek and Cox-Ingersoll-Ross models, including the market price of risk parameters. To do so, maximum likelihood techniques for dependent observations are used, along with a method proposed by Duan for dealing with the unobservable nature of the spot rates.

The article of Merton on jump-diffusion processes and option pricing is covered in Chapter 6 and is used as a pretext for the introduction of Lévy processes and their financial applications, including path properties, change of measure, option pricing, and parameter estimation.

Using the famous article of Duan on GARCH models and option pricing, the properties and parameter estimation of GARCH models are presented in Chapter 7. The chapter also covers the goodness-of-fit tests, using the Khmaladze transform and parametric bootstrap. I show that as a limiting case, one obtains stochastic volatility models, in particular the model studied by Hull & White. The well-known Heston model is also discussed.

In Chapter 8, weak replication of hedge funds and simple credit risk models are used to illustrate the tremendous importance of dependence models. This issue is discussed at great length with the use of copulas. All aspects pertaining to these models are covered: properties, simulation, dependence measures, estimation, and goodness-of-fit. Because of the inherent serial dependence observed in financial time series, I also show how to deal with residuals of stochastic volatility models.

Finally, in Chapters 9 and 10, I cover the topic of filtering and its financial applications, when unobservable factors have to be predicted. Following the insights of Schwartz on filtering in a commodities context, the famous Kalman filter is introduced. Two other methods, the IMM and particle filters are then studied. The latter is a class of Monte Carlo methods for solving the general filtering problem. Estimation of the parameters of the underlying models is also discussed. Then, filtering is applied in three contexts, namely estimation of ARMA models, estimation and prediction of Hidden Markov models using the powerful EM algorithm, and hedge funds replication.

This book, written over such a long period of time, has benefited from the valuable help and feedback of many people. I first wish to thank Matt Davidson, professor at University of Western Ontario, and Hugues Langlois, a Ph.D. student at McGill University, for their helpful comments and suggestions on an earlier version of this book. I would also like to thank my colleague at HEC Jean-François Plante, for his detailed and valuable comments on the chapter on copulas, and HEC Montréal for their financial support. Finally, I would like to thank the students in the Financial Engineering program at HEC Montréal, especially Frédéric Godin, for their comments and suggestions along with their understanding throughout the years when they often found themselves in the position to be the first to test all the new material. Finally, a special thanks to Alexis Constantineau for his help in the preparation of exercises and to David-Shaun Guay for converting my MATLAB programs into R programs.

Bruno Rémillard
Montréal, October 3rd, 2012.

List of Figures

1.1 Graph of strikes vs call prices on Apple with a 24-day maturity, on January 14th, 2011. 17

2.1 Graph of $\tanh(\alpha) = \frac{e^\alpha - e^{-\alpha}}{e^\alpha + e^{-\alpha}}$, $\alpha \in [-5, 5]$. 37

3.1 Dependogram of the P-values for the returns of Apple, computed with $N = 1000$ replications and $p = 6$. 65
3.2 Nonparametric density estimation of the returns of Apple and graphical comparisons with the Gaussian, Johnson SU, and Student distributions. 68
3.3 Graph of C_0 and φ_1 for a call option under the Black-Scholes model, using optimal hedging and delta hedging. . . . 80
3.4 Graph of the estimated densities of the hedging errors for a call option under the Black-Scholes model, using optimal hedging and delta hedging. 10000 portfolios were simulated. 81
3.5 Graph of C_0 and φ_1 for a call option under the Variance Gamma process, using optimal hedging and delta hedging. 82
3.6 Graph of the estimated densities of the hedging errors for a call option under the Variance Gamma process, using optimal hedging and delta hedging. 10000 portfolios were simulated. . 83
3.7 Graph of C_0 and φ_1 for a call option under a regime-switching model with 3 Gaussian regimes, using optimal hedging and delta hedging. 87
3.8 Graph of the estimated densities of the hedging errors for a call option under a regime-switching model with 3 Gaussian regimes, using optimal hedging and delta hedging. 1000000 portfolios were simulated. 88
3.9 Graph of the value of Apple from January 14th to February 18th, 2011, and the associated hedging portfolio under optimal hedging. 88

4.1 Nonparametric estimation of the density and distribution function of the losses. 112
4.2 Standard deviation of the estimation of the VaR in the standard Gaussian case for the parametric and nonparametric methods. Here, $\sigma = 1$. 113

4.3 Standard deviation of the estimation of the expected shortfall in the standard Gaussian case for the parametric and nonparametric methods. Here, $\sigma = 1$. 115
4.4 $\Omega(r)$ as a function of the Sharpe ratio S in the Gaussian case. 129

5.1 Implied and true spot rates for the Vasicek model with $\alpha = 0.5$, $\beta = 2.55$, $\sigma = 0.365$, $q_1 = 0.3$ and $q_2 = 0$. 160
5.2 Implied and true spot rates for the CIR model with $\alpha = 0.5$, $\beta = 2.55$, $\sigma = 0.365$, $q_1 = 0.3$ and $q_2 = 0$. 170

6.1 Densities of weighted-symmetric Gaussian distributions. . . . 187
6.2 Nonparametric estimation of the density and Merton densities with true and estimated parameters (cumulants), for the simulated data set *DataMerton*. 208
6.3 Nonparametric estimation of the density and Merton densities with true and estimated parameters (mle), for the simulated data set *DataMerton*. 210
6.4 Nonparametric estimation of the density and Merton density with estimated parameters (mle) for the returns of Apple. . . 211
6.5 Nonparametric estimation of the density and Merton density with estimated parameters (cumulants) for the returns of Apple. 212
6.6 Nonparametric estimation of the density and Variance Gamma density with the true and estimated parameters (mle) for the simulated data set *DataVG*. 213
6.7 Nonparametric estimation of the density and Variance Gamma density with the true and estimated parameters (combined) for the simulated data set *DataVG*. 213
6.8 Nonparametric estimation of the density and Variance Gamma density with estimated parameters (mle) for the returns of Apple. 214
6.9 Nonparametric estimation of the density and Variance Gamma density with estimated parameters (combined) for the returns of Apple. 214

7.1 Graphs of the prices, returns, and volatilities for a GARCH(1,1) model starting from different points. Here $x_1 = 0$, $y_1 = 0.01$, $\sigma_1 = 0$ and $v_1 = 0.1$. 225
7.2 Trajectory of the process W for the innovations of a GARCH(1,1) model and uniform 95% confidence band for a Brownian motion path. 232
7.3 Trajectory of the process W for the innovations of a GARCH(1,1) model for the returns of Apple and uniform 95% confidence band for a Brownian motion path. 233

7.4 Empirical distribution D_n and uniform 95% confidence band around D. 235

8.1 Default probabilities on a 100-year horizon. 265
8.2 Estimation of the densities of (X_1, X_2) using 10000 pairs of observations. 268
8.3 Graph of 10000 pairs of points (X_{i1}, X_{i2}). 268
8.4 (a) Graph of 1000 pairs (X_{i1}, X_{i2}) of independent exponential observations; (b) Graph of 1000 pairs of their normalized ranks $\left(\frac{R_{i1}}{1001}, \frac{R_{i2}}{1001}\right)$. 269
8.5 (a) Graph of 1000 pairs (X_{i1}, X_{i2}) of independent Gaussian observations; (b) Graph of 1000 pairs of their normalized ranks $\left(\frac{R_{i1}}{1001}, \frac{R_{i2}}{1001}\right)$. 270
8.6 (a) Graph of 1000 pairs (X_{i1}, X_{i2}) of independent Cauchy observations; (b) Graph of 1000 pairs of their normalized ranks $\left(\frac{R_{i1}}{1001}, \frac{R_{i2}}{1001}\right)$. 270
8.7 Graphs of the pseudo-observations for the data of Table 8.6. . 279
8.8 Graphs of the estimated copula for the data of Table 8.6. . . 279
8.9 Estimation of K for the data of Table 8.6. 281
8.10 Variance of $\mathbb{K}$ for the independence copula. 282
8.11 Graph of K_n and 95% confidence band around Kendall function for the independence copula. 284
8.12 Graph of K_n and 95% confidence band around Kendall function for the independence copula of returns of Apple and Microsoft. 285
8.13 Graph of 5000 pairs of points from a Student copula with $\nu = 1$ degree of freedom (Cauchy copula) with $\tau = -0.5$ (panel a) and $\tau = 0$ (panel b). 300
8.14 Gaussian, Laplace, and Pareto margins with (a) Gaussian copula, (b) Clayton copula, (c) Gumbel copula, and $\tau \in \{.1, .5, .9\}$. 302
8.15 Autocorrelograms of the residuals of GARCH(1,1) models for Apple (panel a) and Microsoft (panel b). 321
8.16 Empirical distribution function D_n and 95% confidence band about the uniform distribution function for the innovations of Apple. 322
8.17 Empirical distribution function D_n and 95% confidence band about the uniform distribution function for the innovations of Microsoft. 322
8.18 Graph of the empirical Kendall function and 95% confidence band about $K_\perp$. $N = 1000$ bootstrap samples were generated to compute the P-values and the 95% quantile of the Kolmogorov-Smirnov statistic. 323
8.19 95% confidence band about K_{AMH}. 324
8.20 95% confidence band about $K_{Clayton}$. 324
8.21 Rosenblatt transform of the pseudo-observations using the Student copula with parameters $\hat{\rho} = 0.51$ and $\hat{\nu} = 3.51$. 326

9.1 Observed log-prices and their predictions, using the Kalman filter starting at $(0, 1)$. 351
9.2 Graphs of the signal and out-of-sample prediction, with 95% confidence intervals. 352

10.1 Values and log-returns of the S&P 500 from April 17th, 2007 to December 31st, 2008. 386
10.2 Forecasted densities for the log-returns of the S&P 500, as of December 31st, 2008. 387
10.3 Forecasted densities for the log-returns of Apple as of January 14th, 2011. 388
10.4 Returns of the replication portfolio when using the regression method. 392
10.5 Returns of the replication portfolio when using the Kalman filter. 393
10.6 Compounded values of \$1 for the returns of the index and the replication portfolios. 393
10.7 Evolution of the weights β_i of the replication portfolio, computed with a rolling window of 24 months. 394
10.8 Evolution of the weights β_i of the replication portfolio, computed with the Kalman filter. 395

List of Tables

1.1 95% confidence intervals for the average return per annum (μ) and the volatility per annum (σ). The estimation of the covariance V is also given. 8
1.2 Some market values of call options on Apple with a 24-day maturity, on January 14th, 2011. 16
1.3 Call option market values and their estimations for Apple (aapl) on January 14th, 2011, for a strike price $K = 350$. 18
1.4 Libor rates (%) around January 14th, 2011. 18
1.5 Implied volatility for call option values for Apple (aapl) on January 14th, 2011, for a strike price $K = 350$. 19

2.1 Explicit estimation of the parameters of model (2.1) for Apple and Microsoft on an annual time scale, using the adjusted prices from January 13th, 2009, to January 14th, 2011. 40
2.2 95% confidence intervals for the estimation of the volatility vector v on an annual time scale, for Apple and Microsoft, using the prices from January 13th, 2009, to January 14th, 2011. . 41

3.1 Tests statistics and P-values for the goodness-of-fit tests of the null hypothesis of normality for the returns of Apple. 67
3.2 Statistics for hedging errors under the Black-Scholes model. . 81
3.3 Statistics for hedging errors under the Variance Gamma process. 83
3.4 C_0 and φ_1 for each regime, together with the estimated probability η of each regime. 86
3.5 Statistics for the hedging errors in a Gaussian regime-switching model, where we chose the most probable regime. 86

4.1 Estimation of the VaR and estimation errors for a 95% level of confidence. 114
4.2 Estimation of the expected shorfall and estimation errors for a 95% level of confidence. 115
4.3 95% confidence intervals for the performance ratios of Apple and Microsoft. 131

5.1 95% confidence intervals for the estimation of the parameters of the Vasicek model. Here $\phi = e^{-\alpha h}$, with $h = 1/360$. 159

5.2 95% confidence intervals for the estimation of parameters of the CIR model. Here $\phi = e^{-\alpha h}$, with $h = 1/360$. 170

6.1 95% confidence intervals for the parameters of the Merton model using the cumulant matching method, applied to the simulated data set *DataMerton*. 208

6.2 95% confidence intervals for the parameters of the Merton model using the maximum likelihood method, applied to the simulated data set *DataMerton*. 210

6.3 95% confidence intervals for the parameters of the Merton model applied to the returns of Apple. 211

6.4 95% confidence intervals for the parameters of the Variance Gamma model applied to the simulated data set *DataVG*. . . 212

6.5 95% confidence intervals for the parameters of the Variance Gamma model applied to the returns of Apple. 214

7.1 95% confidence intervals for the maximum likelihood estimation of the parameters of a GARCH(1,1) model, using the simulated data set *DataGARCH*. 232

7.2 95% confidence intervals for the maximum likelihood estimation of the parameters of a GARCH(1,1) model, using the data set *ApplePrices*. 233

7.3 95% confidence intervals for the maximum likelihood estimation of the parameters of a GARCH(1,1) model with GED innovations, using the data set *ApplePrices*. 234

7.4 Estimation of the parameters of Apple on an annual time scale, using Gaussian and GED innovations. 238

8.1 Transition probabilities from the state at period n to the state at period $n+1$. 260

8.2 Transition probabilities from the states at period n to the states at period $n+1$. 261

8.3 Example of a transition matrix. 263

8.4 VaR for the loss of the portfolio, based on 10^5 simulations. . . 263

8.5 Premia (in basis points) for the nth to default based on different families of copula 267

8.6 Data set. 277

8.7 Ranks and pseudo-observations for the data set. 277

8.8 95% confidence intervals and P-values for tests of independence based on classical statistics (Pearson rho, Kendall tau, Spearman rho, van der Waerden rho), and on Kolmogorov-Smirnov and Cramér-von Mises statistics using the empirical Kendall function and the empirical copula. The quantiles and the P-values were computed using $N = 10000$ iterations. 284
8.9 95% confidence intervals and P-values for tests of independence based on classical statistics (Pearson rho, Kendall tau, Spearman rho, van der Waerden rho), and on Kolmogorov-Smirnov and Cramér-von Mises statistics for the returns of Apple and Microsoft. The quantiles and the P-values were computed using $N = 10000$ iterations. 285
8.10 95% confidence intervals for the parameters of GARCH(1,1) models with GED innovations, applied to the returns of Apple and Microsoft. 320
8.11 Tests of the hypothesis of GED innovations. The P-values were computed with $N = 1000$ bootstrap samples. 321
8.12 Tests of goodness-of-fit based on the Cramér–von Mises (S_n) and Kolmogorov–Smirnov (T_n) statistics computed with K_n. $N = 1000$ bootstrap samples were generated. 325
8.13 Tests of goodness-of-fit based on $S_n^{(B)}$, using Rosenblatt transform. The P-values were estimated with $N = 1000$ parametric bootstrap replications. 325

9.1 Estimation of the parameters of the signal and observations for Apple, using Kalman filter. 350
9.2 95% confidence intervals for the estimated parameters for the first 250 observations. The out-of-sample RMSE is 0.0045. . . 352
9.3 95% confidence intervals for the estimated parameters, using quasi-maximum likelihood. 354
9.4 95% confidence intervals for the estimated parameters, using a smooth particle filter with $N = 1000$ particles. The out-of-sample RMSE is 0.3580. 366

10.1 Parameter estimations for 3 regimes. 386
10.2 P-values of the goodness-of-fit tests for Apple, using $N = 10000$ replications. 387
10.3 Parameter estimations for 3 regimes. 388
10.4 In-sample and out-of-sample statistics for the tracking error (TE), Pearson correlation (ρ), Kendall tau (τ), mean (μ), volatility (σ), skewness (γ_1), and kurtosis (γ_2). 392

A.1 Generators of Pearson type distributions. 428

Symbol Descriptions

$C_\perp$	Independence copula.
C_+	Fréchet-Hoeffding upper bound.
C_-	Fréchet-Hoeffding lower bound.
$\mathcal{N}$	Standard Gaussian distribution function
S_d^+	The set of all symmetric and positive definite $d \times d$ matrices.
$\lfloor x \rfloor$	Integer part of x, that is it is the largest integer, smaller or equal to x.
x^+	Positive part of x, i.e., $x^+ = \max(0, x)$.
$o(h)$	One writes $f(h) = o(h)$ when $f(h)/h \to 0$ as $h \to 0$.
$X_n \rightsquigarrow X$	X_n converges in law to X.

Chapter 1

Black-Scholes Model

In this chapter, we introduce a first model for the dynamical behavior of an asset, the so-called Black-Scholes model. One will learn how to estimate its parameters and how to compute the estimation errors. Then we will state the famous Black-Scholes formula for the price of a European call option, together with the general Black-Scholes equation for the value of a European option, and its representation as an expectation. The concept of implied volatility is then introduced. Finally, we conclude this chapter by estimating the sensitivity parameters of the option value, called "Greeks," using Monte Carlo methods. One interesting application of greeks is to measure the impact of estimation errors on the value of options.

1.1 The Black-Scholes Model

In Black and Scholes [1973], the authors introduced their famous model for the price of a stock, together with a partial differential equation for the value of a European option. To do so, they assumed the following "ideal conditions" on the market:

- The short-term interest rate, also called risk-free rate, is known and constant until the maturity of the option.
- The log-returns of the stock follows a Brownian motion with drift.
- The option is European, i.e., it can only be exercised at maturity.
- There are no transaction costs, nor penalties for short selling.
- It is possible to buy or borrow any fraction of the security.

To make things simpler, we assume that the market is liquid and frictionless, that trading can be done in continuous time, and that the risk-free rate is constant during the life of the option. The latter hypothesis will be weakened in Chapter 3.

1.2 Dynamic Model for an Asset

Before discussing the model proposed by Black and Scholes [1973] for the distribution of an asset, we examine first some properties of financial data.

1.2.1 Stock Exchange Data

What do we do when there are dividends or stock splits? In each case, there is a predictable but abnormal jump in the prices. In order to make historical data comparable, it is necessary to take these jumps into account.

For example, in the case of Apple, there were two 2:1 stock splits (June 21st, 2000, and February 28th, 2005), while the last dividend was reported on November 21st, 1995.

In many textbooks on Financial Engineering, e.g., Hull [2006] and Wilmott [2006], it is suggested to replace the stock price S_i at the ex-dividend period i by $S_i + D_i$, where D_i is the dividend. In fact, it is often more convenient to do just the opposite[1], by subtracting the dividend from the pre-dividend price, i.e., the adjusted price for period $i-1$ is $S_{i-1} - D_i = S_{i-1} f_i$, with $f_i = 1 - D_i/S_{i-1}$. In fact, all pre-dividend values S_j, $j < i$, are then multiplied by the same factor f_i. For splits, one proceeds in a similar way. For example, in the case of a 2:1 stock split, all pre-split prices are multiplied by 0.5. The returns are then calculated from these adjusted closing prices. This is in accordance to the standards of the *Center for Research in Security Prices (CRSP)*. The same method is applied to the adjusted prices available on YAHOO!FINANCE website.

Close vs Adjusted Close

It is important to distinguish between the observed value (closing price) and the real value of the asset (adjusted closing price).

1.2.2 Continuous Time Models

Let $S(t)$ be the (adjusted) value of an asset at time t. Before defining the Black-Scholes model, one needs to define what is a Brownian motion.

Definition 1.2.1 *A stochastic process W is a Brownian motion*[2] *if it is a*

[1] This way, the most recent prices are not adjusted. Otherwise, the most recent prices will be adjusted and could be quite different from the observed closing prices.

[2] It is also called a Wiener process after Norbert Wiener who was the first to give a rigorous definition of the process.

continuous Gaussian process starting at 0*, with zero expectation and covariance function*

$$\mathrm{Cov}\{W(s), W(t)\} = \min(s,t), \; s,t \geq 0.$$

In particular, if $0 = t_0 \leq t_1 \leq \cdots \leq t_n$*, then the increments* $W(t_i) - W(t_{i-1})$ *are independent, have a Gaussian distribution with mean* 0 *and variance* $t_i - t_{i-1}$, $i \in \{1, \ldots, n\}$.

Definition 1.2.2 (Black-Scholes Model) *In the Black-Scholes model, one assumes that the value* S *of the underlying asset is modeled by a geometric Brownian motion, i.e.,*

$$S(t) = se^{\left(\mu - \frac{\sigma^2}{2}\right)t + \sigma W(t)}, \quad t \geq 0, \tag{1.1}$$

where W *is a Brownian motion.*

Note that S *is often defined as the solution to the following stochastic differential equation:*

$$dS(t) = \mu S(t)dt + \sigma S(t)dW(t), \; S(0) = s.$$

Remark 1.2.1 *The Black-Scholes model depends on two unknown parameters:* μ *and* σ*. Therefore they must be estimated. In practice, data are collected at regular time intervals of given length* h*. For example, observations can be collected every 5 seconds, daily, weekly, monthly, etc.*

It is also important to characterize the parameters. For example, is μ *an expectation? If so, it is the expectation of which variable? One can ask the same question for the parameter* σ*.*

In what follows, we will see that μ *and* σ *can be interpreted in terms of the log-returns*

$$X_i = X_i(h) = \ln\{S(ih)\} - \ln[S\{(i-1)h\}], \quad i \in \{1, \ldots, n\}.$$

First, we need to find the joint law of these returns. In addition to being important for the estimation of the unknown parameters, the distribution of the returns is needed in order to find expressions for option prices or when using Monte Carlo methods for pricing complex options, measuring risk or performance, etc.

Counting Time

In applications, the time scale is usually in years, while the data are often collected daily. Here, we follow the advice in Hull [2006] and choose trading days instead of calendar days. According to Hull [2006], many studies show that the trading days account for most of the volatility. Therefore, for daily data, the convention is to use $h = 1/252$ instead of $h = 1/365$ or even $h = 1/360$, as it is the case for interest rates in UK and USA.

1.2.3 Joint Distribution of Returns

Proposition 1.2.1 *Under the Black-Scholes model, for $h > 0$ given, the returns $X_i = \ln\{S(ih)\} - \ln[S\{(i-1)h\}]$, $i \in \{1, \ldots, n\}$, are independent and $X_i \sim N\left(\mu h - \frac{\sigma^2}{2}h, \sigma^2 h\right)$, i.e., X_i has a Gaussian distribution with mean $\left(\mu - \frac{\sigma^2}{2}\right)h$ and variance $\sigma^2 h$.*

PROOF. For $i \in \{1, \ldots, n\}$, one has

$$\begin{aligned} S(ih) \Big/ S\{(i-1)h\} &= \frac{s e^{\mu i h - \frac{ih\sigma^2}{2} + \sigma W(ih)}}{s e^{\mu(i-1)h - \frac{(i-1)h\sigma^2}{2} + \sigma W\{(i-1)h\}}} \\ &= e^{\mu h - \frac{h\sigma^2}{2} + \sigma\{W(ih) - W\{(i-1)h\}\}}, \end{aligned}$$

so

$$\begin{aligned} X_i &= \ln\left[S(ih) \Big/ S\{(i-1)h\}\right] \\ &= \mu h - \frac{\sigma^2}{2}h + \sigma\{W(ih) - W\{(i-1)h\}\} \\ &\sim N\left(\mu h - \frac{\sigma^2}{2}h, \sigma^2 h\right). \end{aligned}$$

The increments $W(ih) - W\{(i-1)h\}$, $i \in \{1 \ldots n\}$, being independent, by definition of Brownian motion, it follows that the returns X_i, $i \in \{1 \ldots n\}$, are also independent.

■

Remark 1.2.2 *Since σ appears to be a measure of variability, it is also called volatility in financial applications and it is often reported in percentage. For example, a volatility of 20% per annum means that $\sigma = 0.2$, on an annual time scale.*

1.2.4 Simulation of a Geometric Brownian Motion

The following algorithm is a direct application of Proposition 1.2.1.

Algorithm 1.2.1 *To generate $S(h), \ldots, S(nh)$, one proceeds the following way:*

- *Generate independent Gaussian variables $Z_1, \ldots, Z_n$, with mean 0 and variance 1.*

- *For $i \in \{1 \ldots n\}$, set $X_i = \mu h - \frac{\sigma^2}{2}h + \sigma\sqrt{h}Z_i$.*

- *For $i \in \{1 \ldots n\}$, set $S(ih) = s e^{\sum_{j=1}^{i} X_j}$.*

This algorithm is implemented in the MATLAB function *SimBS1d*.

1.2.5 Joint Law of Prices

One of the most efficient method for estimating parameters is the maximum likelihood principle, described in Appendix B.1. As shown in Remark B.1.2, the maximum likelihood principle can be used with prices or returns. Since the returns are independent and identically distributed in the Black-Scholes model, we will estimate the parameters using the returns instead of the prices.

However, for sake of completeness, we also give the conditional law of the prices, together with their joint law.

Proposition 1.2.2 *The conditional distribution of $S\{(i+1)h\}$, given $S(0) = s, \ldots, S(ih) = s_i$, depends only on $S(ih)$, and its density is*

$$f_{S\{(i+1)h\}|S(ih)}(x|s_i) = \frac{e^{-\frac{1}{2\sigma^2 h}\left\{\ln(x/s_i)-\left(\mu-\sigma^2/2\right)h\right\}^2}}{x\sigma\sqrt{2\pi h}}\mathbb{I}(x>0), \quad x \in \mathbb{R}.$$

PROOF. It follows from Proposition 1.2.1 that for all $i \geq 0$, $S\{(i+1)h\} = S(ih)e^{X_{i+1}}$, and $S(ih) = se^{X_1+\cdots+X_i}$ is independent of $e^{X_{i+1}}$. Therefore, the conditional distribution of $S\{(i+1)h\}$ given $S(0) = s_0$, ..., $S(ih) = s_i$, is a log-normal distribution (see A.6.8), being the exponential of a Gaussian variate with mean $\ln(s_i) + \left(\mu - \sigma^2/2\right) h$ and variance $\sigma^2 h$.

■

Remark 1.2.3 *As a by-product of Proposition 1.2.2 and the multiplication formula* (A.19), *the joint law of $S(h), \ldots, S(nh)$, given $S(0) = s$, is*

$$\begin{aligned} f_{S(h),\ldots,S(nh)|S(0)}(s_1,\ldots,s_n|s) &= \prod_{i=1}^n f_{S((ih)|S(0),\ldots,S((i-1)h)}(s_i|s,\ldots,s_{i-1}) \\ &= \prod_{i=1}^n f_{S(ih)|S((i-1)h)}(s_i|s_{i-1}) \\ &= \prod_{i=1}^n \frac{e^{-\frac{\left(\ln(s_i)-\ln(s_{i-1})-\left(\mu-\sigma^2/2\right)h\right)^2}{2\sigma^2 h}}}{s_i\sqrt{2\pi\sigma^2 h}}. \end{aligned} \quad (1.2)$$

1.3 Estimation of Parameters

The maximum likelihood principle described in Appendix B.1 will now be used to estimate parameters μ and σ, using the returns as the observations. Usually this method of estimation is more precise than any other one. Since the data are Gaussian, the method of moments, described in Appendix B.8, could also be used. However these two methods yield different results in general.

Recall that the original data set are the prices $S(0) = s_0$, $S(h) = s_1, \ldots, S(nh) = s_n$, from which we compute the returns $x_i = \ln(s_i) - \ln(s_{i-1})$, $i \in \{1, \ldots, n\}$.

Proposition 1.3.1 *The estimations of μ and σ, obtained by the maximum likelihood principle, are given by*

$$\begin{aligned} \hat{\mu}_n &= \frac{\bar{x}}{h} + \frac{s_x^2}{2h}, \\ \hat{\sigma}_n &= \frac{s_x}{\sqrt{h}}, \end{aligned}$$

where $\bar{x} = \frac{1}{n}\sum_{i=1}^{n} x_i$ *and* $s_x^2 = \frac{1}{n}\sum_{i=1}^{n}(x_i - \bar{x})^2$.

The proof is given in Appendix 1.C.1.

Remark 1.3.1 *In practice, σ^2 is estimated by*

$$\frac{1}{(n-1)h}\sum_{i=1}^{n}(x_i - \bar{x})^2,$$

the reason being that this estimator is unbiased, while being also quite close to $\frac{s_x^2}{h}$. For example, with MATLAB, the function std(x) returns

$$\sqrt{\frac{1}{n-1}\sum_{i=1}^{n}(x_i - \bar{x})^2}$$

instead of

$$\sqrt{\frac{1}{n}\sum_{i=1}^{n}(x_i - \bar{x})^2}.$$

Remark 1.3.2 *When parameters are estimated using the maximum likelihood principle and the observations are not Gaussian, one rarely finds explicit expressions for the estimators. Therefore, one has to use numerical algorithms to maximize the likelihood. In this case, domain constraints on the parameters must be taken into account. For example, in the Black-Scholes model, $\sigma > 0$. One can easily replace this sign constraint by setting $\sigma = e^{\alpha}$, with $\alpha \in \mathbb{R}$.*

1.4 Estimation Errors

Once the maximum likelihood estimations have been computed, we should concern ourselves with the properties of the estimation errors, i.e., find the asymptotic behavior of $\sqrt{n}\,(\hat{\mu}_n - \mu)$ and $\sqrt{n}\,(\hat{\sigma}_n - \sigma)$. In general, the limiting distribution will be a multivariate Gaussian distribution (see Appendix A.6.18).

The next proposition follows from the results of Appendix B.5.1. To denote "convergence in law," we use the symbol $\rightsquigarrow$, as defined in Appendix B.

Proposition 1.4.1 *As n tends to ∞, $\begin{pmatrix} \sqrt{n}\,(\hat{\mu}_n - \mu) \\ \sqrt{n}\,(\hat{\sigma}_n - \sigma) \end{pmatrix}$ converges in law to a centered bivariate Gaussian distribution with covariance matrix V, denoted*

$$\begin{pmatrix} \sqrt{n}\,(\hat{\mu}_n - \mu) \\ \sqrt{n}\,(\hat{\sigma}_n - \sigma) \end{pmatrix} \rightsquigarrow N_2(0, V),$$

where $V = \frac{\sigma^2}{2}\begin{pmatrix} \frac{2}{h} + \sigma^2 & \sigma \\ \sigma & 1 \end{pmatrix}$. A consistent estimator of V is given by

$$\hat{V} = \frac{\hat{\sigma}_n^2}{2}\begin{pmatrix} \frac{2}{h} + \hat{\sigma}_n^2 & \hat{\sigma}_n \\ \hat{\sigma}_n & 1 \end{pmatrix}. \tag{1.3}$$

In particular, we obtain that $\sqrt{n}\,(\hat{\mu}_n - \mu)/\sqrt{\hat{V}_{11}} \rightsquigarrow N(0,1)$ and $\sqrt{n}\,(\hat{\sigma}_n - \sigma)/\sqrt{\hat{V}_{22}} \rightsquigarrow N(0,1)$.

The proof is given in Appendix 1.C.2.

1.4.1 Estimation of Parameters for Apple

For an example of application of the previous results, consider the MATLAB file *ApplePrices* [3] containing the adjusted values of Apple stock on Nasdaq (aapl), from January 13th, 2009, to January 14th, 2011.

The results of the estimation of parameters μ and σ on an annual time scale, are given in Table 1.1. They have been obtained with the MATLAB function *EstBS1dExp*, using the results in Propositions 1.3.1 and 1.4.1. According to our comments on time scale and trading days at the end of Section 1.2.2, we have taken $h = 1/252$.

Remark 1.4.1 *When using a transformation like $\sigma = \exp(\alpha)$ to work with unconstrained parameters, we obtain an estimation V_0 of the limiting covariance matrix for the estimation error on the parameter $\theta = (\mu, \alpha)$. It follows from the delta method (Theorem B.3.4.1) that the estimation $\hat{V}$ of the limiting covariance matrix for the estimation error on the parameter (μ, σ) is*

$$\hat{V} = \hat{J} V_0 \hat{J},$$

where $\hat{J} = \mathrm{diag}(1, \hat{\sigma}_n)$, i.e., $\hat{J} = \begin{pmatrix} 1 & 0 \\ 0 & \hat{\sigma}_n \end{pmatrix}$. This is because $(\mu, \sigma) =$

[3] For example, these prices can be obtained at the YAHOO!FINANCE website.

TABLE 1.1: 95% confidence intervals for the average return per annum (μ) and the volatility per annum (σ). The estimation of the covariance V is also given.

Parameter	Last Year ($n = 254$)	Last Two Years ($n = 506$)
μ	0.5347 ± 0.5185	0.7314 ± 0.4120
σ	0.2656 ± 0.0231	0.2978 ± 0.0183
V	$\begin{pmatrix} 17.7747 & 0.0094 \\ 0.0094 & 0.0353 \end{pmatrix}$	$\begin{pmatrix} 22.3568 & 0.0132 \\ 0.0132 & 0.0444 \end{pmatrix}$

$H(\mu, \alpha) = (\mu, e^{\alpha})$, so the Jacobian matrix of H is $J = \begin{pmatrix} 1 & 0 \\ 0 & \sigma \end{pmatrix}$, which is estimated by $\hat{J}$.

This approach is implemented in the MATLAB function EstBS1dNum. *For more details on the transformation of parameters and the corresponding limiting covariance matrix, see Remark B.3.3.*

Measurement Scale

Note that since σ is a measure of scale, we should take that into account when choosing the starting point of the minimization algorithm, otherwise the algorithm might not converge, even for simple models. Here, we chose $\mu = 0$ and $\alpha = \ln(\sigma) = \ln(1)$.

Using properties of the Gaussian distribution, we can define the following unbiased and consistent estimator for σ, often used in statistical quality control applications:

$$\hat{\sigma}_n = \sqrt{\frac{\pi}{4h}} \times \frac{1}{n-1} \sum_{i=2}^{n} |X_i - X_{i-1}|.$$

In this case, we obtain that $\sqrt{n}\,(\hat{\sigma}_n - \sigma) \rightsquigarrow N\left(0, 0.8264\sigma^2\right)$. See Exercise 1.9 for details. This estimator is not used in practice even if its estimation error is just a little bit larger than the estimation error of the maximum likelihood estimator.

In the next section we will present the famous Black-Scholes formula for the value of a European call option[4] under the Black-Scholes model.

[4] A call option gives its owner the right of buying an asset for a fixed price, called the strike price.

1.5 Black-Scholes Formula

After presenting the Black-Scholes formula, we will see that in general, the value of a European option can be expressed as the solution to a partial differential equation. It can also be expressed as an expectation, implying the possibility of using Monte Carlo simulations for valuing the option.

1.5.1 European Call Option

Suppose that the price S of the underlying asset is modeled by a geometric Brownian motion (1.1), and that r is the risk-free rate, assumed to be constant on the period $[0, T]$. One also assumes that there is no dividend on that period. It can be shown [Black and Scholes, 1973] that the value at time t, of a European call option of strike price K and maturity T, depends only on $S(t) = s$ and on the volatility σ, and is given by

$$C(t,s) = s\mathcal{N}(d_1) - Ke^{-r(T-t)}\mathcal{N}(d_2), \tag{1.4}$$

where $\mathcal{N}$[5] is the distribution function of a standard Gaussian variable, and

$$\begin{aligned} d_1 &= \frac{\ln(s/K) + r\tau + \sigma^2\tau/2}{\sigma\sqrt{\tau}}, \\ d_2 &= d_1 - \sigma\sqrt{\tau} = \frac{\ln(s/K) + r\tau - \sigma^2\tau/2}{\sigma\sqrt{\tau}}, \end{aligned}$$

where $\tau = T - t$ is the time to maturity. Note that as $t \to T$, $\tau \to 0$, and one can check that $\lim_{t\to T} C(t,s) = \max(s - K, 0)$. Furthermore, as expected, $C(t,s)/s \to 0$, as $s \to 0$, and $C(t,s)/s \to 1$, as $s \to \infty$.

Risk-Free Rate

For short-term maturities, the risk-free rate r is often deduced from the Libor rate (%) with the closest maturity. Since the Libor is not a compounded rate, the relationship between r and Libor is

$$r = \ln\left(1 + \frac{Libor}{100}\right). \tag{1.5}$$

When the maturity is expressed in (trading) days, then $\tau = \frac{maturity}{252}$ on an annual scale. This is the convention that we adopt throughout the book.

[5] In MATLAB, the function corresponding to $\mathcal{N}$ is *normcdf*.

Remark 1.5.1 *It is remarkable that the value of the option does not depend on the average return μ, only on the volatility σ. It does not mean however that μ is not important. For example, if one wants to characterize the behavior (mean, variance, VaR, etc.) of the future value at time t of a portfolio containing call options based on asset S, one needs to consider $C\{t, S(t)\}$, which in turn depends on both parameters μ and σ.*

1.5.1.1 Put-Call Parity

For a European put[6] option on an asset paying no dividends (during the pricing period), its value $\tilde{C}$ must satisfy

$$C(t, s) - \tilde{C}(t, s) = s - Ke^{-r(T-t)}. \tag{1.6}$$

By a non-arbitrage argument, it is easy to check that the value at time t of an option with payoff $\Phi(s) = s - K$ is simply $S(t) - Ke^{-r(T-t)}$, since a portfolio consisting, at time t, of 1 share of the asset and $Ke^{-r(T-t)}$ borrowed at (continuous) rate r will have value $S(T) - K$ at maturity T.

Now, a portfolio consisting of a long position in 1 call option and a short position in 1 put option has value $C(t, s) - \tilde{C}(t, s)$ at time t, while its value at maturity is

$$\max\{S(T) - K, 0\} - \max\{K - S(T), 0\} = S(T) - K.$$

As a result, one must have $C(t, s) - \tilde{C}(t, s) = s - Ke^{-r(T-t)}$.

1.5.1.2 Early Exercise of an American Call Option

When there are no dividends, it is never optimal to exercise an American call option before maturity and its value will be the same as a European call option. To see why, note that an American call option is a European call option with the additional feature that it can be exercised before maturity. Therefore, it must be as valuable as a European call option, i.e.,

$$C_{American}(t, s) \geq C_{European}(t, s),$$

where $C_{American}(t, s)$ and $C_{European}(t, s)$ are respectively American and European call options on the same underlying asset with the same strike price and maturity. Now, we can rewrite the put-call parity relationship (1.6) as

$$\begin{aligned} C_{European}(t, s) &= s - Ke^{-r(T-t)} + \tilde{C}_{European}(t, s) \\ &= s - K + K\left\{1 - e^{-r(T-t)}\right\} + \tilde{C}_{European}(t, s) > s - K. \end{aligned}$$

Putting these two observations together and remarking that $C(t, s) > 0$, one obtains that

$$C_{American}(t, s) \geq C_{European}(t, s) > \max(s - K, 0).$$

[6] A put option gives its owner the right of selling an asset for a fixed price.

Therefore, the value of an American call option prior to maturity is always higher than its immediate exercise value.

A similar reasoning shows that this is not true for put options. In this case, if the underlying price falls enough and the call option's value is low enough, then it might become optimal to exercise a put option prior to maturity.

1.5.2 Partial Differential Equation for Option Values

In Black and Scholes [1973] and Merton [1974], it is shown that the value C of a European call option with maturity T satisfies the following partial differential equation:

$$\frac{\partial C}{\partial t} + rs\frac{\partial C}{\partial s} + \sigma^2 \frac{s^2}{2}\frac{\partial^2 C}{\partial s^2} = rC, \tag{1.7}$$

with boundary condition $C(T, s) = \max(s - K, 0)$, for all $s \geq 0$ and for all $t \in (0, T)$. It can also be shown that the equation also holds for a general payoff Φ.

A popular method for proving the validity of (1.7) is to construct a self-financing portfolio $\Pi_t = \psi_{t1} S_t + \psi_{t0} e^{rt}$, so that at maturity $\Pi_T = \Phi\{S(T)\}$, where $\Phi\{S(T)\}$ is the payoff. In fact, $\psi_{t1} = \frac{\partial}{\partial s} C(t, s)\big|_{s=S_t}$. The justification of (1.7) is given in Appendix 1.A.

In simple cases, numerical methods used for solving partial differential equations can be used to solve (1.7); see, e.g., Wilmott [2006]. However, as we will see in later chapters, derivatives can depend on several risk factors or even depend on the path taken by the underlying asset. In such cases, numerical solution to partial differential equation can become cumbersome. Therefore, we now turn to a representation of the derivative's price which is easier to handle.

1.5.3 Option Value as an Expectation

Under the absence of arbitrage, there exists an equivalent probability measure under which the discounted value of an option is a martingale (see Appendix 1.B). Such a measure is called an *equivalent martingale measure*[7] or risk-neutral measure, and one can show that it is unique for the Black-Scholes model. In this case, the actual value of an option is simply the expected discounted value of the option at a later date, for example at maturity, under the equivalent martingale measure Q. The value of the option at time t is thus given by

$$C(t, s) = e^{-r(T-t)} E\left[\Phi\{\tilde{S}(T)\}\middle|\, \tilde{S}(t) = s\right], \tag{1.8}$$

[7]Under an equivalent martingale measure, the discounted value of any tradable security is a martingale.

where, under Q, $\tilde{W}$ is a Brownian motion and

$$\tilde{S}(u) = se^{\left(r-\frac{\sigma^2}{2}\right)(u-t)+\sigma\left\{\tilde{W}(u)-\tilde{W}(t)\right\}}, \quad u \in [t,T].$$

Equivalently, one has

$$d\tilde{S}(u) = r\tilde{S}(u)du + \sigma\tilde{S}(u)d\tilde{W}(u), \quad t \le u \le T,$$

with $\tilde{S}(t) = s$.

Note that the law of $\tilde{S}$ is not the same as the law of S, μ being replaced by r. In practice, only the process S is observed, not $\tilde{S}$.

Using the Feynman-Kac formula in Proposition 7.4.1, one can show that (1.8) is the solution of the partial differential equation (1.7).

Example 1.5.1 *For example, for a European call option with strike price K, we have $\Phi(s) = \max(s-K,0)$. One can then recover the Black-Scholes formula* (1.4) *using Proposition A.6.3 and the expectation formula* (1.8).

Remark 1.5.2 *Formula* (1.8) *can be extended to path-dependent options like Asian options, lookback options, etc. One has to estimate or evaluate the discounted payoff of the option under the dynamics $\tilde{S}$. Monte Carlo methods are then more than appropriate in this context since a Brownian motion is easy to simulate.*

1.5.3.1 Equivalent Martingale Measures and Pricing of Options

The important notion of non-arbitrage is related to the existence of an equivalent martingale measure Q under which the discounted asset $e^{-rt}S_t$ is a martingale. This, in turns, is equivalent to the existence of a positive martingale M under the underlying probability measure P of the model (called the objective measure), such that $e^{-rt}S_tM_t$ is a martingale. The equivalent martingale measure Q is then defined by its density M with respect to P. In fact, for any security X, we can compute its expectation under law Q by the formula

$$E_Q(X) = E_P(XM).$$

For pricing purposes, the value of any derivative is computed with respect to Q, not under P. Under the new law Q, the distribution of the underlying asset will change in general, as exemplified in the Black-Scholes model. Finally, note that when there are (infinitely) many equivalent martingales measures, the "right one" is the one determined by the market. It could be called the implied distribution. See, e.g., Section 1.6.2. We will come back to these important notions in the next chapters.

1.5.4 Dividends

Suppose that during the period of validity of the option, dividends $D_1, \ldots, D_m$ are paid at periods $t_1, \ldots, t_m$, with $0 = t_0 < t_1 < \cdots < t_m < T$. For simplicity, set $t_{m+1} = T$. By analogy with the Black-Scholes model, we assume that during periods $[t_{k-1}, t_k)$, $k \in \{1, \ldots, m+1\}$, we have

$$S_t = S_{t_{k-1}} e^{\left(\mu - \frac{\sigma^2}{2}\right)(t - t_{k-1}) + \sigma\{W(t) - W(t_{k-1})\}}, \quad t \in [t_{k-1}, t_k),$$

while there is a jump at the dividend period t_k. More precisely,

$$S_{t_k} = S_{t_k-} - D_k, \quad k \in \{1, \ldots, m\}.$$

To obtain the value on an option with payoff Φ at maturity, one can still use the expectation formula (1.8), taking the jumps into account.

In fact, if $t \in [t_m, T]$, then (1.8) yields

$$C(t, s) = e^{-r(T-t)} E\left[\Phi\left\{\tilde{S}(T)\right\}\middle| \tilde{S}(t) = s\right].$$

Next, if $t \in [t_{m-1}, t_m)$, then using the properties of conditional expectations, we have

$$\begin{aligned} C(t, s) &= e^{-r(t_m - t)} E\left[C\left\{t_m, \tilde{S}(t_m)\right\}\middle| \tilde{S}(t) = s\right] \\ &= e^{-r(t_m - t)} E\left[C\left\{t_m, \tilde{S}(t_m-) - D_m\right\} |\tilde{S}(t) = s\right]. \end{aligned}$$

As a result, if $t \in [t_{k-1}, t_k)$, $k \in \{1, \ldots, m\}$, then

$$\begin{aligned} C(t, s) &= e^{-r(t_k - t)} E\left[C\left\{t_k, \tilde{S}(t_k)\right\}\middle| \tilde{S}(t) = s\right] \\ &= e^{-r(t_k - t)} E\left[C\left\{t_k, \tilde{S}(t_k-) - D_k\right\}\middle| \tilde{S}(t) = s\right]. \end{aligned}$$

1.5.4.1 Continuously Paid Dividends

If one assumes that the dividends are paid continuously at a constant rate δ, then, under the equivalent martingale measure, we have

$$\tilde{S}(u) = \tilde{S}(t) e^{\left(r - \delta - \frac{\sigma^2}{2}\right)(u - t) + \sigma\{\tilde{W}(u) - \tilde{W}(t)\}}, \quad u \in [t, T].$$

As a result we obtain the following extension of the Black-Scholes formula: The value of a European call option on a stock paying continuously dividends at a rate δ is given by

$$C_\delta(t, s) = s e^{-\delta\tau} \mathcal{N}(d_1) - K e^{-r(T-t)} \mathcal{N}(d_2), \tag{1.9}$$

where

$$\begin{aligned} d_1 &= \frac{\ln(s/K) + (r - \delta)\tau + \sigma^2\tau/2}{\sigma\sqrt{\tau}}, \\ d_2 &= d_1 - \sigma\sqrt{\tau} = \frac{\ln(s/K) + (r - \delta)\tau - \sigma^2\tau/2}{\sigma\sqrt{\tau}}, \end{aligned}$$

and where $\tau = T - t$ is the time to maturity.

Note that the value of a call option with continuously paid dividends at rate δ, can be expressed in terms of the value of a call option without dividends. In fact, it is easy to check that

$$C_\delta(t,s,r) = C_0\left(t, se^{-\delta\tau}\right) = e^{-\delta\tau} C_0(t,s,r-\delta). \tag{1.10}$$

For the put-call parity, we have a similar result. In fact, if $\tilde{C}_\delta$ is the value of a European put with continuously paid dividends at rate δ, then

$$\begin{aligned} \tilde{C}_\delta(t,s) &= C_\delta(t,s) - se^{-\delta\tau} + Ke^{-r\tau} & (1.11) \\ &= \tilde{C}_0\left(t, se^{-\delta\tau}\right) = e^{-\delta\tau}\tilde{C}_0(t,s,r-\delta). & (1.12) \end{aligned}$$

1.6 Greeks

It is often important to measure the sensitivity of the option value with respect to the variables t, s, r, and σ. The so-called *greeks* are measures of sensitivity based on partial derivatives with respect to those parameters. Explicit formulas for greeks are known only in few cases, in particular the European call option [Wilmott, 2006]. In general, since there is no explicit expression for the option value, the greeks must be approximated. This will be done in Section 1.7. Here are the main definitions and interpretations for these useful parameters.

- The sensitivity of the option value with respect to the underlying asset price, called *delta*, is defined by

$$\Delta = \frac{\partial C}{\partial s}. \tag{1.13}$$

  The delta of an option is quite useful in hedging since it corresponds to the number of shares needed to create a risk-free portfolio replicating the value of the option at maturity; see Appendix 1.A.

- The sensitivity of the option value with respect to time, called *theta*, is defined by

$$\Theta = \frac{\partial C}{\partial t}. \tag{1.14}$$

  Note that $-\Theta$, evaluated at $\tau = T - t$, yields the sensitivity with respect to the time to maturity τ.

- The sensitivity of the option value with respect to the interest rate r, called *rho*, is defined by

$$\rho = \frac{\partial C}{\partial r}. \tag{1.15}$$

- The sensitivity of the option value with respect to the volatility, called *vega*, is defined by

$$\mathcal{V} = \frac{\partial C}{\partial \sigma}. \tag{1.16}$$

As shown next in Section 1.6.3, the vega is also important in determining the error on the option price due to the estimation of the volatility.

- A measure of convexity, the second order derivative of the option value with respect to the underlying asset prices, called *gamma*, is defined by

$$\Gamma = \frac{\partial^2 C}{\partial s^2}. \tag{1.17}$$

Γ is useful in some approximations.

1.6.1 Greeks for a European Call Option

Using the Black-Scholes formula (1.4), it is easy to check that

- $\Delta = \dfrac{\partial C}{\partial s} = \mathcal{N}(d_1) > 0.$

- $\Theta = \dfrac{\partial C}{\partial t} = -\dfrac{\sigma s}{2\sqrt{T-t}} \dfrac{e^{-d_1^2/2}}{\sqrt{2\pi}} - Kre^{-r(T-t)}\mathcal{N}(d_2) < 0.$

- $\rho = \dfrac{\partial C}{\partial r} = K(T-t)e^{-r(T-t)}\mathcal{N}(d_2) > 0.$

- $\mathcal{V} = \dfrac{\partial C}{\partial \sigma} = s\sqrt{T-t}\, \dfrac{e^{-d_1^2/2}}{\sqrt{2\pi}} > 0.$

  Since the vega is positive, it means that the value of the option is an increasing function of the volatility. This property is essential in determining the so-called implied volatility.

- $\Gamma = \dfrac{\partial^2 C}{\partial s^2} = \dfrac{1}{s\sigma\sqrt{T-t}} \dfrac{e^{-d_1^2/2}}{\sqrt{2\pi}} > 0.$

  Since the gamma is positive, it means that the value of the option is a convex function of the underlying asset value.

Remark 1.6.1 *For continuously paid dividends at rate δ, using formula (1.10), it is easy to check that $\Delta_\delta(t,s) = e^{-\delta\tau}\Delta_0\left(t, se^{-\delta\tau}\right)$. Also $\Gamma_\delta(t,s) = e^{-2\delta\tau}\Gamma_0\left(t, se^{-\delta\tau}\right)$. Next, $\Theta_\delta(t,s) = \Theta_0\left(t, se^{-\delta\tau}\right) + s\Delta_\delta(t,s)$. Finally, $\rho_\delta(t,s) = \rho_0\left(t, se^{-\delta\tau}\right)$ and $\mathcal{V}_\delta(t,s) = \mathcal{V}_0\left(t, se^{-\delta\tau}\right)$.*

1.6.2 Implied Distribution

One might ask why there is no sensitivity parameter corresponding to the partial derivative with respect to the strike price. In fact, there is one and it is related to the implied distribution [Breeden and Litzenberger, 1978]. Assuming that the value of a European call option is given by the expectation formula (1.8), and using the properties of expectations, namely (A.2), we obtain

$$C(t,s) = E_Q\left[\max\{\tilde{S}(T) - K, 0\}|\tilde{S}(t) = s\right] = \int_K^\infty Q\{\tilde{S}(T) > y\}dy, \quad (1.18)$$

where Q denotes the equivalent martingale measure. As a result,

$$\frac{\partial C}{\partial K} = -Q\{\tilde{S}(T) > K\} = \tilde{F}(K) - 1,$$

where $\tilde{F}$ is the distribution function of $\tilde{S}(T)$ given $\tilde{S}(t) = s$, under the equivalent martingale measure Q. As a result $\frac{\partial C}{\partial K}$ is non-decreasing and it follows that $\frac{\partial^2 C}{\partial K^2} = \tilde{f}(K) \geq 0$, where $\tilde{f}$ is the associated density, provided it exists. It also shows that the value of a call option is always a convex function of the strike. Note that in the case of the Black-Scholes model, the implied distribution is the log-normal, since $\ln\{\tilde{S}(T)\}$ has a Gaussian distribution with mean $\ln(s) + \left(r - \frac{\sigma^2}{2}\right)\tau$ and variance $\sigma^2\tau$, under the equivalent martingale measure. Since (1.18) is assumed to be always valid, not only for the Black-Scholes model, the implied distribution function can be approximated from the market prices of the calls if there are enough strike prices available. See, e.g., Aït-Sahalia and Lo [1998].

As an example, consider the values of call options on Apple, on January 14th, 2011, with a 24-day maturity. The first data are shown in Table 1.2; the complete data set is in the MATLAB structure *AppleCalls* containing the strikes and call market values for four different maturities. The graph is displayed in Figure 1.1. One can notice that the value of the call for a strike $K = \$210$ seems too low, while the call values for strikes $K = \$160$ and $K = \$170$ are too close, destroying the (theoretical) convexity of the curve.

TABLE 1.2: Some market values of call options on Apple with a 24-day maturity, on January 14th, 2011.

Strike	160.00	170.00	200.00	210.00	220.00	240.00
Call	169.62	169.60	147.85	**112.25**	125.70	108.00

1.6.3 Error on the Option Value

In practice, the value of the option is estimated by $\hat{C}_n = C(\hat{\sigma}_n)$, where $\hat{\sigma}_n$ is the estimation of the volatility σ. As a result, an error in the estimation

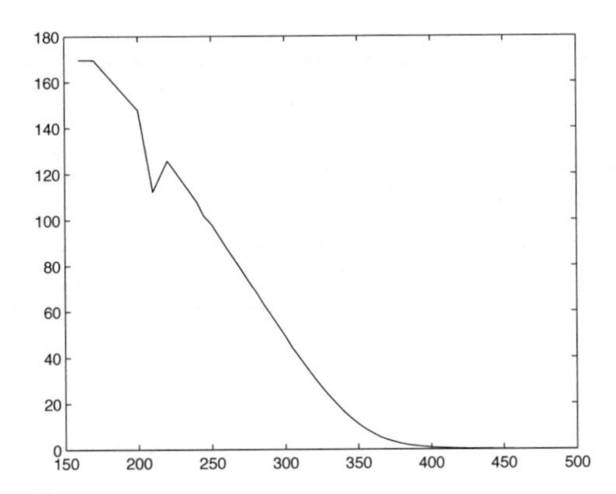

FIGURE 1.1: Graph of strikes vs call prices on Apple with a 24-day maturity, on January 14th, 2011.

of σ converts into an error on the estimation of the value of the option. The following interesting result can be found in Lo [1986].

Proposition 1.6.1 *As $n \to \infty$,*

$$\frac{\sqrt{n}\,(\hat{C}_n - C)}{\left|\hat{\mathcal{V}}_n\right| \sqrt{\hat{V}_{22}}} \rightsquigarrow N(0,1),$$

where $\hat{V}_{22}$ is given by (1.3), and $\hat{\mathcal{V}}_n = \frac{\partial C}{\partial \sigma}(\hat{\sigma}_n) = \mathcal{V}(\hat{\sigma}_n)$ is the estimation of the vega of the option. In other terms, the error on the option value is the error on the volatility multiplied by the absolute value of the vega.

The proof is given in Appendix 1.C.3.

Example 1.6.1 (Call Option on Apple) *The closing price for Apple, on January 14th, 2011, is* \$348.48. *Recall from Table 1.1 that the estimation of the volatility using the last $n = 254$ days is* 0.2656 ± 0.0231, *while it is* 0.2978 ± 0.0183 *for the last $n = 506$ days.*

In Table 1.3, we compared the market values of call options with strike $K = 350$ and their estimations, for three maturities: February 2011 (24 trading days), April 2011 (63 trading days), and January 2012 (255 trading days). Note that options usually expire on the third Friday of the month.

It seems that the market values corresponding to smaller maturities belong to the confidence interval computed with fewer data, while for longer maturities, one needs more historical data for the confidence interval to cover the market price. Here is a possible explanation: It is obvious that the uncertainty about the value of the asset increases with the maturity, yielding an increased

TABLE 1.3: Call option market values and their estimations for Apple (aapl) on January 14th, 2011, for a strike price $K = 350$.

Maturity (days)	Value	95% C.I. ($n = 254$)	95% C.I. ($n = 506$)
24	11.10	10.7115 ± 0.9907	12.0957 ± 0.7873
63	18.65	17.8574 ± 1.6040	20.0979 ± 1.2739
255	43.17	37.5775 ± 3.1950	42.0363 ± 2.5329

uncertainty about the value of the option, since its vega is positive. Therefore, for the market value to belong to the confidence interval, the latter must be larger as well, which can be achieved by having a larger estimated volatility; also, since the estimated volatility is likely to increase with the length of the historical series, chances are that the market value will belong to a confidence interval computed with more data. On the other hand, for small maturities, the recent past is more likely to better explain the future behavior of the asset than a longer series, so the market value is more likely to belong to a confidence interval based on fewer historical data.

The call value estimation $\hat{C} = 12.0957$, *corresponding to 24 (trading) days to maturity and estimated volatility* $\hat{\sigma} = 0.2978$, *was obtained using the MATLAB function* FormulaBS[8] *with the following parameters:* $S = 348.48$, $K = 350$, $\tau = 24/252$, $r = \ln(1 + 0.0026) = 0.0026$, *and* $\sigma = 0.2978$, *i.e.,*

$$\hat{C} = FormulaBS(S, K, r, \tau, \sigma).$$

The estimation $\hat{\mathcal{V}} = 42.9035$ *was obtained using the MATLAB function* FormulaBSGreeks[9] *with the same parameters. Detailed calculations are given in the MATLAB script* AppleCallEst. *The Libor rates of Table 1.4 corresponding to the 1-month, 3-month, and 1-year maturities were used to compute the risk-free rate* r, *according to formula* (1.5) *and the comments on the risk-free rate in Section 1.5.1.*

TABLE 1.4: Libor rates (%) around January 14th, 2011.

1-month	3-month	6-month	1-year
0.26	0.30	0.46	0.78

[8] In the Financial Toolbox of MATLAB, the corresponding function is *blsprice*.

[9] In the Financial Toolbox of MATLAB, the corresponding functions are *blsdelta*, *blsgamma*, and *blsvega*.

1.6.4 Implied Volatility

The implied volatility of a call option corresponds to the unique volatility value yielding the observed call price using the Black-Scholes formula (1.4). Its existence is based on the following result.

Proposition 1.6.2 *For every observed value C_{obs} of an option price in the open interval*

$$\left(\max\left\{0, S(t) - Ke^{-r(T-t)}\right\}, S(t)\right)$$

corresponds a unique value σ_{imp}, called the implied volatility, such that

$$C_{obs} = C\{t, S(t), K, T, r, \sigma_{imp}\}.$$

PROOF. Since

$$\begin{aligned} \lim_{\sigma \to 0} C(t, s, K, T, r, \sigma) &= \max\left\{0, s - Ke^{-r(T-t)}\right\}, \\ \lim_{\sigma \to \infty} C(t, s, K, T, r, \sigma) &= s, \end{aligned}$$

and $\mathcal{V} = \frac{\partial C}{\partial \sigma} > 0$, one may conclude that C is an increasing function with respect to σ, and there exists a unique value σ_{imp} such that

$$C_{obs} = C(t, S(t), K, T, r, \sigma_{imp}),$$

whenever $C_{obs} \in \left(\max\left\{0, S(t) - Ke^{-r(T-t)}\right\}, S(t)\right)$.

■

Example 1.6.2 (Call Option on Apple (continued)) *Taking the same parameters as in Example 1.6.1, one ends up with the implied volatilities given in Table 1.5. These values have been computed with the MATLAB function* ImpliedVolBS.[10] *Recall from Table 1.1 that the estimation of the volatility using the last $n = 254$ days is 0.2656 ± 0.0231, i.e., $\sigma \in [0.2425 \pm 0.2887]$, while for the last $n = 506$ days, the estimation is 0.2978 ± 0.0183 i.e., $\sigma \in [0.2795, 0.3161]$.*

TABLE 1.5: Implied volatility for call option values for Apple (aapl) on January 14th, 2011, for a strike price $K = 350$.

Maturity (days)	Implied Volatility
24	$0.2746 \in [0.2425, 0.2887]$
63	$0.2770 \in [0.2425, 0.2887]$
255	$0.3060 \in [0.2795, 0.3161]$

One can conclude that the implied volatility for 1-month and 3-month maturities are not significantly different from the volatility estimated by using

[10]In the Financial Toolbox of MATLAB, the corresponding function is *blsimpv*.

the last year of historical data, while the implied volatility for the 1-year maturity is not significantly different from the volatility estimated by using the last two years of historical data. One could restate the argument on the larger uncertainty about the option price that we made in Example 1.6.1, by saying that the implied volatility increases with the maturity. This is reflected in the so-called volatility smile effect discussed in Chapter 3.

1.6.4.1 Problems with Implied Volatility

Given $S(t) = s$, for different strike prices K_1, K_2 or different maturity dates T_1, T_2, one cannot solve in general the following set of equations with respect to σ:

$$\begin{aligned} C_1 &= C(t, s, K_1, T_1, r, \sigma), \\ C_2 &= C(t, s, K_2, T_2, r, \sigma). \end{aligned}$$

To overcome this problem, Gouriéroux and Jasiak [2001] suggested to add random error terms with null expectation. However, that would imply an incomplete market.

Note that even if one does not believe in the Black-Scholes, implied volatility is often used to compare option prices. In fact, the implied volatility surface, corresponding to the implied volatility for different values of strikes and maturities, is often used in practice. See, e.g., Carr and Madan [2001].

1.7 Estimation of Greeks using the Broadie-Glasserman Methodologies

While it is generally impossible to find explicit expressions for the option value, we can however fairly easily estimate them with a Monte Carlo approximation of the expected value in (1.8). On a similar note, expressions for the greeks are often not available. An easy way to circumvent this problem, which is often used in practice, is to estimate them with a finite difference approximation. For example, the delta could be approximated as

$$\Delta \approx \frac{C(t, s + \epsilon) - C(t, s)}{\epsilon}$$

where ϵ is a small positive scalar. However, such procedures are plagued by an inevitable tradeoff; a large ϵ will produce biased estimations of the greeks, while small ϵ values will results in high estimation variance.

Fortunately, Broadie and Glasserman [1996] proposed methods to estimate an option's value, together with unbiased estimations of the greeks. They considered several models, including the Black-Scholes model.

According to formula (1.8), the value of a European option with payoff Φ at maturity is

$$C(t,s) = e^{-r\tau} E\left[\Phi\left\{s e^{\left(r-\frac{\sigma^2}{2}\right)\tau+\sigma\sqrt{\tau}Z}\right\}\right] \tag{1.19}$$

$$= e^{-r\tau}\int_{-\infty}^{+\infty}\Phi\left\{s e^{\left(r-\frac{\sigma^2}{2}\right)\tau+\sigma\sqrt{\tau}z}\right\}\frac{e^{-z^2/2}}{\sqrt{2\pi}}dz$$

$$= e^{-r\tau}\int_{0}^{+\infty}\Phi(x)\frac{e^{-\frac{1}{2\sigma^2\tau}\left\{\ln(x/s)-\left(r-\frac{\sigma^2}{2}\right)\tau\right\}^2}}{x\sigma\sqrt{2\pi\tau}}dx, \tag{1.20}$$

where $Z \sim N(0,1)$ and $\tau = T - t$ is the time to maturity.

Suppose that one generates $Z_1, \ldots, Z_N \sim N(0,1)$, with N large enough. Further set $\tilde{S}_i = s e^{\left(r-\frac{\sigma^2}{2}\right)\tau+\sigma\sqrt{\tau}Z_i}$, $i \in \{1\ldots, N\}$.

Then, an unbiased and consistent estimation of $C(t,s)$ is given by

$$\hat{C} = \frac{1}{N}\sum_{i=1}^{N} e^{-r\tau}\Phi\left(\tilde{S}_i\right).$$

This Monte Carlo approach was proposed a long time ago by Boyle [1977]. However, no unbiased estimation of the greeks was proposed until Broadie and Glasserman [1996]. In their article, the authors proposed in fact two methodologies to estimate greeks, based respectively on representations (1.19) and (1.20). These methodologies have the advantage of being computed in parallel with the option price, not sequentially.

1.7.1 Pathwise Method

The first methodology, called *pathwise method*, is based on representation (1.19). To be applicable, one has to assume that the payoff function Φ is differentiable "almost everywhere," i.e., everywhere but possibly at a countable set of points[11]. However, note that the partial derivatives of any order for $\tilde{S}_i$ exist for any possible parameter $\theta \in \{s, r, t, \sigma\}$.

Proposition 1.7.1 *Suppose that Φ is differentiable almost everywhere. Then simultaneous unbiased estimations of the option value and its first order derivatives are given respectively by*

$$\hat{C} = \frac{1}{N}\sum_{i=1}^{N} e^{-r\tau}\Phi\left(\tilde{S}_i\right) \tag{1.21}$$

[11] The right mathematical formulation would that it is differentiable everywhere but at a set of Lebesgue measure 0. A countable set has Lebesgue measure 0.

and

$$\widehat{\partial_\theta C} = \partial_\theta \hat{C} = -\hat{C}\partial_\theta(r\tau) + e^{-r\tau}\frac{1}{N}\sum_{i=1}^{N}\Phi'(\tilde{S}_i)\partial_\theta \tilde{S}_i. \tag{1.22}$$

Estimation of the Gamma

For the Black-Scholes model, we have $\Gamma = \frac{\mathcal{V}}{s^2\sigma\tau}$. Therefore, since $\hat{\mathcal{V}}$ is an unbiased estimation of $\mathcal{V}$, it follows that $\hat{\Gamma} = \frac{\hat{\mathcal{V}}}{s^2\sigma\tau}$ is an unbiased estimation of Γ.

This is important since for a European call option, the second derivative of the payoff with respect to s is 0 almost everywhere.

Remark 1.7.1 *Since these estimations are averages of independent and identically distributed random vectors $X_1, \ldots, X_N$ with mean $\mathcal{G} \in \mathbb{R}^p$, one can determine the asymptotic behavior of the estimation errors. In fact, the central limit theorem (Theorem B.4.1) applies to yield*

$$\sqrt{N}(\bar{X} - \mathcal{G}) \rightsquigarrow N_p(0, V),$$

where V is estimated by

$$\frac{1}{N-1}\sum_{i=1}^{N}\left(X_i - \bar{X}\right)\left(X_i - \bar{X}\right)^\top.$$

Example 1.7.1 *For a European call option, $\Phi(s) = \max(s-K, 0)$, so $\Phi'(s) = \mathbb{I}(s > K)$ almost everywhere. As a result,*

$$\hat{\Delta} = e^{-r\tau}\frac{1}{N}\sum_{i=1}^{N}\frac{\tilde{S}_i}{s}\mathbb{I}\left(\tilde{S}_i > K\right)$$

and

$$\hat{\mathcal{V}} = e^{-r\tau}\frac{1}{N}\sum_{i=1}^{N}(Z_i\sqrt{\tau} - \sigma\tau)\tilde{S}_i\mathbb{I}\left(\tilde{S}_i > K\right).$$

Therefore $\mathcal{G} = (C, \Delta, \mathcal{V})^\top$ can be estimated as the mean of the 3-dimensional random vectors

$$X_i = e^{-r\tau}\mathbb{I}\left(\tilde{S}_i > K\right)\left(\tilde{S}_i - K, \frac{\tilde{S}_i}{s}, \tilde{S}_i(Z_i\sqrt{\tau} - \sigma\tau)\right)^\top,$$

$i \in \{1, \ldots, N\}$.

1.7.2 Likelihood Ratio Method

The second method proposed by Broadie and Glasserman [1996] is based on representation (1.20). For $x > 0$, set

$$f(x) = \frac{e^{-\frac{1}{2\sigma^2\tau}\left\{\ln(x/s)-\left(r-\frac{\sigma^2}{2}\right)\tau\right\}^2}}{x\sigma\sqrt{2\pi\tau}}.$$

Then f is the density of $\tilde{S}(T)$ given $\tilde{S}(t) = s$. Note that f is differentiable with respect to every parameter $\theta \in \{s, r, \sigma, t\}$.

Proposition 1.7.2 *Simultaneous unbiased estimations of the value of the option and derivatives of order 1 are given by*

$$\hat{C} = \frac{1}{N}\sum_{i=1}^{N} e^{-r\tau}\Phi\left(\tilde{S}_i\right) \tag{1.23}$$

and

$$\widehat{\partial_\theta C} = -\hat{C}\partial_\theta\left(r\tau\right) + e^{-r\tau}\frac{1}{N}\sum_{i=1}^{N}\Phi(\tilde{S}_i)\,\partial_\theta\left[\ln\{f(x)\}\right]\big|_{x=\tilde{S}_i}. \tag{1.24}$$

In particular

$$\hat{\Delta} = e^{-r\tau}\frac{1}{N}\sum_{i=1}^{N}\frac{Z_i}{s\sigma\sqrt{\tau}}\Phi(\tilde{S}_i)$$

and

$$\hat{\mathcal{V}} = e^{-r\tau}\frac{1}{N}\sum_{i=1}^{N}\frac{\left(Z_i^2 - 1 - Z_i\sigma\sqrt{\tau}\right)}{\sigma}\Phi(\tilde{S}_i).$$

Moreover, an unbiased estimation of the gamma is given by

$$\hat{\Gamma} = e^{-r\tau}\frac{1}{N}\sum_{i=1}^{N}\frac{\left(Z_i^2 - 1 - Z_i\sigma\sqrt{\tau}\right)}{s^2\sigma^2\tau}\Phi(\tilde{S}_i) = \frac{\hat{\mathcal{V}}}{s^2\sigma\tau}. \tag{1.25}$$

1.7.3 Discussion

In Broadie and Glasserman [1996], the authors concluded that the pathwise method is more precise than the likelihood ratio method. However their conclusion is only based on a few examples. One cannot conclude that it will be true in general for all kind of options. It is yet to be proven.

1.8 Suggested Reading

For the Black-Scholes model, I recommend Black and Scholes [1973], [Hull, 2006, Chapter 13] and [Wilmott, 2006, Chapters 5–8]. Lo [1986] also proposes interesting empirical tests for the model. Finally, for the greeks, the interested reader should consult Broadie and Glasserman [1996].

1.9 Exercises

Exercise 1.1

State the main assumptions of the Black-Scholes model. Do you think they are met in practice?

Exercise 1.2

Suppose that the daily price S of an asset satisfies the Black-Scholes model with parameters μ and σ (on a daily time scale). With the daily data, we first compute the log-returns x_t, $t = 1, \ldots, 506$. Then with the MATLAB function *fminunc*, we find the following estimations for $\theta = (\mu, \ln(\sigma))$.

```
theta =

    0.0029   -3.9769

hessian =

  1.0e+006 *

    1.4404   -0.0007
   -0.0007    0.0010
```

(a) Compute 95% confidence intervals for μ and σ.

(b) Compute 95% confidence intervals for the mean and volatility per annum.

Exercise 1.3

Can you explain the difference between the results of EstBS1Dexp with the explicit method vs the results of EstBS1DNum with the numerical method?

Exercise 1.4

Consider that the Black-Scholes model is valid for asset S, and assume it has a 36% volatility per annum. The risk-free rate is 2% and the actual value of the asset is $71.

(a) We buy 100 3-month maturity European calls and 50 2-month maturity European puts on the underlying asset. If both options have a strike price of $70, how much money do we need to borrow to get this portfolio?

(b) One month later, the value of S has grown up to $74.27, while the volatility raised to 47%. Compute the P&L according to this new information.

Exercise 1.5

Let C be the value of a European call option given by the Black-Scholes formula. Prove that each of the following limits exist and give a financial interpretation of each result.

(a) $\lim_{\sigma\to 0} C(t, s, K, T, r, \sigma) = \max\left\{s - K^{-r(T-t)}, 0\right\}$.

(b) $\lim_{\sigma\to\infty} = C(t, s, K, T, r, \sigma) = s$.

(c) $\lim_{t\to T} C(t, s) = \max(s - K, 0)$.

(d) Set $d_{1,t} = \frac{\ln\{S(t)/K\}+r(T-t)+\sigma^2(T-t)/2}{\sigma\sqrt{T-t}}$, and set $d_{2,t} = d_{1,t} - \sigma\sqrt{T-t}$. Deduce from (c) and the Black-Scholes formula that at time t, if one has $N(d_{1,t})$ shares of the asset and if one borrows $Ke^{-r(T-t)}N(d_{2,t})$ at the risk-free rate, then the final value of the portfolio is the payoff, i.e., one can replicate exactly the payoff of the option.

Exercise 1.6

Suppose that $S \in [\$50, \$150]$, $K = \$100$, $\sigma = 0.25$, $r = 1.5\%$, and $\tau = 1$.

(a) Plot the graph of Δ for the call and the put, as a function of the underlying asset.

(b) Plot the graph of Γ for the call and the put, as a function of the underlying asset.

(c) Plot the graph of $\mathcal{V}$ for the call and the put, as a function of the underlying asset.

Exercise 1.7

(a) Compute the greeks for a European binary call option.

(b) Compute the greeks for a European put option.

(c) For a European option with payoff $\Phi\{S(T)\}$, show that $\Gamma = \frac{\mathcal{V}}{s^2 \sigma \tau}$. Hint: Use representation (1.20).

Exercise 1.8

(a) Explain what is the implied volatility.

(b) Find the implied volatility of a 9-month maturity European call option with a market value of \$1.12, where the actual price of the asset is $S_0 = \$73$, the strike is \$75, and the annual risk-free rate is constant at 1.5%.

Exercise 1.9

Let $X_1, X_2, \ldots, X_n$ be i.i.d. Gaussian with mean μ and variance σ^2.

(a) Show that the mean and variance of $Y_1 = |X_1 - X_2|$ are $\sigma\sqrt{\frac{4}{\pi}}$ and $2\sigma^2\left(1 - \frac{2}{\pi}\right)$ respectively.

(b) Prove that the covariance between $Y_1 = |X_1 - X_2|$ and $Y_2 = |X_2 - X_3|$ is $2\sigma^2\left\{\frac{1}{6} - \frac{(2-\sqrt{3})}{\pi}\right\}$. This exercise is difficult.

(c) Consider the estimation of σ given by

$$\hat{\sigma}_n = \sqrt{\frac{\pi}{4}} \times \frac{1}{n-1}\sum_{i=2}^{n} |X_i - X_{i-1}|.$$

Deduce from (a) and (b) that $E\left(\hat{\sigma}_n\right) = \sigma$ and $n\mathrm{Var}\left(\hat{\sigma}_n\right) \to \frac{2\pi}{3} + \sqrt{3} - 3 \approx .8264$, as $n \to \infty$.

Exercise 1.10

For a European put, write the MATLAB expressions for its value and the greeks $(\Delta, \Gamma, \mathcal{V})$, by using the likelihood ratio method of Broadie-Glasserman, as a function of the following variables:

- s : initial value of the underlying asset;
- K : strike price;
- r : risk-free rate;
- τ : time to maturity, i.e., $(T - t)$;

- σ : volatility;
- N : number of simulations.

To write the function, use the following intermediate variables:

- Z : Nx1 vector of standard Gaussian random variables;
- Stilde : Nx1 vector of final values under risk-neutral measure of the underlying asset;
- Phi: payoff of the put option at maturity.

Using $N = 1000000$, compare the estimated values with the true values, if $s = 100$, $K = 100$, $r = 1.5\%$, $\tau = 1$, and $\sigma = 0.25$.

1.10 Assignment Questions

1. Write a MATLAB function for estimating the greeks (delta, gamma, theta, rho, vega) of a European put option, using the two methods proposed by Broadie and Glasserman [1996]. Using $N = 100000$, compare the estimated values with the true values, if $s = 100$, $K = 100$, $r = 1.5\%$, $\tau = 1$, and $\sigma = 0.25$. Which method is the most precise?

2. Use the MATLAB functions *NumJacobian*, *NumHessian*, and *FormulaBS* to compute the greeks of a call option. For a wide range of strikes, maturities, and stock prices, compare the numerical values with the theoretical values.

3. Construct a MATLAB function for the computation of the implied volatility surface. The input must be a structure like the one of *AppleCalls*. The number of maturities must be deduced from the size of the structure.

1.A Justification of the Black-Scholes Equation

The following justification is based on Björk [1999] and uses stochastic calculus. The idea is to construct a self-financing portfolio $\Pi_t = \psi_{t1} S(t) + \psi_{t0} e^{rt}$ so that at maturity, $\Pi_T = \Phi\{S(T)\}$, where $\Phi\{S(T)\}$ is the payoff function. Since the portfolio if self-financing, one must have

$$d\Pi_t = \psi_{t1} dS(t) + \psi_{t0} d(e^{rt}) = \left\{\mu\psi_{t1} S(t) + r\psi_{t0} e^{rt}\right\} dt + \sigma\psi_{t1} S(t) dW(t).$$

On the other hand, the option value process $C\{t, S(t)\}$ can be obtained using Itô's formula:

$$\begin{aligned} dC\{t, S(t)\} &= \partial_t C\{t, S(t)\}dt + \partial_s C\{t, S(t)\}dS(t) + \frac{S^2(t)\sigma^2}{2}\partial_s^2 C\{t, S(t)\}dt \\ &= \{\partial_t C\{t, S(t)\} + \mu S(t)\partial_s C\{t, S(t)\}\}\, dt \\ &\quad + \frac{\sigma^2 S^2(t)}{2}\partial_s^2 C\{t, S(t)\}dt + \sigma S(t)\partial_s C\{t, S(t)\}dW(t). \end{aligned}$$

Let the self-financing portfolio be formed in such a way that it replicates the option's value at any time t. Since the two representations must agree, they must have the same diffusion term and it follows that

$$\psi_{t1} = \partial_s C\{t, S(t)\},$$

which implies that

$$\psi_{t0} = e^{-rt}\left[C\{t, S(t)\} - \psi_{t1} S(t)\right].$$

They must also have the same drift term, that is

$$\begin{aligned} \mu\psi_{t1} S(t) + r\psi_{t0} e^{rt} &= (\mu - r)S(t)\partial_s C\{t, S(t)\} + rC\{t, S(t)\} \\ &= \partial_t C\{t, S(t)\} + \mu S(t)\partial_s C\{t, S(t)\} \\ &\quad + \frac{\sigma^2 S^2(t)}{2}\partial_s^2 C\{t, S(t)\}. \end{aligned}$$

As a result, $C(t, s)$ is the unique solution of the following partial differential equation (1.7), i.e.,

$$\frac{\partial C}{\partial t} + rs\frac{\partial C}{\partial s} + \sigma^2 \frac{s^2}{2}\frac{\partial^2 C}{\partial s^2} = rC,$$

with boundary condition $C(T, s) = \Phi(s)$, for all $s \geq 0$ and for all $t \in (0, T)$.

1.B Martingales

Martingales are important stochastic processes that are used throughout the book. But first, we need to define the notion of a filtration, since a martingale is always defined with respect to a given filtration.

A filtration is a collection of increasing sigma-algebras $\mathcal{F}_t$, $t \geq 0$. In applications, $\mathcal{F}_t$ is often generated by the past values of a stochastic process X_s, $s \in [0, t]$, together with exogenous variables. If there are no exogenous variables, the filtration is called the natural filtration of the process X.

Definition 1.B.1 (Martingale) *A stochastic process M is a martingale with respect to a filtration $\mathbb{F} = \{\mathcal{F}_t; t \geq 0\}$, if for all $0 \leq s \leq t$, $E(|M_t|) < \infty$, M_t is $\mathcal{F}_t$-measurable and for all $s \in [0,t]$,*

$$E(M_t|\mathcal{F}_s) = M_s.$$

For example, consider the Brownian motion W. It is easy to see that with respect to its natural filtration $\mathbb{F}$, W is a martingale since $W(t) - W(s)$ is independent of $\mathcal{F}_s$. For the same reason, $e^{\sigma W(t) - \sigma^2 t/2}$ is also a martingale.

1.C Proof of the Results

1.C.1 Proof of Proposition 1.3.1

Following Proposition 1.2.1, the returns are independent and Gaussian, with mean $\left(\mu - \frac{\sigma^2}{2}\right) h$ and variance $\sigma^2 h$. Hence the likelihood function that we need to maximize is

$$L(\mu,\sigma) = \prod_{i=1}^{n} \frac{1}{\sqrt{2\pi\sigma^2 h}} e^{-\frac{\left\{x_i - h\left(\mu - \sigma^2/2\right)\right\}^2}{2\sigma^2 h}},$$

for $\mu \in \mathbb{R}$ and $\sigma > 0$. This is equivalent to minimizing

$$\begin{aligned} g(\mu,\sigma) &= -\ln\left(L(\mu,\sigma)\right) \\ &= n\ln(\sigma) + \frac{1}{2\sigma^2 h}\sum_{i=1}^{n}\left(x_i - \mu h + \frac{\sigma^2 h}{2}\right)^2 + \frac{n}{2}\ln(2\pi h). \end{aligned}$$

Next,

$$\begin{aligned} \frac{\partial}{\partial\mu} g &= \tfrac{n}{\sigma^2}(\mu h - \sigma^2 h/2 - \bar{x}), \\ \frac{\partial}{\partial\sigma} g &= \tfrac{n}{\sigma} - \tfrac{1}{\sigma^3 h}\textstyle\sum_{i=1}^{n}(x_i - \mu h + \sigma^2 h/2)^2 - \tfrac{n}{\sigma}(\mu h - \sigma^2 h/2 - \bar{x}). \end{aligned}$$

The two derivatives are zero when $\hat{\mu}_n$ and $\hat{\sigma}_n$ are given by

$$\begin{aligned} \hat{\mu}_n &= \tfrac{\bar{x}}{h} + \tfrac{(\hat{\sigma}_n)^2}{2} = \tfrac{\bar{x}}{h} + \tfrac{s_x^2}{2h}, \\ \hat{\sigma}_n &= \sqrt{\tfrac{1}{nh}\textstyle\sum_{i=1}^{n}(x_i - \bar{x})^2} = \tfrac{s_x}{\sqrt{h}}. \end{aligned}$$

It is easy to check that these two values yield the minimum value of g.

■

1.C.2 Proof of Proposition 1.4.1

Following Example B.5.1, $\sqrt{n}(\bar{x} - \mu h + \sigma^2 h/2) \rightsquigarrow N(0, \sigma^2 h)$, and $\sqrt{n}(s_x - \sigma\sqrt{h}) \rightsquigarrow N(0, \sigma^2 h/2)$. Moreover, the two statistics are asymptotically independent. Hence $\sqrt{n}\left(\frac{s_x}{\sqrt{h}} - \sigma\right) \rightsquigarrow N(0, \sigma^2/2)$. In addition,

$$\sqrt{n}(\hat{\mu}_n - \mu) = \frac{1}{h}\sqrt{n}\left(\bar{x} - \mu h + \sigma^2 h/2\right) + \frac{1}{2h}\sqrt{n}\left(s_x^2 - \sigma^2 h\right).$$

As a result, $\sqrt{n}(\hat{\mu}_n - \mu) \rightsquigarrow N\left(0, \frac{\sigma^2}{h} + \frac{\sigma^4}{2}\right)$, and the limiting covariance between the two statistics is $\frac{\sigma^3}{2}$.

■

1.C.3 Proof of Proposition 1.6.1

It follows from the delta method (B.3.4.1) and Proposition 1.4.1 that

$$\begin{aligned}
\sqrt{n}\,(\hat{C} - C) &= \sqrt{n}\,(C(\hat{\sigma}_n) - C(\sigma)) \\
&\approx \frac{\partial C}{\partial \sigma}(\hat{\sigma}_n) \times \sqrt{n}\,(\hat{\sigma}_n - \sigma) \\
&\rightsquigarrow N\left(0, (\hat{\mathcal{V}})^2 \hat{V}_{22}\right).
\end{aligned}$$

Hence the result.

■

Bibliography

Y Aït-Sahalia and A. W. Lo. Nonparametric estimation of state-price densities implicit in financial asset prices. *Journal of Finance*, 53(2):499–547, 1998.

T. Björk. *Arbitrage Theory in Continuous Time.* Oxford University Press, 1999.

F. Black and M. Scholes. The pricing of options and corporate liabilities. *Journal of Political Economy*, 81:637–654, 1973.

P. Boyle. Options: A Monte Carlo approach. *Journal of Financial Economics*, 4:323–338, 1977.

D. T. Breeden and R. H. Litzenberger. Prices of state-contingent claims implicit in option prices. *The Journal of Business*, 51(4):621–651, 1978.

M. Broadie and P. Glasserman. Estimating security price derivatives using simulation. *Management Science*, 42:260–285, 1996.

P. Carr and D. Madan. Determining volatility surfaces and option values from an implied volatility smile. In *Quantitative Analysis in Financial Markets*, pages 163–191. World Sci. Publ., River Edge, NJ, 2001.

C. Gouriéroux and J. Jasiak. *Financial Econometrics: Problems, Models, and Methods.* Princeton Series in Finance. Princeton University Press, 2001.

J. C. Hull. *Options, Futures, and Other Derivatives.* Prentice-Hall, sixth edition, 2006.

A. W. Lo. Statistical tests of contingent claims asset-pricing models. *Journal of Financial Economics*, 17:143–173, 1986.

R. C. Merton. On the pricing of corporate debt: The risk structure of interest rates. *Journal of Finance*, 29:449–470, 1974.

P. Wilmott. *Paul Wilmott on Quantitative Finance*, volume 1. John Wiley & Sons, second edition, 2006.

Chapter 2

Multivariate Black-Scholes Model

In this chapter, we will study properties of the Black-Scholes model for multiple assets. We will show how to estimate the parameters of the model and how to measure the estimation errors. Examples of options on several assets will be given, including quanto options, where some underlying assets are traded in foreign currencies. Finally, we will extend the Monte Carlo approach of Broadie-Glasserman to estimate the associated *greeks* and see how to measure the impact of estimation errors on the value of options.

2.1 Black-Scholes Model for Several Assets

In many cases, an option's payoff may depend on several assets, e.g., swap options, quantos, basket options, etc. Therefore one has to model the dynamical behavior of several assets.

Definition 2.1.1 *$W = (W_1, \ldots, W_d)$ is a d-dimensional Brownian motion with correlation matrix R if it is a continuous random vector starting at 0, with independent increments, and such that $W(t) - W(s) \sim N_d(0, R(t-s))$, whenever $0 \le s \le t$.*

Note that it follows that the components $W_1, \ldots, W_d$ are correlated Brownian motions with

$$\mathrm{Cov}\left\{W_j(s), W_k(t)\right\} = R_{jk} \min(s,t), \quad s,t, \ge 0, \quad j,k \in \{1,\ldots,d\}.$$

One can now state the extension of the Black-Scholes model to several assets.

Definition 2.1.2 *The values $S_1, \ldots, S_d$ of assets are modeled by geometric Brownian motions if*

$$S_j(t) = S_j(0) e^{\left(\mu_j - \frac{\sigma_j^2}{2}\right)t + \sigma_j W_j(t)}, \quad t \ge 0, \quad j \in \{1,\ldots,d\}, \tag{2.1}$$

where the Brownian motions W_i are correlated, with correlation matrix R.

One can also say that $S = (S_1, \ldots, S_d)$ are correlated geometric Brownian *motions with parameters μ and Σ, where the covariance matrix Σ is defined by*

$$\Sigma_{jk} = \sigma_j \sigma_k R_{jk}, \quad j,k \in \{1,\ldots,d\}. \tag{2.2}$$

Remark 2.1.1 *As in the one-dimensional case, the geometric Brownian motions satisfy the following system of stochastic differential equations:*

$$dS_j(t) = \mu_j S_j(t)dt + \sigma_j S_j(t) dW_j(t), \quad j \in \{1,\ldots,d\}. \tag{2.3}$$

2.1.1 Representation of a Multivariate Brownian Motion

Using the properties of Gaussian random vectors (Section A.6.18), a d-dimensional Brownian motion W with correlation matrix R can be constructed as a linear combination of d independent univariate Brownian motions $Z = (Z_1,\ldots,Z_d)^\top$ by setting $W(t) = b^\top Z(t)$, where $b^\top b = R$. We can then rewrite (2.1) as

$$\frac{S_j(t)}{S_j(0)} = e^{\left(\mu_j - \frac{\sigma_j^2}{2}\right)t + \sigma_j \sum_{k=1}^d b_{kj} Z_k(t)}, \quad t \geq 0, \quad j \in \{1,\ldots,d\}.$$

An interesting decomposition of $R = b^\top b$ is when b is an upper triangular matrix, as in Cholesky decomposition. In this case,

$$\frac{S_j(t)}{S_j(0)} = e^{\left(\mu_j - \frac{\Sigma_{jj}}{2}\right)t + \sum_{k=1}^j a_{kj} Z_k(t)}, \quad t \geq 0, \quad j \in \{1,\ldots,d\}, \tag{2.4}$$

where the Brownian motions $Z_1,\ldots,Z_d$ are independent, and $a_{jk} = \sigma_k b_{jk}$. The matrix a is upper triangular, contains all the information on the dependence, and is uniquely determined by the Cholesky decomposition $a^\top a = \Sigma$, i.e.,

$$\Sigma_{jk} = (a^\top a)_{jk} = \sum_{l=1}^{\min(j,k)} a_{lj} a_{lk} = \sigma_j \sigma_k R_{jk}, \quad j,k \in \{1,\ldots,d\},$$

under the constraints $a_{jj} > 0$, $j \in \{1,\ldots,d\}$. In the bivariate case, one can check that $a = \begin{pmatrix} \sigma_1 & R_{12}\sigma_2 \\ 0 & \sigma_2\sqrt{1 - R_{12}^2} \end{pmatrix}$.

2.1.2 Simulation of Correlated Geometric Brownian Motions

The following algorithm is a direct application of (2.4) and the independence of the increments of Brownian motions.

Algorithm 2.1.1 *To generate $S(h),\ldots,S(nh)$ with parameters μ, Σ, we can proceed in the following way:*

- *Find the Cholesky decomposition a of Σ.*

- *Generate independent Gaussian variables* $Z_{11}, \ldots, Z_{nd}$, *with mean* 0 *and covariance* 1.
- *For* $i \in \{1, \ldots, n\}$ *and* $j \in \{1, \ldots, d\}$, *compute*

$$X_{ij} = \left(\mu_j - \frac{\sigma_j^2}{2}\right) h + \sqrt{h} \sum_{k=1}^{d} a_{kj} Z_{ik}.$$

- *For* $i \in \{1, \ldots, n\}$ *and* $j \in \{1, \ldots, d\}$, *set* $S_j(ih) = s_j e^{\sum_{k=1}^{i} X_{kj}}$.

This algorithm is implemented in the MATLAB function *SimBS*.

2.1.3 Volatility Vector

Definition 2.1.3 *The volatility vector* v, *based on model* (2.4), *is defined by*

$$v = \vec{a} = (a_{11}, \ldots, a_{1d}, a_{22}, \ldots, a_{dd})^\top. \tag{2.5}$$

The term "volatility vector" is appropriate because we recover the volatility parameter when $d = 1$, these values multiply independent Brownian motions, and as we will see later, option values depend only on the volatility vector, as in the univariate Black-Scholes model.

The model defined by (2.1) contains many unknown parameters: vectors $\mu = (\mu_1, \ldots, \mu_d)^\top$ and $\sigma = (\sigma_1, \ldots, \sigma_d)^\top$, and correlation matrix R. For model (2.4), the parameters are μ and the upper triangular matrix a (or the volatility vector v). We must estimate these parameters. To do so, we will assume, as it is generally the case in practice, that the data are observed at regular time intervals of length h.

Since we will use the maximum likelihood principle to estimate the parameters, we need to find the joint distribution of the returns.

2.1.4 Joint Distribution of the Returns

Proposition 2.1.1 *Suppose that* S *is a* d*-dimensional geometric Brownian motion with parameters* μ *and* Σ. *Then, for a given* $h > 0$, *the returns* $X_i = (X_{i1}, \ldots, X_{id})^\top$, $i \in \{1, \ldots, n\}$, *defined by*

$$X_{ij} = \ln\{S_j(ih)\} - \ln[S_j\{(i-1)h\}], \quad j \in \{1, \ldots, d\},$$

are independent and identically distributed, with $X_i \sim N_d(mh, \Sigma h)$, *where*

$$m_j = \mu_j - \sigma_j^2/2, \qquad j \in \{1, \ldots, d\},$$

and Σ *is defined by* (2.2).

The proof is given in Appendix 2.B.1.

2.2 Estimation of Parameters

The maximum likelihood principle will now be used to estimate parameters μ, σ, and R of correlated geometric Brownian motions with representation (2.1).

Suppose that we observe prices $S(0) = s_0$, $S(h) = s_1, \ldots, S(nh) = s_n$, from which one compute the returns x_i, where

$$x_{ij} = \ln(s_{ij}) - \ln(s_{i-1,j}), \quad j \in \{1, \ldots, d\}, \quad i \in \{1, \ldots, n\}.$$

Consider the statistics

$$\bar{x} = \frac{1}{n}\sum_{i=1}^{n} x_i \quad \text{and} \quad S_x = \frac{1}{n}\sum_{i=1}^{n}(x_i - \bar{x})(x_i - \bar{x})^\top.$$

2.2.1 Explicit Method

Proposition 2.2.1 *For correlated geometric Brownian motions with representation* (2.1), *the maximum likelihood estimates of μ, σ and R are given by*

$$\begin{aligned} \hat{\mu}_{n,j} &= \frac{\bar{x}_j}{h} + \frac{(S_x)_{jj}}{2h}, & j \in \{1, \ldots, d\}, \\ \hat{\sigma}_{n,j} &= \frac{\sqrt{(S_x)_{jj}}}{\sqrt{h}}, & j \in \{1, \ldots, d\}, \\ \hat{R}_{n,jk} &= \frac{(S_x)_{jk}}{h\hat{\sigma}_{n,j}\hat{\sigma}_k}, & j, k \in \{1, \ldots, d\}. \end{aligned}$$

The proof is given in Appendix 2.B.2.

Remark 2.2.1 *It follows from Proposition 2.2.1 that $\hat{\Sigma} = \frac{S_x}{h}$, where Σ satisfies* (2.2), *and $\hat{m} = \frac{\bar{x}}{h}$, where $m_j = \mu_j - \frac{\Sigma_{jj}}{2}$, $j \in \{1, \ldots, d\}$.*

Corollary 2.2.1 *For correlated geometric Brownian motions with representation* (2.4), *the maximum likelihood estimates of μ and a are given by*

$$\hat{\mu}_{n,j} = \frac{\bar{x}_j}{h} + \frac{(S_x)_{jj}}{2h}, \quad j \in \{1, \ldots, d\},$$

and $\hat{a}_n$ is the Cholesky decomposition of $\frac{S_x}{h}$, i.e., $\hat{a}_n^\top \hat{a}_n = \frac{S_x}{h}$.

Remark 2.2.2 *In practice, one uses*

$$\tilde{S}_x = \frac{1}{n-1}\sum_{i=1}^{n}(x_i - \bar{x})(x_i - \bar{x})^\top,$$

which is very close to S_x when n is not small. With MATLAB, one has $\tilde{S}_x = \mathrm{cov}(X)$, if X is the $n \times d$ matrix with i-th line defined by $x_i^\top$, $i \in \{1, \ldots, n\}$. The associated correlation matrix is obtained by the MATLAB function corrcoef, while $\hat{a}_n$ can be computed by $\mathrm{chol}(\tilde{S}_x/h)$.

2.2.2 Numerical Method

In general, we cannot compute explicitly the maximum likelihood estimates of a model, so we have to rely on numerical methods for optimization. One of the main problems encountered with numerical methods for optimization is the problem of constraints on the parameters.

For example, for model (2.1), the matrix R must be positive definite and symmetric. In two dimensions, this condition is simple since the correlation matrix R is determined by the unique number $\rho = R_{12}$. In this case, one needs to assume that the correlation coefficient ρ is in the interval $(-1, 1)$. To get rid of this constraint, we can set $\rho = \tanh(\alpha)$. Its inverse, called the Fisher transformation, is $\alpha = \frac{1}{2}\ln\left(\frac{1+\rho}{1-\rho}\right)$. See Figure 2.1.

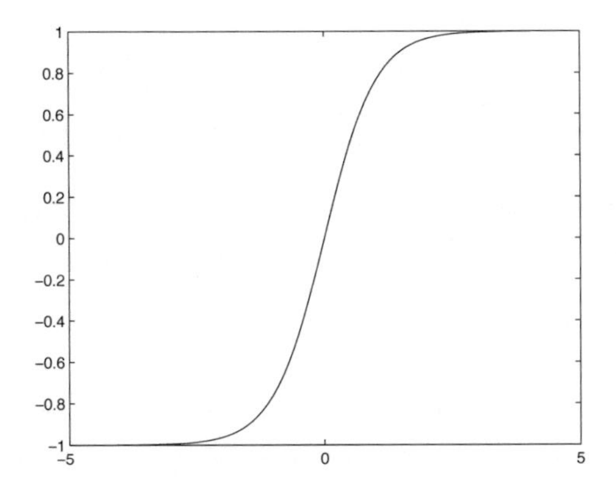

FIGURE 2.1: Graph of $\tanh(\alpha) = \frac{e^{\alpha}-e^{-\alpha}}{e^{\alpha}+e^{-\alpha}}$, $\alpha \in [-5, 5]$.

For higher dimensions, it is impossibly difficult to find explicit constraints on a matrix coefficients to ensure that it is positive definite. This is where representation (2.4) becomes interesting. The only constraint on the upper triangular matrix a is that its diagonal is positive. As parameters are expressed by vectors, it is therefore natural to work with the volatility vector v.

In addition, as we will see later, it is relatively easy to compute the estimator error on option prices in terms of the error on v.

2.3 Estimation Errors

We will now compute the estimation errors for representations (2.1) and (2.4), and we will illustrate these results using real data.

2.3.1 Parametrization with the Correlation Matrix

The proof of the following proposition in given in Appendix 2.B.3.

Proposition 2.3.1 *Let $\hat{\mu}_n, \hat{\sigma}_n, \hat{\Sigma}_n, \hat{R}_n$ be the maximum likelihood estimators of μ, σ, Σ, R respectively, as given in Proposition 2.2.1. Then the behavior of the errors for the estimation of μ_j and σ_j, $j \in \{1, \ldots, d\}$, can be obtained from Proposition 1.4.1. Also* $\text{Cov}\left\{\sqrt{n}\left(\hat{\sigma}_{n,i} - \sigma_i\right), \sqrt{n}\left(\hat{\sigma}_{n,j} - \sigma_j\right)\right\} \to R_{ij}^2 \frac{\sigma_i \sigma_j}{2}$, *as $n \to \infty$, for all $1 \le i \le j \le d$.*

In addition, as $n \to \infty$, $\sqrt{n}\left(\hat{\Sigma}_{n,ij} - \Sigma_{ij}\right) \rightsquigarrow N\left(0, \sigma_i^2\sigma_j^2 + \Sigma_{ij}^2\right)$, and for all $i, j, k, l \in \{1, \ldots, d\}$,

$$\text{Cov}\left\{\sqrt{n}\left(\hat{\Sigma}_{n,ij} - \Sigma_{ij}\right), \sqrt{n}\left(\hat{\Sigma}_{n,kl} - \Sigma_{kl}\right)\right\} \to \Sigma_{ik}\Sigma_{jl} + \Sigma_{kj}\Sigma_{il}. \tag{2.6}$$

Finally, for all $1 \le i < j \le d$, $\sqrt{n}\left(\hat{R}_{n,ij} - R_{ij}\right) \rightsquigarrow N\left(0, \left(1 - R_{ij}^2\right)^2\right)$, as $n \to \infty$, and $\text{Cov}\left\{\sqrt{n}\left(\hat{R}_{n,ij} - R_{ij}\right), \sqrt{n}\left(\hat{R}_{n,kl} - R_{kl}\right)\right\}$ *converges to*

$$\begin{aligned} & R_{ik}R_{jl} + R_{jk}R_{il} - R_{ij}(R_{ik}R_{il} + R_{jk}R_{jl}) - R_{kl}(R_{ik}R_{jk} + R_{il}R_{jl}) \\ & -\frac{R_{ij}R_{kl}}{2}\left(R_{ik}^2 + R_{il}^2 + R_{jk}^2 + R_{jl}^2\right), \end{aligned}$$

for all $i, j, k, l \in \{1, \ldots, d\}$.

Remark 2.3.1 *As a by-product of Proposition 2.3.1, we obtain the well-known result of Fisher [1921] for his correlation transform. For all $1 \le i < j \le d$ for which $|R_{ij}| < 1$, set $\hat{\alpha}_{n,ij} = \frac{1}{2}\ln\left(\frac{1+\hat{R}_{n,ij}}{1-\hat{R}_{n,ij}}\right)$, and $\alpha_{ij} = \frac{1}{2}\ln\left(\frac{1+R_{ij}}{1-R_{ij}}\right)$, so $R = \tanh(\alpha)$ and $\hat{R}_n = \tanh(\alpha_n)$ componentwise. Then for $1 \le i < j \le d$, $\sqrt{n}\left(\hat{\alpha}_{n,ij} - \alpha_{ij}\right) \rightsquigarrow N(0,1)$, as $n \to \infty$, whenever $|R_{ij}| < 1$. This follows from the delta method (Theorem B.3.1) since*

$$\frac{d}{dR}\frac{1}{2}\ln\left(\frac{1+R}{1-R}\right) = \frac{1}{1-R^2}.$$

2.3.2 Parametrization with the Volatility Vector

Define the $d(d+3)/2$-dimensional vector θ by $\theta_j = \mu_j$, $j \in \{1, \ldots, d\}$, and

$$a = \begin{pmatrix} \exp\left(\theta_{d+1}\right) & \theta_{d+2} & \cdots & \theta_{2d} \\ 0 & \exp\left(\theta_{2d+1}\right) & \cdots & \theta_{3d-1} \\ \vdots & \vdots & \cdots & \vdots \\ 0 & 0 & \cdots & \exp\left(\theta_{\frac{d(d+3)}{2}}\right) \end{pmatrix}.$$

Then θ contains all the values needed to compute μ and a. Moreover, there are no constraints on θ. Note that $a_{jj} = \exp\left(\theta_{jd - \frac{(j-1)(j-2)}{2} + 1}\right)$, $j \in \{1, \ldots, d\}$, and $a_{jk} = \theta_{jd-\frac{j(j-1)}{2}+k}$, $j,k \in \{1,\dots,d\}$, with $j<k$. The likelihood function to minimize is

$$g(\theta) = \frac{nd}{2}\ln(2\pi h) + \frac{n}{2}\ln(|\Sigma|) + \frac{1}{2h}\sum_{i=1}^{n}(x_i - mh)^\top \Sigma^{-1}(x_i - mh),$$

where $\Sigma = a^\top a$ and $m_j = \mu_j - \Sigma_{jj}/2$, $j \in \{1,\dots,d\}$.

Denote by $\hat{\theta}_n$ the value of θ minimizing $g(\theta)$. Recall that for any function $L : \mathbb{R}^d \mapsto \mathbb{R}$, the Hessian matrix H of L is defined by $H_{jk} = \partial_{\theta_j}\partial_{\theta_k} L(\theta)$, $j,k \in \{1,\dots,d\}$.

The following proposition gives the behavior of the estimation errors. Its proof is given in Appendix 2.B.4.

Proposition 2.3.2 $\sqrt{n}\,(\hat{\theta}_n - \theta) \rightsquigarrow N_{d(d+3)/2}\left(0, \hat{\mathcal{I}}^{-1}\right)$, *and*

$$\begin{pmatrix} \sqrt{n}\,(\hat{\mu}_n - \mu) \\ \sqrt{n}\,(\hat{v}_n - v) \end{pmatrix} \rightsquigarrow N_{d(d+3)/2}(0, \hat{V}),$$

where v is the volatility vector defined by (2.5), $\hat{V} = \hat{J}\hat{\mathcal{I}}^{-1}\hat{J}$, and the Fisher information matrix $\mathcal{I}$ is estimated by $\hat{\mathcal{I}} = H(\hat{\theta}_n)/n$, $H(\hat{\theta}_n)$ being the Hessian matrix[1] *of g evaluated at $\hat{\theta}_n$, or it is estimated by the numerical gradient, as explained in Appendix B.5.1. Here $\hat{J}$ is the estimation of the Jacobian matrix of the transformation $(\mu, v) \mapsto \theta$, i.e., $\hat{J}$ is a diagonal matrix with $\hat{J}_{jj} = 1$ everywhere but at the indices j corresponding to diagonal elements a_{kk}, i.e., when $j = kd - \frac{(k-1)(k-2)}{2} + 1$, for $k \in \{1,\dots,d\}$. In the latter case,*

$$\hat{J}_{jj} = \hat{a}_{n,kk} = \exp\left(\hat{\theta}_{n,\; kd-\frac{(k-1)(k-2)}{2}+1}\right).$$

Remark 2.3.2 *Using Proposition 2.3.2, we can recover the error of estimation of σ_j, for $j \in \{1,\dots,d\}$. In fact, since $\sigma_j^2 = \sum_{k=1}^{j} a_{kj}^2$, the estimation of σ_j^2 is given by $\hat{\sigma}_{n,j}^2 = \sum_{k=1}^{j} \hat{a}_{n,kj}^2$. If $\sqrt{n}\,(\hat{a}_{n,kj} - a_{kj}) \rightsquigarrow Z_{kj}$, then an application of the delta method (Theorem B.3.1) yields that for all $j \in \{1,\dots,d\}$,*

$$\sqrt{n}\,(\hat{\sigma}_{n,j} - \sigma_j) \rightsquigarrow \frac{1}{\sigma_j}\sum_{k=1} a_{kj} Z_{kj}, \tag{2.7}$$

which has a Gaussian distribution with mean 0 and variance

$$\frac{1}{\sigma_j^2}\sum_{k=1}^{j}\sum_{l=1}^{j} a_{kj} a_{lj} \operatorname{Cov}(Z_{kj}, Z_{lj}).$$

According to our parametrization, the variance of Z_{jj} is estimated by

---

[1]Depending on the minimization algorithm used, e.g., the quasi-newton method used by the MATLAB function *fminunc*, one could obtain $H(\hat{\theta}_n)$ directly as an output of the minimization procedure.

$\hat{V}_{jd-\frac{(j-1)(j-2)}{2}+1,\ jd-\frac{(j-1)(j-2)}{2}+1}$. *Moreover, for any* $k,l \in \{1,\ldots,j-1\}$, $\mathrm{Cov}(Z_{kj}, Z_{jj})$ *is estimated by* $\hat{V}_{jd-\frac{(j-1)(j-2)}{2}+1,\ kd-\frac{k(k-1)}{2}+j}$, *while* $\mathrm{Cov}(Z_{kj}, Z_{lj})$ *is estimated by* $\hat{V}_{kd-\frac{k(k-1)}{2}+j,\ ld-\frac{l(l-1)}{2}+j}$.

2.3.3 Estimation of Parameters for Apple and Microsoft

The MATLAB function *EstBS2dExp* can be used to estimate explicitly the parameters μ and Σ on an annual time scale for the bivariate Black-Scholes model with representation (2.1).

The MATLAB file *MicrosoftPrices* contains the adjusted prices of Microsoft[2] from January 13th, 2009, to January 14th, 2011. The results of the explicit estimation for Apple and Microsoft are given in Table 2.1.

Multiple Time Series

Always check that the dates coincide when working with multiple time series. To do so, one can use the MATLAB function *intersect*.

TABLE 2.1: Explicit estimation of the parameters of model (2.1) for Apple and Microsoft on an annual time scale, using the adjusted prices from January 13th, 2009, to January 14th, 2011.

Apple	$\mu = 0.7314 \pm 0.4116$	$\sigma = 0.2978 \pm 0.0183$
Microsoft	$\mu = 0.2441 \pm 0.4142$	$\sigma = 0.2998 \pm 0.0185$
$\rho = 0.4589 \pm 0.0687$		

The MATLAB function *EstBS2dNumCor*[3] can be used to find the maximum likelihood estimators by numerical methods of μ, σ, $\rho = R_{12}$ of model (2.1) in the bivariate case, while the MATLAB function *EstBS2dVol* can be used to find the maximum likelihood estimators of μ and v of model (2.4), in addition to the covariance matrix V of Proposition 2.3.2.

Note that the results obtained with the MATLAB functions *EstBS2dNumCor* (with the Hessian method) are quite similar to the ones obtained with the explicit method with MATLAB function *EstBS2dExp*. In addition, with *EstBS2dNumVol* the errors on the volatility vector v are given in Table 2.2.

[2]For example, these prices can also be obtained at the YAHOO!FINANCE website.

[3]Here, we used the Fisher transformation $\rho = \tanh(\theta)$, so to compute the error on ρ, we have to use the results of Proposition 2.3.1.

TABLE 2.2: 95% confidence intervals for the estimation of the volatility vector v on an annual time scale, for Apple and Microsoft, using the prices from January 13th, 2009, to January 14th, 2011.

$v_1 = a_{11}$	$= 0.2975 \pm 0.0184$
$v_2 = a_{12}$	$= 0.1374 \pm 0.0247$
$v_3 = a_{22}$	$= 0.2661 \pm 0.0247$

Starting Points

Even for a model as simple as the one considered here, the numerical algorithms might not converge to the global optimum, depending on the starting point of the algorithm. For example, for the data set we used here, a starting point corresponding to $\mu_1 = \mu_2 = 0 = \rho$ and $\sigma_1 = \sigma_2 = .001$ does not work, while $\mu_1 = \mu_2 = 0 = \rho$ and $\sigma_1 = \sigma_2 = 1$ or $\mu_1 = \mu_2 = 0 = \rho$ and $\sigma_1 = \sigma_2 = .1$ do work. The latter two starting points make sense since we are estimating parameters on an annual time scale.

2.4 Evaluation of Options on Several Assets

In this section, we first give the value of a European type options on several underlying assets, when they are modeled by geometric Brownian motions according to (2.1) or (2.4). It is also assumed that the "ideal conditions" of the univariate Black-Scholes model are satisfied. In particular, the risk-free interest rate r is constant for the maturity of the option.

As in the univariate case, the value of the option is related to the solution of a partial differential equation, which could be solved numerically. It can also be represented as an expectation, under an equivalent martingale measure. Therefore the value of the option can be estimated by Monte Carlo methods.

2.4.1 Partial Differential Equation for Option Values

In this section, we write the price of a derivative as the solution of a partial differential equation. The justification is very similar to the one given in Appendix 1.A, and can be found in Björk [1999].

In fact, for an option with payoff $\Phi\{S(T)\}$ at maturity T, the value of the option at time t is given by $C\{t, S(t)\}$, where C satisfies the following partial differential equation:

$$\frac{\partial C}{\partial t} + r\sum_{j=1}^{d} s_j \partial_{s_j} C + \frac{1}{2}\sum_{j=1}^{d}\sum_{k=1}^{d} s_j s_k \Sigma_{jk} \partial_{s_j}\partial_{s_k} C = rC, \tag{2.8}$$

for $(t, s) \in (0, T) \times (0, \infty)^d$, with the boundary condition $C(T, s) = \Phi(s)$, $s \in [0, \infty)^d$.

2.4.2 Option Value as an Expectation

Assume that the dynamics of the underlying assets $S = (S_1, \ldots, S_d)$ are given by (2.4). Then the value of the option with payoff $\Phi\{S(T)\}$ at maturity T is

$$C(t, s) = e^{-r(T-t)} E\left[\Phi\{\tilde{S}(T)\} | S(t) = s\right], \tag{2.9}$$

where, under the equivalent martingale measure, $\tilde{Z}_1, \ldots, \tilde{Z}_d$ are independent Brownian motions, and

$$\tilde{S}_j(u) = s_j e^{\left(r - \frac{\Sigma_{jj}}{2}\right)(u-t) + \sum_{k=1}^{j} a_{kj}\left(\tilde{Z}_k(u) - \tilde{Z}_k(t)\right)}, \quad u \in [t, T], \tag{2.10}$$

with a being the Cholesky decomposition of Σ and $S(t) = s = (s_1, \ldots, s_d)^\top$.

Alternatively, one can write $\tilde{S}$ as the solution of a stochastic differential equation viz.,

$$d\tilde{S}_j(u) = r\tilde{S}_j(u)du + \tilde{S}_j(u)\sum_{k=1}^{j} a_{kj} d\tilde{Z}_k(u), \quad t < u \le T, \tilde{S}_t = S(t) = s,$$

$j \in \{1, \ldots, d\}$.

Remark 2.4.1 *In (2.9), the payoff function Φ may depend on the whole trajectory, not just on $S(T)$. For example, for an Asian type option on a basket,*

$$\Phi(S) = \max\left\{\frac{1}{m}\sum_{i=1}^{m}\sum_{j=1}^{d} \omega_j S_j(t_i) - K, 0\right\},$$

for given weights $\omega_1, \ldots, \omega_d$ and time periods $t_1, \ldots, t_m$.

Its continuous time analog could be given by the payoff

$$\Phi(S) = \max\left\{\frac{1}{T}\sum_{j=1}^{d} \omega_j \int_0^T S_j(u)du - K, 0\right\}.$$

2.4.2.1 Vanilla Options

In the simple case where the payoff function Φ only depends on the value of the underlying assets at maturity, i.e., $\Phi = \Phi\{S(T)\}$, the expectation formula (2.9) can be written in the simpler form

$$C(t,s) = e^{-r\tau} E\left\{\Phi(\tilde{S})\right\}, \tag{2.11}$$

where $\tau = T - t$, and

$$\tilde{S}_j = s_j e^{\left(r - \frac{\Sigma_{jj}}{2}\right)\tau + \sqrt{\tau}\sum_{k=1}^{j} a_{kj} Z_k}, \quad j \in \{1,\ldots,d\}, \tag{2.12}$$

where $\tilde{Z}_1,\ldots,\tilde{Z}_d$ are independent $N(0,1)$ random variables.

Remark 2.4.2 *Formulas* (2.9) *and* (2.11) *are quite useful since we can simulate quite easily Brownian motions or standard Gaussian variates.*

Note also that because of (2.9)*, the value of an option depends only on its past values up to time* t*, the short-term interest rate* r*, the time to maturity* $\tau = T - t$*, and the upper diagonal matrix* a *(or equivalently on the volatility vector* v *defined by* (2.5)*).*

We will now give some examples where the value of the options can be computed explicitly.

2.4.3 Exchange Option

An exchange option gives its owner the right of exchanging a given fraction of the shares of an asset for a fraction of the shares of another given asset. For example, one could replace 2 shares of the first asset with 3 shares of the second asset. More generally, the payoff at maturity of this option can be written as

$$\Phi(s_1,s_2) = \max(q_2 s_2 - q_1 s_1, 0),$$

where q_1 and q_2 are both positive. Using formula (2.11), one can find explicitly the value of the option.

Proposition 2.4.1 *In terms of the parameters of model* (2.1)*, the value* $C(t,s_1,s_2)$ *of the exchange option, given* $S_1(t) = s_1$, $S_2(t) = s_2$*, is*

$$q_2 s_2 \mathcal{N}(D_1) - q_1 s_1 \mathcal{N}(D_2), \tag{2.13}$$

where

$$D_1 = \frac{\ln\left(\frac{q_2 s_2}{q_1 s_1}\right) + \frac{1}{2}(T-t)\sigma^2}{\sigma\sqrt{T-t}}, \tag{2.14}$$

$$D_2 = \frac{\ln\left(\frac{q_2 s_2}{q_1 s_1}\right) - \frac{1}{2}(T-t)\sigma^2}{\sigma\sqrt{T-t}}, \tag{2.15}$$

$$\sigma^2 = \sigma_1^2 + \sigma_2^2 - 2R_{12}\sigma_1\sigma_2. \tag{2.16}$$

The proof is given in Appendix 2.B.5. Note that the value of the option is independent of the risk-free rate. Can you explain why?

2.4.4 Quanto Options

In this section, we assume that the underlying assets are priced in foreign currencies. An option based on foreign assets is called a quanto. In addition to their dynamics, we also have to model the exchange rates of the foreign currencies with respect to the domestic currency, i.e., the value in the domestic currency of 1 unit of the foreign currency. For example, if the Euro/CAD exchange rate is 1.35, it means that €1 is worth \$1.35.

Suppose that the underlying assets $S = (S_1, \ldots, S_d)$, together with the associated exchange rates $\mathcal{C} = (\mathcal{C}_1, \ldots, \mathcal{C}_d)$ are modeled by geometric Brownian motions, as in (2.1). More precisely, for $j \in \{1, \ldots, d\}$,

$$\begin{aligned} S_j(t) &= S_j(0)e^{\left(\mu_j - \frac{1}{2}\sigma_j^2\right)t + \sigma_j W_j(t)}, \\ \mathcal{C}_j(t) &= \mathcal{C}_j(0)e^{\left(\alpha_j - \frac{1}{2}\beta_j^2\right)t + \beta_j Z_j(t)}, \end{aligned}$$

where $W = (W_1, \ldots, W_d)$ and $Z = (Z_1, \ldots, Z_d)$ are correlated Brownian motions, with R_{ww} being the correlation matrix for the Brownian motion W, R_{zz} being the correlation matrix for the Brownian motion Z, and R_{wz} being the correlation matrix between the Brownian motions W and Z, i.e.,

$$(R_{wz})_{jk} = \text{Cor}\left\{W_j(t), Z_k(t)\right\}, \quad j, k \in \{1, \ldots, d\}.$$

For simplicity, set $R_j = (R_{wz})_{jj}$, $j \in \{1, \ldots, d\}$. Finally, assume that the risk-free interest rates in the foreign currencies are given respectively by $r_1, \ldots, r_d$, while the risk-free interest rate in the domestic currency is r. If for some fixed j, the asset S_j is already expressed in the domestic currency, one simply set $r_j = r$, $\alpha_j = \beta_j = 0$, so $\mathcal{C}_j \equiv 1$.

To determine the value of a quanto option, we have to express the value of all risky assets (including the exchange rates) into the domestic currency. These risky assets, when expressed into the domestic currency, are given by

$$\begin{aligned} &\bar{S}_1(t) = \mathcal{C}_1(t)S_1(t), \ldots, \bar{S}_d(t) = \mathcal{C}_d(t)S_d(t), \\ &\bar{S}_{d+1}(t) = e^{r_1 t}\mathcal{C}_1(t), \ldots, \bar{S}_{2d}(t) = e^{r_d t}\mathcal{C}_d(t). \end{aligned}$$

It is obvious that any option based on S and $\mathcal{C}$ can also be written as a function of the transformed assets $\bar{S} = \left(\bar{S}_1, \ldots, \bar{S}_{2d}\right)$. Therefore, an option depending on foreign assets is always equivalent to a certain option depending on domestic assets. The next step is to find the dynamics of $\bar{S}$.

Proposition 2.4.2 *$\bar{S}$ are correlated geometric Brownian motions. In fact, for all $j \in \{1, \ldots, d\}$, one has*

$$\begin{aligned} \bar{S}_j(t) &= S_j(0)\mathcal{C}_j(0)e^{\left(\mu_j + \alpha_j + R_j\sigma_j\beta_j - \frac{1}{2}\bar{\sigma}_j^2\right)t + \bar{\sigma}_j \bar{W}_j(t)}, \\ \bar{S}_{d+j}(t) &= \mathcal{C}_j(0)e^{\left(r_j + \alpha_j - \frac{1}{2}\beta_j^2\right)t + \beta_j Z_j(t)}, \end{aligned}$$

where $\bar{W} = (\bar{W}_1, \ldots, \bar{W}_d)$ *and* Z *are correlated Brownian motions,*

$$\bar{\sigma}_j^2 = \sigma_j^2 + 2R_j\sigma_j\beta_j + \beta_j^2,$$

and

$$\bar{R}_j = \mathrm{Cor}(\bar{W}_j(t), Z_j(t)) = \frac{\beta_j + R_j\sigma_j}{\bar{\sigma}_j}.$$

Moreover, for $j, k \in \{1, \ldots, d\}$*,*

$$\begin{aligned}
(R_{\bar{w}\bar{w}})_{jk} &= \mathrm{Cor}(\bar{W}_j(t), \bar{W}_k(t)) \\
&= \frac{\sigma_j\sigma_k (R_{ww})_{jk} + \sigma_j\beta_k (R_{wz})_{jk}}{\bar{\sigma}_j\bar{\sigma}_k} \\
&\quad + \frac{\sigma_k\beta_j (R_{wz})_{kj} + \beta_j\beta_k (R_{zz})_{jk}}{\bar{\sigma}_j\bar{\sigma}_k}, \\
(R_{\bar{w}z})_{jk} &= \mathrm{Cor}(\bar{W}_j(t), Z_k(t)) = \frac{\sigma_j (R_{wz})_{jk} + \beta_j (R_{zz})_{jk}}{\bar{\sigma}_j}.
\end{aligned}$$

The proof is given in Appendix 2.B.6.

Example 2.4.1 *If* $d = 1$*, then there is only one foreign asset* S *and one foreign currency with exchange rate* $\mathcal{C}$*. Assume that the foreign asset* S *is modeled by a geometric Brownian motion with parameters* μ_1 *and* σ_1*, and the exchange rate* $\mathcal{C}$ *is modeled by a geometric Brownian motion with parameters* α *and* β*, where the correlation between the two Brownian motions is* ρ*. If* r_f *is the foreign short-term rate, then* $\bar{S}_1(t) = S(t)\mathcal{C}(t)$ *and* $\bar{S}_2(t) = e^{r_f t}\mathcal{C}(t)$ *are correlated geometric Brownian motions with mean parameters* $\bar{\mu}_1 = \mu_1 + \alpha + \rho\sigma_1\beta - \frac{\bar{\sigma}_1^2}{2}$*,* $\bar{\mu}_2 = r_f + \alpha - \frac{\beta^2}{2}$*, volatility parameters* $\bar{\sigma}_1 = \sqrt{\sigma_1^2 + 2\rho\sigma_1\beta + \beta^2}$*,* $\bar{\sigma}_2 = \beta$*, and correlation* $\bar{\rho} = \frac{\beta + \rho\sigma_1}{\bar{\sigma}_1}$*.*

To conclude, one can now state the main result for valuing quanto options.

Proposition 2.4.3 *The value of an option based on the foreign assets* S *and exchange rates* $\mathcal{C}$ *can be computed by using formula* (2.9) *when the payoff is expressed in terms of the domestic risky assets* $\bar{S}$*, where under the equivalent martingale measure, the assets* $\bar{S}_j$ *are correlated Brownian motions with parameters* r *and* $\bar{\sigma}_j$*, and assets* $\bar{S}_{d+j}$ *are correlated Brownian motions with parameters* r *and* β_j*, for* $j \in \{1, \ldots, d\}$*. The correlation matrices are the same as in Proposition 2.4.2.*

Example 2.4.2 (Call Option Paying Off in the Foreign Currency) *As a first example of a quanto option, consider a European call option of the foreign asset* S*, modeled by a geometric Brownian motion with parameters* μ_1 *and* σ_1*, where the exchange rate* $\mathcal{C}$ *is modeled by a geometric Brownian motion with parameters* α *and* β*, where* r *is the domestic short-term rate,* r_f *is* *the foreign risk-free rate and the strike price K is expressed into the foreign currency. The correlation between the two Brownian motions is denoted by ρ.*

At maturity, the payment of the option is in the foreign currency, so the payoff (expressed into the domestic currency), is

$$\begin{aligned} \mathcal{C}(T)\max(S(T)-K,0) &= \max\{S(T)\mathcal{C}(T)-K\mathcal{C}(T),0\} \\ &= \max\left\{\bar{S}_1(T)-Ke^{-r_fT}\bar{S}_2(T),0\right\}. \end{aligned}$$

Therefore, this option is an exchange option between the assets $\bar{S}_2, \bar{S}_1$, with $q_1 = Ke^{-r_fT}$ and $q_2 = 1$. Using formula (2.16), we obtain that $\sigma = \sigma_1$. Next, applying formula (2.13) with $\mathcal{C}(t) = c$ and $S(t) = s$, one finds that

$$C(t,s,c) = c\left\{s\mathcal{N}(d_1) - Ke^{-r_f\tau}\mathcal{N}(d_2)\right\},$$

where $\tau = T-t$, $d_1 = \frac{\ln\left(\frac{s}{K}\right)+r_f\tau+\frac{1}{2}\sigma^2\tau}{\sigma\sqrt{\tau}}$, $d_2 = d_1 - \sigma\sqrt{\tau}$.

This is the usual Black-Scholes formula, but computed with the foreign short-term rate r_f as the risk-free rate, strike $\bar{K} = cK$, asset value $\bar{S} = cs$, and volatility $\sigma = \sigma_1$. In fact, this is nothing but the value of the call option in the foreign market, with payment transferred into the domestic currency.

Example 2.4.3 (Call Option Paying Off in the Domestic Currency) *As an example of quanto, Wilmott [2006, Section 11.7] considers the value at maturity of a call on a foreign asset when the payoff is expressed into the domestic currency. For example, if the asset is the NIKKEI index with a value at maturity of 12000 and strike 11000, and if the domestic currency is the USD, then the payoff is* \$1000, *instead of being* 1000 × ExchangeRate, *as in the previous example.*

The solution to this problem is even simpler since it is equivalent to an option on a single asset. In fact, the payoff at maturity is $\max\left\{e^{r_fT}\bar{S}_1(T)\big/\bar{S}_2(T)-K,0\right\}$. *Under the equivalent martingale measure, if $S_1(t) = s$, and $\mathcal{C}(t) = c$, then*

$$\begin{aligned} \frac{\bar{S}_1(T)}{\bar{S}_2(T)} &= \frac{cse^{r_fT}e^{\left(r-\frac{\bar{\sigma}_1^2}{2}\right)\tau+\bar{\sigma}_1\left\{\tilde{W}_1(T)-\tilde{W}_1(t)\right\}}}{ce^{r_ft}e^{\left(r-\frac{\beta^2}{2}\right)\tau+\beta\left\{\tilde{Z}(T)-\tilde{Z}(t)\right\}}} \\ &= se^{(r_f-r-\rho\sigma_1\beta)\tau}e^{\left(r-\frac{\sigma^2}{2}\right)\tau+\sigma\left(\tilde{W}(T)-\tilde{W}(t)\right)}, \end{aligned}$$

where $\tilde{W}_1$ and $\tilde{Z}$ are correlated Brownian motions with correlation $\bar{\rho} = \frac{\beta+\rho\sigma_1}{\bar{\sigma}_1}$, $\tilde{W} = \frac{\bar{\sigma}_1\tilde{W}\_\beta Z}{\sigma}$ is a Brownian motion, and $\tau = T-t$. Here

$$\sigma^2 = \bar{\sigma}_1^2 - 2\bar{\rho}\bar{\sigma}_1\beta + \beta^2 = \sigma_1^2.$$

As a result, one can thus apply directly the Black-Scholes formula (1.4) with $s' = se^{(r_f-r+\rho\sigma_1\beta)\tau}$ and $\sigma = \sigma_1$. Using the extended Black-Scholes formula (1.9), we can interpret the value of this particular quanto option as a European call on a risky asset continuously paying dividends at constant rate $\delta = r - r_f + \rho\sigma_1\beta$.

2.5 Greeks

The greeks are easily extended to the multivariate case, the only potential difficulty being the vega, because of the parametrization of the covariance matrix Σ, as defined by (2.2).

- Δ corresponds to the gradient of the value of the option with respect to the underlying assets, i.e., $\Delta = \nabla_s C$, so $\Delta_j = \partial_{s_j} C$, $j \in \{1, \ldots, d\}$, where $s = (s_1, \ldots, s_d)$ is the value of the underlying assets at time t.

  This value is essential for hedging. It can also be useful for the evaluation of the Value-at-Risk (VaR), as we shall see in Chapter 4.

- Γ is the Hessian matrix of C, seen as a function of s, i.e.,

$$\Gamma_{jk} = \partial_{s_j} \partial_{s_k} C, \quad j, k \in \{1, \ldots, d\}.$$

  We shall see in Chapter 4 that the matrix Γ can also be useful in approximating the VaR.

- $\Theta = \partial_t C$.

- Rho $= \partial_r C$.

- To define $\mathcal{V}$, one must choose a good parametrization for the parameters related to the volatility of the assets. We suggest to use the natural parametrization induced by model (2.4).

  Letting $v = \vec{a} = (a_{11}, \ldots, a_{1d}, a_{22}, \ldots, a_{dd})^\top$ be the volatility vector, one defines $\mathcal{V}$ as follows:

$$\mathcal{V} = \nabla_v C,$$

  i.e.,

$$\mathcal{V}_j = \partial_{v_j} C, \quad j \in \{1, \ldots, d(d+1)/2\}.$$

  Note that when $d = 1$, one recovers the previous definition of vega, since $v = \sigma$.

2.5.1 Error on the Option Value

For a d-dimensional asset S, consider the Black-Scholes model with representation (2.4) and volatility vector v defined by (2.5). According to formula (2.9), the value of the option depends only on the unknown value of v, together with the given values r and T. Therefore, the value C of the option is estimated by $\hat{C}_n = C(\hat{v}_n)$. The estimation error on v has consequences on the estimation error on C, as stated in the next proposition.

Proposition 2.5.1

$$\frac{\sqrt{n}\,(\hat{C}-C)}{\sqrt{\hat{\mathcal{V}}^\top \hat{V}_v \hat{\mathcal{V}}}} \rightsquigarrow N(0,1),$$

where $\hat{\mathcal{V}} = \mathcal{V}(\hat{v}_n)$ is the vega computed at $v = \hat{v}_n$, and $\hat{V}_v$ is the matrix obtained from $\hat{V}$, as defined in Proposition 2.3.2, by eliminating the first d rows and columns.

The proof is given in Appendix 2.B.7.

2.5.2 Extension of the Broadie-Glasserman Methodologies for Options on Several Assets

Using representation (2.4), one can generalize the approach of Broadie and Glasserman [1996] to European options on several assets. This can also be done for exotic options.

Recall that according to formula (2.11), if $Z = (Z_1, \ldots, Z_d)^\top \sim N_d(0, I)$, i.e., the components are d independent standard Gaussian random variables, and if the payoff of the option is $\Phi\{S(T)\} = \Phi\{S_1(T), \ldots, S_d(T)\}$, then the value of the option at time t, when $S(t) = s$, is given by

$$C(t,s) = e^{-r\tau} E\left\{\Phi(\tilde{S})\right\},$$

where $\tilde{S} = \left(\tilde{S}_1, \ldots, \tilde{S}_d\right)$, with

$$\tilde{S}_j = s_j e^{\left(r - \frac{\Sigma_{jj}}{2}\right)\tau + \sqrt{\tau}\sum_{k=1}^{j} a_{kj} Z_k}, \quad j \in \{1, \ldots, d\}, \tag{2.17}$$

$\tau = T - t$, and a is the Cholesky decomposition of Σ.

To estimate $C = C(t, s)$ by Monte Carlo simulations, it suffices to generate independent random vectors

$$Z_1 = (Z_{11}, \ldots, Z_{1d})^\top, \ldots, Z_N = (Z_{N1}, \ldots, Z_{Nd})^\top$$

with distribution $N_d(0, I)$. Then, one sets $\tilde{S}_i = \left(\tilde{S}_{i1}, \ldots, \tilde{S}_{id}\right)$, with

$$\tilde{S}_{ij} = \left(\tilde{S}_i\right)_j = s_j e^{\left(r - \frac{\Sigma_{jj}}{2}\right)\tau + \sqrt{\tau}\sum_{k=1}^{j} a_{kj} Z_{ik}},$$

for $i \in \{1 \ldots, N\}$, and $j \in \{1, \ldots, d\}$.

The generalization of the pathwise methodology proposed in Broadie and Glasserman [1996] in the univariate case is given in the next proposition.

Proposition 2.5.2

$$\hat{C} = e^{-r\tau} \frac{1}{N} \sum_{i=1}^{N} \Phi\left(\tilde{S}_i\right)$$

is an unbiased estimation of C. If in addition Φ has partial derivatives $\partial_j \Phi = \partial_{s_j} \Phi$ almost everywhere[4]*, for all $j \in \{1, \ldots, d\}$, then*

$$\widehat{\partial_\theta C} = \partial_\theta \hat{C} = -\hat{C} \partial_\theta (r\tau) + e^{-r\tau} \frac{1}{N} \sum_{i=1}^{N} \sum_{k=1}^{d} \partial_k \Phi(\tilde{S}_i) \partial_\theta \tilde{S}_{ik}.$$

is an unbiased estimator of $\partial_\theta C$, for every parameter

$$\theta \in \{r, t, s_1, \ldots, s_d, v_1, \ldots, v_{d(d+1)/2}\}.$$

To extend the likelihood ratio methodology to the multivariate case, set

$$f(x) = \frac{e^{-\frac{1}{2\tau} \zeta^\top \Sigma^{-1} \zeta}}{x_1 x_2 \cdots x_d a_{11} a_{22} \cdots a_{dd} (2\pi\tau)^{d/2}}$$

where $\zeta = (\zeta_1, \ldots, \zeta_d)^\top$ with

$$\zeta_j = \ln(x_j / s_j) - \left(r - \frac{\Sigma_{jj}}{2} \right) \tau, \quad j \in \{1, \ldots, d\}.$$

Note that the density f is differentiable with respect to all parameters.

Proposition 2.5.3 *Simultaneous unbiased estimations of C and the greeks of order one are given by*

$$\hat{C} = \frac{1}{N} \sum_{i=1}^{N} e^{-r\tau} \Phi\left(\tilde{S}_i\right)$$

and

$$\widehat{\partial_\theta C} = -\hat{C} \partial_\theta (r\tau) + e^{-r\tau} \frac{1}{N} \sum_{i=1}^{N} \Phi(\tilde{S}_i)\, \partial_\theta \left\{ \ln(f(x)) \right\} \big|_{x = \tilde{S}_i} . \tag{2.18}$$

In particular, an unbiased estimation of Δ is given by

$$\hat{\Delta}_j = e^{-r\tau} \frac{1}{N} \sum_{i=1}^{N} \frac{\left(a^{-1} Z_i\right)_j}{s_j \sqrt{\tau}} \Phi(\tilde{S}_i), \quad j \in \{1, \ldots, d\}. \tag{2.19}$$

Furthermore, an unbiased estimation of Γ is given, for all $j, k \in \{1, \ldots, d\}$, by

$$\hat{\Gamma}_{jj} = e^{-r\tau} \frac{1}{N} \sum_{i=1}^{N} \Phi(\tilde{S}_i) \frac{\left\{ \left(a^{-1} Z_i\right)_j \right\}^2 - \left(\Sigma^{-1}\right)_{jj} - \left(a^{-1} Z_i\right)_j \sqrt{\tau}}{s_j^2 \tau}, \tag{2.20}$$

[4]This means everywhere but at a set of Lebesgue measure 0.

and

$$\hat{\Gamma}_{jk} = e^{-r\tau}\frac{1}{N}\sum_{i=1}^{N}\Phi(\tilde{S}_i)\frac{\left(a^{-1}Z_i\right)_j\left(a^{-1}Z_i\right)_k - \left(\Sigma^{-1}\right)_{jk}}{s_j s_k \tau}, \quad j \neq k. \tag{2.21}$$

An unbiased estimation of $\text{Rho} = \partial_r C$ *is given by*

$$\widehat{\text{Rho}} = -\tau\hat{C} + \sqrt{\tau}e^{-r\tau}\frac{1}{N}\sum_{i=1}^{N}\Phi(\tilde{S}_i)\mathbf{1}^\top a^{-1}Z_i, \tag{2.22}$$

where $\mathbf{1} = (1,\ldots,1)^\top$.

An unbiased estimation of $\Theta = \partial_t C$ *is given by*

$$\begin{aligned}\widehat{\Theta} &= \left(\frac{d}{2\tau}+r\right)\hat{C} \\ &\quad -\frac{e^{-r\tau}}{2\tau N}\sum_{i=1}^{N}\Phi(\tilde{S}_i)\left(Z_i^\top Z_i + 2\sqrt{\tau}\, m^\top a^{-1}Z_i\right),\end{aligned} \tag{2.23}$$

where $m = r\mathbf{1} - \frac{1}{2}\text{diag}(\Sigma)$ *and* $(\text{diag}(\Sigma))_j = \Sigma_{jj}$, $j \in \{1,\ldots,d\}$. *Finally, an unbiased estimation of* $\mathcal{V}$ *is given, for* $j,k \in \{1,\ldots,d\}$, $j \leq k$, *by*

$$\widehat{\partial_{a_{jj}}C} = e^{-r\tau}\frac{1}{N}\sum_{i=1}^{N}\Phi(\tilde{S}_i)\left\{\left(a^{-1}Z_i\right)_j\left(Z_{ij} - a_{jj}\sqrt{\tau}\right) - \frac{1}{a_{jj}}\right\} \tag{2.24}$$

and

$$\widehat{\partial_{a_{jk}}C} = e^{-r\tau}\frac{1}{N}\sum_{i=1}^{N}\Phi(\tilde{S}_i)\left(a^{-1}Z_i\right)_k\left(Z_{ij} - a_{jk}\sqrt{\tau}\right), \quad j < k. \tag{2.25}$$

Remark 2.5.1 *Note that when* $x = \tilde{S}_i$, *we have* $\zeta = \sqrt{\tau}a^\top Z_i$, *so* $\Sigma^{-1}\zeta = \sqrt{\tau}a^{-1}Z_i$, $i \in \{1,\ldots,N\}$.

The proof is given in Appendix 2.B.8.

2.6 Suggested Reading

For options on several assets, including quantos, I recommend [Björk, 1999, Chapters 9,12] and [Wilmott, 2006, Chapter 11].

2.7 Exercises

Exercise 2.1

If $W_1, ..., W_d$ are correlated Brownian motions, show that

$$Cov\left\{W_j\left(s\right), W_k\left(t\right)\right\} = R_{jk} \min(s,t), \quad s,t \geq 0, \quad j,k \in \{1,\ldots,d\}.$$

Exercise 2.2

Suppose that the covariance matrix Σ of 3 sets of returns is given by

$$\Sigma = \begin{pmatrix} 1.00 & -0.68 & 0.12 \\ -0.68 & 0.58 & -0.09 \\ 0.12 & -0.09 & 0.23 \end{pmatrix}$$

Find the Cholesky decomposition a of Σ, i.e., $a^\top a = \Sigma$.

Exercise 2.3

For a bivariate Black-Scholes model, show that the Cholesky decomposition a of the covariance matrix Σ is given by

$$a = \begin{pmatrix} \sigma_1 & R\sigma_2 \\ 0 & \sigma_2\sqrt{1-R^2} \end{pmatrix},$$

in terms of the volatilities σ_1, σ_2 and the correlation $R = R_{12}$. Can you interpret the coefficients of the volatility vector v?

Exercise 2.4

Suppose that stock A and stock B follow the Black-Scholes model. With the associated daily returns, we computed the following volatility vector $v = (0.011348, 0.005345, 0.013248)$. Estimate the volatilities σ_A and σ_B as well as the correlation R between the underlying Brownian motions.

Exercise 2.5

For a trivariate Black-Scholes model, show that if a is the Cholesky decomposition of the covariance matrix Σ, then a_{ij} has the same values as in the bivariate case for $1 \leq j \leq 2$, while $a_{13} = \sigma_3 R_{13}$, $a_{23} = \sigma_3 \frac{R_{23} - R_{12}R_{13}}{\sqrt{1-R_{12}^2}}$, and

$$a_{33} = \sigma_3 \sqrt{\frac{1 - R_{12}^2 - R_{13}^2 - R_{23}^2 + 2R_{12}R_{13}R_{23}}{1 - R_{12}^2}}.$$

Exercise 2.6

Try to estimate the parameters of the bivariate Black-Scholes model for Apple and Microsoft, using the MATLAB function *EstBS2DCor* with the starting points $\mu_1 = \mu_2 = 0 = \rho$ and $\sigma_1 = \sigma_2 = .001$. Can you explain the results?

Exercise 2.7

Compute the greeks (delta, gamma, vega) for an exchange option. How do they relate to the greeks of a call option on one asset? How much does one have to invest in the risk-free asset to replicate exactly the payoff?

Exercise 2.8

Suppose that two stocks satisfy the Black-Scholes model with parameters $\mu_1 = 0.04$, $\mu_2 = 0.09$, $\sigma_1 = 0.25$, $\sigma_2 = 0.35$, and $R_{12} = 0.25$. We want to price an exchange option whose payoff at maturity is $\max(3S_1 - 2S_2, 0)$.

(a) Why would an investor be interested by this derivative?

(b) Write the expressions for σ, D_1, D_2 and $C(t, s_1, s_2)$.

(c) Today, $S_1 = \$43$ and $S_2 = \$65$. Compute the value of the exchange option if it expires in 6 months. Assume that the risk-free rate is 2%.

(d) Compute the (bivariate) delta of the option.

(d) What is the impact of a change in the risk-free interest rate for this option, from a hedging point of view?

Exercise 2.9

What is the relationship between the value C_{12} of an exchange option with payoff $\max(q_1 s_1 - q_2 s_2, 0)$ and the value C_{21} of an exchange option with payoff $\max(q_2 s_2 - q_1 s_1, 0)$, when the maturity is τ?

Exercise 2.10

Suppose that a stock traded in France satisfies the stochastic differential equation

$$dS(t) = 0.04S(t)dt + 0.15S(t)dW(t),$$

while the EUR/CAD's exchange rate process satisfies

$$dC(t) = 0.025C(t)dt + 0.40C(t)dZ(t).$$

The correlation between the Brownian motions W and Z is 0.65. The annual risk-free rate in France and Canada are respectively 4% and 2%, and the 1-year forward contract on EUR/CAD is \$1.34. The spot exchange rate c_0 on EUR/CAD is determined by the covered interest rate parity formula:

$$F = c_0 e^{r_{EUR} - r_{CAD}}.$$

We want to price a European put on the asset S paying off in CAD.

(a) Why would an investor buy a quanto call or a quanto put paying off in the domestic currency?

(b) Write the expressions of $\bar{\sigma_1}^2$, $\bar{\sigma_2}^2$, and σ^2, and compute their values.

(c) Write the expression of the value of the put $P(t, S_0, c_0)$, including the values of D_1 and D_2.

(d) What is the exchange rate c_0 today?

(e) Today, $S = \$27$. Compute the value of a 3-month maturity put if the strike price is \$28.

2.8 Assignment Questions

1. Construct a MATLAB function for the estimation of the parameters of a Black-Scholes model with representation (2.4). The input values are a $n \times d$ matrix S representing the prices of the assets over time, and the scaling parameter h, representing the time between observations, on an annual time scale. The number of assets must be determined from the data S. The output values are the estimation of the parameters μ and v on an annual basis, together with their estimation errors and the estimation of the covariance matrix V.

2. Construct a MATLAB program for the estimation of the value of an exchange option together with its greeks, using the generalization of the likelihood method of Broadie-Glasserman. The number of simulations N must be an input. The output must also contain the exact values of the option and the greeks.

3. Construct a MATLAB program for the estimation of the value of a call-on-max option on the cumulative returns of two assets modeled by correlated Brownian motions, i.e., the payoff at maturity is $\max\left[\max\left[\ln\{S_1(T)/S_1(0)\}, \ln\{S_2(T)/S_2(0)\}\right] - K, 0\right]$. One also wants to estimate the greeks $(\Delta, \mathcal{V})$ using the generalization of the pathwise method of Broadie-Glasserman. The inputs are: the strike K, the risk-free rate r, the maturity T, the volatilities σ_1, σ_2, the correlation $R = R_{12}$, and the number of simulations N.

2.A Auxiliary Result

The following result is needed for valuing some options.

2.A.1 Evaluation of $E\left\{e^{aZ}\mathcal{N}(b+cZ)\right\}$

Proposition 2.A.1 *If* $Z \sim N(0,1)$, *then*

$$E\left\{e^{aZ}\mathcal{N}(b+cZ)\right\} = e^{\frac{a^2}{2}}\mathcal{N}\left(\frac{b+ac}{\sqrt{1+c^2}}\right). \tag{2.26}$$

PROOF.

$$\begin{aligned}
E\left\{e^{aZ}\mathcal{N}(b+cZ)\right\} &= \int_{-\infty}^{\infty}\int_{-\infty}^{b+cz} e^{az}\frac{1}{2\pi}e^{-\frac{z^2}{2}-\frac{w^2}{2}}dwdz \\
&= e^{\frac{a^2}{2}}\int_{-\infty}^{\infty}\int_{-\infty}^{b+cz}\frac{1}{2\pi}e^{-\frac{(z-a)^2}{2}-\frac{w^2}{2}}dwdz \\
&= e^{\frac{a^2}{2}}\int_{-\infty}^{\infty}\int_{-\infty}^{b+ac+cy}\frac{e^{-\frac{y^2}{2}-\frac{w^2}{2}}}{2\pi}dwdy \\
&\qquad \text{(by setting } y = z-a\text{)} \\
&= e^{\frac{a^2}{2}}P(Z_2 \le b+ac+cZ_1) \\
&= e^{\frac{a^2}{2}}P(Z_2 - cZ_1 \le b+ac) \\
&= e^{\frac{a^2}{2}}\mathcal{N}\left(\frac{b+ac}{\sqrt{1+c^2}}\right),
\end{aligned}$$

since $Z_2 - cZ_1 \sim N(0, 1+c^2)$.

■

2.B Proofs of the Results

2.B.1 Proof of Proposition 2.1.1

According to model (2.1), we have,

$$\begin{aligned}
\frac{S_j(ih)}{S_j\{(i-1)h\}} &= \frac{s_j e^{\mu_j ih - \frac{ih\sigma_j^2}{2} + \sigma_j W_j(ih)}}{s_j e^{\mu_j (i-1)h - \frac{(i-1)h\sigma_j^2}{2} + \sigma_j W_j\{(i-1)h\}}} \\
&= e^{m_j h + \sigma_j [W_j(ih) - W_j\{(i-1)h\}]},
\end{aligned}$$

for any $i \in \{1,\ldots,n\}$ and any $j \in \{1,\ldots,d\}$. Hence

$$\begin{aligned} X_{ij} &= \ln\{S_j(ih)\} - \ln[S_j\{(i-1)h\}] \\ &= m_j h + \sigma_j [W_j(ih) - W_j\{(i-1)h\}] \sim N\left(m_j h, \sigma_j^2 h\right). \end{aligned}$$

Since the increments $\Delta W_i = W(ih) - W\{(i-1)h\}$, $i \in \{1,\ldots,n\}$, are independent and since $W(ih) - W\{(i-1)h\} \sim N_d(0, hR)$, it follows that X_i, $i \in \{1,\ldots,n\}$ are also independent, with

$$\mathrm{Cov}(X_{ij}, X_{ik}) = \mathrm{Cov}(\sigma_j \Delta W_{ij}, \sigma_k \Delta W_{ik}) = \sigma_j \sigma_k R_{jk} h = h\Sigma_{jk},$$

for any $j,k \in \{1,\ldots,d\}$.

■

2.B.2 Proof of Proposition 2.2.1

It follows from Proposition 2.1.1 that the returns X_j are independent and identically distributed Gaussian vectors with mean mh and covariance Σh. Hence the joint density of the returns, given $S(0)$, is

$$L(m,\Sigma) = \prod_{i=1}^{n} \frac{1}{(2\pi h)^{d/2}|\Sigma|^{1/2}} e^{-\frac{1}{2h}(x_i - mh)^\top \Sigma^{-1}(x_i - mh)},$$

which we want to maximize, for $m \in \mathbb{R}^d$ and $\Sigma \in S_d^+$, where S_d^+ is the set of all positive definite and symmetric $d \times d$ matrices. Since one can omit the constant term $\frac{nd}{2}\ln(2\pi h)$, the problem is equivalent to minimizing

$$\begin{aligned} g(m,\Sigma) &= -\ln\left(L(m,\Sigma)\right) - \frac{nd}{2}\ln(2\pi h) \\ &= \frac{n}{2}\ln(|\Sigma|) + \frac{1}{2h}\sum_{i=1}^{n}(x_i - mh)^\top \Sigma^{-1}(x_i - mh) \\ &= \frac{n}{2}\ln(|\Sigma|) + \frac{1}{2h}\sum_{i=1}^{n}(x_i - \bar{x})^\top \Sigma^{-1}(x_i - \bar{x}) \\ &\quad + n\frac{1}{2h}(\bar{x} - mh)^\top \Sigma^{-1}(\bar{x} - mh) \\ &= \frac{n}{2}\ln(|\Sigma|) + \frac{n}{2h}\mathrm{Tr}(S_x \Sigma^{-1}) + n\frac{1}{2h}(\bar{x} - mh)^\top \Sigma^{-1}(\bar{x} - mh) \\ &\geq \frac{n}{2}\ln(|S_x/h|) + \frac{nd}{2h} = g(\bar{x}/h, S_x/h), \end{aligned}$$

since for any $B \in S_d^+$, one has $\mathrm{Tr}(B) - \ln(|B|) \geq d$.

It follows from the last chain of inequalities that $\hat{\Sigma} = S_x/h$, and $\hat{m} = \bar{x}/h$. The proof of Corollary 2.2.1 is then a consequence of the uniqueness of the Cholesky decomposition.

■

2.B.3 Proof of Proposition 2.3.1

Using the delta method (Theorem B.3.1) and the fact that for two standard Gaussian random variables Z_1, Z_2, with correlation ρ, $\mathrm{Cov}\left(Z_1^2, Z_2^2\right) = 2\rho^2$, one can deduce that as $n \to \infty$,

$$\mathrm{Cov}\left\{\sqrt{n}\left(\hat{\sigma}_{n,i} - \sigma_i\right), \sqrt{n}\left(\hat{\sigma}_{n,j} - \sigma_j\right)\right\} \to R_{ij}^2 \frac{\sigma_i \sigma_j}{2},$$

for all $1 \le i \le j \le d$. Moreover, for standard Gaussian random variables $Z_1, \ldots, Z_d$, we also have

$$\mathrm{Cov}(Z_i Z_j, Z_k Z_l) = R_{ik} R_{jl} + R_{il} R_{kj}, \quad i, j, k, l \in \{1, \ldots, d\}. \tag{2.27}$$

As a result, as $n \to \infty$, $\sqrt{n}\left(\hat{\Sigma}_{n,ij} - \Sigma_{ij}\right) \rightsquigarrow N\left(0, \sigma_i^2 \sigma_j^2 + \Sigma_{ij}^2\right)$, and for all $i, j, k, l \in \{1, \ldots, d\}$,

$$\mathrm{Cov}\left\{\sqrt{n}\left(\hat{\Sigma}_{n,ij} - \Sigma_{ij}\right), \sqrt{n}\left(\hat{\Sigma}_{n,kl} - \Sigma_{kl}\right)\right\} \to \Sigma_{ik}\Sigma_{jl} + \Sigma_{kj}\Sigma_{il}.$$

Next, since $\hat{R}_{n,ij} = \frac{\hat{\Sigma}_{n,ij}}{\hat{\sigma}_{n,i}\hat{\sigma}_{n,j}}$, it is easy to check that

$$\begin{aligned}
\sqrt{n}\left(\hat{R}_{n,ij} - R_{ij}\right) &= \frac{\sqrt{n}\left(\hat{\Sigma}_{n,ij} - \Sigma_{ij}\right)}{\sigma_i \sigma_j} - R_{ij}\frac{\sqrt{n}\left(\hat{\sigma}_{n,i}^2 - \sigma_i^2\right)}{2\sigma_i^2} \\
&\quad - R_{ij}\frac{\sqrt{n}\left(\hat{\sigma}_{n,j}^2 - \sigma_j^2\right)}{2\sigma_j^2} + o_P(1).
\end{aligned}$$

Using (2.B.3), one can then conclude that as $n \to \infty$, $\sqrt{n}\left(\hat{R}_{n,ij} - R_{ij}\right) \rightsquigarrow N\left(0, \left(1 - R_{ij}^2\right)^2\right)$, for all $1 \le i < j \le d$. As a by-product, we also obtain that as $n \to \infty$, $\mathrm{Cov}\left\{\sqrt{n}\left(\hat{R}_{n,ij} - R_{ij}\right), \sqrt{n}\left(\hat{R}_{n,kl} - R_{kl}\right)\right\}$ converges to

$$\begin{aligned}
&R_{ik}R_{jl} + R_{jk}R_{il} - R_{ij}(R_{ik}R_{il} + R_{jk}R_{jl}) - R_{kl}(R_{ik}R_{jk} + R_{il}R_{jl}) \\
&- \frac{R_{ij}R_{kl}}{2}\left(R_{ik}^2 + R_{il}^2 + R_{jk}^2 + R_{jl}^2\right).
\end{aligned}$$

■

2.B.4 Proof of Proposition 2.3.2

It follows from Proposition B.5.1 that $\sqrt{n}\left(\hat{\theta}_n - \theta\right) \rightsquigarrow N_{d(d+3)/2}\left(0, \hat{\mathcal{I}}^{-1}\right)$, where $\hat{\mathcal{I}}$ is the estimation of the Fisher information matrix.

Now, since $g(\theta) = -\sum_{i=1}^n \ln(f(x_i; \theta))$, the Hessian matrix $H = H(\theta)$ of g, defined by $H_{jk} = \partial_{\theta_j}\partial_{\theta_k} g(\theta)$, $1 \le j, k \le d$, satisfies

$$\begin{aligned}
\frac{H_{jk}(\hat{\theta}_n)}{n} &= \frac{1}{n}\sum_{i=1}^n (y_i)_j (y_i)_k - \frac{1}{n}\sum_{i=1}^n \frac{\partial_{\theta_j}\partial_{\theta_k} f(x_i; \hat{\theta}_n)}{f(x_i; \hat{\theta})} \\
&\stackrel{n\to\infty}{\longrightarrow} (\mathcal{I}_\theta)_{jk},
\end{aligned}$$

where $(y_i)_j = \frac{\partial_{\theta_j} f(x_i;\hat{\theta}_n)}{f(x_i;\hat{\theta}_n)}$, $j \in \{1,\ldots,d\}$ and $i \in \{1,\ldots,n\}$. Hence, the Fisher information matrix can be estimated by $\hat{\mathcal{I}} = \frac{H(\hat{\theta}_n)}{n}$. Finally, for $j \in \{1,\ldots,d\}$, $a_{jj} = \exp\left(\theta_{jd-\frac{(j-1)(j-2)}{2}+1}\right)$. Hence, from the delta method (Theorem B.3.1), if we set $j = kd - \frac{(k-1)(k-2)}{2} + 1$, we obtain

$$\sqrt{n}\left(\hat{a}_{n,kk} - a_{kk}\right) = \sqrt{n}\left(e^{\hat{\theta}_{n,j}} - e^{\theta_j}\right) \approx \hat{J}_{jj}\sqrt{n}\left(\hat{\theta}_{n,j} - \theta_j\right).$$

Consequently, the limiting covariance matrix of the estimation errors $\begin{pmatrix} \sqrt{n}\left(\hat{\mu}_n - \mu\right) \\ \sqrt{n}\left(\hat{v}_n - v\right) \end{pmatrix}$ can be estimated by $\hat{J}\hat{\mathcal{I}}^{-1}\hat{J}$.

■

2.B.5 Proof of Proposition 2.4.1

Set $\tau = T - t$. Using representation (2.4), we have

$$a = \begin{pmatrix} \sigma_1 & R\sigma_2 \\ 0 & \sigma_2\sqrt{1-R^2} \end{pmatrix},$$

since

$$a^\top a = \Sigma = \begin{pmatrix} \sigma_1^2 & R\sigma_1\sigma_2 \\ R\sigma_1\sigma_2 & \sigma_2^2 \end{pmatrix}.$$

As a result, from formula (2.12), we can write

$$q_1\tilde{S}_1(T) = q_1 s_1 e^{\left(r-\frac{\sigma_1^2}{2}\right)\tau + \sigma_1\sqrt{\tau}Z_1}$$

and

$$\begin{aligned} q_2\tilde{S}_2(T) &= q_2 s_2 e^{\left(r-\frac{\sigma_2^2}{2}\right)\tau + \sigma_2 R\sqrt{\tau}Z_1 + \sigma_2\sqrt{1-R^2}\sqrt{\tau}Z_2} \\ &= \left(q_2 s_2 e^{\sigma_2 R\sqrt{\tau}Z_1 - \frac{1}{2}\sigma_2^2 R^2\tau}\right) \times e^{\left(r-\frac{\sigma_2^2(1-R^2)}{2}\right)\tau + \sigma_2\sqrt{1-R^2}\sqrt{\tau}Z_2} \\ &= \bar{s} e^{\left(r-\frac{\bar{\sigma}^2}{2}\right)\tau + \bar{\sigma}\sqrt{\tau}Z_2} \end{aligned}$$

where $Z_1 \sim N(0,1)$ is independent of $Z_2 \sim N(0,1)$, $\bar{s} = q_2 s_2 e^{\sigma_2 R\sqrt{\tau}Z_1 - \frac{1}{2}\sigma_2^2 R^2\tau}$, and $\bar{\sigma} = \sigma_2\sqrt{1-R^2}$. Set

$$
\begin{aligned}
\bar{K} &= q_1\tilde{S}_1(T) = q_1 s_1 e^{\left(r-\frac{\sigma_1^2}{2}\right)\tau+\sigma_1\sqrt{\tau}Z_1}, \\
\bar{d}_1 &= \frac{\ln(\bar{s}/\bar{K}) + r\tau + \bar{\sigma}^2\tau/2}{\bar{\sigma}\sqrt{\tau}} \\
&= \frac{\ln\left(\frac{q_2 s_2}{q_1 s_1}\right) + \sqrt{\tau}(R\sigma_2-\sigma_1)Z_1 + \frac{1}{2}\tau\left(\sigma_1^2+\sigma_2^2(1-2R^2)\right)}{\sigma_2\sqrt{1-R^2}\sqrt{\tau}}, \\
\bar{d}_2 &= \frac{\ln(\bar{s}/\bar{K}) + r\tau - \bar{\sigma}^2\tau/2}{\bar{\sigma}\sqrt{\tau}} \\
&= \frac{\ln\left(\frac{q_2 s_2}{q_1 s_1}\right) + \sqrt{\tau}(R\sigma_2-\sigma_1)Z_1 + \frac{1}{2}\tau\left(\sigma_1^2-\sigma_2^2\right)}{\sigma_2\sqrt{1-R^2}\sqrt{\tau}}.
\end{aligned}
$$

Then, since $\bar{S}(T) = \bar{s}e^{\left(r-\frac{\bar{\sigma}^2}{2}\right)\tau+\bar{\sigma}\sqrt{\tau}Z_2}$ is independent of $\bar{K}$, we have, according to the Black-Scholes formula (1.4),

$$
\begin{aligned}
C &= E\left[e^{-r\tau}\max\left\{q_2\tilde{S}_2(T) - q_1\tilde{S}_1(T), 0\right\}\right] \\
&= E\left[E\left[e^{-r\tau}\max\left\{\bar{S}(T)-\bar{K},0\right\}\right]\right] \\
&= E\left\{\bar{s}\mathcal{N}(\bar{d}_1) - \bar{K}e^{-r\tau}\mathcal{N}(\bar{d}_2)\right\}.
\end{aligned}
$$

To evaluate the last expression, we have to compute an expectation of the form $E\left\{e^{aZ}\mathcal{N}(b+cZ)\right\}$, where $Z \sim N(0,1)$. This is done in Appendix 2.A.1. Using the expression of $E\left\{e^{aZ}\mathcal{N}(b+cZ)\right\}$ given by (2.26) with

$$
a = \sigma_2 R\sqrt{\tau}, \qquad b = \frac{\ln\left(\frac{q_2 s_2}{q_1 s_1}\right) + \frac{1}{2}\tau\left\{\sigma_1^2+\sigma_2^2(1-2R^2)\right\}}{\sigma_2\sqrt{1-R^2}\sqrt{\tau}},
$$

and $c = \frac{R\sigma_2-\sigma_1}{\sigma_2\sqrt{1-R^2}}$, one finds

$$
D_1 = \frac{b+ac}{\sqrt{1+c^2}} = \frac{\ln\left(\frac{q_2 s_2}{q_1 s_1}\right) + \frac{1}{2}\tau\sigma^2}{\sigma\sqrt{\tau}},
$$

where σ is given by (2.16), thus obtaining (2.14). Hence,

$$
E\left\{\bar{s}\mathcal{N}(\bar{d}_1)\right\} = q_2 s_2 e^{-\frac{1}{2}\sigma_2^2R^2\tau} e^{\frac{(\sigma_2 R\sqrt{\tau})^2}{2}}\mathcal{N}(D_1) = q_2 s_2\mathcal{N}(D_1).
$$

Finally, applying (2.26) with

$$
a = \sigma_1\sqrt{\tau}, \qquad b = \frac{\ln\left(\frac{q_2 s_2}{q_1 s_1}\right) + \frac{1}{2}\tau\left(\sigma_1^2-\sigma_2^2\right)}{\sigma_2\sqrt{1-R^2}\sqrt{\tau}},
$$

and $c = \frac{R\sigma_2 - \sigma_1}{\sigma_2\sqrt{1-R^2}}$, one ends up with

$$D_2 = \frac{b+ac}{\sqrt{1+c^2}} = \frac{\ln\left(\frac{q_2 s_2}{q_1 s_1}\right) - \frac{1}{2}\tau\sigma^2}{\sigma\sqrt{\tau}},$$

which corresponds to (2.15). As a result,

$$e^{-r\tau} E\left\{\bar{K}\mathcal{N}(\bar{d}_2)\right\} = e^{-r\tau} q_1 s_1 e^{\left(r - \frac{\sigma_1^2}{2}\right)\tau} e^{\frac{(\sigma_1\sqrt{\tau})^2}{2}} \mathcal{N}(D_2) = q_1 s_1 \mathcal{N}(D_2).$$

Combining the last expressions, we obtain that the value of the option is (2.13), which completes the proof.

■

2.B.6 Proof of Proposition 2.4.2

Set $\bar{W}_j(t) = \frac{1}{\bar{\sigma}_j}\{\sigma_j W_j(t) + \beta_j Z_j(t)\}$, $j \in \{1, \ldots, d\}$. Then $\bar{W}$ and Z are correlated Brownian motions, and it is easy to check that the corresponding correlation matrices $R_{\bar{w}\bar{w}}$ and $R_{\bar{w}z}$ are the ones stated in the proposition. In addition, for $j \in \{1, \ldots, d\}$,

$$\begin{aligned}
\bar{S}_j(t) &= S_j(t)\mathcal{C}_j(t) = S_j(0)\mathcal{C}_j(0) e^{\left(\mu_j - \frac{1}{2}\sigma_j^2\right)t + \sigma_j W_j(t)} \times e^{\left(\alpha_j - \frac{1}{2}\beta_j^2\right)t + \beta_j Z_j(t)} \\
&= S_j(0)\mathcal{C}_j(0) e^{\left(\mu_j + \alpha_j - \frac{1}{2}\sigma_j^2 - \frac{1}{2}\beta_j^2\right)t + \sigma_j W_j(t) + \beta_j Z_j(t)} \\
&= S_j(0)\mathcal{C}_j(0) e^{\left(\mu_j + \alpha_j + R_j\sigma_j\beta_j - \frac{1}{2}\bar{\sigma}_j^2\right)t + \bar{\sigma}_j \bar{W}_j(t)}.
\end{aligned}$$

The rest of the proof is easy.

■

2.B.7 Proof of Proposition 2.5.1

According to the delta method (Theorem B.3.1) and Proposition 2.3.2, we have

$$\begin{aligned}
\sqrt{n}\left(\hat{C}_n - C\right) &= \sqrt{n}\left(C(\hat{v}_n) - C(v)\right) \approx \left(\nabla_v C(\hat{v}_n)\right)^\top \times \sqrt{n}\left(\hat{v}_n - v\right) \\
&= \hat{\mathcal{V}}^\top \times \sqrt{n}\left(\hat{v}_n - v\right) \rightsquigarrow N\left(0, \mathcal{V}^\top \hat{V}_v \hat{\mathcal{V}}\right).
\end{aligned}$$

■

2.B.8 Proof of Proposition 2.5.3

In the general case, under integrability conditions, the proof is a consequence of

$$\begin{aligned}
\partial_\theta C &= \int \Phi(x)\partial_\theta f(x)dx = \int \Phi(x)\frac{\partial_\theta f(x)}{f(x)} f(x)dx \\
&= E\left\{\Phi(\tilde{S})\, \partial_\theta \ln(f(x))\big|_{x=\tilde{S}}\right\}.
\end{aligned}$$

One also has

$$\partial_{s_j}\partial_{s_k} C = E\left\{\Phi(\tilde{S})\frac{\partial_{s_j}\partial_{s_k} f(x)\big|_{x=\tilde{S}}}{f(\tilde{S})}\right\}.$$

For the case of Δ, it suffices to remark that $\partial_{s_j} f = f\frac{\left(\Sigma^{-1}\zeta\right)_j}{s_j\tau}$. Finally,

$$\frac{\partial_{s_j}\partial_{s_k} f}{f} = \frac{\left(\Sigma^{-1}\zeta\right)_j\left(\Sigma^{-1}\zeta_j\right)_k - \tau\left(\Sigma^{-1}\right)_{jk}}{s_j s_k \tau^2} - \begin{cases} \frac{\left(\Sigma^{-1}\zeta_j\right)_j}{s_j^2\tau} & j = k \\ 0 & j \neq k \end{cases}.$$

In the case of $\mathcal{V}$, since $\Sigma = a^\top a$, we obtain that for every $i, j \in \{1\ldots, d\}$, $i \leq j$, we have

$$\partial_{a_{ij}}\Sigma_{kl} = \begin{cases} 2a_{ij} & k = j = l \\ a_{il} & k = j < l \text{ or } i \leq l < k = j, \\ a_{ik} & l = j < k \text{ or } i \leq k < l = j \\ 0 & \text{otherwise} \end{cases}.$$

In particular

$$\partial_{a_{ij}}\Sigma_{kk} = \begin{cases} 2a_{ij} & k = j \\ 0 & k \neq j \end{cases}, \quad i, j \in \{1\ldots, d\}, \quad i \leq j. \tag{2.28}$$

Next, $\partial_\theta\Sigma^{-1} = -\Sigma^{-1}\left(\partial_\theta\Sigma\right)\Sigma^{-1}$, so one may conclude that for all $i, j \in \{1\ldots, d\}$, $i \leq j$, we have

$$\partial_{a_{ij}}\left(\Sigma^{-1}\right)_{kl} = -\left(a^{-1}\right)_{ki}\left(\Sigma^{-1}\right)_{lj} - \left(a^{-1}\right)_{li}\left(\Sigma^{-1}\right)_{kj}. \tag{2.29}$$

Using (2.28) and (2.29), one finds that

$$\partial_{a_{ij}}\left\{\frac{1}{2}\zeta^\top\Sigma^{-1}\zeta\right\} = \tau a_{ij}\left(\Sigma^{-1}\zeta\right)_j - \left(a\Sigma^{-1}\zeta\right)_i\left(\Sigma^{-1}\zeta\right)_j.$$

So we may conclude that for $i, k \in \{1\ldots, d\}$, $i < k$,

$$\widehat{\partial_{a_{ik}} C} = e^{-r\tau}\frac{1}{N}\sum_{j=1}^N \Phi(\tilde{S}_j)\left(a^{-1}Z_j\right)_k\left(Z_{ji} - a_{ik}\sqrt{\tau}\right)$$

and

$$\widehat{\partial_{a_{ii}} C} = e^{-r\tau}\frac{1}{N}\sum_{j=1}^N \Phi(\tilde{S}_j)\left\{\left(a^{-1}Z_j\right)_i\left(Z_{ji} - a_{ii}\sqrt{\tau}\right) - \frac{1}{a_{ii}}\right\}$$

In the case of Rho, it suffices to remark that $\partial_r\zeta = -\tau\mathbf{1}$, so

$$\partial_r \ln(f) = -\frac{1}{2\tau}\left(\partial_r\zeta^\top\Sigma^{-1}\zeta + \zeta^\top\Sigma^{-1}\partial_r\zeta\right) = \mathbf{1}^\top\Sigma^{-1}\zeta.$$

Finally, for Θ, $\partial_t \ln(f) = -\partial_\tau \ln(f)$, and

$$\begin{aligned} \partial_t \ln(f) &= -\partial_\tau \ln(f) \\ &= -\frac{1}{2\tau^2}\zeta^\top\Sigma^{-1}\zeta + \frac{d}{2\tau} - \frac{1}{\tau}\left(r\mathbf{1} - \text{diag}(\Sigma)/2\right)^\top\Sigma^{-1}\zeta. \end{aligned}$$

■

Bibliography

T. Björk. *Arbitrage Theory in Continuous Time.* Oxford University Press, 1999.

M. Broadie and P. Glasserman. Estimating security price derivatives using simulation. *Management Science*, 42:260–285, 1996.

R. A. Fisher. On the 'probable error' of a coefficient of correlation deduced from a small sample. *Metron*, 1:3–32, 1921.

P. Wilmott. *Paul Wilmott on Quantitative Finance*, volume 1. John Wiley & Sons, second edition, 2006.

Chapter 3

Discussion of the Black-Scholes Model

In this chapter we will examine the main weaknesses of the Black-Scholes model and we will see how to check the underlying hypotheses of the model about the returns. Then we present some extensions to the Black-Scholes model. Finally, we complete the chapter with a preliminary study of hedging errors and hedging techniques in discrete time.

3.1 Critiques of the Model

According to Propositions 1.2.1 and 2.1.1, an important property of the Black-Scholes model is that the returns $X_i = \ln\left[\frac{S(ih)}{S\{(i-1)h\}}\right]$, $i \in \{1, \ldots, n\}$, are independent and identically distributed, with a Gaussian distribution. We now show how to test these two hypotheses, starting with the serial independence.

3.1.1 Independence

Consider (univariate) observations $X_1, X_2, \ldots$ which are stationary, i.e., for any $m \geq 1$ and any $i \geq 1$, the law of m consecutive observations $(X_i, \ldots, X_{i+m-1})$ does not depend on i. For such sequences, the null hypothesis of serial independence can take the form $H_0 : X_1, \ldots, X_p$ are independent, for a given $p \geq 2$. In fact, the latter null hypothesis is a little bit weaker than the one we would like to test, which corresponds to $p = \infty$.

To perform this test, we could compute the autocorrelations and use them to build a test statistic. For example, one could use the well-known Ljung-Box test statistic defined by

$$L_{n,p} = n(n+2)\sum_{j=1}^{p} \frac{r_{n,j}^2}{n-j},$$

where $r_{n,j}$ is the sample correlation between the pairs (X_i, X_{i+j}). Under the null hypothesis of independence, $\sqrt{n}\, r_{n,j} \rightsquigarrow N(0,1)$, and the limiting variables are all independent, so $L_{n,p} \rightsquigarrow \chi^2(p)$.

However, by using autocorrelations, we are not doing a real test of independence. We can define a time series having zero (theoretical) autocorrelations while having a very strong serial dependence, as shown next. For such models, a test based on autocorrelations will be inconsistent in general, i.e., the probability of rejecting the null hypothesis does not always tend to 1 as $n \to \infty$.

An Interesting Time Series

For example, consider the tent map series defined by $X_{i+1} = 1 - |2X_i - 1| = 2\min(X_i, 1 - X_i)$, with $X_1 \sim \text{Unif}(0,1)$. It then follows that $X_i \sim \text{Unif}(0,1)$ for all $i \geq 1$. The series is also strongly dependent, X_{i+1} being a deterministic function of X_i. However, the autocorrelation of lag j, i.e., the correlation between X_i and X_{i+j} is 0 for all $j \geq 1$. It can be shown that the power of the Ljung-Box test, for any p, does not tend to 1 as $n \to \infty$, proving that the test is not consistent, even for such a strongly dependent series. See Exercise 3.2.

There exists powerful and consistent statistical tests that can be used to verify the hypothesis of serial independence implied by the Black-Scholes model, whether the law of the returns is known or not, as in Genest and Rémillard [2004] or Genest et al. [2007]. These tests are based on distribution functions instead of correlations.

Following Genest and Rémillard [2004], for all $A \in \mathcal{A}_p = \{B \subset \{1, \ldots, p\}; B \ni 1, |B| > 1\}$, we can define nonparametric Cramér-von Mises statistics $T_{n,A}$ that converge jointly in law, under the null hypothesis of serial independence, to independent random variables T_A. See Appendix 3.A for more details.

Because these statistics are based on ranks, they are distribution-free, i.e., they do not depend on the underlying distribution of the observations. Therefore tables of quantiles and also P-values can be easily computed via Monte Carlo methods. Denoting by $p_{n,A}$ the associated P-values, one can then construct a *dependogram*, i.e., a graph of the P-values $p_{n,A}$ of $T_{n,A}$, for all $A \in \mathcal{A}_p$. P-values below the dashed horizontal line at 5% suggest dependence for the given indices. An example of a dependogram, obtained by using the MATLAB function *SerIndGRTest* with the returns of Apple, is given in Figure 3.1.1.

Being a graphical tool like the correlogram, the dependogram is not a good way to test independence. As proposed in Genest and Rémillard [2004], one can also define a powerful test of serial independence by combining the P-values of the test statistics $T_{n,A}$ by setting

$$\mathfrak{F}_{n,p} = -2 \sum_{A \subset \mathcal{A}_p} \ln(p_{n,A}).$$

Under the null hypothesis of serial independence, $\mathfrak{F}_{n,p} \rightsquigarrow \mathfrak{F}_p \sim \chi^2\left(2^p - 2\right)$.

Example 3.1.1 *Using the MATLAB function* SerIndGRTest *with the returns* *of Apple from January 14th, 2009, to January 14th, 2011, we obtain* $\mathfrak{I}_{n,2} = 8.27$*, for a P-value of 1.6%, while* $\mathfrak{I}_{n,6} = 148.79$*, for a P-value of* 0%*, both tests leading to the rejection of the null hypothesis of serial independence for the returns of Apple, during the period 2009–2011. One may conclude that the serial independence assumption implied by the Black-Scholes model is not met.*

The fact that we reject the null hypothesis using only the pairs (X_i, X_{i+1}) *indicates a significant serial dependence. Note that using the Ljung-Box test with the same amount of information, i.e., with the first 5 lags, we end up with a P-value of 21.9%, and the first autocorrelation is not significant. Even by considering the 20 first lags, we do not reject the null hypothesis, the P-value being 5.8%. This is another indication of the lack of power of the Ljung-Box test for detecting serial dependance.*

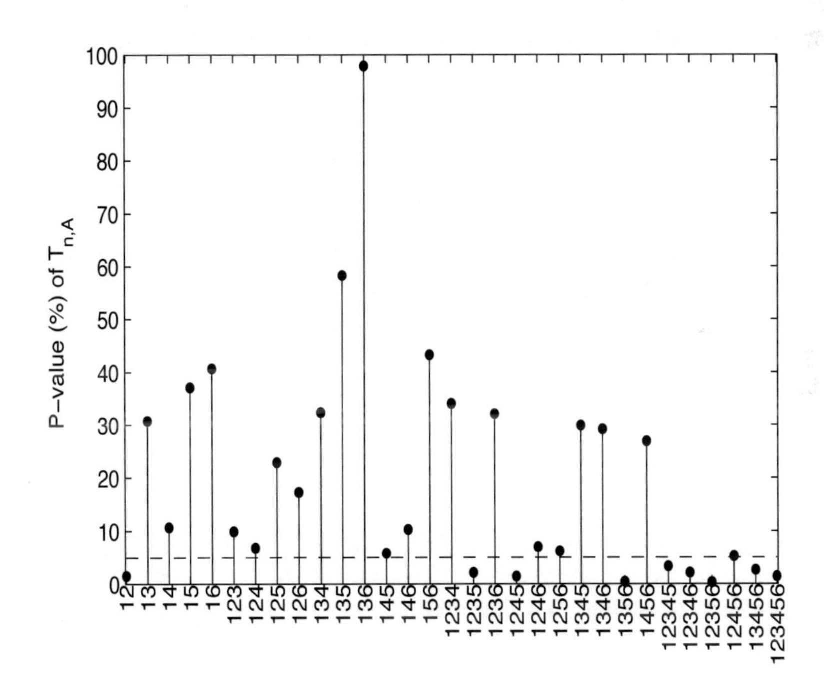

FIGURE 3.1: Dependogram of the P-values for the returns of Apple, computed with $N = 1000$ replications and $p = 6$.

3.1.2 Distribution of Returns and Goodness-of-Fit Tests of Normality

One can also test the Gaussian distribution hypothesis implied by the Black-Scholes model, assuming the returns are independent. A simple way to test the normality of data is to use the empirical skewness coefficient $\hat{\gamma}_{n,1}$, defined by formula (B.7), together with the empirical kurtosis coefficient $\hat{\gamma}_{n,2}$, defined by formula (B.8). It follows from Proposition B.4.1 that $\sqrt{n}\hat{\gamma}_1 \rightsquigarrow N(0,6)$ and $\sqrt{n}(\hat{\gamma}_{n,2} - 3) \rightsquigarrow N(0,24)$, the two statistics being asymptotically independent. The associated P-values are respectively

$$\begin{aligned} P_{n,1} &= 2\left\{1 - \mathcal{N}\left(\sqrt{n}\,|\hat{\gamma}_{n,1}|\,/\sqrt{6}\right)\right\}, \\ P_{n,2} &= 2\left\{1 - \mathcal{N}\left(\sqrt{n}\,|\hat{\gamma}_{n,2} - 3|\,/\sqrt{24}\right)\right\}. \end{aligned}$$

The Jarque-Berra test[1], used mainly in the econometric literature as a goodness-of-fit test of normality, is based on the empirical skewness and kurtosis coefficients viz.

$$JB_n = n\left\{\frac{\hat{\gamma}_{n,1}^2}{6} + \frac{(\hat{\gamma}_{n,2} - 3)^2}{24}\right\}.$$

It follows from Proposition B.4.1 that $JB_n \rightsquigarrow \chi^2(2)$.

Another test, which we call the "revisited" Jarque-Berra test, can be constructed by combining the P-values $P_{n,1}$ and $P_{n,2}$ viz.

$$JBR_n = -2\ln\left(P_{n,1}\right) - 2\ln\left(P_{n,2}\right).$$

Under the null hypothesis, $JBR_n \rightsquigarrow \chi^2(4)$, since $P_{n,1}$ and $P_{n,2}$ are asymptotically independent and uniform over $(0,1)$. According to Littell and Folks [1973], the revisited Jarque-Berra test should be more powerful than the original Jarque-Berra test.

Remark 3.1.1 *Neither the Jarque-Berra test nor its improved version are consistent tests, since even with very large sample sizes, it is possible that the null hypothesis of a Gaussian distribution would not be rejected even if it is false.*

Fortunately, there exists consistent test of goodness-of-fit. They are usually based on the empirical distribution function. Examples of such test statistics are the well-known Kolmogorov-Smirnov test and the Cramér-von Mises test. However, the classical versions of these tests appearing in statistical packages[2] are designed for a simple null hypothesis, i.e., a distribution with known

[1] The test is implemented in the MATLAB function *jbtest*, available in the Statistics Toolbox.

[2] This includes the MATLAB function *kstest* of the Statistics Toolbox.

parameters. In general, as it is the case for the Black-Scholes model, we are interested in testing a composite hypothesis, i.e., where the distribution depends on unknown parameters.

Dealing with Estimated Parameters

When the parameters of the distribution have to be estimated, like it is the case for the Gaussian distribution, the classical versions of the Kolmogorov-Smirnov and the Cramér-von Mises tests yield incorrect results. They have to be replaced by adequate versions that take into account parameter uncertainty.

For example, the corrected version of the Kolmogorov-Smirnov test is the Lilliefors test[3]. It is implemented in the MATLAB function *GofGauss1dKS*. A corrected version of the Cramér-von Mises test is implemented in the MATLAB function *GofGauss1dCVM*. Both tests are described in Appendices 3.B.1 and 3.B.2.

In the financial community, it has become widely acknowledged that returns display more extreme values than should be observed with a Gaussian distribution. That is why models with fatter tails have been proposed, e.g., Student, Variance Gamma, generalized Pareto, etc. Doing so, however, creates other problems such as market incompleteness.

One can also use a non-formal graphical test, by comparing the estimated density of the observations (see Appendix 3.C) with the density of the Gaussian distribution having the same mean and variance. All these calculations are implemented in the MATLAB function *EstPdfGK*. For the particular case of the returns of Apple, from January 14th, 2009, to January 14th, 2011, we may conclude from Table 3.1 that for all goodness-of-fit tests, we reject the hypothesis that the daily returns are Gaussian. Figure 3.2 reinforces this conclusion.

TABLE 3.1: Tests statistics and P-values for the goodness-of-fit tests of the null hypothesis of normality for the returns of Apple.

Test	Statistic	P-value (%)
Jarque-Berra	33.63	0.00
Improved Jarque-Berra	39.21	0.00
Lilliefors	1.36	0.00
Cramer-von Mises	0.50	0.00

[3]The test is implemented in the MATLAB function *lillietest*, available in the Statistics Toolbox.

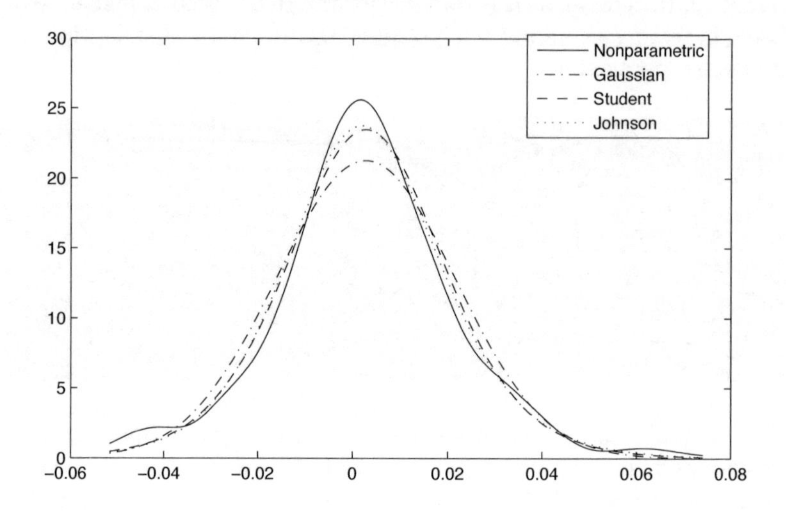

FIGURE 3.2: Nonparametric density estimation of the returns of Apple and graphical comparisons with the Gaussian, Johnson SU, and Student distributions.

3.1.3 Volatility Smile

While the Black-Scholes model assumes that the volatility is constant, implied volatilities computed from market prices appear to depend on the strike level and the time to maturity. Implied volatility plotted against either the strike price or maturity often shows a convex pattern, and this phenomenon has been coined the "volatility smile." To address this issue, one can choose models with non-constant or stochastic volatility, or regime-switching models. In Ané and Geman [2000] and Geman et al. [2002], the authors also considered models of the form

$$S(t) = B(\alpha_t),$$

where α_t is an increasing process independent of the geometric Brownian motion B. The special case $\alpha_t = \sigma^2 t$ corresponds to the Black-Scholes model.

3.1.4 Transaction Costs

Another unrealistic hypothesis of the Black-Scholes model is the absence of transaction costs. Unfortunately, explicit formulas are very hard to obtain when they are taken into account. This problem is still explored in the literature. For an excellent introduction on the subject, see, e.g., Boyle and Vorst [1992].

3.2 Some Extensions of the Black-Scholes Model

One can extend the Black-Scholes model along several directions.

3.2.1 Time-Dependent Coefficients

In this case, for any $s > t$, the conditional distribution of $\ln\{S(s)\}$ given the past values $S(u)$, $u \in [0,t]$, is Gaussian with mean

$$\ln\{S(t)\} + \int_t^s \left\{\mu(u) - \frac{\sigma^2(u)}{2}\right\} du$$

and variance

$$\int_t^s \sigma^2(u)du.$$

This model can also be characterized as the solution to the stochastic differential equation

$$dS(t) = \mu(t)S(t)dt + \sigma(t)S(t)dW(t),$$

where W is a Brownian motion.

Remark 3.2.1 *As in the traditional Black-Scholes model with constant coefficients of Section 1.1, $\mu(t)$ and $\sigma(t)$ generally depend on parameters that must be estimated.*

If one assumes that the short-term interest rate is not constant (but deterministic), one can still obtain explicitly the value of a European call option, using the formula

$$C(t,s) = \frac{B(t)}{B(T)} E\left[\max\left\{0, \tilde{S}(T) - K\right\}\middle|\, \tilde{S}(t) = s\right],$$

where

$$B(u) = \exp\left\{\int_0^u r(v)dv\right\},$$

and the conditional distribution of $\ln\{\tilde{S}(T)\}$ given the past values up to time t, is Gaussian with mean $\ln\{S(t)\} + \int_t^T \left\{r(u) - \frac{\sigma^2(u)}{2}\right\} du$ and variance $\int_t^T \sigma^2(u)du$.

3.2.1.1 Extended Black-Scholes Formula

To obtain $C(t,s)$, it suffices to apply the Black-Scholes formula (1.4) with the average risk-free rate $\bar{r}$ and the average volatility $\bar{\sigma}$ defined by

$$\bar{r} = \frac{1}{T-t}\int_t^T r(u)du = \frac{1}{T-t}\ln\left\{\frac{B(T)}{B(t)}\right\}$$

and

$$(\bar{\sigma})^2 = \frac{1}{T-t}\int_t^T \sigma^2(u)du.$$

This follows from the fact that under the equivalent martingale measure, the conditional distribution of $\ln\{\tilde{S}(T)\}$ given the past values up to time t, is Gaussian with mean $\ln\{S(t)\} + \left(\bar{r} - \frac{\bar{\sigma}^2}{2}\right)\tau$ and variance $\bar{\sigma}^2\tau$, $\tau = T - t$.

Remark 3.2.2 *One can also check that* $C(t,s)$ *is the unique solution of the partial differential equation*

$$\frac{\partial C}{\partial t} + r(t)s\frac{\partial C}{\partial s} + \sigma^2(t)\frac{s^2}{2}\frac{\partial^2 C}{\partial s^2} = r(t)C, \tag{3.1}$$

with boundary conditions $C(T,s) = \max(s-K,0)$.

In Wilmott [2006a], it is suggested to transform (3.1) *into a partial differential equation with constant coefficients constants, similar to* (1.7). *However, the above approach can be implemented more easily by adjusting the parameters in the Black-Scholes model.*

3.2.2 Diffusion Processes

Another extension of the Black-Scholes model is through diffusion processes with coefficients depending on S, defined by a stochastic differential equation of the form

$$dS(t) = \mu\{t,S(t)\}S(t)dt + \sigma\{t,S(t)\}S(t)dW(t), \tag{3.2}$$

where W is a Brownian motion. One can also write

$$S(t) = S(0) + \int_0^t \mu\{u,S(u)\}S(u)du + \int_0^t \sigma\{u,S(u)\}S(u)dW(u).$$

For those with no prior knowledge of stochastic differential equations, just think of a diffusion process as the continuous time limit of a sequence of discrete time Markov processes of the form

$$S_k^{(n)} = S_{k-1}^{(n)} + \mu_{n,k}\left(S_{k-1}^{(n)}\right) + \sigma_{n,k}\left(S_{k-1}^{(n)}\right)\epsilon_k^{(n)}, \quad k \in \{1,\ldots,n\}, \tag{3.3}$$

representing observations at periods $\frac{k}{n}T$, where the innovations $\epsilon_k^{(n)}$ are independent of the past values $S_j^{(n)}$, for $j \in \{0, \ldots, k-1\}$. See, e.g., Labbé et al. [2012]. In general, (3.2) has no explicit solution, which makes estimation more difficult. This is why approximations by Markov models of the form (3.3) can be useful, in addition to being more realistic.

For diffusion models, the value $C(t, s)$ of a European option with payoff Φ at maturity is the unique solution of the partial differential equation

$$\frac{\partial C}{\partial t} + r(t)s\frac{\partial C}{\partial s} + \sigma^2(t, s)\frac{s^2}{2}\frac{\partial^2 C}{\partial s^2} = r(t)C, \tag{3.4}$$

with the boundary condition $C(T, s) = \Phi(s)$. One also has a representation in terms of an expectation, viz.

$$C(t, s) = \frac{B(t)}{B(T)} E\left[\Phi\left\{\tilde{S}(T)\right\}\middle|\, \tilde{S}(t) = s\right], \tag{3.5}$$

where, for a Brownian motion $\tilde{W}$, $\tilde{S}$ is a diffusion process satisfying

$$\tilde{S}(t) = s, \quad d\tilde{S}(u) = r(u)\tilde{S}(u)du + \sigma\{u, S(u)\}\tilde{S}(u)d\tilde{W}(u), \quad u \in (t, T].$$

Equation (3.5) allows estimation by Monte Carlo methods of $C(t, s)$, if one can generate $\tilde{S}(T)$ given $\tilde{S}(t) = s$. The main difficulty is to generate exact values of $\tilde{S}(T)$. In most cases, one has to find approximations of the solution of the stochastic differential equation or construct variables X_n such that the distribution of X_n is close to the distribution of $\tilde{S}(T)$ by using approximations such as (3.3). Examples of application of this methodology include the well-known binomial and multinomial tree methods. See Labbé et al. [2012] for a review on this topic together with new methodologies for approximating (3.5).

Continuous Time vs Discrete Time

Even if discrete time models are more realistic, continuous time limits such as diffusion processes are not without interest since they can provide useful approximations, especially when they can be simulated exactly. In fact, I like to think of a continuous time model not as a real model, but instead, as an approximation to a real life discrete time model. This can prove very useful in practice for the pricing of options and discrete time hedging.

3.3 Discrete Time Hedging

Another weakness of the Black-Scholes model is the hypothesis of continuous time hedging, which is unrealistic. Therefore, one must consider hedging in discrete time, say at times $t_0 = 0, \ldots, t_n = T$.

Suppose that for $k \in \{0, \ldots, n\}$, S_k represents the values of the underlying assets at period t_k, β_k is the discount factor at period t_k, and the information (σ-algebra) available up to period t_k is denoted by $\mathcal{F}_k$; in particular, S_k is $\mathcal{F}_k$-measurable,. It is assumed that β is predictable, in the sense that β_k is $\mathcal{F}_{k-1}$-measurable, $k \in \{1, \ldots, n\}$. For simplicity, set $\Delta_k = \beta_k S_k - \beta_{k-1} S_{k-1}$, $k \in \{1, \ldots, n\}$.

Finally, let $(\pi_0, \vec{\varphi}) = (\pi_0, \varphi_1, \ldots, \varphi_n)$ be the self-financing strategy, where π_0 is the initial amount invested in the portfolio, and $\varphi_k^{(j)}$ represents the number of shares of asset j in the hedging portfolio, during period $[t_{k-1}, t_k)$, $j \in \{1, \ldots, d\}$, $k \in \{1, \ldots, n\}$.

The discounted value $\breve{\pi}_n$ of the hedging portfolio at maturity then satisfies

$$\breve{\pi}_n = \beta_n \pi_n = \pi_0 + \sum_{k=1}^{n} \varphi_k^\top \Delta_k. \tag{3.6}$$

Finally, the (discounted) hedging error G_n associated with the strategy $(\pi_0, \vec{\varphi})$ is given by

$$G_n = \beta_n \Phi - \breve{\pi}_n = \beta_n \Phi - \pi_0 - \sum_{k=1}^{n} \varphi_k^\top \Delta_k, \tag{3.7}$$

where Φ is the payoff at maturity.

There exists many ways to choose the strategy $(\pi_0, \vec{\varphi})$. For example, one could use delta hedging, which consists in assuming that the Black-Scholes model is valid and then discretize the corresponding Δ. This will be covered in the next section.

One could also want to minimize a given loss function of the hedging error G_n. In Section 3.4, we will present the optimal solution when the loss is the quadratic mean.

Remark 3.3.1 *Except for a very simple model (binomial tree), the hedging error G_n will never be zero. In particular, there is no hedging strategy in discrete time so that one can replicate exactly the value of non-trivial options like European call options. Following Boyle and Emanuel [1980] and Wilmott [2006b], when delta hedging is used, the first order of the hedging error has approximately a gamma distribution.*

3.3.1 Discrete Delta Hedging

From now on, suppose that one observes the assets S at times $t_k = \frac{k}{n}T$, i.e., $S_k = S(t_k)$, $k \in \{0, \ldots, n\}$, and that these periods also correspond to the rebalancing periods of the portfolio. For example, one could change the composition of the portfolio at the beginning of every trading day.[4]

The delta hedging methodology simply consists in implementing the Black-Scholes model in discrete time, even if that model is incorrect. More precisely, set

$$\alpha = \frac{1}{T}\text{Cov}\left[\ln\{S(T)\}\right].$$

Then α is the covariance matrix associated with the Black-Scholes model (2.4). It is then estimated by $\frac{n}{T}\Sigma_x$, where Σ_x is the sample covariance matrix of the returns

$$x_{ij} = \ln\left(S_i^{(j)}/S_{i-1}^{(j)}\right), \quad j \in \{1, \ldots, d\}, \quad i \in \{, \ldots, n\}.$$

One could also use other estimations for α, e.g., implied volatilities, etc.

Suppose that $V^{BS} = V^{BS}(t, s)$ is the unique solution of the Black-Scholes partial differential equation

$$\partial_t V^{BS} + r\sum_{j=1}^{d} s_j \partial_{s_j} V^{BS} + \frac{1}{2}\sum_{j=1}^{d}\sum_{k=1}^{d} s_j s_k \alpha_{jk} \partial_{s_j} \partial_{s_k} V^{BS} = rV^{BS}, \qquad (3.8)$$

for all $(t, s) \in (0, T) \times (0, \infty)^d$, with the boundary conditions $V^{BS}(T, s) = \Phi(s)$, $s \in [0, \infty)^d$. The delta hedging strategy consists in setting $\pi_0 = V^{BS}(0, s_0)$, and choosing

$$\varphi_k = \nabla_s V^{BS}\left(t_{k-1}, S_{k-1}\right), \quad k \in \{1, \ldots, n\}.$$

In practice, the delta of the option is rarely known explicitly so one can use one of the Broadie-Glasserman methodologies described in Sections 1.7 and 2.5.2.

As a result, the discounted value of the delta-hedged portfolio at maturity is

$$\check{\pi}_n^{BS} = \beta_n \pi_n^{BS} = V^{BS}(0, S_0) + \sum_{k=1}^{n} \Delta_k^\top \nabla_s V^{BS}\left(t_{k-1}, S_{k-1}\right).$$

What happens when the number of trading periods tends to infinity? If the Black-Scholes model is correct, then the hedging error tends to 0. If the Black-Scholes model is incorrectly specified, e.g., the volatility is incorrect, then the hedging error could tend to a nonzero value, which could be catastrophic on the long run. For general models, the behavior can be quite complicated, as it is shown in Appendix 3.D. In practice it can be quite instructive to generate a large number of trajectories under different models and analyze the hedging errors.

[4] Even this assumption can be too optimistic, e.g., if the market is not liquid enough. It could take all day to liquidate a position.

3.4 Optimal Quadratic Mean Hedging

We will now use the quadratic mean as a loss function, i.e., we are interested in finding π_0 and a predictable investment strategy $\vec{\varphi} = (\varphi_k)_{k=1}^n$ that minimize the expected quadratic hedging error $E\left[\{G_n(\pi_0, \vec{\varphi})\}^2\right]$, where

$$G_n = G_n(\pi_0, \vec{\varphi}) = \beta_n \Phi - \check{\pi}_n,$$

and the discounted value of the portfolio at period k is

$$\check{\pi}_k = \pi_0 + \sum_{j=1}^{k} \varphi_j^\top \Delta_j, \quad k = 0, \ldots, n.$$

To simplify notations, we express the discounted values $\check{S}_k$ in terms of the excess returns R_k, i.e., for $k = 0, \ldots, n$, and $j \in \{1, \ldots, d\}$, $\check{S}_{kj} = \beta_k S_{kj} = \check{S}_{k-1,j} e^{R_{kj}}$. Also let $D(s)$ be the diagonal matrix with components s_j, $j = 1, \ldots, d$. One can then write $\check{S}_k = D(\check{S}_{k-1}) e^{R_k}$, where e^x stands for the vector with components e^{x_j}, $j \in \{1, \ldots, d\}$. Further let $\mathbf{1}$ be the vector with all components equal to 1.

3.4.1 Offline Computations

Once a dynamical model is chosen for the asset prices, one must start with some computations that are necessary for the implementation, whatever the payoff of the option. Unfortunately, they do not seem to have an obvious interpretation.

Set $\mathcal{P}_{n+1} = 1$, and for $k = n, \ldots, 1$, define

$$\begin{aligned}
\gamma_{k+1} &= E(\mathcal{P}_{k+1} | \mathcal{F}_k), \\
\mathfrak{a}_k &= E\left\{ \left(e^{R_k} - \mathbf{1}\right) \left(e^{R_k} - \mathbf{1}\right)^\top \mathcal{P}_{k+1} | \mathcal{F}_{k-1} \right\} \\
&= E\left\{ \left(e^{R_k} - \mathbf{1}\right) \left(e^{R_k} - \mathbf{1}\right)^\top \gamma_{k+1} | \mathcal{F}_{k-1} \right\}, \\
\mathfrak{b}_k &= E\left\{ \left(e^{R_k} - \mathbf{1}\right) \gamma_{k+1} | \mathcal{F}_{k-1} \right\}, \\
\rho_k &= \mathfrak{a}_k^{-1} \mathfrak{b}_k, \\
\mathcal{P}_k &= \prod_{j=k}^{n} \left\{ 1 - \rho_j^\top \left(e^{R_j} - \mathbf{1}\right) \right\}.
\end{aligned}$$

Lemma 3.4.1 *Suppose that* $E(\gamma_{k+1} | \mathcal{F}_{k-1}) \mathfrak{a}_k - \mathfrak{b}_k \mathfrak{b}_k^\top$ *is positive definite* P*-a.s. Then* $\gamma_k \in (0, 1]$ *and* $\mathfrak{a}_k$ *is invertible. If this is true for every* $k \in \{1, \ldots, n\}$*, then* $(\gamma_{k+1})_{k=0}^n$ *is a positive submartingale.*

Note that an alternative expression for γ is

$$\gamma_k = E\left[\gamma_{k+1}\left\{1 - \rho_k^\top\left(e^{R_k} - \mathbf{1}\right)\right\}|\mathcal{F}_{k-1}\right] = E\left(\gamma_{k+1}|\mathcal{F}_{k-1}\right) - \rho_k^\top \mathfrak{b}_k.$$

If the discounted prices are a martingale, then $\mathcal{P} = \gamma \equiv 1$, $\mathfrak{b} = \rho \equiv 0$, and $\mathfrak{a}_k = E\left\{\left(e^{R_k} - \mathbf{1}\right)\left(e^{R_k} - \mathbf{1}\right)^\top |\mathcal{F}_{k-1}\right\}$, $k \in \{1, \ldots, n\}$. In this case, the conditions of Lemma 3.4.1 mean that S is a genuine d-dimensional process.

Remark 3.4.1 *In the univariate case, Schweizer [1995] stated sufficient conditions for the validity of the assumptions of Lemma 3.4.1. It is not obvious how they could be generalized to the multivariate case. In practice, for a given dynamical model, the conditions of Lemma 3.4.1 must be verified, often using brute force calculations.*

3.4.2 Optimal Solution of the Hedging Problem

Theorem 3.4.1 *Under the assumptions of Lemma 3.4.1, the solution* $(\pi_0, \vec{\varphi})$ *of the minimization problem is* $\pi_0 = E(\beta_n \Phi \mathcal{P}_1)/\gamma_1$, *and* $\varphi_k = D^{-1}\left(\breve{S}_{k-1}\right)\left(\alpha_k - \breve{\pi}_{k-1}\rho_k\right)$, *where*

$$\alpha_k = \mathfrak{a}_k^{-1} E\left\{\beta_n \Phi\left(e^{R_k} - \mathbf{1}\right)\mathcal{P}_{k+1}|\mathcal{F}_{k-1}\right\},$$

for all $k \in \{1, \ldots, n\}$. *Also,* $E(G_n) = 0$.

Note that $\varphi_k^\top \Delta_k = \left(\alpha_k - \breve{\pi}_{k-1}\rho_k\right)^\top\left(e^{R_k} - \mathbf{1}\right)$, for every $k \in \{1, \ldots, n\}$.

Remark 3.4.2 *Theorem 3.4.1 has been proven in Schweizer [1995] for a single asset, under more restrictive conditions. The more general case stated here comes from Remillard and Rubenthaler (2013). It was first proven and stated in a different way by Černý and Kallsen [2007].*

Option Value

Let C_k be the optimal investment at period k, so that the value of the portfolio at period n is as close as possible to the payoff Φ, in terms of mean square error. Then $C_n = \Phi$, and it follows from Theorem 3.4.1 that C_k is given by

$$\beta_k C_k = \frac{E(\beta_n \Phi \mathcal{P}_{k+1}|\mathcal{F}_k)}{E(\mathcal{P}_{k+1}|\mathcal{F}_k)}, \qquad k = 0, \ldots, n. \tag{3.9}$$

Even if it seems natural to interpret C_k as the value of the option at period k, one must be cautious since C_k could be negative. Also, this "option value" depends on the number of hedging periods. However, is it less realistic than the equivalent martingale price which is obtained with an infinite number of hedging periods?

3.4.3 Relationship with Martingales

Corollary 3.4.1 *Under the assumptions of Lemma 3.4.1, set* $U_k = \frac{E(\mathcal{P}_k|\mathcal{F}_k)}{E(\mathcal{P}_k|\mathcal{F}_{k-1})} = \frac{\gamma_{k+1}}{\gamma_k}\left\{1-\rho_k^\top\left(e^{R_k}-\mathbf{1}\right)\right\}$, $k \in \{1,\ldots,n\}$. *If* $M_0 = 1$ *and* $M_k = U_k M_{k-1}$, $k \in \{1,\ldots,n\}$, *then* $(M_k, \mathcal{F}_k)_{k=0}^n$ *is a martingale (not necessarily positive) and*

$$\beta_{k-1}C_{k-1} = \beta_k E(C_k U_k|\mathcal{F}_{k-1}), \quad \beta_k E(S_k U_k|\mathcal{F}_{k-1}) = \beta_{k-1}S_{k-1}. \tag{3.10}$$

In particular $(\beta_k C_k M_k, \mathcal{F}_k)_{k=0}^n$ *and* $(\beta_k S_k M_k, \mathcal{F}_k)_{k=0}^n$ *are martingales, and* $\pi_0 = C_0 = E(\beta_n \Phi M_n|\mathcal{F}_0)$.

An alternative expression for α_k *is*

$$\alpha_k = \mathfrak{a}_k^{-1} E\left\{\beta_k C_k \left(e^{R_k}-\mathbf{1}\right)\gamma_{k+1}|\mathcal{F}_{k-1}\right\}.$$

Remark 3.4.3 *If* M *is a positive martingale, then the probability measure* Q *defined by*

$$E_Q(X) = E_P(M_n X), \quad X \in \mathcal{F}_n,$$

is a measure equivalent to P *for which* $\beta_k S_k$ *is a* Q*-martingale. Hence,* Q *is an equivalent martingale measure, and* C_k *can now be interpreted as the option price since it follows from* (3.9) *that* $\beta_k C_k = E_Q(\beta_n \Phi|\mathcal{F}_k)$, $k \in \{0,\ldots,n\}$.

3.4.3.1 Market Price vs Theoretical Price

Suppose $\mathcal{M}_0$ is the market price of the option and let $C_0 = \pi_0$ be the theoretical price of the option, as obtained in Theorem 3.4.1. Consider the following strategy:

- If $\mathcal{M}_0 = C_0$, do nothing.
- If $\mathcal{M}_0 > C_0$, short the option, invest $\mathcal{M}_0 - C_0$ in the risk-free asset and long the optimal portfolio with starting value C_0.
- If $\mathcal{M}_0 < C_0$, borrow $\mathcal{M}_0$, buy the option and short the optimal portfolio with starting value C_0.

The discounted P&L of the strategy is then given by

$$\mathcal{G} = \begin{cases} 0, & \text{if } \mathcal{M}_0 = C_0, \\ \mathcal{M}_0 - C_0 - G_n, & \text{if } \mathcal{M}_0 > C_0, \\ C_0 - \mathcal{M}_0 + G_n, & \text{if } \mathcal{M}_0 < C_0. \end{cases} \tag{3.11}$$

Since $E(G_n) = 0$, we have $E(\mathcal{G}) = |C_0 - \mathcal{M}_0|$. As a result, on average, it could be possible to make money if the option is "mispriced" according to our theoretical price C_0.

Note that (3.11) holds for any hedging strategy. However, since in general the expected value $E(G_n)$ is not necessarily 0, there is no guarantee that the expected P&L will be positive. In fact it could be negative! See, e.g., Example 3.4.5.

In the next sections, we give examples of implementation of the optimal hedging methodology for interesting dynamical models.

3.4.4 Markovian Models

Suppose that the discounted price process $\check{S}$ is a time homogeneous Markov chain such that the conditional distribution of R_k given $\check{S}_{k-1} = s$ is $\mu(s;dx)$. Then $\gamma_k = \gamma_k(\check{S}_{k-1})$, $\mathfrak{a}_k = \mathfrak{a}_k(\check{S}_{k-1})$, and $\rho_k = \rho_k(\check{S}_{k-1})$, i.e, γ_k, $\mathfrak{a}_k$, and ρ_k are deterministic functions of $\check{S}_{k-1}$. If in addition, $\Phi = \check{\Phi}\left(\check{S}_n\right)$, and β is deterministic, then $\beta_k C_k(S_k) = \check{C}_k(\check{S}_k)$ and $\alpha_k(S_k) = \check{a}_k(\check{S}_{k-1})$, i.e, $\check{C}_k$ and $\check{a}_k$ are deterministic functions of $\check{S}_k$ and $\check{S}_{k-1}$.

More precisely, for $k = n, \ldots, 1$, we have

$$\begin{aligned}
\mathfrak{a}_k(s) &= \int (e^x - \mathbf{1})(e^x - \mathbf{1})^\top \gamma_{k+1}\{D(s)e^x\}\mu(s;dx),\\
\rho_k(s) &= \mathfrak{a}_k^{-1}(s)\int (e^x - \mathbf{1})\gamma_{k+1}\{D(s)e^x\}\mu(s;dx),\\
\gamma_k(s) &= \int \gamma_{k+1}\{D(s)e^x\}\mu(s;dx) - \rho_k^\top(s)\mathfrak{a}_k(s)\rho_k(s),\\
\check{C}_{k-1}(s) &= \frac{1}{\gamma_k(s)}\int \check{C}_k\{D(s)e^x\}\gamma_{k+1}\{D(s)e^x\}\\
&\qquad \times\left\{1 - \rho_k^\top(s)(e^x - \mathbf{1})\right\}\mu(s;dx),\\
\check{a}_k(s) &= \mathfrak{a}_k^{-1}(s)\int \check{C}_k\{D(s)e^x\}\gamma_{k+1}\{D(s)e^x\}\\
&\qquad \times (e^x - \mathbf{1})\mu(s;dx).
\end{aligned}$$

Here are some simple examples.

3.4.5 Application to Geometric Random Walks

For this model, the excess returns $R_1, \ldots, R_n$ are i.i.d. Set $\mathfrak{b} = E\left(e^{R_k} - 1\right)$ and $\mathfrak{a} = E\left\{\left(e^{R_k} - 1\right)\left(e^{R_k} - 1\right)^\top\right\}$. Assume that the genuine dimension of the process is d, which implies that the covariance matrix of e^{R_k} is invertible. Then, by Lemma 3.F.1, the conditions of Lemma 3.4.1 are met and $\mathfrak{a}$ is invertible. Set $\rho = \mathfrak{a}^{-1}\mathfrak{b}$ and $c = 1 - \rho^\top\mathfrak{b}$. Then $c > 0$, and for all $k \in \{1, \ldots, n\}$,

$$\begin{aligned}
\mathfrak{a}_k &= \mathfrak{a}c^{n-k},\\
\rho_k &= \rho,\\
\mathcal{P}_k &= \prod_{j=k}^{n}\left\{1 - \rho^\top\left(e^{R_j} - \mathbf{1}\right)\right\},\\
\gamma_k &= c^{n+1-k},\\
U_k &= \frac{1 - \rho^\top\left(e^{R_k} - \mathbf{1}\right)}{c}.
\end{aligned}$$

In particular, using (3.10), we obtain that for all $k \in \{1, \ldots, n\}$,

$$\beta_{k-1} C_{k-1} = E\left[\beta_k C_k \frac{\{1 - \rho^\top (e^{R_k} - 1)\}}{c} \middle| \mathcal{F}_{k-1} \right].$$

Moreover, if $\Phi = \check{\Phi}(\check{S}_n)$, then $\beta_k C_k = \check{C}_k(\check{S}_k)$, where

$$\check{C}_{k-1}(s) = E\left[\check{C}_k \left\{ D(s) e^{R_1} \right\} \frac{\{1 - \rho^\top (e^{R_1} - 1)\}}{c} \right], \tag{3.12}$$

and

$$\check{a}_k(s) = \mathfrak{a}^{-1} E\left[\check{C}_k \left\{ D(s) e^{R_1} \right\} \left(e^{R_1} - 1 \right) \right], \tag{3.13}$$

$k = 1, \ldots, n$. Here we have used the fact that the law of R_k is the same as the law of R_1. Also, φ_k can be easily expressed in terms of discounted values viz.

$$\varphi_k = D^{-1}(\check{S}_{k-1}) \left\{ \check{a}_k(\check{S}_{k-1}) - \rho^\top \check{\pi}_{k-1} \right\}. \tag{3.14}$$

Note that $\check{C}_{k-1}(0) = \check{C}_n(0)$ and $\check{a}_k(0) = \rho \check{C}_n(0)$, for all $k = 1, \ldots, n$. For example, for a call option, we will have $\check{C}_k(0) = 0$ for all k, while for a put option with strike K, $\check{C}_k(0) = \beta_n K$, for all $k \in \{0, \ldots, n\}$.

Because (3.12)–(3.13) are simple expectations, they can be easily approximated. The simplest approximation method is by using Monte Carlo simulations and interpolation. Another one is by linear approximation, provided one can evaluate $E\left\{ e^{\theta^\top R_1} \mathbb{I}(R_1 \in \mathcal{C}) \right\}$ for a suitable class of subsets $\mathcal{C}$. This can be done for example for the univariate Black-Scholes model, i.e., when the returns are Gaussian. An example of implementation is given in the MATLAB function *HedgingLI* for call and put options.

There is one interesting example where computations can be done explicitly and this is the binomial tree model. It is shown in Appendix 3.E that the optimal hedging leads to the usual solution with no hedging error, and the prices coincide with the prices obtained under an equivalent martingale measure, as in Cox et al. [1979].

Example 3.4.1 (Monte Carlo Approximation) *For simplicity, assume that $d = 1$. If $\check{C}_k(s) = s$, then $\check{C}_{k-1}(s) = s$. To preserve this property for the approximation, one should take the expectations with respect to the same empirical measure, i.e., if $(x_i)_{i=1}^N$ is a sample from R_1, then $\hat{\mathfrak{b}} = \frac{1}{N} \sum_{i=1}^N (e^{x_i} - 1)$ and $\hat{\mathfrak{a}} = \frac{1}{N} \sum_{i=1}^N (e^{x_i} - \mathbf{1})^2$. Further set $\hat{\rho} = \hat{\mathfrak{b}} / \hat{\mathfrak{a}}$ and $\hat{c} = 1 - \hat{\rho} \hat{\mathfrak{b}}$.*

Fix grid points $0 \le s_1 < \cdots < s_m$. Starting at $k = n$, compute, for all $j \in \{1, \ldots, m\}$,

$$\begin{aligned} \widehat{\check{C}}_{k-1}(s_j) &= \frac{1}{N} \sum_{i=1}^N \left\{ \frac{1 - \hat{\rho}(e^{x_i} - 1)}{\hat{c}} \right\} \widehat{\check{C}}_k\left(s_j e^{x_i}\right), \\ \hat{\check{a}}_k(s_j) &= \frac{1}{N \hat{\mathfrak{a}}} \sum_{i=1}^N (e^{x_i} - 1) \widehat{\check{C}}_k\left(s_j e^{x_i}\right). \end{aligned}$$

Repeat these steps for $k = n-1,\ldots,1$, interpolating $\hat{\check{C}}_{k-1}$ and $\hat{\mathfrak{a}}_k$ for values $s \notin \mathcal{S} = \{s_j; 1 \le j \le m\}$ when needed.

Example 3.4.2 (Semi-Exact Calculations) *For simplicity, take $d = 1$. It is assumed here that one can evaluate $E\left\{e^{\theta R_1}\mathbb{I}(R_1 \le x)\right\}$ for all x and $\theta = 0,1,2$. Again we need to define a grid, defined by $0 = s_0 < s_1 < \cdots < s_m < s_{m+1} = \infty$.*

$\check{C}_k$ is approximated by the continuous piecewise linear function

$$\sum_{j=0}^{m} \mathbb{I}(s_j \le s < s_{j+1})(A_{kj} + sB_{kj}),$$

with $B_{kj} = \frac{\hat{\check{C}}_k(s_{j+1}) - \hat{\check{C}}_k(s_j)}{s_{j+1}-s_j}$, $A_{kj} = \hat{\check{C}}_k(s_j) - s_jB_{kj}$, $j = 0,\ldots,m-1$, and $A_{km} = A_{k,m-1}$, $B_{km} = B_{k,m-1}$.

For $\theta = 0,1,2$, $j = 0,\ldots,m$ and $l = 1,\ldots,m$, set

$$I_{\theta,j,l} = E\left\{e^{\theta R_1}\mathbb{I}\left(s_j \le s_l e^{R_1} < s_{j+1}\right)\right\}.$$

Then, for $l = 1,\ldots,m$,

$$\hat{\check{C}}_{k-1}(s_l) = \sum_{j=0}^{m} A_{kj}\frac{\{(1+\rho)I_{0jl} - \rho I_{1jl}\}}{c} + s_l\sum_{j=0}^{m} B_{kj}\frac{\{(1+\rho)I_{1jl} - \rho I_{2jl}\}}{c}$$

and

$$\hat{\mathfrak{a}}_k(s_l) = \frac{1}{\mathfrak{a}}\sum_{j=0}^{m}\{A_{kj}(I_{1jl} - I_{0jl}) + s_lB_{kj}(I_{2jl} - I_{1jl})\}.$$

Linear interpolation is used for the values s not on the grid. Recall that $\check{C}_{k-1}(0) = \check{C}_n(0)$ and $a_k(0) = \rho\check{C}_n(0)$, for all $k = 1,\ldots,n$.

Note that in the Gaussian case, if $R \sim N(\mu,\sigma^2)$, then for any $u \in \mathbb{R}$,

$$E\left\{e^{uR}\mathbb{I}(R \le x)\right\} = e^{u\mu + \frac{u^2\sigma^2}{2}}\mathcal{N}\left(\frac{x-\mu}{\sigma} - u\sigma\right).$$

3.4.5.1 Illustrations

Example 3.4.3 (Black-Scholes Model) *As a first illustration, consider the following Black-Scholes model: The excess returns R_k are i.i.d. and*

$$R_k = \left(\mu - \frac{\sigma^2}{2} - r\right)\frac{T}{n} + \sigma\sqrt{\frac{T}{n}}Z_k,$$

where the Z_ks are i.i.d. $N(0,1)$, $\mu = 0.0918$, $\sigma = 0.06$, $T = 1$, and the annual rate is 5%. We want to value a call option with strike $K = 100$ and maturity 1 year, using 22 replication periods and 2000 grid points. Note that delta hedging is optimal in the continuous time limit.

The Monte Carlo method with a sample size of $N = 10000$ *simulated returns was used to approximate the functions* $\check{C}$ *and* $\check{a}$. *The results for the value of the call together with* φ_1 *appear in Figure 3.4.3.*

Then, to compute the hedging errors, 10000 trajectories of 22 *periods were simulated. The estimated density of hedging errors and related statistics are given in Figure 3.4 and Table 3.2 respectively.*

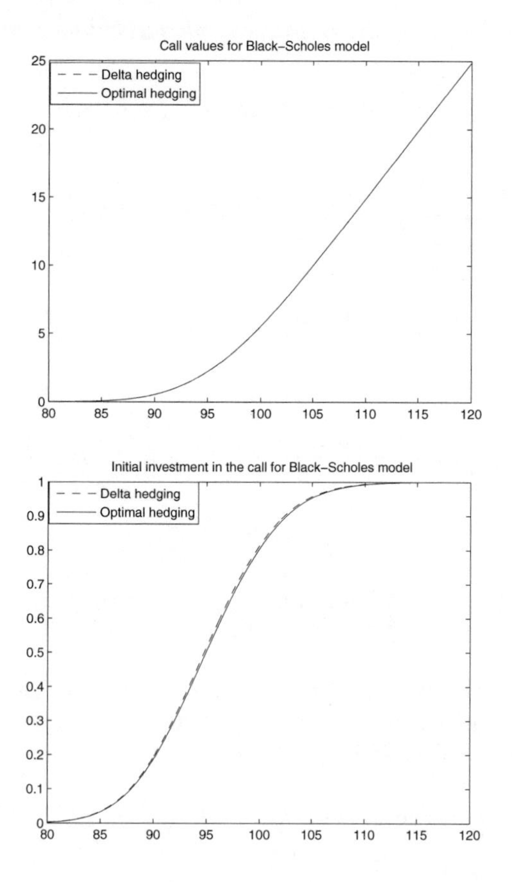

FIGURE 3.3: Graph of C_0 and φ_1 for a call option under the Black-Scholes model, using optimal hedging and delta hedging.

Note that the values of C_0 *and* φ_1 *under optimal hedging and delta hedging are quite close, even with 22 hedging periods. As the number of hedging periods increases, the RMSE decreases to zero.*

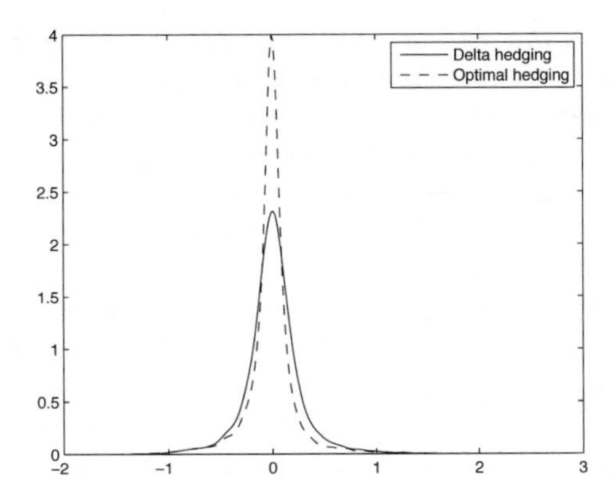

FIGURE 3.4: Graph of the estimated densities of the hedging errors for a call option under the Black-Scholes model, using optimal hedging and delta hedging. 10000 portfolios were simulated.

TABLE 3.2: Statistics for hedging errors under the Black-Scholes model.

Statistic	Optimal	Delta
Average	-0.0065	0.0076
Median	-0.0014	0.0029
Volatility	0.2537	0.2774
Skewness	1.3781	0.5515
Kurtosis	22.0975	8.8391
Minimum	-1.7986	-1.7578
Maximum	3.5389	2.2978
VaR(99%)	0.8231	0.9244
RMSE	0.2538	0.2775

Example 3.4.4 (Variance Gamma Model) *As a second illustration, consider the following (symmetric) Variance Gamma process [Madan et al., 1998]: The excess returns R_k are i.i.d. and*

$$R_k = \left(\mu - \frac{\sigma^2}{2} - r\right)\frac{T}{n} + \frac{\sigma}{\sqrt{2\alpha}}(Y_{1,k} - Y_{2,k}),$$

$Y_{1,k}$, $Y_{2,k}$ are i.i.d. with gamma distribution of parameters $\alpha\frac{T}{n}$ and $\beta = 1$. Also $\mu = 0.0918$, $\sigma = 0.06$, $T = 1$, and the annual rate is 5%.

We want to value a call option with strike $K = 100$ and maturity 1 year, using 22 replication periods and 2000 grid points. Note that $(Y_{1,k} - Y_{2,k})/\sqrt{2}$ has the same law as $\sqrt{Y_{1,k}}Z_k$, where $Z_k \sim N(0,1)$, Z_k independent of $Y_{1,k}$.

For more details on the Variance Gamma process, see Section 6.4.4. The parameters are chosen so that the returns have the same mean and variance as in the previous example. In the present case, we take $\alpha = 1$ so that the annual returns have a Laplace distribution, being the difference of two independent variables with the same exponential distribution. Note that delta hedging is not optimal in the continuous time limit, as shown in Rémillard and Rubenthaler [2009].

The Monte Carlo method with a sample size of $N = 10000$ simulated returns was used to approximate the functions $\check{C}$ and $\check{a}$. The results for the value of the call together with φ_1 appear in Figure 3.5. Then 10000 trajectories of 22 periods were simulated to compute the hedging errors. The estimated density of hedging errors and related statistics are given in Figure 3.6 and Table 3.3 respectively.

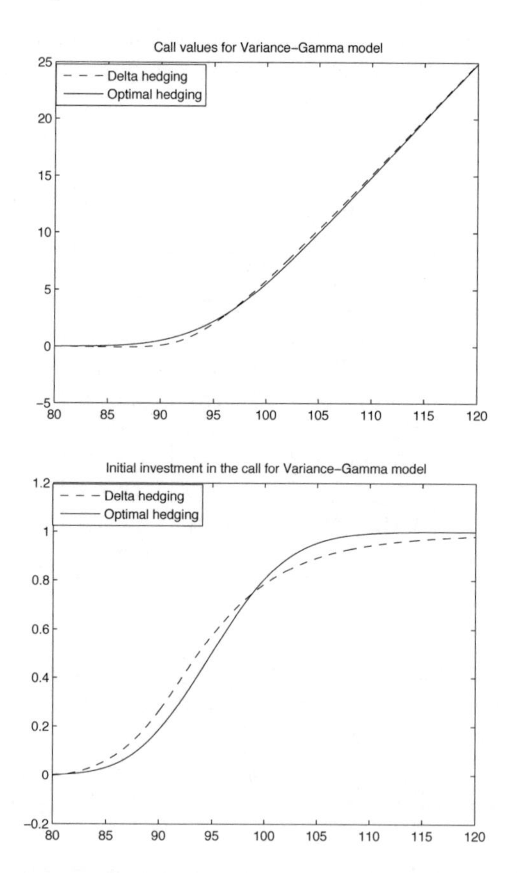

FIGURE 3.5: Graph of C_0 and φ_1 for a call option under the Variance Gamma process, using optimal hedging and delta hedging.

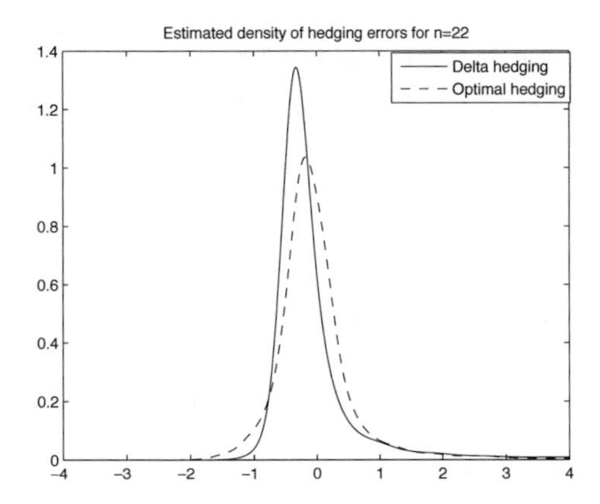

FIGURE 3.6: Graph of the estimated densities of the hedging errors for a call option under the Variance Gamma process, using optimal hedging and delta hedging. 10000 portfolios were simulated.

TABLE 3.3: Statistics for hedging errors under the Variance Gamma process.

Statistic	Optimal	Delta
Average	0.0005	0.0344
Median	-0.1584	-0.2821
Volatility	0.9798	1.1411
Skewness	6.8904	5.5994
Kurtosis	84.3995	47.4645
Minimum	-1.7918	-1.4656
Maximum	20.7638	16.6249
VaR(99%)	4.1281	5.4253
RMSE	0.9798	1.1416

In contrast with the previous example, the values of C_0 and φ_1 under optimal hedging and delta hedging are not that close. Also, the RMSE does not decrease to zero as $n \to \infty$.

3.4.6 Incomplete Markovian Models

Suppose that $(\check{S}_k, h_k)$ is a Markov process, where the conditional distribution of (R_k, h_k) given $\check{S}_{k-1} = s$ and $h_{k-1} - h$ is $\mu(s, h; dx, dp)$. In practice, it happens often that h_k is not observable. This is the case for example for GARCH models since the volatility in these models depends on unknown parameters. See Chapter 7 for details. Then $\mathfrak{a}_k$, $\mathfrak{b}_k$, ρ_k, and γ_k are deterministic functions of $(\check{S}_{k-1}, h_{k-1})$, for all $k \in \{1, \ldots, n\}$.

If in addition $\Phi = C_n(S_n) = \check{C}_n(\check{S}_n)$, and we set $\beta_k \gamma_{k+1} C_k = \check{C}_k(\check{S}_k, h_k)$ and $\alpha_k = \check{a}_k(\check{S}_{k-1}, h_{k-1})$, we have, for $k = n, \ldots, 1$,

$$\begin{aligned}
\mathfrak{a}_k(s,h) &= \int (e^x - \mathbf{1})(e^x - \mathbf{1})^\top \gamma_{k+1}\{D(s)e^x, p\}\,\mu(s,h;dx,dp),\\
\mathfrak{b}_k(s,h) &= \int (e^x - \mathbf{1})\gamma_{k+1}\{D(s)e^x, p\}\,\mu(s,h;dx,dp),\\
\rho_k(s,h) &= \mathfrak{a}_k^{-1}(s,h)\mathfrak{b}(s,h)\\
\gamma_k(s,h) &= \int \gamma_{k+1}\{D(s)e^x, p\}\,\mu(s,h;dx,dp) - \rho_k^\top(s,h)\mathfrak{b}_k(s,h),\\
\check{C}_{k-1}(s,h) &= \int \check{C}_k\{D(s)e^x, p\}\left\{1 - \rho_k^\top(s,h)(e^x - \mathbf{1})\right\}\mu(s,h;dx,dp),\\
\check{a}_k(s,h) &= \mathfrak{a}_k^{-1}(s,h)\int \check{C}_k\{D(s)e^x, p\}(e^x - \mathbf{1})\,\mu(s,h;dx,dp).
\end{aligned}$$

These functions can also be approximated quite accurately by Monte Carlo methods. See, e.g., Rémillard et al. [2010].

Example 3.4.5 (Simple Hidden Markov Model) *We consider here a regime-switching geometric random walk defined as follows: Suppose that $(\tau_k)_{k\geq 0}$ is an homogeneous Markov chain with values in $\{1, \ldots, l\}$, and with transition matrix Q. Here τ_k represents the (non observable) regime at period k.*

Given $\tau_1 = j_1, \ldots, \tau_n = j_n$, we assume that the excess returns $R_1, \ldots, R_n$ are independent, where $R_i \sim \mathbb{P}_{j_i}$, $i \in \{1, \ldots, n\}$, with $E\left\{\left(e^{R_i} - \mathbf{1}\right)|\tau_i = j\right\} = \mathfrak{b}^{(j)}$ and

$$E\left\{\left(e^{R_i} - \mathbf{1}\right)\left(e^{R_i} - \mathbf{1}\right)^\top |\tau_i = j\right\} = \mathfrak{a}^{(j)},$$

$j \in \{1, \ldots, l\}$. We assume that the covariance matrix of e^{R_1} under $\mathbb{P}_j$ is invertible for all $j \in \{1, \ldots, l\}$, so the conditions of Lemma 3.4.1 are met.

In general, S is not a Markov process, unless the regimes are independent. However, (S, τ) is always a Markov process.

One can easily prove that if $\tau_{k-1} = i$, then for $k = n, \ldots, 1$,

$$
\begin{aligned}
\mathfrak{a}_k &= \mathfrak{a}_k(i) = \sum_{j=1}^{l} Q_{ij} \gamma_{k+1}(j) \mathfrak{a}^{(j)}, \\
\mathfrak{b}_k &= \mathfrak{b}_k(i) = \sum_{j=1}^{l} Q_{ij} \gamma_{k+1}(j) \mathfrak{b}^{(j)}, \\
\rho_k &= \rho_k(i) = \mathfrak{a}_k^{-1}(i) \mathfrak{b}_k(i), \\
\gamma_k &= \gamma_k(i) = \sum_{j=1}^{l} Q_{ij} \gamma_{k+1}(j) \left\{ 1 - \rho_k^\top(i) \mathfrak{b}^{(j)} \right\}, \\
U_k &= \frac{\gamma_{k+1}(\tau_k)}{\gamma_k(i)} \left\{ 1 - \rho_k^\top(i) \left(e^{R_k} - \mathbf{1} \right) \right\}.
\end{aligned}
$$

In addition, if $C_n = \Phi(s)$, and if we set $\beta_k C_k \gamma_{k+1} = \check{C}_k(\check{S}_k, \tau_k)$ and $\alpha_k = \check{a}_k(\check{S}_{k-1}, \tau_{k-1})$, then, for $k = n, \ldots, 1$, and $i \in \{1, \ldots, l\}$, we have

$$
\check{C}_{k-1}(s, i) = \sum_{j=1}^{l} Q_{ij} \int \check{C}_k \{D(s)e^x, j\} \left\{ 1 - \rho_k^\top(i) \left(e^x - \mathbf{1} \right) \right\} \mathbb{P}_j(dx),
$$

and

$$
\check{a}_k(s, i) = \mathfrak{a}_k^{-1}(i) \sum_{j=1}^{l} Q_{ij} \int \check{C}_k \{D(s)e^x, j\} \left(e^x - \mathbf{1} \right) \mathbb{P}_j(dx).
$$

Since $\check{C}_k$ and $\check{a}_k$ are weighted expectations, they can be approximated by using Monte Carlo simulations or semi-exact calculations, coupled with interpolations.

Remark 3.4.4 *To be able to implement it in practice however, one needs to be able to predict the regimes, given the observations. This is a filtering problem that is tackled in Section 10.2.2.*

As an example, suppose that we have a 3-regime model with Gaussian distributions under each regime. To be realistic, the parameters[5] *have been taken from Example 10.2.2, where a regime-switching model is fitted to the data of Apple, from January 13th, 2009, to January 14th, 2011.*

We want to value a call option on the asset, with $S_0 = 348.38$ (closing price, January 14th, 2011), $K = 350$, risk-free rate $r = \ln(1 + 0.0026)$, and $n = 24$ hedging periods, i.e., we rebalance the hedging portfolio every day. For the grid, we use $m = 2000$ equidistant points, from $s_1 = 270$ to $s_m = 420$.

[5] Here the parameters are rescaled to take into account the daily time scale, i.e, we take $\mu_j h$ and $\sigma_j \sqrt{h}$, with $h = 1/252$.

The estimations of $\check{C}_k$ *and* $\check{a}$ *are computed with the semi-exact method. The results for the value of the call together with* φ_1 *appear in Figure 3.5. For sake of comparisons, the 3 options prices* $C_0(i)$ *for* $S_0 = 348.48$ *were computed together with the associated* $\varphi_1(i)$. *These results appear in Table 3.4, as well as the corresponding values under the Black-Scholes model and the estimated probability of each regime. The market price value was* 11.10 *on January 14*th, *2011.*

Then 1000000 trajectories of 24 *periods were simulated to compute the hedging errors. The estimated density of hedging errors and related statistics are given in Figure 3.6 and Table 3.3 respectively.*

TABLE 3.4: C_0 and φ_1 for each regime, together with the estimated probability η of each regime.

		Optimal Hedging		Delta Hedging	
Regime	η	Call Value	φ_1	Call Value	φ_1
1	0.0193	15.4679	0.5187	12.0957	0.5005
2	0.5412	9.9883	0.5147	12.0957	0.5005
3	0.4395	10.2309	0.5216	12.0957	0.5005

TABLE 3.5: Statistics for the hedging errors in a Gaussian regime-switching model, where we chose the most probable regime.

Stats	Optimal	Delta
Average	0.0291	-1.9684
Median	-0.2122	-2.2109
Volatility	2.9399	3.2114
Skewness	2.0237	1.2246
Kurtosis	12.0312	7.1825
Minimum	-21.2869	-11.7354
Maximum	44.9545	32.8553
VaR(99%)	11.3826	8.9104
RMSE	2.9400	3.7667

Finally, using the investment strategy described in Section 3.4.3.1 and using the averages in Table 3.5, we find that starting with the most probable regime, the average discounted P&L for the optimal hedging is $11.10 - 9.9883 - 0.0291 = 1.0826$, *while for the delta hedging, it is given by* $12.0957 - 11.10 + (-1.9684) = -0.9727$.

Obviously, in terms of P&L, the optimal hedging strategy is much more attractive than the delta hedging strategy, even if the latter produces a portfolio

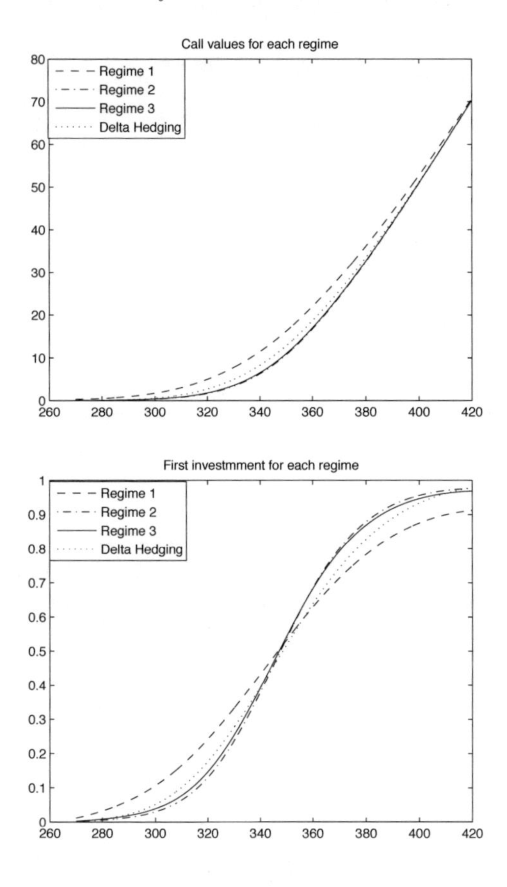

FIGURE 3.7: Graph of C_0 and φ_1 for a call option under a regime-switching model with 3 Gaussian regimes, using optimal hedging and delta hedging.

value much larger than the payoff of the option on average. The problem is just that according to the Black-Scholes model, the option value is too expensive. On the other hand, according to the regime-switching model, it is the market value of the option that is too expensive.

With real data, the optimal hedging strategy looks even better in terms of P&L. For example, for the call option with 24-day maturity on Apple, the hedging errors are respectively -6.1084 *and* -8.3838 *for the optimal and delta hedging. As a result, the discounted P&L is* $11.10 - 9.9883 - (-6.1084) = 9.4955$ *for the optimal hedging method, while it is* $12.0957 - 11.10 + (-8.3838) = -5.117$ *for the delta hedging method. The problem with the optimal hedging is that* φ_k *can be large, so here for example, at some point the discounted value of the portfolio almost reached* -800 *to finish at* 8.9437*, as shown in Figure 3.9. Note that the payoff of the option at maturity was* 0.56*.*

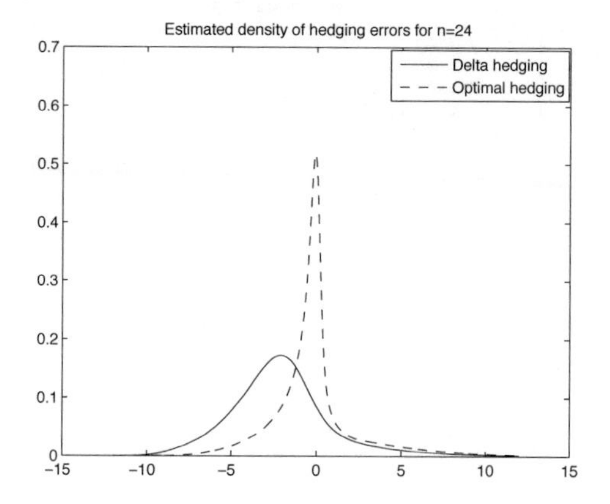

FIGURE 3.8: Graph of the estimated densities of the hedging errors for a call option under a regime-switching model with 3 Gaussian regimes, using optimal hedging and delta hedging. 1000000 portfolios were simulated.

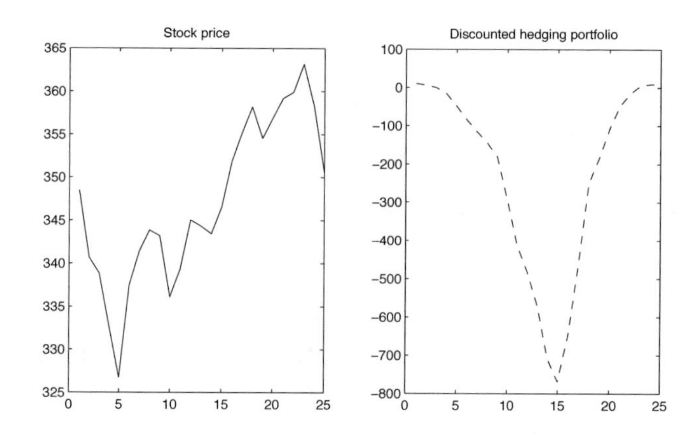

FIGURE 3.9: Graph of the value of Apple from January 14th to February 18th, 2011, and the associated hedging portfolio under optimal hedging.

Example 3.4.6 (GARCH Type Models) *For these models, we assume that*

$$\begin{aligned} R_k &= \pi_1(h_{k-1}, \epsilon_k) \\ h_k &= \pi_2(h_{k-1}, \epsilon_k), \end{aligned}$$

where the innovations ϵ_k are independent and identically distributed with probability law $\mathbb{P}$. Also ϵ_k is independent of h_{k-1}. It is immediate that (S_k, h_k) is *a Markov process. Furthermore, almost all known GARCH(1,1) models can be written in that way; see Chapter 7 for more details on GARCH models.*

It is easy to check that for all $k = n, \ldots, 1$,

$$
\begin{aligned}
\mathfrak{a}_k &= \mathfrak{a}_k(h) = \int \left\{ e^{\pi_1(h,y)} - \mathbf{1} \right\} \left\{ e^{\pi_1(h,y)} - \mathbf{1} \right\}^\top \gamma_{k+1}\{\pi_2(h,y)\}\mathbb{P}(dy), \\
\mathfrak{b}_k &= \mathfrak{b}_k(h) = \int \{ e_1^\pi(h,y) - \mathbf{1} \} \gamma_{k+1} \{\pi_2(h,y)\} \mathbb{P}(dy), \\
\rho_k &= \rho_k(h) = \mathfrak{a}_k^{-1}(h)\mathfrak{b}_k(h), \\
\gamma_k &= \gamma_k(h) = \int \left[1 - \rho_k(h) \left\{ e^{\pi_1(h,y)} - \mathbf{1} \right\} \right] \gamma_{k+1} \{\pi_2(h,y)\} \mathbb{P}(dy).
\end{aligned}
$$

Further assume that $\Phi = C_n(S_n)$. Then setting $\beta_k C_k \gamma_{k+1} = \check{C}_k(\check{S}_k, h_k)$, $\alpha_k = \check{a}_k(\check{S}_k, h_k)$, we have

$$
\check{C}_{k-1}(s,h) = \int \check{C}_k \left\{ D(s) e^{\pi_1(h,y)}, \pi_2(h,y) \right\} \left[1 - \rho_k^\top(h) \left\{ e^{\pi_1(h,y)} - \mathbf{1} \right\} \right] \mathbb{P}(dy),
$$

and

$$
\check{a}_k(s,h) = \mathfrak{a}_k^{-1}(h) \int \check{C}_k \left\{ D(s) e^{\pi_1(h,y)}, \pi_2(h,y) \right\} \left\{ e^{\pi_1(h,y)} - \mathbf{1} \right\} \mathbb{P}(dy).
$$

3.4.7 Limiting Behavior

As in the case of the delta hedging, one can let the number of trading periods tend to infinity. For the Black-Scholes model, the limiting hedging error is 0, while in general, the hedging error is the terminal value of a martingale with mean zero.

For more details, see Rémillard and Rubenthaler [2009], where the case of regime-switching Lévy processes is studied.

3.5 Suggested Reading

For an interesting discussion on the Black-Scholes model, discrete time hedging, and transaction costs, I recommend [Wilmott, 2006b, Chapters 46,47,48].

For another point of view on dynamic hedging and the Black-Scholes model, see the pamphlets of Derman and Taleb [2005] and Haug and Taleb [2008].

3.6 Exercises

Exercise 3.1

Consider the tent map series defined by $X_{i+1} = 1 - |2X_i - 1| = 2\min(X_i, 1 - X_i)$, with $X_1 \sim \text{Unif}(0,1)$. Show that $X_i \sim \text{Unif}(0,1)$ for all $i \geq 1$. Also prove that the correlation between X_i and X_{i+j} is 0 for all $i, j \geq 1$.

Hint: X_{i+j} is a function of $|X_i - 1/2|$ for all $j \geq 1$, and $\epsilon_i = \text{sgn}(X_i - 1/2) = \begin{cases} +1, & X_i > 1/2 \\ -1, & X_i < 1/2 \end{cases}$, is independent of $|X_i - 1/2|$.

Exercise 3.2

The MATLAB data set *DataTentMap* contains 1000 samples of size $n = 500$ generated from the Tent Map series. Compute the percentage of rejection of the Ljung-Box test, using *IndLBTest* with 20 lags. What is the conclusion?

Exercise 3.3

(a) Which moment-based coefficients could we use to test the Gaussian distribution hypothesis? Are there other ways to verify the normality of a sample?

(b) Are the following tests reliable to confirm the Gaussian distribution? If not, explain why.

  (i) Jarque-Berra;

  (ii) Revisited Jarque-Berra;

  (iii) Lilliefors;

  (iv) Cramer-von Mises.

(c) What are the advantage(s) of the test(s) identified in (b)?

Exercise 3.4

Test the serial independence hypothesis for the log-returns of Microsoft, using $p = 6$ consecutive observations and $N = 1000$ Monte Carlo samples. Also plot the dependogram.

Exercise 3.5

Give some reasons why option prices on markets are different from the ones obtained with the Black-Scholes model. For each reason, indicate

(a) which assumption is violated in the model;

(b) what kind of models we could use to address the problem.

Exercise 3.6

For a data set of $n = 506$ observations, one finds that the estimation of the skewness is -0.15 and the estimation of the kurtosis is 3.4. Test the normality hypothesis using the Jarque-Berra test and its revisited version. Find the P-values of both statistics.

Exercise 3.7

What are the main advantages of optimal hedging over delta hedging? Can you find any negative points against optimal hedging?

Exercise 3.8

Test the normality hypothesis for the Microsoft data set.

Exercise 3.9

If $U_1, \ldots, U_n$ are i.i.d. uniform variables over $(0, 1)$, find the distribution of

$$F = -2 \sum_{i=1}^{n} \ln(U_i).$$

Exercise 3.10

For the simulation of hedging errors in discrete time for the Black-Scholes model, Wilmott [2006b] uses the following approximation of the standard Gaussian distribution:

$$Z = \sum_{j=1}^{12} U_j - 6,$$

where $U_1, \ldots, U_{12}$ are independent and uniformly distributed over $(0, 1)$.

(a) Compute the theoretical mean, variance, skewness, and kurtosis of Z.

(b) Using 100000 samples of size 500, estimate the 95% quantile of the distribution of the Cramér-von-Mises statistic for testing the null hypothesis of normality (without specifying the parameters).

(c) Using the quantile found in (b), generate 10000 samples of size 500 of the variable Z and compute the percentage of rejection of the null hypothesis of normality for the Cramér-von-Mises statistic. Is it significantly different from 5%? Note that using 10000 samples, the error of estimation of the percentage of rejection should be smaller than 1%. Compare the results with real samples from a Gaussian distribution.

(d) Using 100000 samples of size 500, estimate the 95% quantile of the distribution of the Kolmogorov-Smirnov statistic for testing the null hypothesis of normality (without specifying the parameters).

(e) Using the quantile found in (d), generate 10000 samples of size 500 of the variable Z and compute the percentage of rejection of the null hypothesis of normality for the Kolmogorov-Smirnov statistic. Is it significantly different from 5%? Compare the results with real samples from a Gaussian distribution.

(f) Generate 10000 samples of size 500 of the variable Z and compute the percentage of rejection of the null hypothesis of normality for the Jarque-Berra test. Is it significantly different from 5%? Compare the results with real samples from a Gaussian distribution.

(g) Generate 10000 samples of size 500 of the variable Z and compute the percentage of rejection of the null hypothesis of normality for the revisited Jarque-Berra test. Is it significantly different from 5%? Compare the results with real samples from a Gaussian distribution.

Exercise 3.11

Compute the mean and variance of

$$T_k = \frac{6^k}{\pi^{2k}} \sum_{i_1=1}^{\infty} \cdots \sum_{i_k=1}^{\infty} \frac{Z_{i_1,\ldots,i_k}^2}{(i_1 \cdots i_k)^2},$$

where the random variables $Z_{i_1,\ldots,i_k}$ are i.i.d. standard Gaussian.

3.7 Assignment Questions

1. Construct a MATLAB function to compute the hedging error of a portfolio with the delta hedging method for replicating a call or a put option, under the Black-Scholes model. The input values are:

   - S: $(n+1) \times N$ matrix of prices at times $\frac{k}{n}T$, $k \in \{0,\ldots,n\}$; the columns represents different trajectories. The rebalancing periods of the portfolio are $\frac{k}{n}T$, $k \in \{0,\ldots,n-1\}$.
   - K: strike price;
   - r: annual risk-free rate;
   - T: time to maturity (on an annual time scale);
   - vol: annual volatility;

- put: 1 if put, zero if call;

Then, using the MATLAB function *HedgingLI*, try to replicate the results of Example 3.4.3.

2. Construct a MATLAB function for approximating the functions $\check{C}_k$ and $\check{a}_k$ for a call or a put option, when the underlying process is the symmetric Variance Gamma process. This function must include a method for generating trajectories of the process at rebalancing times $\frac{k}{n}T$, $k \in \{0, \dots, n\}$. Then try to reproduce the results of Example 3.4.4.

3.A Tests of Serial Independence

Suppose $X_1, \dots, X_n$ are identically distributed observations. Following Genest and Rémillard [2004], if $p \geq 2$, for any $A \in \mathcal{A}_p = \{B \subset \{1, \dots, p\}; |B| > 1, B \ni 1\}$, set

$$\mathbb{G}_{n,A}(x) = \frac{1}{\sqrt{n}} \sum_{i=1}^{n} \prod_{j \in A} \{\mathbb{I}(X_{i+j-1} \leq x_j) - F_n(x_j)\}, \quad x = (x_1, \dots, x_p) \in \mathbb{R}^p,$$

where $|A|$ is the cardinality of A, and $F_n(y) = \frac{1}{n} \sum_{i=1}^{n} \mathbb{I}(X_i \leq y)$, $y \in \mathbb{R}$, is the empirical distribution function estimating the unknown common distribution of $X_1, \dots, X_n$. Here we set $X_{n+i} = X_i$, for all $i \in \{1, \dots, p\}$. It is shown in Genest and Rémillard [2004] that under the null hypothesis, the Cramér-von Mises statistics

$$T_{n,A} = \frac{6^{|A|}}{n} \sum_{i=1}^{n} \mathbb{G}^2_{n,A}(X_i, \dots, X_{i+p-1}) \tag{3.15}$$

converge jointly in law to T_A and the latter are independent. Moreover, if $|A| = k$, the T_A has the same law as

$$T_k = \frac{6^k}{\pi^{2k}} \sum_{i_1=1}^{\infty} \cdots \sum_{i_k=1}^{\infty} \frac{Z^2_{i_1,\dots,i_k}}{(i_1 \cdots i_k)^2},$$

where the random variables $Z_{i_1,\dots,i_k}$ are i.i.d. standard Gaussian. It follows that $E(T_A) = 1$, for all $A \in \mathcal{A}_p$. Note that these limiting random variables are distribution-free in the sense that their distribution is independent of the distribution function F. This is due to the fact that $T_{n,A}$ is based on the ranks of $X_1, \dots, X_n$.

Tables for the distribution of T_A can be constructed. One can also compute P-values for the statistics. To do so, one can use the following algorithm.

Algorithm 3.A.1 *For N large (say $N = 1000$), repeat the following steps for each $k \in \{1,\ldots,N\}$:*

- *Generate independent uniform random variables $U_1^{(k)},\ldots,U_n^{(k)}$;*
- *Compute the associated statistics $T_{n,A}^{(k)}$, $A \in \mathcal{A}_p$, using formula (3.15).*

Then one can approximate the P-value of $T_{n,A}$ by

$$p_{n,A} = \frac{1}{N}\sum_{k=1}^{k} \mathbb{I}\left(T_{n,A}^{(k)} > T_{n,A}\right), \quad A \in \mathcal{A}_p.$$

With these asymptotically independent statistics, one can construct a *dependogram*, i.e., a graph of the P-values $p_{n,A}$, for all $A \in \mathcal{A}_p$. An example of a dependogram, obtained by using the MATLAB function *SerIndGRTest* with the returns of Apple, is given in Figure 3.1.1.

One can also define a powerful test of serial independence by combining the P-values of the test statistics $T_{n,A}$, as proposed in Genest and Rémillard [2004]. More precisely, set

$$\mathfrak{F}_{n,p} = -2\sum_{A\subset\mathcal{A}_p} \ln(p_{n,A}).$$

Under the null hypothesis of serial independence, the P-values $p_{n,A}$ converge to independent uniform variables, so $\mathfrak{F}_{n,p} \rightsquigarrow \mathfrak{F}_p$, where $\mathfrak{F}_p$ has a chi-square distribution with $2^p - 2$ degrees of freedom. This test is also implemented in *SerIndGRTest*.

3.B Goodness-of-Fit Tests

Suppose that $X_1,\ldots,X_n$ are independent and identically distributed random variables from a continuous distribution F, and suppose one wants to test the hypothesis H_0 : F is Gaussian, against the hypothesis H_1 : F is not Gaussian.

Durbin [1973] proposed to base test statistics on functions of the empirical distribution function D_n, where

$$D_n(u) = \frac{1}{n}\sum_{i=1}^{n} \mathbb{I}(U_{n,i} \le u), \quad u \in [0,1],$$

with $U_{n,i} = \mathcal{N}\left(\frac{X_i-\bar{x}}{s_x}\right)$, $i \in \{1,\ldots,n\}$, $\mathcal{N}$ being the distribution function of the standard Gaussian distribution. In fact, if the null hypothesis is true, then the pseudo-observations $U_{n,i}$, $i \in \{1, \ldots, n\}$, should be approximately uniformly distributed on $(0, 1)$, so the empirical distribution function should be close to the uniform distribution function.

One could also test the hypothesis that F belongs to some other parametric family $\{F_\theta; \theta \in \Theta\}$. Then we just set $U_{n,i} = F_{\hat{\theta}_n}(X_i)$, where $\hat{\theta}_n$ is an estimation of θ.

3.B.1 Cramér-von Mises Test

To test $\mathcal{H}_0$ against $\mathcal{H}_1$, we suggest to use a Cramér-von Mises type statistics given by

$$\begin{aligned} S_n &= T_n(U_{n,1}, \ldots, U_{n,n}) = n \int_0^1 \{D_n(u) - u\}^2 du \\ &= \sum_{i=1}^n \left\{ U_{n:i} - \frac{(i - 1/2)}{n} \right\}^2 + \frac{1}{12n}, \end{aligned}$$

where $U_{n:1} < \cdots < U_{n:n}$ are the ordered values of the pseudo-observations $U_{n,1}, \ldots, U_{n,n}$. Large values of the statistic S_n should lead to the rejection of the null hypothesis.

Durbin [1973] showed that statistics based on D_n, such has S_n, have an asymptotic distribution which does not depend on the estimated parameters μ and σ of the assumed Gaussian distribution. It is also true for the exponential distribution. However, this is not true for general parametric families of distributions, so we need a way to estimate the P-value.

3.B.1.1 Algorithms for Approximating the P-Value

To approximate the P-value of S_n, or any other statistic, one can use the following Monte Carlo approach, called parametric bootstrap. We state it first in the Gaussian case and then in the general case.

Algorithm 3.B.1 (P-Value for S_n in the Gaussian Case) *For N large (say $N = 1000$), repeat the following steps for each $k \in \{1, \ldots, N\}$:*

(i) Generate a sample $Y_1^{(k)}, \ldots, Y_n^{(k)}$ from a standard Gaussian distribution and compute the mean $\bar{y}_k$ and the standard deviation s_k of the sample.

(ii) Compute $U_{n,i}^{(k)} = \mathcal{N}\left(\frac{Y_i^{(k)} - \bar{y}_k}{s_k}\right)$, for each $i \in \{1, \ldots, n\}$. Then compute $S_n^{(k)} = T_n\left(U_{n,1}^{(k)}, \ldots, U_{n,n}^{(k)}\right)$.

An approximate P-value for the test is given by

$$\frac{1}{N} \sum_{k=1}^N \mathbb{I}\left(S_n^{(k)} > S_n\right).$$

In the general case, i.e., for testing the goodness-of-fit to a parametric family $\{F_\theta; \theta \in \Theta\}$, the approach is similar. Here $\hat{\theta}_n$ is the estimation of θ from our sample.

Algorithm 3.B.2 (P-Value for S_n in the General Case) *For N large (say $N = 1000$), repeat the following steps for each $k \in \{1, \ldots, N\}$:*

(i) Generate a sample $Y_{n,1}^{(k)}, \ldots, Y_{n,n}^{(k)}$ from distribution $F_{\hat{\theta}_n}$ and estimate θ by $\hat{\theta}_{n,k}$.

(ii) Compute $U_{n,i}^{(k)} = F_{\hat{\theta}_n^{(k)}}\left(Y_i^{(k)}\right)$, for each $i \in \{1, \ldots, n\}$. Then compute $S_n^{(k)} = T_n\left(U_{n,1}^{(k)}, \ldots, U_{n,n}^{(k)}\right)$.

In this case, an approximate P-value for the test is given by

$$\frac{1}{N}\sum_{k=1}^{N} \mathbb{I}\left(S_n^{(k)} > S_n\right).$$

A justification of the parametric bootstrap in the general case in given in Genest and Rémillard [2008].

3.B.2 Lilliefors Test

This test is defined by the usual Kolmogorov-Smirnov statistic

$$\begin{aligned} K_n &= \sup_{u \in [0,1]} n^{1/2} |D_n(u) - u| \\ &= n^{1/2} \max_{1 \le k \le n} \max\left(|U_{n:k} - k/n|, |U_{n:k} - (k-1)/n|\right). \end{aligned}$$

The difference with the classical Kolmogorov-Smirnov test is that its P-value can be computed using the Monte Carlo methods described in Section 3.B.1.1, either for the Gaussian case (Algorithm 3.B.1) or the general case (Algorithm 3.B.2).

3.C Density Estimation

The nonparametric estimation of the density function is usually based on the kernel method, of which the well-known histogram is a particular case. In general, the estimator of the density is defined by

$$f_n(x) = \frac{1}{nh_n}\sum_{i=1}^{n} K\left(\frac{x - x_i}{h_n}\right), \quad x \in \mathbb{R},$$

where K, called the kernel, is a density function and the bandwidth h_n is a sequence of positive numbers converging to 0. Note that if the kernel is continuous, then so is the estimated density f_n.

The choice of the kernel is not that important, compared to the choice of the bandwidth.

3.C.1 Examples of Kernels

- Histogram: $K(x) = \mathbb{I}_{(-0.5,0.5]}(x)$. It corresponds to the density of a uniform distribution over the interval $(-0.5, 0.5]$.

- Gaussian kernel: $K(x) = \frac{e^{-x^2/2}}{\sqrt{2\pi}}$. The optimal bandwidth is $h_n = sn^{-1/5}$, where s is the standard deviation of the observations. This is the kernel used in the MATLAB functions *EstPdfGK* and *EstPdfGKpt*.

- Epanechnikov kernel: $K(x) = \frac{3}{4}\max(0, 1 - x^2)$. The optimal bandwidth is $h_n = sn^{-1/5}$, where s is the standard deviation of the observations.

3.D Limiting Behavior of the Delta Hedging Strategy

This appendix can be omitted. It is based on delicate calculations that are detailed in Rémillard and Rubenthaler [2009] and uses stochastic calculus.

Suppose that for all $k \in \{0, \ldots, n\}$, $S_k = S(t_k)$, where S is continuous semimartingale (martingale + a differentiable process). Further assume that where $\beta_k = e^{-rt_k}$, $k \in \{0, \ldots, n\}$. Then, as the number of trading periods $n \to \infty$, π_n^{BS} should tend to

$$
\begin{aligned}
e^{-rT}\pi_T^{BS} &= V^{BS}(0, S_0) + \int_0^T e^{-ru}\nabla_s V^{BS}\{u, S(u)\}^\top dS(u) \\
&\quad - r\int_0^T e^{-ru}\nabla_s V^{BS}\{u, S(u)\}^\top S(u)du.
\end{aligned}
$$

Using Itô's formula, we then have

$$
\begin{aligned}
e^{-rT}\Phi\{S(T)\} &= V^{BS}(0,S_0) - r\int_0^T e^{-ru}V^{BS}\{u,S(u)\}du \\
&\quad + \int_0^T e^{-ru}\partial_u V^{BS}\{u,S(u)\}du \\
&\quad + \frac{1}{2}\sum_{j=1}^d\sum_{k=1}^d \int_0^T e^{-ru}\partial_{s_j}\partial_{s_k}V^{BS}\{u,S(u)\}d\left[S^{(j)},S^{(k)}\right]_u \\
&\quad + \int_0^T e^{-ru}\nabla_s V^{BS}\{u,S(u)\}^\top dS(u).
\end{aligned}
$$

As a result, using (3.8), we find that

$$
\begin{aligned}
G_\infty^{BS} &= \frac{1}{2}\sum_{j=1}^d\sum_{k=1}^d \int_0^T e^{-ru}\partial_{s_j}\partial_{s_k}V^{BS}\{u,S(u)\}d\left[S^{(j)},S^{(k)}\right]_u \\
&\quad - \frac{1}{2}\sum_{j=1}^d\sum_{k=1}^d \int_0^T e^{-ru}S^{(j)}(u)S^{(k)}(u)\alpha_{jk}\partial_{s_j}\partial_{s_k}V^{BS}\{u,S(u)\}du,
\end{aligned}
$$

where $[\cdot,\cdot]$ denotes the quadratic covariation between semimartingales. For more details, see Protter [2004].

For example, if the correct model is the Black-Scholes model with the same covariance matrix Σ, then $G_\infty^{BS} = 0$, as predicted. However, even in the one-dimensional case, if the volatilities are not equal, that is $\Sigma \neq \alpha$, then

$$
G_\infty^{BS} = \frac{(\Sigma - \alpha)}{2}\int_0^T e^{-ru}S^2(u)\partial_s^2 V^{BS}\{u,S(u)\}du.
$$

Finally, for a one-dimensional time-homogeneous diffusion process, as defined in (3.2), one has

$$
G_\infty^{BS} = \frac{1}{2}\int_0^T e^{-ru}S^2(u)\left[A\{S(u)\} - \alpha\right]\partial_s^2 V^{BS}\{u,S(u)\}du,
$$

where $A(s) = \sigma^2(s)$.

From the last two cases, one can see that the hedging error associated with the delta hedging strategy can be enormous.

3.E Optimal Hedging for the Binomial Tree

The famous binomial tree model of Cox et al. [1979] is a particular case of the geometric random walk model described previously. In this case, $d = 1$ and

$S_k = S_{k-1}\zeta_k$, $\beta_k = (1+R)^{-k}$, where $P(\zeta_k = U) = p$ and $P(\zeta_k = D) = 1-p$, with $D < 1+R < U$. Then $e^{R_k} - 1 = \xi_k = \frac{\zeta_k}{1+R} - 1$, $\mathfrak{b} = p\frac{U-D}{1+R} - \frac{1+R-D}{1+R}$ and $\mathfrak{a} = p(1-p)\frac{(U-D)^2}{(1+R)^2} + \mathfrak{b}^2$. Moreover, setting $q = \frac{1+R-D}{U-D}$, one finds that $U_k = \frac{q}{p}$ with probability p and $U_k = \frac{1-q}{1-p}$ with probability $1-p$.

As a result,

$$C_{k-1}(s) = \frac{1}{1+R}\left\{qC_k(sU) + (1-q)C_k(sD)\right\}, \quad k \in \{1,\ldots,n\}.$$

In addition, $\alpha_k = \alpha_k(S_{k-1})$, where

$$\alpha_k(s) = \frac{U-D}{s\mathfrak{a}(1+R)^2}\left\{p(1-q)C_k(sU) - q(1-p)C_k(sD)\right\}.$$

We also have $\varphi_1 = \frac{C_1(sU)-C_1(sD)}{s(U-D)}$, showing that $\pi_1 = C_1$. We will now prove by induction that $\pi_k = C_k$ for all $k \in \{0,\ldots,n\}$.

We know that it is true for $k = 0, 1$. Suppose it is true for $k-1$. Then $\varphi_k = \varphi_k(S_{k-1})$, where

$$\begin{aligned}
\varphi_k(s) &= \alpha_k(s) - \beta_{k-1}\pi_{k-1}\rho_k \\
&= \frac{(U-D)}{s\mathfrak{a}(1+R)^2}\left\{p(1-q)C_k(sU) - q(1-p)C_k(sD)\right\} \\
&\qquad -(p-q)C_{k-1}(s)\frac{(U-D)}{s\mathfrak{a}(1+R)} \\
&= \frac{(U-D)}{s\mathfrak{a}(1+R)^2}\left\{C_k(sU) - C_k(sD)\right\}\left\{p(1-p) + (p-q)^2\right\} \\
&= \frac{C_k(sU) - C_k(sD)}{s(U-D)},
\end{aligned}$$

since $\mathfrak{b} = \frac{(U-D)(p-q)}{1+R}$. Thus

$$\begin{aligned}
\beta_k\pi_k &= \beta_{k-1}\pi_{k-1} + \varphi_k\left(e^{R_k} - 1\right) = \beta_{k-1}C_{k-1}(s) + \varphi_k(s)\beta_{k-1}s\left(\frac{\zeta_k}{1+R} - 1\right) \\
&= \beta_k\left[qC_k(sU) + (1-q)C_k(sD) + \left\{C_k(sU) - C_k(sD)\right\}\frac{(\zeta_k - 1 - R)}{(U-D)}\right].
\end{aligned}$$

Now, if $\zeta_k = U$, then $\pi_k = C_k(sU)$ since $\frac{(\zeta_k-1-R)}{(U-D)} = 1-q$; if $\zeta_k = D$, then $\pi_k = C_k(sD)$ since $\frac{(\zeta_k-1-R)}{(U-D)} = -q$. This proves the desired equality $\pi_k = C_k(S_k)$. In particular, $G \equiv 0$, proving that we have perfect hedging and we recover the well-known results of Cox et al. [1979].

3.F A Useful Result

Lemma 3.F.1 *Suppose* $A = \Sigma + bb^\top$ *where* Σ *is symmetric and invertible. Then* A *is invertible, and* $A^{-1} = \Sigma^{-1} - \frac{\Sigma^{-1}bb^\top\Sigma^{-1}}{1+b^\top\Sigma^{-1}b}$. *Moreover,* $1 - b^\top A^{-1}b = \frac{1}{1+b^\top\Sigma^{-1}b} > 0$.

PROOF. Since $A\left(\Sigma^{-1} - \frac{\Sigma^{-1}bb^\top\Sigma^{-1}}{1+b^\top\Sigma^{-1}b}\right) = I$, A is invertible and $A^{-1} = \Sigma^{-1} - \frac{\Sigma^{-1}bb^\top\Sigma^{-1}}{1+b^\top\Sigma^{-1}b}$. Setting $c = b^\top\Sigma^{-1}b$, one gets $1-b^\top A^{-1}b = 1-c+\frac{c^2}{1+c} = \frac{1}{1+c} > 0$.
■

Bibliography

T. Ané and H. Geman. Order flow, transaction clock, and normality of asset returns. *Journal of Finance*, 55:2259–2284, 2000.

P. P. Boyle and D. Emanuel. Discretely adjusted option hedges. *Journal of Financial Economics*, 8:259–282, 1980.

P. P. Boyle and T. Vorst. Option replication in discrete time with transaction costs. *The Journal of Finance*, XLVII(2):271–293, 1992.

A. Černý and J. Kallsen. On the structure of general mean-variance hedging strategies. *Ann. Probab.*, 35(4):1479–1531, 2007.

J. C. Cox, S. A. Ross, and M. Rubinstein. Option pricing: A simplified approach. *Journal of Financial Economics*, 7:229–263, 1979.

E. Derman and N. N. Taleb. The illusion of dynamic delta replication. *Quant. Finance*, 5:323–326, 2005.

J. Durbin. Weak convergence of the sample distribution function when parameters are estimated. *Ann. Statist.*, 1(2):279–290, 1973.

H. Geman, D. B. Madan, and M. Yor. Stochastic volatility, jumps and hidden time changes. *Finance Stoch.*, 6:63–90, 2002.

C. Genest and B. Rémillard. Tests of independence or randomness based on the empirical copula process. *Test*, 13:335–369, 2004.

C. Genest and B. Rémillard. Validity of the parametric bootstrap for goodness-of-fit testing in semiparametric models. *Ann. Inst. H. Poincaré Sect. B*, 44:1096–1127, 2008.

C. Genest, K. Ghoudi, and B. Rémillard. Rank-based extensions of the Brock Dechert Scheinkman test for serial dependence. *J. Amer. Statist. Assoc.*, 102:1363–1376, 2007.

E. G. Haug and N. N. Taleb. Why we have never used the Black-Scholes-Merton option pricing formula. *Wilmott Magazine*, pages 72–79, 2008.

C. Labbé, B. Rémillard, and J.-F. Renaud. A Simple Discretization Scheme for Nonnegative Diffusion Processes, with Applications to Option Pricing. *Journal of Computational Finance*, 15:3–35, 2012.

R. C. Littell and J. L. Folks. Asymptotic optimality of Fisher's method of combining independent tests. II. *J. Amer. Statist. Assoc.*, 68:193–194, 1973.

D. B. Madan, P. P. Carr, and E. C. Chang. The variance gamma process and option pricing. *European Finance Review*, 2:79–105, 1998.

P. E. Protter. *Stochastic Integration and Differential Equations*, volume 21 of *Applications of Mathematics (New York)*. Springer-Verlag, Berlin, second edition, 2004. Stochastic Modelling and Applied Probability.

B. Rémillard and S. Rubenthaler. Optimal hedging in discrete and continuous time. Technical Report G-2009-77, Gerad, 2009.

B. Rémillard and S. Rubenthaler. Optimal hedging in discrete time. Quantitative Finance, To appear, (2013).

B. Rémillard, A. Hocquard, and N. A. Papageorgiou. Option Pricing and Dynamic Discrete Time Hedging for Regime-Switching Geometric Random Walks Models. Technical report, SSRN Working Paper Series No. 1591146, 2010.

M. Schweizer. Variance-optimal hedging in discrete time. *Math. Oper. Res.*, 20(1):1–32, 1995.

P. Wilmott. *Paul Wilmott on Quantitative Finance*, volume 1. John Wiley & Sons, second edition, 2006a.

P. Wilmott. *Paul Wilmott on Quantitative Finance*, volume 3. John Wiley & Sons, second edition, 2006b.

Chapter 4

Measures of Risk and Performance

We first discuss properties that should be satisfied by risk measures associated with loss variables. Next, we show how Monte Carlo methods can be used to estimate the distribution function of the loss and other measures of risk such as the VaR and the expected shortfall. We also show how to compute the estimation error in each case. Then, we cover some approximation methods for losses, following Jaschke [2002]. We begin by approximating the loss by a quadratic form of the returns of the underlying assets. Next, assuming that the returns have a joint Gaussian distribution, we discuss four approximation methods: Monte Carlo simulations, Edgeworth and Cornish-Fisher expansions, saddlepoint approximations, and inversion of characteristic functions. Finally, we also state properties that performance measures should satisfy. Estimation and precision of three popular performance measures (Sharpe, Sortino, and Omega ratios) are discussed.

4.1 Measures of Risk

Since the first Basel Capital Accord in 1998[1], financial institutions must compute some measures of risk. In Basel I, for measuring market risk, the banks were allowed to use either the standardized Basel model or use their internal Value at Risk (VaR) models. Nowadays, all the financial industry players need to measure risk.

In what follows, we will focus on the following measures of (market) risk: VaR, expected shortfall, and more generally on coherent risk measures. We will also see how these measures can be estimated.

4.1.1 Portfolio Model

Let $S_t = \left(S_t^{(1)}, \ldots, S_t^{(m)}\right)$ be m risk factors entering in the composition of a portfolio. For example, S_t could represent the value at time t of m assets.

Suppose that the value V of a portfolio at time t is of the form $V = V(t, S_t)$. For example, the portfolio could be composed of 20 options written on $m = 10$ underlying assets. Since V is not necessarily a linear function of the assets, it

[1] `http://www.bis.org`

is almost impossible to find the distribution of $V(t, S_t)$ even if we know the distribution of S_t.

We are interested in computing measures of risk for the discounted loss of the portfolio at time t, i.e., we are interested in the distribution function F of the random variable

$$X = V_0 - e^{-rt}V(t, S_t),$$

where r is the risk-free interest rate, assumed to be constant up to time t, and $V_0 = V(0, S_0)$.

4.1.2 VaR

Since the random variable X represent possible losses for an investment, its quantile[2] of order 95%, $F^{-1}(0.95)$, is often called the Value-at-Risk (VaR). So the VaR is characterized by the equation

$$P(X > \text{VaR}) = 0.05.$$

The last equation means that 5% of the time, the losses should be greater than the VaR(95%); 1% of the time the losses will be greater than VaR(99%), etc. More generally, for $p \in (0, 1)$, the VaR of order $100p\%$, denoted by $\text{VaR}(100p\%)$, is the quantile of order p, that is $\text{VaR}(100p\%) = F^{-1}(p)$.

4.1.3 Expected Shortfall

Another widely used measure of risk is the expected shortfall, denoted by $ES(p)$, and defined by

$$\begin{aligned} ES(p) &= E\{X|X > \text{VaR}(100p\%)\} \\ &= \text{VaR}(100p\%) + e\{\text{VaR}(100p\%)\}, \end{aligned} \tag{4.1}$$

where e is the mean excess function defined in Appendix 4.C. The expected shortfall can be interpreted as the average loss when it exceeds the corresponding VaR(100p%). It is also sometimes called the "tail conditional expectation" or TailVaR.

Remark 4.1.1 *Note that the function $x \mapsto E(X|X > x)$ is non-decreasing, so $ES(p)$ is also non-decreasing. This property will be proven in Appendix 4.I.2.*

One could ask if there are other measures of risk. This question led several people to define properties that should be satisfied by risk measures. These axioms are defined next.

---

[2]See Section 4.B.

4.1.4 Coherent Measures of Risk

We begin by presenting the Axioms proposed by Artzner et al. [1999] for coherent measures of risk.

Let X and Y be random variables representing discounted losses. According to Artzner et al. [1999], a coherent measure of risk ρ should satisfy the following properties:

A1: $\rho(X + a) = \rho(X) + a$, for any $a \in \mathbb{R}$.

A2: $X \leq Y \quad \Rightarrow \quad \rho(X) \leq \rho(Y)$.

A3: $\rho(\lambda X + (1 - \lambda)Y) \leq \lambda\rho(X) + (1 - \lambda)\rho(Y)$, for all $\lambda \in [0, 1]$.

A4: $\rho(\lambda X) = \lambda\rho(X)$ for any $\lambda \geq 0$.

Note that Axioms A1–A4 are equivalent to Axioms A1, A2, A4, and

A3': $\rho(X + Y) \leq \rho(X) + \rho(Y)$.

In general, the VaR and the expected shortfall do not respect the Axioms, so they are not coherent measures of risk.[3] The question is: Is there any coherent measure of risk? The answer is yes. The expectation is one! Here is another example.

Example 4.1.1 (Artzner et al. [1999]) *For a given $p \in (0, 1)$, define*

$$WCE_p(X) = \sup_{A;\ P(A)>1-p} E(X|A).$$

Then WCE_p, called the Worst Conditional Expectation (of order p), is a coherent measure of risk satisfying

$$ES_p(X) \leq WCE_p(X).$$

Remark 4.1.2 *It is easy to check if the WCE is a coherent measure of risk but it is impossibly difficult to compute in general. The authors succeeded in computing it in the particular case when the probability space Ω is finite and when the underlying probability is the uniform distribution, i.e., every possible outcome has the same probability. In that case, they proved that $ES_p(X) = WCE_p(X)$, whenever X takes* card(Ω) *different values!*

They also proved the following result.

Lemma 4.1.1 *If Ω is finite, then ρ satisfies Axioms A1–A4 if and only if there exists a set of probability measures $\mathcal{P}$ on Ω so that $\rho = \rho_{\mathcal{P}}$, where*

$$\rho_{\mathcal{P}}(X) = \sup_{P \in \mathcal{P}} E_P(X). \tag{4.2}$$

[3] However, one can modify the expected shortfall so that it becomes a coherent measure of risk. See, e.g., Tasche [2002].

Of course, on a general probability space Ω, (4.2) still defines a coherent measure of risk. In fact, all measures of risk that have some kind of continuity properties are of this form.

Going back to our problem, i.e., measures of risk associated with a portfolio, one could choose $\mathcal{P}$ as the set of possible distributions for the risk factors S_t. For example, the risk associated with $X = V_0 - e^{-rt}V(t, S_t)$ would then be defined by

$$\rho_{\mathcal{P}}(X) = \sup_{P \in \mathcal{P}} E_P \left\{ V_0 - e^{-rt} V(t, S_t) \right\}.$$

Remark 4.1.3 *If the set $\mathcal{P}$ contains only one probability distribution, then $\mathcal{P}$ can be associated with a "pure scenario." Otherwise, $\mathcal{P}$ is called the set of "generalized scenarios." The associated risk measure of X comes from the worst scenario.*

Axiom A4 is clearly the most debatable one since it implies that the risk is a linear function of the leverage effect. Considering the recent financial crisis, its realism is questionable. If one rejects Axiom 4, then a risk measure satisfying Axiom A1–A3 is called a convex risk measure.

In Artzner et al. [1999], they were also able to characterize the set of convex risk measures on a finite probability space Ω.

Lemma 4.1.2 *If Ω is finite, then ρ satisfies Axioms A1–A3 if and only if there exists a set of probability measures $\mathcal{P}$ on Ω and a real-valued function c on $\mathcal{P}$ so that*

$$\rho(X) = \sup_{P \in \mathcal{P}} \left\{ c(P) + E_P(X) \right\}. \tag{4.3}$$

4.1.4.1 Comments

My personal view of the axiomatic definition of risk measures is that these Axioms are quite debatable, particulary A3 and A4. Even A1 in my opinion should not be considered. Adding a constant value to an investment does not make it more risky, unless you consider also the risk of default. Also, the mean of a random variable is considered a coherent measure of risk, which is counterintuitive and hardly justifiable. Perhaps it is the use of the term "risk" that is incorrect. These axioms define more a preference measure than a risk measure. In such a case, Axioms A1–A4 make more sense.

Finally, as a single number cannot characterize the distribution of a random variable, a single number cannot characterize the risk associated with an investment. In fact, the only way to measure correctly the risk of an investment is by knowing its distribution function. In other words, collapsing a distribution function into a single number constitute an over-simplification.

4.1.5 Coherent Measures of Risk with Respect to a Stochastic Order

Instead of trying to measure the risk associated with an investment, one could just try to compare investments, to check if one investment is less risky than another one. Such comparison methods have been used widely in Actuarial science and Statistics under the term 'stochastic orders," for which there is a huge literature.

Following Yamai and Yoshiba [2002], if $\preceq$ is a given stochastic order, we say that a risk measure ρ is coherent with respect to that order if $X \preceq Y$ implies $\rho(X) \leq \rho(Y)$.

One of the major drawbacks of such a definition is that since the stochastic orders are necessarily partial orders, one could not always compare investments. Is this really important? It may just be normal not to be able to compare two investments as one cannot always compare two distribution functions.

4.1.5.1 Simple Order

The following stochastic order, called the simple order, is due to Lehmann [1951].

Definition 4.1.1 (Simple order) $X \preceq_{st} Y$ *if and only if*

$$P(X > x) \leq P(Y > x), \text{ for all } x \in \mathbb{R}.$$

Remark 4.1.4 *If* $X \preceq_{st} Y$, *then* $E\{f(X)\} \leq E\{f(Y)\}$, *for all increasing functions* f *such that* $E\{|f(X)|\}$ *and* $E\{|f(Y)|\}$ *are both finite.*

A risk measure is then said to be coherent with respect to the simple order if $X \preceq_{st} Y$ entails $\rho(X) \leq \rho(Y)$.

Proposition 4.1.1 $X \preceq_{st} Y$ *if and only if* $\mathrm{VaR}_X(p) \leq \mathrm{VaR}_Y(p)$ *for all* $p \in (0,1)$.

In particular, for any given $p \in (0,1)$, $\mathrm{VaR}_X(p)$ *is a coherent measure of risk with respect to the simple order.*

The proof is given in Appendix 4.I.1.

According to Proposition 4.1.1, for two investments to be comparable with respect to the simple order, we have to be able to compare their VaR of any order.

4.1.5.2 Hazard Rate Order

One may ask when, for all x, the conditional distribution of X given $X > x$ is stochastically smaller, in the sense of the simple order, than the conditional distribution of Y given $Y > x$. The following order answers precisely this question.

Definition 4.1.2 (Hazard rate order) $X \preceq_{hr} Y$ *if and only if* $x \mapsto \frac{P(Y>x)}{P(X>x)}$ *is non-decreasing. The hazard rate order is also called the concave order.*

Proposition 4.1.2 *The hazard rate order is stronger than the simple order, i.e., if* $X \preceq_{hr} Y$ *then* $X \preceq_{st} Y$.

PROOF. Since $\lim_{x\to-\infty} \frac{P(Y>x)}{P(X>x)} = 1$, then $X \preceq_{hr} Y$ entails that $x \mapsto \frac{P(Y>x)}{P(X>x)}$ is non decreasing, so for all x, $\frac{P(Y>x)}{P(X>x)} \geq 1$. Hence, $X \preceq_{st} Y$.

■

Here is another characterization of the hazard rate order. When X and Y have densities, it follows that $X \preceq_{hr} Y$ if and only if $r_X(x) = \frac{f_X(x)}{P(X>x)} \geq r_Y(x) = \frac{f_Y(x)}{P(Y>x)}$ for all x such that $P(X > x)P(Y > x) > 0$. The function $r_X(x) = \frac{f_X(x)}{P(X>x)}$ is called the hazard rate function of X, hence the name "hazard rate order."

Proposition 4.1.3 *If* $X \preceq_{hr} Y$, *then* $ES_X(p) \leq ES_Y(p)$, *for any* $p \in (0,1)$.

In particular, the expected shortfall is coherent with respect to the hazard rate order.

The proof is given in Appendix 4.I.2.

Having shown that the VaR and the expected shortfall are not worthless measures of risk, being coherent with respect to the simple order and the hazard rate order, we will now focus on methods to estimate measures of risk.

4.2 Estimation of Measures of Risk by Monte Carlo Methods

Recall that we are interested in measures of risk for the loss

$$X = V_0 - e^{-rt}V(t, S_t),$$

where r is the risk-free rate, $V_0 = V(0, S_0)$, and $S_t = \left(S_t^{(1)}, \ldots, S_t^{(m)}\right)$ are the risk factors.

In this section, we assume that we can simulate the risk factors and that the function $V(t, s)$ can be computed or approximated with sufficient precision. To compute the VaR as well as any other measure of risk, we will find a way to estimate the distribution function of the random variable X.

The first method that we propose is the Monte Carlo method. It is not the fastest method but it has the great advantage of being precise, when the number of simulations is large enough. This is why the Monte Carlo method often serves as a benchmark when comparing other evaluation methods for measuring risk.

4.2.1 Methodology

The methodology can be described simply. Conditionally on S_0, we generate n independent values $S_{1,t}$, ..., $S_{n,t}$ of S_t, and we compute the associated losses

$$X_j = V_0 - e^{-rt}V(t, S_{j,t}), \qquad j \in \{1, \ldots, n\}.$$

We therefore have at our disposal n independent values of X, from which we could estimate any measure of risk. We will also be able to estimate the unknown distribution function F of X.

4.2.2 Nonparametric Estimation of the Distribution Function

Suppose that $X_1, \ldots, X_n$ are independent and identically distributed observations from a distribution function F. The empirical distribution function, denoted by F_n, is

$$F_n(x) = \frac{1}{n} \sum_{j=1}^{n} \mathbb{I}(X_j \leq x), \quad x \in \mathbb{R}. \tag{4.4}$$

$F_n(x)$ is just the proportion of the observations in the sample with values not exceeding x. In fact, $nF_n(x)$ has a binomial distribution with parameters n and $p = F(x)$. Note that the empirical distribution can be computed by using the MATLAB function *ecdf* of the *Statistics* Toolbox.

When the distribution function F is continuous, $P(X_i = X_j) = 0$ whenever $i \neq j$, so the ordered values $X_{n:1} < \cdots < X_{n:n}$ are called the order statistics of the sample. In this case, $F_n(X_{n:k}) = k/n$, and the graph of F_n is determined by the points

$$(X_{n:1}, 1/n), (X_{n:2}, 2/n), \ldots, (X_{n:n}, n/n).$$

As discussed later in Remark 4.2.1, most of the time in financial applications, we may assume that the distribution function of the losses of a portfolio is continuous.

4.2.2.1 Precision of the Estimation of the Distribution Function

It follows from an easy application of the central limit theorem for the Bernoulli distribution that for x given,

$$\sqrt{n}\, \{F_n(x) - F(x)\} \rightsquigarrow N\left(0, F(x)\{1 - F(x)\}\right).$$

It means that when n is large enough, for a fixed x,

$$F_n(x) \pm \frac{1.96}{\sqrt{n}} \sqrt{F_n(x)\{1 - F_n(x)\}}$$

defines a 95% confidence interval for $F(x)$. However, we can do much better. Instead of estimating F at one point, we can estimate it at every point. More precisely, as a random function of x, we can prove that

$$\sqrt{n}\,\{F_n(x) - F(x)\} \rightsquigarrow B\{F(x)\}, \tag{4.5}$$

for all values of x for which F is continuous, where B is a Brownian bridge (see Appendix 4.A for more details). In fact, the convergence in (4.5) means that for all continuous functions Ψ of the sample path space of continuous functions on $[-\infty, \infty]$,

$$\Psi\left\{\sqrt{n}\,(F_n - F)\right\} \rightsquigarrow \Psi\left\{B \circ F\right\}. \tag{4.6}$$

In particular, if F is continuous, then taking $\Psi(h) = \sup_{x\in\mathbb{R}} |h(x)|$ in (4.6), we may conclude that

$$\sqrt{n}\,\sup_x |F_n(x) - F(x)| \rightsquigarrow \sup_{x\in\mathbb{R}} |B\{F(x)\}| = \sup_{0\le t\le 1} |B(t)|. \tag{4.7}$$

Note that (4.7) leads to the famous Kolmogorov-Smirnov goodness-of-fit test (of a simple hypothesis), and there exist tables for the distribution function of the random variable $\sup\limits_{t\in[0,1]} |B(t)|$. In fact, from Shorack and Wellner [1982, page 34], one has

$$P\left\{\sup_{t\in[0,1]} |B(t)| > x\right\} = 2\sum_{n=1}^{\infty} (-1)^{n+1} e^{-2n^2x^2}, \tag{4.8}$$

for all $x \in (0,\infty)$. For example, with formula (4.8), we find that the 95% quantile of $\sup_{t\in[0,1]} |B(t)|$ is 1.358. It means that in 95% of the time, the following inequality is true:

$$\sup_x |F_n(x) - F(x)| \le \frac{1.358}{\sqrt{n}}.$$

Hence, 95% of the time, the graph $\{(x, F(x)); x \in \mathbb{R}\}$ of the unknown distribution function F will lie completely inside the uniform confidence band

$$\left\{(x,y) : F_n(x) - \frac{1.358}{\sqrt{n}} \le y \le F_n(x) + \frac{1.358}{\sqrt{n}}, x \in \mathbb{R}\right\}. \tag{4.9}$$

Remark 4.2.1 *In most financial applications, either F is continuous or the set of all $\{F(x); F \text{is continuous at } x\}$ is included is a closed interval $[a,b] \subset [0,1]$. The latter occurs when the losses are bounded or the gains are bounded, and that one can reach at least one of either bounds with positive probability. It then follows that*

$$\sqrt{n}\,\sup_x |F_n(x) - F(x)| \rightsquigarrow \sup_{t\in[a,b]} |B(t)| \le \sup_{t\in[0,1]} |B(t)|.$$

This leads to an overestimation of the estimation error, i.e., the confidence interval will be wider than necessary.

For an example of a bounded loss X, suppose for instance that we have a long position in a call (bought at value V_0) and that X represents the value of the loss at maturity, that is $t = T$. Then $P(X \leq V_0) = 1$ and there is a positive probability that the loss is V_0; in fact $P(X = V_0) = P(S_T < K) > 0$, where K is the strike price. However, if X represents the value of the loss at a period t before the maturity T of the option, then X is still bounded above by V_0 but reaching the bound has probability 0, so then the distribution function of X is continuous.

Example 4.2.1 *Suppose we bought 1000 call options on Apple (strike price of \$350) with a 6-month maturity on January 14th, 2011, when the value of the asset is \$348, and we shorted 1000 shares of Microsoft currently at \$28. We want to estimate the distribution of losses in 1 month, assuming that the Black-Scholes assumptions hold. We use the parameters in Tables 2.1 and 2.2, i.e., the estimated parameters for μ and the volatility vector v are $\mu = (0.7314, 0.2441)^\top$ and $v = (0.2975, 0.1374, 0.2661)^\top$. Also, according to Table 1.4, the 6-month rate is 0.46%.*

Objective Measure vs Equivalent Martingale Measure

Contrary to what we do for option pricing, we need to simulate the prices with respect to the real measure, not the equivalent martingale measure.

Using the MATLAB program SimLosses, *we simulated 10^6 losses, saved in the MATLAB file* Losses. *The corresponding nonparametric estimations of the density and the distribution function are displayed in Figure 4.1, together with a 95% uniform confidence band, as given by* (4.9). *The estimation of the distribution function was done with MATLAB function* EstCdf.

4.2.3 Nonparametric Estimation of the VaR

For simplicity, assume that the distribution function F is continuous. Recall that $X_{n:1} <, \ldots, < X_{n:n}$ are the order statistics of the sample. Since $F_n(X_{n:k}) = k/n$, using the definition of quantiles (Appendix 4.B), we have

$$\widehat{\mathrm{VaR}}_n(100p\%) = F_n^{-1}(p) = X_{n:k}, \quad \text{for } p \in \left(\frac{k-1}{n}, \frac{k}{n}\right], \; i \in \{1, \ldots, n\}.$$

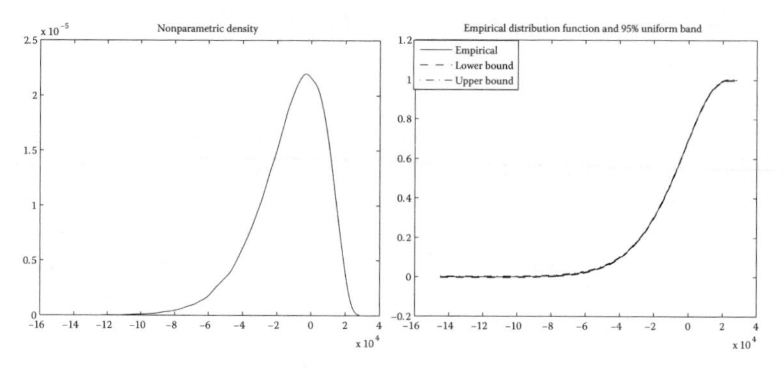

FIGURE 4.1: Nonparametric estimation of the density and distribution function of the losses.

In applications, one should have $n(1-p) \geq 100$. For example, if one wants to estimate the VaR(99.9%), one should choose $n \geq 100000$.

The precision of the estimation of the VaR is given in the following proposition which is proven in Appendix 4.I.3.

Proposition 4.2.1 *If F has a continuous positive density f on the set $\{x; 0 < F(x) < 1\}$, then for a fixed $p \in (0,1)$,*

$$\sqrt{n}\left\{\widehat{\mathrm{VaR}}_n(100p\%) - \mathrm{VaR}(100p\%)\right\} \rightsquigarrow N\left(0, \sigma^2_{\mathrm{VaR}(p)}\right), \tag{4.10}$$

where

$$\sigma^2_{\mathrm{VaR}(p)} = \frac{p(1-p)}{f^2\{\mathrm{VaR}(100p\%)\}}, \quad 0 < p < 1. \tag{4.11}$$

Remark 4.2.2 *To estimate the variance $\sigma^2_{\mathrm{VaR}(p)}$, we can estimate $f\{\mathrm{VaR}(100p\%)\}$ in formula (4.11) by $\hat{f}_n\left\{\widehat{\mathrm{VaR}}_n(100p\%)\right\}$, where $\hat{f}_n$ is the density estimation obtained from the kernel method described in Appendix 3.C. Another way of estimating σ_p^2, without using formula (4.11), is to use the bootstrap methodology described in Appendix 4.D.*

Example 4.2.2 *If the portfolio loss X has a Gaussian distribution with mean μ and variance σ^2, then the VaR can be simply estimated by*

$$\widehat{\mathrm{VaR}}_n(100p\%) = \bar{x} + s_x z_p,$$

where $\bar{x}$ and s_x^2 are respectively the sample mean and variance, and $z_p = \mathrm{VaR}_{N(0,1)}(100p\%)$, since $\mathrm{VaR}_{N(\mu,\sigma^2)}(100p\%) = \mu + \sigma z_p$. It then follows from Example B.5.1 that

$$\sqrt{n}\left\{\widehat{\mathrm{VaR}}_n(100p\%) - \mathrm{VaR}(100p\%)\right\} \rightsquigarrow N\left(0, \sigma^2(1 + z_p^2/2)\right). \tag{4.12}$$

Note that from formula (4.11), the estimation of the VaR *using the nonparametric method has variance* $2\sigma^2\pi p(1-p)e^{z_p^2}$, *compared to* $\sigma^2\left(1+\frac{z_p^2}{2}\right)$ *in the parametric case. In particular, for the* 95% VaR, *the variances are respectively* $1.5339\sigma^2$ *and* $2.1132\sigma^2$. *The graphs of the two standard deviations as a function of* p *are displayed in Figure 4.2 for* $\sigma = 1$. *As expected, the parametric method is more precise than the nonparametric one.*

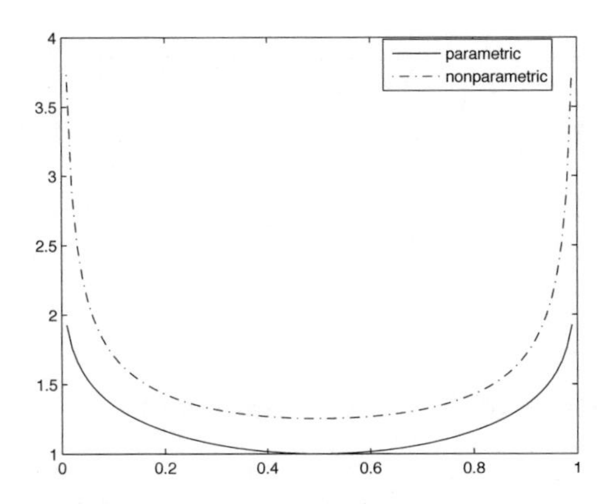

FIGURE 4.2: Standard deviation of the estimation of the VaR in the standard Gaussian case for the parametric and nonparametric methods. Here, $\sigma = 1$.

4.2.3.1 Uniform Estimation of Quantiles

There also exists an analog for the uniform approximation result of the distribution function given by (4.7) for quantiles. Here is the result: For $0 < a \le b < 1$ given, one can prove that

$$\sqrt{n} \sup_{a\le p\le b} \left|\widehat{\text{VaR}}_n(100p\%) - \text{VaR}(100p\%)\right| \rightsquigarrow \sup_{a\le p\le b} \left\{\frac{|B(p)|}{f\{F^{-1}(p)\}}\right\} = \mathbb{Q}_{F,a,b}.$$

Unfortunately, the distribution of $\mathbb{Q}_{F,a,b}$ is unknown but it can be easily simulated, so one can find its quantiles. See Appendix 4.E for details. However, since the distribution clearly depends on F through f and F^{-1}, it is not possible to tabulate $\mathbb{Q}_{F,a,b}$.

If c is an estimation of 95% quantile for $\mathbb{Q}_{F,a,b}$, then 95% of the time, the graph $\{(p, \text{VaR}(100p\%))\,; p \in [a,b]\}$ will lie completely inside the uniform confidence band

$$\left|\widehat{\text{VaR}}_n(100p\%) - \text{VaR}(100p\%)\right| \le \frac{c}{\sqrt{n}}, \qquad \text{for all } p \in [a,b].$$

Example 4.2.3 (Example 4.2.1 (continued)) *We want to estimate the* VaR*(95%),* VaR*(99%), and* VaR*(99.9%) for the losses of the portfolio created in Example 4.2.1. Using the MATLAB function* EstVaR*, we obtain the results displayed in Table 4.1. The bootstrap method (using 1000 bootstrap samples) seems to be more precise than the density method.*

TABLE 4.1: Estimation of the VaR and estimation errors for a 95% level of confidence.

Level	VaR	95% Error	
		$\hat{f}_n$	Bootstrap
95%	1404490	122.88	3.82
99%	19352.24	138.45	4.002
99.9%	23066.08	155.10	6.39

4.2.4 Estimation of Expected Shortfall

Suppose that $X_1, \ldots, X_n$ are independent and identically distributed observations from a random variable X with distribution function F. Suppose also that $E(|X|) < \infty$. Recall that

$$\begin{aligned} ES(100p\%) &= E\{X|X > \mathrm{VaR}(100p\%)\} \\ &= \mathrm{VaR}(100p\%) + e\left\{\mathrm{VaR}(100p\%)\right\}, \end{aligned}$$

so we can estimate $ES(p)$ by

$$\widehat{ES}_n(100p\%) = \widehat{\mathrm{VaR}}_n(100p\%) + \frac{1}{1-p}\,\frac{1}{n}\sum_{i=1}^{n}\left\{X_i - \widehat{\mathrm{VaR}}_n(100p\%)\right\}^+ . \quad (4.13)$$

The estimator is also consistent and asymptotically unbiased. The precision of the estimation is given by the next proposition, which is proven in Appendix 4.I.4.

Proposition 4.2.2 *If $E(X^2) < \infty$ and if its distribution function F has a continuous density f on the set $\{x; 0 < F(x) < 1\}$, with $f > 0$ on that set, then for a fixed $p \in (0,1)$, we have*

$$\sqrt{n}\left\{\widehat{ES}_n(100p\%) - ES(100p\%)\right\} \rightsquigarrow N\left(0, \sigma^2_{ES(p)}\right),$$

where $\sigma^2_{ES(p)}$ can be estimated by the sample variance of the pseudo-observations

$$\frac{\left\{X_1 - \widehat{\mathrm{VaR}}_n(100p\%)\right\}^+}{1-p}, \ldots, \frac{\left\{X_n - \widehat{\mathrm{VaR}}_n(100p\%)\right\}^+}{1-p}.$$

Example 4.2.4 (Example 4.2.1 (continued)) *We want to estimate the* Expected Shortfall *for levels 95%, 99%, and 99.9% for the losses of the portfolio created in Example 4.2.1. Using the MATLAB function* EstES, *we obtain the results displayed in Table 4.2.*

TABLE 4.2: Estimation of the expected shorfall and estimation errors for a 95% level of confidence.

Level	Expected Shortfall	95% Error
95%	17267.68	109.81
99%	21100.11	139.6
99.9%	24163.99	299

Remark 4.2.3 *If the portfolio loss X has a Gaussian distribution with mean μ and variance σ^2, then the expected shortfall can be simply estimated by the parametric estimate*

$$\widehat{ES}_n(100p\%) = \bar{x} + \frac{s_x}{(1-p)} \frac{e^{-\frac{z_p^2}{2}}}{\sqrt{2\pi}},$$

where $\bar{x}$ and s_x^2 are respectively the sample mean and variance, and $z_p =$ $\text{VaR}_{N(0,1)}(100p\%)$. As illustrated in Figure 4.3, the precision of the estimation is better in the parametric than in the nonparametric case.

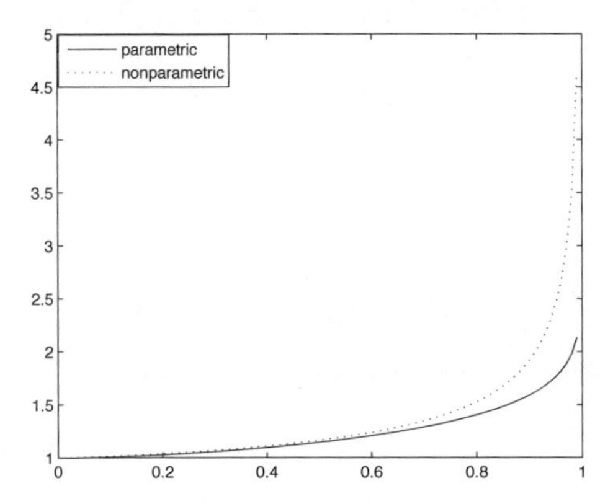

FIGURE 4.3: Standard deviation of the estimation of the expected shortfall in the standard Gaussian case for the parametric and nonparametric methods. Here, $\sigma = 1$.

4.2.5 Advantages and Disadvantages of the Monte Carlo Methodology

The Monte Carlo methodology is relatively easy to implement if one has a fast algorithm to simulate the risk factors. It can also be implemented with a lot of models, not only with factors following the Black-Scholes dynamics. For some interesting risk measures, including the distribution function, we can compute confidence intervals. Estimations are done in a nonparametric way, so they are usually less precise than using parametric methods; however, this is not really a disadvantage since in practice, one cannot use the parametric approach, the distribution of X being unknown.

4.3 Measures of Risk and the Delta-Gamma Approximation

The basic idea behind the so-called Delta-Gamma approximation is to replace the loss X by an approximation based on a second order Taylor expansion of $V(t, S_t)$ about $V(t, S_0)$. See, e.g., Mina and Ulmer [1999] and Jaschke [2002]. Doing so, the approximation will only involve a quadratic function of the returns of the risk factors. This approximation is not proposed necessarily as an alternative to the Monte Carlo method. It could also be used to generate control variables needed for variance reduction when using the Monte Carlo method. See, e.g., Glasserman et al. [2000]. For t and S_0 given, set

$$
\begin{aligned}
\delta_j &= S_0^{(j)} \partial_j V(t, S_0) = S_0^{(j)} \partial_{s^{(j)}} V(t, S_0), \quad j \in \{1, \ldots, m\}, \\
\gamma_{jk} &= S_0^{(j)} S_0^{(k)} \partial_j \partial_k V(t, S_0), \quad j, k \in \{1, \ldots, m\}.
\end{aligned}
$$

As a result, the loss X of the investment is approximated by $\tilde{X}$, where

$$
\tilde{X} = \alpha - e^{-rt} \left(\delta^\top R + \frac{1}{2} R^\top \gamma R \right) = \alpha - e^{-rt} Y, \tag{4.14}
$$

with $\alpha = V_0 - e^{-rt}V(t,S_0)$, and $R^{(j)} = \ln\left(S_t^{(j)}\right) - \ln\left(S_0^{(j)}\right)$, $j \in \{1,\ldots,m\}$. In fact, since $\left(S_t^{(j)} - S_0^{(j)}\right)/S_0^{(j)} \approx R_j = \ln\left(S_t^{(j)}\right) - \ln\left(S_0^{(j)}\right)$, it follows that

$$
\begin{aligned}
V(t,S_t) &\approx V(t,S_0) + \sum_{j=1}^{m}\left(S_t^{(j)} - S_0^{(j)}\right)\partial_j V(t,S_0) \\
&\quad + \frac{1}{2}\sum_{1\le j,k\le m}\left(S_t^{(j)} - S_0^{(j)}\right)\left(S_t^{(k)} - S_0^{(k)}\right)\partial^2_{jk}V(t,S_0) \\
&\approx V(t,S_0) + \sum_{j=1}^{m}\delta_j R^{(j)} + \frac{1}{2}\sum_{1\le j,k\le m}\gamma_{jk}R^{(j)}R^{(k)} \\
&= V(t,S_0) + \delta^\top R + \frac{1}{2}R^\top \gamma R = V(t,S_0) + Y,
\end{aligned}
$$

from which one may conclude that

$$
\begin{aligned}
X &= V_0 - e^{-rt}V(t,S_0) - e^{-rt}\left\{V(t,S_t) - V(t,S_0)\right\} \\
&\approx V_0 - e^{-rt}V(t,S_0) - e^{-rt}Y = \alpha - e^{-rt}Y = \tilde{X}.
\end{aligned}
$$

4.3.1 Delta-Gamma Approximation

The Delta-Gamma approximation consists in replacing the real loss variable by its second order approximation $\tilde{X} = \alpha - e^{-rt}Y$ defined by formula (4.14), and then compute measures of risk for $\tilde{X}$ instead of X. For example, the VaR of X will be estimated by the VaR of $\tilde{X}$, and the expected shortfall of X will be estimated by expected shortfall of $\tilde{X}$. Moreover, since $\tilde{X}$ is an affine function of $Y = \delta^\top R + \frac{1}{2}R^\top\gamma R$, it follows that

$$
\mathrm{VaR}_{\tilde{X}}(p) = \alpha + e^{-rt}\mathrm{VaR}_{-Y}(p) = \alpha - e^{-rt}\mathrm{VaR}_Y(1-p),
$$

and

$$
ES_{\tilde{X}}(p) = \alpha + e^{-rt}ES_{-Y}(p).
$$

Even if R has a Gaussian distribution, the distribution of Y is not known in general.

4.3.2 Delta-Gamma-Normal Approximation

From now on, we assume that $R \sim N_m(\mu,\Sigma)$. In this case, the approximation is called the Delta-Gamma-normal approximation and the goal is to find approximately the distribution of Y.

Using the Cholesky decomposition of Σ given by $\Sigma = C^\top C$, we end up with a simplified formula for Y, in terms of independent Gaussian standard variables. In fact, since R has the same law as $\mu + C^\top Z$, with $Z \sim N_m(0, I)$, it follows that

$$\begin{aligned} Y &\stackrel{Law}{=} \delta^\top \mu + \frac{1}{2}\mu^\top \gamma\mu + (\delta + \gamma\mu)^\top C^\top Z + \frac{1}{2} Z^\top (C\gamma C^\top) Z \\ &= \alpha' + \beta^\top Z + \frac{1}{2} Z^\top (C\gamma C^\top) Z \\ &= \alpha' + Q, \end{aligned} \tag{4.15}$$

where $\alpha' = \delta^\top \mu + \frac{1}{2}\mu^\top \gamma\mu$, $\beta = C(\delta + \gamma\mu)$, and

$$Q = \beta^\top Z + \frac{1}{2} Z^\top (C\gamma C^\top) Z. \tag{4.16}$$

As a result, from (4.15) we have

$$\mathrm{VaR}_Y(p) = \alpha' + \mathrm{VaR}_Q(p) \text{ and } ES_Y(p) = \alpha' + ES_Q(p).$$

Note also that

$$\tilde{X} = \alpha - e^{-rt}\alpha' - e^{-rt} Q, \tag{4.17}$$

so if G is the distribution function of Q, the distribution function F of $\tilde{X}$ is

$$F(x) = 1 - G\left(\frac{\alpha - e^{-rt}\alpha' - x}{e^{-rt}},\right), \quad x \in \mathbb{R}.$$

Over the years, many people worked on finding the distribution of Q or at least finding a good approximation of it. Unfortunately, the distribution function and the density of Q are still unknown in general. However, we can easily find its moment generating function. This is done in the next section.

4.3.3 Moment Generating Function and Characteristic Function of Q

The moment generating function of Q, defined by $M_Q(u) = E\left(e^{uQ}\right)$, $u \in [-u_0, u_0]$, for some $u_0 > 0$, is given in the next proposition. It can be used for example to compute the cumulants of Q (see Section A.5.1) which will be used to find approximately the distribution of Q. It is also needed in the saddlepoint approximation in Section 4.3.6.

Suppose that A is the matrix formed with the normalized eigenvectors of $C\gamma C^\top$ and let Λ be the diagonal matrix of the corresponding eigenvalues $\lambda_1, \ldots, \lambda_m$, so that $A\Lambda A^\top = C\gamma C^\top$. Set $h = A^\top \beta$. Note that the decomposition is possible because $C\gamma C^\top$ is symmetric; furthermore, the entries of A and Λ are real numbers. Note also that both A and Λ can be obtained from the MATLAB function *eig*.

Proposition 4.3.1 *For all* $|u| < \dfrac{1}{\max_{1\le j\le m}|\lambda_j|}$*, we have*

$$M_Q(u) = \prod_{j=1}^{m}\left(\frac{e^{\frac{1}{2}\frac{u^2h_j^2}{1-\lambda_j u}}}{\sqrt{1-\lambda_j u}}\right). \tag{4.18}$$

Moreover, the cumulants of Q *are given by*

$$\kappa_n = \begin{cases} \frac{1}{2}\sum_{j=1}^m \lambda_j = \frac{1}{2}Trace(\Lambda), & n=1, \\ \frac{(n-1)!}{2}\sum_{j=1}^m \lambda_j^{n-2}\left(\lambda_j^2 + nh_j^2\right), & n\ge 2. \end{cases} \tag{4.19}$$

The proof is given in Appendix 4.I.5. It follows easily from (4.18) that Q is a weighted sum of m independent non-central chi-square random variables (Appendix A.6.12), up to a centering constant. In fact, it is the law of $\frac{1}{2}\sum_{j=1}^m \lambda_j\left(W_j + \frac{h_j}{\lambda_j}\right)^2 - \frac{1}{2}\sum_{j=1}^m \frac{h_j^2}{\lambda_j}$, where $W_1,\ldots,W_m$ are i.i.d. $N(0,1)$. As a corollary to Proposition 4.3.1, we also obtain the characteristic function of Q by replacing u with iu in formula (4.18). This function will play a crucial role later on.

Corollary 4.3.1 *The characteristic function* ϕ *of* Q *is given by*

$$\phi(u) = E\left(e^{iuQ}\right) = \prod_{j=1}^{m}\left\{\frac{e^{\frac{-u^2h_j^2}{2(1-iu\lambda_j)}}}{\sqrt{1-iu\lambda_j}}\right\}, \quad u\in\mathbb{R}. \tag{4.20}$$

We are now in a position to describe four ways to approximate the distribution of Q and the distribution of $\tilde{X}$.

4.3.4 Partial Monte Carlo Method

Here we simply use the fact that by hypothesis $R\sim N_m(\mu,\Sigma)$. The partial Monte Carlo methods consists in generating a large number n of independent returns $R_1,\ldots,R_n$ and compute $Y_i = \delta^\top R_i + \frac{1}{2}R_i^\top\gamma R_i$, $i\in\{1,\ldots,n\}$. One can then use the estimation techniques developed in Section 4.2. Note that we can also use (4.15) and simulate $Z_i\sim N_m(0,I)$ and compute Q_i, $i\in\{1,\ldots,n\}$.

Remark 4.3.1 *The (full) Monte Carlo method presented in Section 4.2 can also be implemented in parallel to the partial Monte Carlo method since* $S_t^{(j)} = S_0^{(j)}e^{R^{(j)}}$*, for all* $j\in\{1,\ldots,m\}$ *and we assumed that the distribution of* R *is Gaussian.*

4.3.4.1 Advantages and Disadvantages of the Methodology

The methodology is easy to implement but it depends a lot on the precision of the approximation of X by $\tilde{X}$. One could ask why use an approximate method when the full Monte Carlo approach is available if V can be computed. At the very least, it provides us with a way to compare the precision of the estimation of X by $\tilde{X}$. Also, it could be used as a control variate to reduce the variance of risk measures for X.

4.3.5 Edgeworth and Cornish-Fisher Expansions

These expansions serve as approximations of the distribution function and quantiles of a sum of independent random variables, using their cumulants. In our framework, according to (4.39), Q is a sum of independent random variables and we know that its cumulants are given by (4.19).

The Edgeworth expansion is an approximation of the distribution function based on its cumulants, so we will be able to approximate the distribution function of $\tilde{X} = \alpha - e^{-rt}\alpha' - e^{-rt}Q$ as well. Therefore, we can use it to compute measures of risk. There exists also an approximation of the density using cumulants. The Edgeworth expansion, well-known in Statistics, seems to have been used in a financial engineering context for the first time in Jarrow and Rudd [1982].

4.3.5.1 Edgeworth Expansion for the Distribution Function

Recall that $\mathcal{N}$ stands for the distribution function of a standard Gaussian variable. Denote the associated density by $\phi(x) = \mathcal{N}'(x) = \frac{e^{-x^2/2}}{\sqrt{2\pi}}$. The relationship

$$\frac{(-1)^k}{\phi(x)} \frac{d^k \phi(x)}{dx^k} = (-1)^k \frac{\phi^{(k)}(x)}{\phi(x)} = H_k(x)$$

define polynomials H_n of degree n; their expression is

$$H_n(x) = \sum_{j=0}^{[n/2]} \frac{(-1)^j}{2^j} \frac{n!}{j!(n-2j)!} x^{n-2j}, \quad x \in \mathbb{R}.$$

These polynomials play a major role in the Edgeworth expansion. They are related to the famous Hermite polynomials h_n viz.

$$H_n(x) = 2^{-n/2} h_n(x/\sqrt{2}).$$

For a general distribution function F of a random variable X, let F_{X_0} be the distribution function of the standardized random variable $X_0 = \frac{X-\mu}{\sigma}$, i.e., $F_X(x) = F_{X_0}\left(\frac{x-\mu}{\sigma}\right)$. Assume that the first six moments (or cumulants) of F exist and denote these cumulants by $\kappa_1, \ldots, \kappa_6$. In particular, recall that the mean μ is given $\mu = \kappa_1$ and the variance σ^2 is given by $\sigma^2 = \kappa_2$. Set

$\gamma_r = \frac{\kappa_{r+2}}{\sigma^{r+2}}$, $r \in \{1,2,3,4\}$. Then, according to Abramowitz and Stegun [1972], the Edgeworth expansion of F is given by

$$
\begin{aligned}
F_{X_0}(x) \quad \approx \quad & N(x) - \phi(x)\left\{\frac{\gamma_1}{6}H_2(x) + \frac{\gamma_2}{24}H_3(x) + \frac{\gamma_1^2}{72}H_5(x)\right.\\
& +\frac{\gamma_3}{120}H_4(x) + \frac{\gamma_1\gamma_2}{144}H_6(x) + \frac{\gamma_1^3}{1296}H_8(x)\\
& +\frac{\gamma_4}{720}H_5(x) + \frac{\gamma_2^2}{1152}H_7(x) + \frac{\gamma_1\gamma_3}{720}H_7(x)\\
& \left.+\frac{\gamma_1^2\gamma_2}{1728}H_9(x) + \frac{\gamma_1^4}{31104}H_{11}(x)\right\} + \cdots,
\end{aligned}
$$

This Edgeworth expansion is implemented in the MATLAB function *EdgeworthCdf*, while the Edgeworth expansion for the density is implemented in *EdgeworthPdf*.

Example 4.3.1 (Chi-Square Distribution) *If $X \sim \chi^2(\nu)$, then its moment generating function is $M_X(u) = (1-2u)^{-\nu/2}$, for $u < 1/2$, so*

$$\kappa_n = \nu(n-1)!2^{n-1}, \quad n \geq 1.$$

The MATLAB script DemoEdgeworth *illustrates the accuracy of the Edgeworth expansion for a chi-square distribution as a function of its degrees of freedom. After running the script, why do you think the approximation gets better as ν increases?*

4.3.5.2 Advantages and Disadvantages of the Edgeworth Expansion

- The Edgeworth expansion is very easy to compute and requires only the first six cumulants.
- Computation is extremely fast.
- The approximation is good when the law is close to the law of a Gaussian random variable, as it is the case for a sum of a relatively large number of independent random variables.
- When the distribution is not close to a Gaussian, the approximation can be very poor, even if all cumulants exist.
- Often the approximation can be poor in the tails, leading to poor estimations of VaR(100p%) values of p close to 1.

4.3.5.3 Cornish-Fisher Expansion

The Cornish-Fisher expansion is like the inverse of the Edgeworth expansion since it is used to approximate the inverse distribution function F, i.e., the quantile function or VaR function. It is also based on cumulants. More precisely, according to Abramowitz and Stegun [1972], the Cornish-Fisher Expansion is given by $F_X^{-1}(p) = \mu + \sigma F_{X_0}^{-1}(p)$, where

$$\begin{aligned} F_{X_0}^{-1}(p) &\approx x + \{\gamma_1 V_1(x)\} + \left\{\gamma_2 V_2(x) + \gamma_1^2 V_{11}(x)\right\} \\ &\quad + \left\{\gamma_3 V_3(x) + \gamma_1\gamma_2 V_{12}(x) + \gamma_1^3 V_{111}(x)\right\} \\ &\quad + \left\{\gamma_4 V_4(x) + \gamma_2^2 V_{22}(x) + \gamma_1\gamma_3 V_{13}(x) \right. \\ &\quad \left. + \gamma_1^2\gamma_2 V_{112}(x) + \gamma_1^4 V_{1111}(x)\right\} + \cdots, \end{aligned}$$

with $x = N^{-1}(p)$, $V_1(x) = \frac{1}{6}H_2(x)$, and

$$\begin{aligned} V_2(x) &= \frac{1}{24}H_3(x), \quad V_{11}(x) = -\frac{1}{36}\{2H_3(x) + H_1(x))\}, \\ V_3(x) &= \frac{1}{120}H_4(x), \quad V_{12}(x) = -\frac{1}{24}\{H_4(x) + H_2(x)\}, \\ V_{111}(x) &= \frac{1}{324}\{12H_4(x) + 19H_2(x)\}, \\ V_4(x) &= \frac{1}{720}H_5(x), V_{22} = -\frac{1}{384}\{3H_5(x) + 6H_3(x) + 2H_1(x)\}, \\ V_{13} &= -\frac{1}{180}\{2H_5(x) + 3H_3(x)\}, \\ V_{112} &= \frac{1}{288}\{14H_5(x) + 37H_3(x) + 8H_1(x)\}, \\ V_{1111} &= -\frac{1}{7776}\{252H_5(x) + 832H_3(x) + 227H_1(x)\}. \end{aligned}$$

This expansion is implemented in the MATLAB function *CornishFisher*.

Example 4.3.2 (Chi-Square Distribution) *The MATLAB script* DemoCornish-Fisher *illustrates the accuracy of the Cornish-Fisher expansion for a chi-square distribution as a function of its degrees of freedom.*

4.3.5.4 Advantages and Disadvantages of the Cornish-Fisher Expansion

The Cornish-Fisher expansion has basically the same qualities and weaknesses as the Edgeworth expansion.

4.3.6 Saddlepoint Approximation

Another approximation method that has been recently used in financial engineering [Aït-Sahalia and Yu, 2006, Carr and Madan, 2009] is the so-called *saddlepoint approximation*. It was introduced in Statistics by Daniels [1954] and it is based on the Legendre transform of the cumulant generating function $K = \ln(M)$. So we have to assume that the moment generating function M

exists in an open interval $(u_0, u_1) \ni 0$. This is much stronger than simply assuming that the first 6 moments exist.

By Remark A.5.1, $K''(u) > 0$ for all $u \in (u_0, u_1)$, so K is convex, K' is increasing and its Legendre transform is

$$L(x) = \sup_{u \in (u_0, u_1)} xu - K(u) \geq 0.$$

This function is finite for all points in the support of the law given by $\mathcal{R} = \{K'(u); u \in (u_0, u_1)\}$. It then follows that for all $x \in \mathcal{R}$, there is a unique $u_x \in (u_0, u_1)$ defined by $K'(u_x) = x$, and $L(x) = xu_x - K(u_x)$. In particular, if μ is the mean of the distribution, then $L(\mu) = 0$ since $K'(0) = \mu$. Note that since K' is increasing, it is relatively easy to invert. Also note that $L(x) > 0$ whenever $x \neq \mu$, $L'(x) = u_x$, and $L''(x) = 1/K''(u_x) > 0$.

4.3.6.1 Approximation of the Density

If the density f associated with K is continuous on $\chi = \mathcal{R} \cap \{x; f(x) > 0\}$, then $f(x)$ is approximated by

$$\hat{f}(x) = \frac{e^{-L(x)}}{\sqrt{2\pi K''(u_x)}} = \sqrt{\frac{L''(x)}{2\pi}} e^{-L(x)}, \quad x \in \chi. \tag{4.21}$$

Remark 4.3.2 *The saddlepoint density $\hat{f}$ defined by (4.21) is not necessarily a density but can be normalized to become one. It is then called the normalized saddlepoint density. Note that for a location and scale family defined by $f_{\mu,\sigma}(x) = \frac{1}{\sigma} f\left(\frac{x-\mu}{\sigma}\right)$, we have $\hat{f}_{\mu,\sigma}(x) = \frac{1}{\sigma}\hat{f}\left(\frac{x-\mu}{\sigma}\right)$, since $L_{\mu,\sigma}(x) = L\left(\frac{x-\mu}{\sigma}\right)$.*

Example 4.3.3 *For a Gaussian distribution with parameters μ and σ^2, $K(u) = \mu u + \frac{1}{2}u^2\sigma^2$, for all u. As a result, $K'(u) = \mu + \sigma^2 u$, $u_x = (x - \mu)/\sigma^2$ and $\chi = (-\infty, \infty)$. It is then easy to check that $L(x) = \frac{1}{2}\left(\frac{x-\mu}{\sigma}\right)^2$, so $\hat{f}$ is indeed the Gaussian density. In this case, the saddlepoint approximation is exact!*

Another interesting case is the gamma distribution with parameters (α, β). Then $K(u) = -\alpha \ln(1 - \beta u)$, which exists for all $u < 1/\beta$. Then $u_x = \frac{1}{\beta} - \frac{\alpha}{x}$, so for any $x > 0$, $L(x) = \frac{x}{\beta} - \alpha - \alpha \ln\left(\frac{x}{\alpha\beta}\right)$ and $L''(x) = \alpha/x^2$. As a result,

$$\hat{f}(x) = \frac{1}{c_\alpha} \frac{x^{\alpha-1}}{\beta^\alpha} e^{-x/\beta} \mathbb{I}(x > 0),$$

where $c_\alpha = \sqrt{2\pi}\alpha^{\alpha-1/2}e^{-\alpha}$. So in this case, the saddlepoint approximation is almost correct, up to a constant. In other words, the normalized saddlepoint density is the density of the gamma distribution with parameters (α, β). Note that by Stirling formula [Abramowitz and Stegun, 1972],

$$c_\alpha = \sqrt{2\pi}\alpha^{\alpha-1/2}e^{-\alpha} \approx \Gamma(\alpha), \tag{4.22}$$

in the sense that $c_\alpha/\Gamma(\alpha) \to 1$, as $\alpha \to \infty$. In fact $c_\alpha/\Gamma(\alpha)$ is increasing, with $c_\alpha/\Gamma(\alpha) \to 0$, as $\alpha \to 0$.

The saddlepoint approximation for the density of the non-central chi-square distribution is implemented in the MATLAB function *SaddleNChi2Pdf*.

4.3.6.2 Approximation of the Distribution Function

One can also approximate the distribution function with respect to the Gaussian distribution or with respect to any other distribution having a nice cumulant generating function. It was first proposed by Lugannani and Rice [1980]. For the general case, the methodology is described in Appendix 4.F; see Butler [2007, Chapter 16] for more details. For an application in option pricing, see Carr and Madan [2009].

Using the same notations as before, the saddlepoint approximation of a continuous distribution function, with respect to the Gaussian base, is

$$\hat{F}(x) = \begin{cases} \mathcal{N}(w_x) + \mathcal{N}'(w_x)\left(\frac{1}{w_x} - \frac{1}{v_x}\right), & x \neq \mu, \\ \frac{1}{2} + \frac{\kappa_3}{6\kappa_2^{3/2}\sqrt{2\pi}}, & x = \mu, \end{cases} \tag{4.23}$$

where $w_x = \text{sign}(u_x)\sqrt{2L(x)}$ and $v_x = u_x\sqrt{K''(u_x)}$.

Remark 4.3.3 *From a practical point of view, the approximation of F at μ is not good in general. It is much better to interpolate between close values. The proposed approximation is particularly bad for non-symmetric distributions like the Gamma or the non-central chi-square.*

The saddlepoint approximation of the distribution function for the non-central chi-square distribution is implemented in the MATLAB function *SaddleNChi2Cdf*.

Note that in the case of the Gaussian distribution, the approximation is exact in the sense that $\hat{F} = F$. However, even after rescaling, the saddlepoint approximation is no longer exact for the gamma distribution. This is illustrated in the MATLAB script *DemoSaddlepoint*, which compares the approximation to the real function for the non-central chi-square distribution with 5 values for the degrees of freedom and 3 values for the non-centrality parameter. As one can guess, the fit is better as the degrees of freedom increase, the distribution becoming closer to the Gaussian one.

4.3.6.3 Advantages and Disadvantages of the Methodology

The methodology is harder to implement than the Edgeworth expansion, since we cannot have explicit expressions in general. However, for a given distribution class, for example quadratic forms of Gaussian variables, it can be done in full generality and the approximation seems to be quite satisfactory [Kuonen, 1999].

In the case of the density approximation, we end up with an un-normalized density, but at least it is non-negative. In that sense, it is much better than the Edgeworth expansion. Another advantage is that there is no fine tuning of parameters, as it is the case for the inversion of the characteristic function discussed next. Finally, having an approximation of the density, we could use it for parameter estimation, when there is no explicit density but one can compute the cumulant generating function. See, e.g., Aït-Sahalia and Yu [2006].

4.3.7 Inversion of the Characteristic Function

If ϕ is the characteristic function of a continuous distribution F, then a particular case of the Gil-Pelaez formula stated in Appendix 4.H expresses the distribution function in terms of its characteristic function ϕ viz.

$$F(x) = \frac{1}{2} - \int_{-\infty}^{\infty} \frac{\phi(u)}{2\pi i u} e^{-iux} du, \quad x \in \mathbb{R}. \tag{4.24}$$

If ϕ is known, as it is the case for Q in the Gaussian case, according to (4.20), then one can obtain good approximations of the VaR as well as for other measures of risk. Note that since Q is a quadratic function of independent Gaussian random variables according to (4.39), its distribution function F is necessarily continuous. It remains to compute the integral in (4.24), which involves working with complex numbers. See Appendix 4.G for pitfalls to avoid when using complex numbers in MATLAB.

4.3.7.1 Davies Approximation

If ϕ is the characteristic function of F, then Davies [1973] showed that

$$F(x) \approx \frac{1}{2} - \frac{1}{\pi} \sum_{k=0}^{K} \mathrm{Im} \left\{ \frac{\phi\left(\left(k+\frac{1}{2}\right)\Delta\right) e^{-i\left(k+\frac{1}{2}\right)\Delta x}}{\left(k+\frac{1}{2}\right)} \right\}, \tag{4.25}$$

where K is large and Δ is small. In addition, if the density f of F exists, then differentiating (4.25) yields

$$f(x) \approx \frac{\Delta}{\pi} \sum_{k=0}^{K} \mathrm{Re} \left\{ \phi\left(\left(k+\frac{1}{2}\right)\Delta\right) e^{-i\left(k+\frac{1}{2}\right)\Delta x} \right\}.$$

4.3.7.2 Implementation

An interesting application of (4.25) is when ϕ is given by (4.20), so we can approximate the distribution function of Q. However, to implement formula (4.25), one must choose K and Δ. Unfortunately, there is no universal way to find good choices for K and Δ, one of the problems being related to the scale. For example, changing the scale of measurement changes the scale of u in $\phi(u)$, which in turns affects the choice of Δ. As a rule, one should begin by taking values of K between 10 and 100 and Δ about 0.5 and then plot the graph of F according to formula (4.25). Then one has to change the parameters until the results are satisfactory, i.e., the resulting plot looks like the graph of a distribution function.

Remark 4.3.4 *Decreasing Δ and increasing K often improve the approximation.*

The MATLAB script *DemoDavies* illustrates some choices of K and Δ for selected distributions, including discrete distributions like the Poisson distribution.

Remark 4.3.5 *The inversion of characteristic functions is also important for option pricing, as exemplified in Heston [1993] and Carr and Madan [1999]. Therefore, one can use Davies approximation not only in a context of risk measures but also for pricing derivatives.*

4.4 Performance Measures

The most popular measure of performance is the so-called Sharpe ratio [Sharpe, 1966]. Another example of a performance measure is the Sortino ratio. Also, [Keating and Shadwick, 2002] proposed the Omega function or Omega ratio. In what follows, we will study the properties and estimation of these three measures of performance. But first, as it has been done for measures of risk, one could define properties that any "good" measure of performance should satisfy.

4.4.1 Axiomatic Approach of Cherny-Madan

According to Cherny and Madan [2009], a performance measure α defined for any random variable X (representing an excess return) with a finite first moment, should satisfy the following axioms:

B1: For any $x \geq 0$, the set $\mathcal{A}_x = \{X; \alpha(X) \geq x\}$ is convex.

B2: If $X \leq Y$, then $\alpha(X) \leq \alpha(Y)$.

B3: $\alpha(\lambda X) = \alpha(X)$ for any $\lambda > 0$.

B4: For any $x \geq 0$, the set $\mathcal{A}_x = \{X; \alpha(X) \geq x\}$ is closed for the convergence in mean[4], meaning that if $\alpha(X_n) \geq x$ and $X_n \stackrel{L_1}{\to} X$, then $\alpha(X) \geq x$.

B5: If $X \stackrel{Law}{=} Y$, then $\alpha(X) = \alpha(Y)$.

[4]That is not the exact description of their axiom but since they assumed that the sequence is bounded, convergence in probability is equivalent to convergence in L_1.

B6: If $X \preceq_{hr} Y$, then $\alpha(X) \le \alpha(Y)$.

B7: $X \ge 0$ almost surely if and only if $\alpha(X) = +\infty$.

B8: If $E(X) < 0$, then $\alpha(X) = 0$; if $E(X) > 0$, then $\alpha(X) > 0$.

Note that Axiom B8 is not really necessary. Further note that the Sharpe ratio does not satisfy all the Axioms B1–B7, while the Omega ratio does.

4.4.2 The Sharpe Ratio

Suppose that $X_1, \ldots, X_n$ are independent and identically distributed observations of a random variable X, with mean μ and standard deviation σ. In an investment context, $X_1, \ldots, X_n$ represent successive observations of investment returns. If r is some target rate, e.g., the periodic risk-free rate, then $X_i - r$ represents the excess return. Taking a constant rate is not really a restriction. For monthly returns for example, we could subtract the risk-free rate for the corresponding month and work directly with the excess returns. In that case, we would simply take $r = 0$.

For a given r, the Sharpe ratio is defined by $\mathcal{S}_1 = (\mu - r)/\sigma$. It can be estimated by $\hat{\mathcal{S}}_{n,1} = (\hat{\mu}_n - r)/\hat{\sigma}_n$, where $\hat{\mu}_n = \bar{x}$ and $\hat{\sigma}_n = s_x$. The next proposition, proven in Appendix 4.I.6, establishes the precision of the estimation.

Proposition 4.4.1 *If X has finite fourth moment, i.e., γ_1 and γ_2 exist, then*

$$\sqrt{n}\left(\hat{\mathcal{S}}_{n,1} - \mathcal{S}_1\right) \rightsquigarrow N(0, \sigma^2_{\mathcal{S}_1}),$$

where $\sigma^2_{\mathcal{S}_1} = 1 + \mathcal{S}_1^2(\gamma_2 - 1)/4 - \mathcal{S}_1\gamma_1$. $\sigma^2_{\mathcal{S}_1}$ can also be estimated by the sample variance of the pseudo-observations

$$W_{n,i} = \frac{X_i - \hat{\mu}_n}{\hat{\sigma}_n} - \frac{1}{2}\hat{\mathcal{S}}_{n,1}\left\{\left(\frac{X_i - \hat{\mu}_n}{\hat{\sigma}_n}\right)^2 - 1\right\}, \quad i \in \{1, \ldots, n\}.$$

In the particular case of a Gaussian distribution, $\gamma_1 = 0$, $\gamma_2 = 3$, and $\sigma^2_{\mathcal{S}_1} = 1 + \mathcal{S}_1^2/2$.

The estimation and computation of the estimation error of the Sharpe ratio are implemented in the MATLAB function *EstSharpe*.

4.4.3 The Sortino Ratio

Suppose that $X_1, \ldots, X_n$ are independent and identically distributed observations of a random variable X, with mean μ and standard deviation σ. Then the Sortino ratio is defined by $\mathcal{S}_2 = (\mu - r)/D$, where $D^2 = E\{\max(r - X, 0)^2\}$. It can be estimated by $\hat{\mathcal{S}}_2 = (\bar{x} - r)/\hat{D}_n$, where $\hat{D}_n^2 = \frac{1}{n}\sum_{i=1}^n \max(r - X_i, 0)^2$. The next proposition deals with the precision of the estimation.

Proposition 4.4.2 *If the fourth moment of X exists, then*

$$\sqrt{n}\left(\hat{\mathcal{S}}_2 - \mathcal{S}_2\right) \rightsquigarrow N(0, \sigma^2_{\mathcal{S}_2}),$$

where $\sigma^2_{\mathcal{S}_2}$ can be estimated by the sample variance of the pseudo-observations

$$W_{n,i} = \frac{X_i - \bar{x}}{\hat{D}_n} - \hat{\mathcal{S}}_2 \frac{(Z_i^2 - \hat{D}_n^2)}{2\hat{D}_n^2}, \text{ with } Z_i = \max(r - X_i, 0), \quad i \in \{1, \ldots, n\}.$$

The proof is given in Appendix 4.I.7. The estimation and computation of the estimation error of the Sortino ratio are implemented in the MATLAB function *EstSortino*.

Remark 4.4.1 *Note that in the Gaussian case, i.e., $X_i \sim N(\mu, \sigma^2)$,*

$$D = \sigma^2 \left\{\mathcal{N}(-\mathcal{S}_1)\left(1 + \mathcal{S}_1^2\right) - \mathcal{S}_1 \mathcal{N}'(\mathcal{S}_1)\right\}.$$

4.4.4 The Omega Ratio

The Ω function [Keating and Shadwick, 2002] or Omega ratio is now used frequently in portfolio selection. Suppose that X is a random variable with distribution function F for which the first moment exists. Set $s_0 = \inf\{x; F(x) > 0\}$. Then $\Omega = \Omega_F$ is defined for $t > s_0$ by

$$\Omega(t) = \frac{E\{\max(X - t, 0)\}}{E\{\max(t - X, 0)\}} = \frac{\int_t^\infty \{1 - F(x)\}dx}{\int_{-\infty}^t F(x)dx}. \tag{4.26}$$

It is also related to the gain-loss ratio defined in Cherny and Madan [2009]. In a financial setting, X is the portfolio return and t is the level of return. The next two sections are adapted from Laroche and Rémillard [2008].

4.4.4.1 Relationship with Expectiles

Let $\alpha \in (0, 1)$ be given. According to Newey and Powell [1987], μ_α is called the expectile of order α of F if

$$\mu_\alpha = \operatorname{argmin}_\theta E\left[|\alpha - \mathbb{I}_{\{X \le \theta\}}|\{(X - \theta)^2 - X^2\}\right].$$

For example, taking $\alpha = 1/2$ yields $\mu_{1/2} = E(X)$. It follows from Newey and Powell [1987] and Abdous and Rémillard [1995] that μ_α is in fact the quantile of order α of the distribution function G, i.e., $G(\mu_\alpha) = \alpha$, where $G = G_F$ is defined for all $t \in \mathbb{R}$ by

$$\begin{aligned} G(t) &= \frac{E\{\max(t - X, 0)\}}{E\{|t - X|\}} & (4.27) \\ &= \frac{\int_{-\infty}^t F(x)dx}{\int_{-\infty}^t F(x)dx + \int_t^\infty \{1 - F(x)\}dx}. \end{aligned}$$

Using (4.27), we obtain that for all $t > s_0$,

$$\Omega(t) = \frac{1 - G(t)}{G(t)} \text{ or } G(t) = \frac{1}{1 + \Omega(t)}.$$

A major advantage of G over Ω is that G exists for any $t \in \mathbb{R}$ and that $0 \leq G(t) \leq 1$. Moreover, $\Omega_{F_1}(t) \geq \Omega_{F_2}(t)$ if and only if $\bar{G}_{F_1}(t) \geq \bar{G}_{F_2}(t)$, where $\bar{G} = 1 - G$ is the associated survival function. Note that $\bar{G}$ can also be used to measure performance, with the value 1 associated with the best possible performance while the value 0 is associated with the worst possible performance.

4.4.4.2 Gaussian Case and the Sharpe Ratio

In the Gaussian case, that is if $X \sim N(\mu, \sigma^2)$, then $S_1 = (\mu - t)/\sigma$, and we have

$$E\{\max(X - t, 0)\} = \sigma\{\mathcal{N}'(S_1) + S_1\mathcal{N}(S_1)\},$$

where $\mathcal{N}$ is the distribution function of a standard Gaussian random variable, and $\mathcal{N}'$ is its density. Furthermore,

$$E\{\max(t - X, 0)\} = \sigma\{\mathcal{N}'(S_1) - S_1\mathcal{N}(-S_1)\},$$

so

$$\Omega(t) = \frac{\mathcal{N}'(S_1) + S_1\mathcal{N}(S_1)}{\mathcal{N}'(S_1) - S_1\mathcal{N}(-S_1)}.$$

In particular, taking $t = r$, we obtain a relationship between the Sharpe and $\Omega(r)$ ratios. See Figure 4.4.

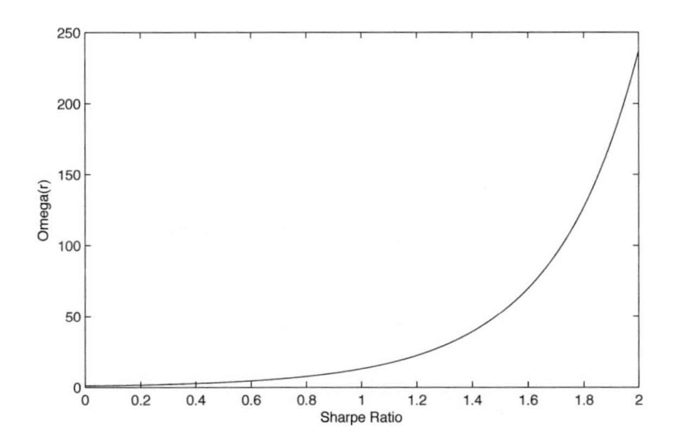

FIGURE 4.4: $\Omega(r)$ as a function of the Sharpe ratio S in the Gaussian case.

4.4.4.3 Relationship with Stochastic Dominance

The following proposition shows that Ω and $\bar{G}$ are monotone with respect to the stochastic order of Lehmann (see Definition 4.1.1).

Proposition 4.4.3 *Suppose that F_1 and F_2 have the same means. Then $\Omega_{F_1}(t) \le \Omega_{F_2}(t)$ for all t, or $\bar{G}_{F_1}(t) \le \bar{G}_{F_2}(t)$ for all t, if and only if $F_1 \preceq_{st} F_2$.*

4.4.4.4 Estimation of Omega and $\bar{G}$

Suppose that $X_1, \ldots, X_n$ are independent observations of a random variable X with a finite first moment. For a given t, set $Y_i = \max(X_i - t, 0)$ and $Z_i = \max(t - X_i, 0)$, $i \in \{1, \ldots, n\}$. Then for any $t > \min_{1 \le i \le n} X_i$, an estimator of $\Omega(t)$ is given by

$$\hat{\Omega}_n(t) = \bar{Y}_n / \bar{Z}_n, \tag{4.28}$$

where $\bar{Y}_n = \frac{1}{n}\sum_{i=1}^n Y_i$ and $\bar{Z}_n = \frac{1}{n}\sum_{i=1}^n Z_i$.

In addition, for all t, $\bar{G}(t)$ can be estimated by

$$\hat{\bar{G}}_n(t) = \bar{Y}_n / \{\bar{Y}_n + \bar{Z}_n\}. \tag{4.29}$$

As it is the case with any statistic, the estimation of Ω is not perfect, contrary to what is stated in Keating and Shadwick [2002]. The asymptotic behavior of the estimation error is described in the next proposition.

Before stating the result, note that $\bar{Y}_n$ and $\bar{Z}_n$ are unbiased and convergent estimators of $\mu_Y = E(\{\max(X - t, 0)\}$ and $\mu_Z = E\{\max(t - X, 0)\}$ respectively. However, $\hat{\Omega}_n(t)$ is a biased estimator of $\Omega(t)$, but it is convergent and asymptotically unbiased. The same properties hold true for $\hat{\bar{G}}_n(t)$.

Proposition 4.4.4 *For every $t > s_0$, $\sqrt{n}\left\{\hat{\Omega}_n(t) - \Omega(t)\right\} \rightsquigarrow N(0, \sigma_\Omega^2)$, where*

$$\sigma_\Omega^2 = \left\{\sigma_Y^2 + \Omega^2(t)\sigma_Z^2\right\} / \mu_Z^2 + 2\Omega^2(t),$$

which can be estimated by

$$s_\Omega^2 = \left\{s_Y^2 + \hat{\Omega}_n^2(t) s_Z^2\right\} / \bar{Z}_n^2 + 2\hat{\Omega}_n^2(t). \tag{4.30}$$

Moreover, for any t, $\sqrt{n}\left\{\hat{\bar{G}}_n(t) - \bar{G}(t)\right\} \rightsquigarrow N(0, \sigma_{\bar{G}}^2)$, where

$$\sigma_{\bar{G}}^2 = \left\{\bar{G}^2(t)\sigma_Z^2 + G^2(t)\sigma_Y^2\right\} / (\mu_Y + \mu_Z)^2 + 2G^2(t)\bar{G}^2(t),$$

which can be estimated by

$$s_G^2 = \frac{\hat{\bar{G}}_n^2(t) s_Z^2 + \hat{G}_n^2(t) s_Y^2}{(\bar{Y}_n + \bar{Z}_n)^2} + 2\hat{G}_n^2(t)\hat{\bar{G}}_n^2(t). \tag{4.31}$$

The proof is given in Appendix 4.I.8. The estimation and computation of the estimation error of the Omega ratio are implemented in the MATLAB function *EstOmega*. For example, a 95% confidence interval for $\Omega(t)$ is $\hat{\Omega}_n(t) \pm 1.96 s_\Omega/\sqrt{n}$, when n is large enough. Note that (4.30) and (4.31) are respectively the empirical variance of the (centered) pseudo-observations $U_{n,i} = \{Y_i - \hat{\Omega}_n(t) Z_i\}/\bar{Z}_n$ and $V_{n,i} = \{\hat{\bar{G}}_n(t)(Y_i + Z_i) - Y_i\}/(\bar{Y}_n + \bar{Z}_n)$, $i \in \{1, \ldots, n\}$.

Example 4.4.1 (Performance Ratios of Apple and Microsoft) *Using the daily prices of Apple and Microsoft from January 13th, 2009, to January 14th 2011, we obtain the following results for the estimation of the three performance ratios, assuming a daily rate* $r = \ln(1 + 0.0026)/252$ *for the period.*

TABLE 4.3: 95% confidence intervals for the performance ratios of Apple and Microsoft.

Ratio	Apple	Microsoft
Sharpe	0.1448 ± 0.0870	0.0413 ± 0.0877
Sortino	0.2325 ± 0.1604	0.0595 ± 0.1332
$\Omega(r)$	1.4803 ± 0.3521	1.1295 ± 0.2909
$\bar{G}(r)$	0.5968 ± 0.0572	0.5304 ± 0.0642

4.5 Suggested Reading

For axiomatic risk measures, the standard reference is Artzner et al. [1999], while it is Cherny and Madan [2009] for axiomatic performance measures. For the delta-gamma approximation, one suggests Jaschke [2002] and Glasserman et al. [2000]. Finally, for applications of approximations to option pricing, I suggest Jarrow and Rudd [1982] and Carr and Madan [2009].

4.6 Exercises

Exercise 4.1

(a) Give an interpretation of the following measures of risk:

(i) $VaR(100p\%)$.

(ii) $ES(100p\%)$.

(b) We want to estimate the VaR of a given portfolio composed of stocks and options on these stocks. Under which measure do we need to simulate the values of the stocks?

Exercise 4.2

Assume that the returns of a portfolio loss X at any time t are correctly modeled by a process of the form $\mu + \sigma W(t)$, where W is a Brownian motion. Using $n = 506$ daily returns, the mean and the standard deviation are respectively 400 and 250. Compute the estimation of the 99% VaR for the value of the portfolio in one year. Also compute the estimation error corresponding to a 95% confidence level. How does this compare to the nonparametric estimation error?

Exercise 4.3

(a) Give a financial interpretation for each of the following axioms:

(i) Translation invariance : $\rho(X + a) = \rho(X) + a$, for any a $\in \mathbb{R}$.

(ii) Monotonicity : $X \leq Y \Rightarrow \rho(X) \leq \rho(Y)$.

(iii) Sub-additivity : $\rho(\lambda X + (1 - \lambda)Y) \leq \lambda\rho(X) + (1 - \lambda)\rho(Y)$.

(iv) Positive homogeneity : $\rho(\lambda X) = \lambda\rho(X)$ for any $\lambda \geq 0$.

(b) Are the following measures of risk coherent with the axioms?

(i) VaR.

(ii) ES.

(iii) WCE.

(c) Give an example of a peculiar coherent measure of risk. Are there better measures of risk?

(d) Give an interpretation of the simple order of stochastic dominance and give an example of a coherent measure of risk with respect to that order.

(e) Give an interpretation of the hazard rate order and give an example of a coherent measure of risk with respect to that order.

Exercise 4.4

Assume, as in Exercise 4.2, that the returns of a portfolio loss X at any time t are modeled by a process of the form $\mu + \sigma W(t)$, where W is a Brownian motion. Using $n = 506$ daily returns, the mean and the standard deviation are respectively 400 and 250. Compute the estimation of the 99% expected shortfall for the value of the portfolio in one year. Also compute the estimation error corresponding to a 95% confidence level. How does this compare to the nonparametric estimation error?

Exercise 4.5

Discuss the advantages and disadvantages of the following approximation methods:

(a) The full Monte Carlo methodology.

(b) The partial Monte Carlo methodology.

(c) The Edgeworth expansion.

(d) The saddlepoint approximation.

(e) The inversion of the characteristic function.

Exercise 4.6

Compute the saddlepoint approximation of the density and the distribution function of a non-central chi-square distribution. Compare the graph of the approximated density with the real one when the degrees of freedom are $\nu = 5$ and the non-centrality parameter is 2.5.

Exercise 4.7

(a) Give an interpretation of the following measures of performance:

  (i) The Sharpe ratio.

  (ii) The Sortino ratio.

  (iii) The Omega ratio.

(b) Are these measures of performance coherent with respect to the Axioms B1–B7 of Cherny and Madan [2009]?

Exercise 4.8

Suppose that over the last 5 years, the monthly returns of a portfolio generated an average return of 0.75% with a 2% volatility. Assuming that the returns are correctly modeled by a Gaussian distribution and that the annual risk-free rate is 2.5%, compute the 95% confidence intervals for each of the following performance measures:

(a) The Sharpe ratio.

(b) The Sortino ratio.

(c) The Omega ratio.

(d) $\bar{G}$.

4.7 Assignment Questions

1. Consider a portfolio of m assets composed of European call options with different strikes and maturities, where the underlying assets are correlated geometric Brownian motions. We want to be able to evaluate the potential loss of the portfolio at a given period t. Construct a MATLAB function to estimate the VaR and the expected shortfall for given levels p using the full Monte Carlo approach, the partial Monte Carlo, and the Cornish-Fisher expansion. We want to be able to compute the estimation errors whenever possible.

2. Construct a MATLAB function to implement the saddlepoint approximation for the random variable Q with cumulant generating function (4.18).

4.A Brownian Bridge

The Brownian bridge is a stochastic process that appears naturally when one studies the asymptotic behavior of empirical distribution functions. It is defined as follows.

Definition 4.A.1 *A Brownian bridge B is a continuous centered Gaussian process on $[0,1]$ so that $B(0) = B(1) = 0$ and $\mathrm{Cov}\{B(s), B(t)\} = \min(s,t) - st$, for any $s, t \in [0,1]$. In particular, the variance of $B(t)$ is $t(1-t)$, $t \in [0,1]$.*

Remark 4.A.1 *A Brownian bridge is often described as a Brownian motion W conditional to $W(1) = 0$. In fact, one can always write B in the following form: $B(t) = W(t) - tW(1) = W(t) - E\{W(t)|W(1)\}$, $t \in [0,1]$.*

4.B Quantiles

Definition 4.B.1 *For a given distribution function F and a fixed $p \in (0,1)$, we have*

$$\{x; F(x) \geq p\} = [q_p, +\infty), \tag{4.32}$$

since F is right-continuous. The value q_p determined by (4.32) defines the quantile of order p of F, denoted by $F^{-1}(p)$.

Remark 4.B.1 *Note that as a function of $p \in (0,1)$, $F^{-1}(p)$ is left-continuous.*

Here are some particular cases of quantiles:

- $F^{-1}(1/2)$ is called the median of the distribution function F;
- $F^{-1}(1/4)$, $F^{-1}(2/4)$, and $F^{-1}(3/4)$ are respectively the first, second, and third quartiles of the distribution function F;
- $F^{-1}(1/10), \ldots, F^{-1}(9/10)$ are called the deciles of the distribution function F;
- $F^{-1}(1/100), \ldots, F^{-1}(99/100)$ are called the centiles of the distribution function F.

4.C Mean Excess Function

Definition 4.C.1 *For a random variable X with distribution function F having a finite first moment, the mean excess function is defined by*

$$\begin{aligned} e(u) &= E(X - u | X > u) \\ &= \frac{1}{1-F(u)} \int_u^\infty \{1 - F(x)\} dx, \end{aligned} \tag{4.33}$$

whenever $P(X > u) = 1 - F(u) > 0$.

In survival analysis, it is also called the mean survival function since $e(u)$ is the mean of a random variable with distribution function F_u defined by

$$F_u(x) = P(X \leq x + u | X > u), \quad x \geq 0,$$

whenever $F(u) < 1$. F_u is called the excess distribution function, hence the term mean excess function for $e(u)$.

4.C.1 Estimation of the Mean Excess Function

Suppose that $X_1, \ldots, X_n$ are independent and identically distributed observations from a random variable X with distribution function F. Suppose also that $E(|X|) < \infty$. Then the mean excess function is defined for all y such that $F(y) < 1$ by

$$e(y) = E\left(X - y | X > y\right).$$

It can be estimated by

$$\hat{e}_n(y) = \frac{\frac{1}{n}\sum_{i=1}^n (X_i - y)^+}{1 - F_n(y)}, \quad y < X_{n:n}, \tag{4.34}$$

where $(z)^+ = \max(0, z)$. The estimator is consistent and asymptotically unbiased. The next result about the estimation error of e is a direct consequence of the delta method (Theorem B.3.1).

Proposition 4.C.1 *Assume that $E(X^2) < \infty$. Then*

$$\sqrt{n}\,\{\hat{e}_n(y) - e(y)\} \rightsquigarrow N(0, \sigma_y^2),$$

where σ_y^2 can be estimated by

$$s_y^2 = \frac{1}{\{1 - F_n(y)\}^2}\left[V_{11} - \hat{e}_n^2(y) F_n(y)\{1 - F_n(y)\}\right],$$

V_{11} being the sample variance of the observations

$$(x_1 - y)^+, \ldots, (x_n - y)^+.$$

4.D Bootstrap Methodology

Suppose that $X_1, \ldots, X_n$ is a sample of independent and identically distributed observations of a random variable X and that we compute a statistic $\hat{\theta}_n = T_n(X_1, \ldots, X_n)$ which estimates a parameter θ. Suppose also that $\Theta_n = \sqrt{n}(\hat{\theta}_n - \theta) \rightsquigarrow N(0, V)$.

In many applications, an explicit formula for V can be impossibly difficult to find. How can we estimate V? One of the easiest and most intuitive ways of estimating V is called the (resampling) bootstrap. This methodology is due to Efron [1979] and it is described next.

4.D.1 Bootstrap Algorithm

For $k \in \{1, \ldots, N\}$, with N large enough, repeat the following steps:

(i) Choose with replacement a sample $X^\star_{1,k},\ldots,X^\star_{n,k}$ from the "population" $\{X_1,\ldots,X_n\}$.

(ii) Compute $\theta^\star_{n,k} = T_n\left(X^\star_{1,k},\ldots,X^\star_{n,k}\right)$.

These resampling steps are implemented in the MATLAB function *bootstrp*.

The main result is that the values $\Theta^\star_{n,k} = \sqrt{n}\left(\theta^\star_{n,k} - \hat{\theta}_n\right)$, $k \in \{1,\ldots,N\}$ are often almost independent copies of Θ_n. In particular, the variance of the bootstrap sample $\theta^\star_{n,k}$, $k \in \{1,\ldots,N\}$ is a good approximation of V.

4.E Simulation of $\mathbb{Q}_{F,a,b}$

To simulate plausible values $\mathbb{Q}_{F,a,b}$, one can use the following Monte Carlo algorithm:

Algorithm 4.E.1 *Estimate the density f of F by $\hat{f}_n$, using the kernel method described in Appendix 3.C.*

For $k \in \{1,\ldots,N = 10000\}$, repeat the following steps:

- *Generate independent observations $U_1,\ldots,U_n$ from the uniform distribution on $(0,1)$ and sort them, i.e., compute the order statistics $U_{n:1},\ldots,U_{n:n}$.*
- *Compute $Q_{n,k} = \max_{k\in\{\lfloor na\rfloor+1,\ldots,\lfloor nb\rfloor\}} \left[\frac{\sqrt{n}\,|U_{n:k} - k/n|}{\hat{f}_n\left\{F_n^{-1}(k/n)\right\}}\right]$, where $\lfloor x\rfloor$ stands for the integer part of x.*

To estimate the 95% quantile of $\mathbb{Q}_{F,a,b}$, just compute the 95% quantile of the pseudo-observations $Q_{n,1},\ldots,Q_{n,N}$.

4.F Saddlepoint Approximation of a Continuous Distribution Function

Let Z be a random variable with cumulant generating function K_Z, distribution function G and density g, called the base distribution. As in Section 4.3.6, we want to approximate the distribution function of a random variable X with cumulant generating function K. To this end, let L be the Legendre transform of K and let L_Z be the Legendre transform of K_Z.

Then the saddlepoint approximation of a continuous distribution function, with respect to the base K_Z is

$$\hat{F}(x) = G(w_x) + g(w_x)\left(\frac{1}{w_x} - \frac{1}{v_x}\right), \tag{4.35}$$

where $K'(u_x) = x$, $L_Z(\xi_x) = L(x)$, w_x solves $\xi_x = K_Z'(w_x)$, $v_x = u_x\sqrt{K''(u_x)/L_Z''(w_x)}$. Since L_Z is convex, there are two solutions to $L_Z(w) = L(x)$. If $x < \mu$, take w_x so that $w_x < K_Z'(0)$, otherwise take $w_x > K_Z'(0)$. Note that this approximation is exact for location and scale transforms of the base, i.e., when $X = a + bZ$.

Finally, one can check that for the Gaussian base, we recover formula (4.23). For, in that case, $L_Z(x) = x^2/2$, $v_x = u_x\sqrt{K''(u_x)}$, so $w_x \in \{\pm\sqrt{L(x)}\}$ has the same sign has u_x.

4.G Complex Numbers in MATLAB

Complex numbers can be used in MATLAB calculations. For example, whenever $a, b \in \mathbb{R}$, the expressions $z = a + bi$ and $z = a + i * b$ make sense and are equal. In particular, the multiplication sign is not necessary in $a + bi$, but one cannot write $a+ib$ in MATLAB. One can also use the MATLAB functions *real* and *imag* to compute the real and imaginary parts of a complex number $z = a + bi$.

Example 4.G.1 *For any $\lambda \in \mathbb{R}$, we have*

$$\begin{aligned}\sqrt{1 - i\lambda} &= (1+\lambda^2)^{1/4}\, e^{\frac{i}{2}\arctan(-\lambda)} \\ &= (1+\lambda^2)^{1/4}\, \cos(\arctan(\lambda)/2) \\ &\quad -i(1+\lambda^2)^{1/4}\, \sin(\arctan(\lambda)/2),\end{aligned}$$

and $\ln(1 - i\lambda) = \frac{1}{2}\ln(1+\lambda^2) + i\arctan(-\lambda) = \frac{1}{2}\ln(1+\lambda^2) - i\arctan(\lambda)$.

With MATLAB, we obtain readily $\ln(1 - 4i) = 1.4166 - 1.3258i$ *and* $\sqrt{1 - 4i} = 1.6005 - 1.2496i$.

Complex Numbers and Indices in MATLAB

When using complex numbers in MATLAB, do not use i or j as indices in loops. In fact, it is suggested to use $1i$ for complex numbers to improve the robustness of your code.

4.H Gil-Pelaez Formula

The Gil-Pelaez formula is used to compute the distribution function F from its characteristic function ϕ. It is much more useful than the expression relating the distribution function with its characteristic function given in Theorem A.4.1. The so-called Gil-Pelaez formula is

$$\frac{F(x)+F(x-)}{2} = \frac{1}{2} - \int_{-\infty}^{\infty} \frac{\phi(u)}{2\pi ui} e^{-uxi} du, \quad x \in \mathbb{R}. \tag{4.36}$$

As a result, if F is continuous at x, then

$$F(x) = \frac{1}{2} - \int_{-\infty}^{\infty} \frac{\phi(u)}{2\pi ui} e^{-uxi} du.$$

Note that using Theorem A.4.1 when F is continuous, one can only compute

$$F(b) - F(a) = \frac{1}{2\pi} \int_{-\infty}^{\infty} \left(\frac{e^{-uai} - e^{-ubi}}{ui} \right) \phi(u)\, du, \quad a < b,$$

so the Gil-Pelaez formula (4.36) is much more interesting.

Example 4.H.1 *Suppose that X is a discrete random variable with $P(X = x_j) = p_j$, with $\sum_j p_j = 1$. Its characteristic function is then given by*

$$\phi(t) = \sum_j p_j e^{itx_j},$$

and we can see that

$$\frac{1}{2} - \int_{-\infty}^{\infty} \frac{\phi(u)}{2\pi iu} e^{-iux} du = \frac{1}{2} + \frac{1}{2} \sum_j p_j \operatorname{sign}(x - x_j) = \tilde{F}(x),$$

using the identity

$$\int_0^{\infty} \frac{\sin(tx)}{t} dt = \operatorname{sgn}(x) = \begin{cases} 1, & \text{if } x > 0, \\ 0, & \text{if } x = 0, \\ -1, & \text{if } x < 0. \end{cases} \tag{4.37}$$

Now, if the F is the distribution function of X, we have $\tilde{F}(x) = \frac{F(x-)+F(x)}{2}$, as stated in (4.36).

4.I Proofs of the Results

4.I.1 Proof of Proposition 4.1.1

Suppose that for all $p \in (0,1)$, $F_X^{-1}(p) = \mathrm{VaR}_X(p) \leq \mathrm{VaR}_Y(p) = F_Y^{-1}(p)$. Let $U \sim U(0,1)$. It follows from Theorem A.6.1 that $X_1 = F_X^{-1}(U) \sim F_X$ and

$Y_1 = F_Y^{-1}(U) \sim F_Y$. By hypothesis, $X_1 \le Y_1$ so for all x,

$$P(X > x) = P(X_1 > x) \le P(Y_1 > x) = P(Y > x),$$

proving that $X \preceq_{st} Y$. Now, if $X \preceq_{st} Y$, then $F_X(x) \ge F_Y(x)$ for all x, so for a given $p \in (0,1)$, by the definition of a quantile (Appendix 4.B), one gets

$$[\mathrm{VaR}_Y(p), \infty) = \{x; F_Y(x) \ge p\} \subset \{x; F_X(x) \ge p\} = [\mathrm{VaR}_X(p), \infty).$$

Hence, $\mathrm{VaR}_X(p) \le \mathrm{VaR}_Y(p)$.

■

4.I.2 Proof of Proposition 4.1.3

Suppose that $x < y$ and $P(X > y) > 0$. Then

$$\begin{aligned} P(X > y)E\{X\mathbb{I}(x < X \le y)\} &\le yP(x < X \le y)P(X > y) \\ &\le P(x < X \le Y)E\{X\mathbb{I}(X > y)\}, \end{aligned}$$

showing that

$$P(X > y)E\{X\mathbb{I}(x < X \le y)\} \le P(x < X \le y)E\{X\mathbb{I}(X > y)\}.$$

Adding $P(X > y)E\{X\mathbb{I}(X > y)\}$ to each side of the last inequality and collecting terms, one obtains

$$P(X > y)E\{X\mathbb{I}(X > x)\} \le P(X > x)E\{X\mathbb{I}(X > y)\},$$

so $E(X|X > x) \le E(X|X > y)$. Next, by Proposition 4.1.2, $X \preceq_{hr} Y$ implies $X \preceq_{st} Y$, which in turn implies that $\mathrm{VaR}_X(p) \le \mathrm{VaR}_Y(p)$ by Proposition 4.1.1. The proof will be completed if we can prove that $E(X|X > x) \le E(Y|Y > x)$, for any x. To this end, note that by the property of expectations and by definition of the mean excess function, one has

$$E(X|X > x) = \frac{1}{P(X > x)} \int_x^\infty P(X > t)dt. \tag{4.38}$$

Since $x \mapsto \frac{P(Y>x)}{P(X>x)}$ is non-decreasing, it follows from (4.38) that

$$\int_x^\infty \frac{P(X > t)dt}{P(X > x)} dt \le \int_x^\infty \frac{P(Y > t)dt}{P(Y > x)} dt,$$

since $t \ge x$. Hence $ES_X(p) \le ES_Y(p)$, for any $p \in (0,1)$.

■

4.I.3 Proof of Proposition 4.2.1

Set $Q_n = \sqrt{n}\left\{F_n^{-1}(p) - q\right\}$, where $q = F^{-1}(p) = \text{VaR}(p)$. Then

$$\begin{aligned}
P(Q_n \le x) &= P\left\{F_n^{-1}(p) \le q + \frac{x}{\sqrt{n}}\right\} = P\left\{F_n\left(q + \frac{x}{\sqrt{n}}\right) \ge p\right\} \\
&= P\left[\sqrt{n}\left\{F_n\left(q + \frac{x}{\sqrt{n}}\right) - F\left(q + \frac{x}{\sqrt{n}}\right)\right\}\right. \\
&\qquad \left. \ge -\sqrt{n}\left\{F\left(q + \frac{x}{\sqrt{n}}\right) - F(q)\right\}\right] \\
&\to P\left[B\{F(q)\} \ge -xf(q)\right] \text{ (as } n \to \infty) = P\left\{-\frac{B(p)}{f(q)} \le x\right\}.
\end{aligned}$$

Since $-B(p) \sim N(0, p - p^2)$, it follows that $Q_n \rightsquigarrow Q \sim N\left(0, \sigma^2_{\text{VaR}(p)}\right)$, where $\sigma^2_{\text{VaR}(p)} = p(1-p)/f^2(q)$.

■

4.I.4 Proof of Proposition 4.2.2

For simplicity, set $q_n = F_n^{-1}(p)$, $q = F^{-1}(p)$, and $Q_n = \sqrt{n}\,(q_n - q)$. By Proposition 4.2.1, we know that $Q_n \rightsquigarrow Q \sim N\left(0, \frac{p(1-p)}{f^2(q)}\right)$. In particular, the sequence Q_n is tight, i.e., for any $\delta > 0$, there exists $M > 0$ so that $P(|Q_n| > M) < \delta$ for all $n \ge 1$. Next, $\widehat{ES}_n(p) = q_n + \frac{1}{1-p}\int_{q_n}^{\infty}\{1 - F_n(x)\}dx$, and $ES(p) = q + \frac{1}{1-p}\int_q^{\infty}\{1 - F(x)\}dx$, so setting $Z_n = \sqrt{n}\left\{\widehat{ES}_n(p) - ES(p)\right\}$ and $\mathbb{F}_n = \sqrt{n}\,(F_n - F)$, we have

$$\begin{aligned}
Z_n &= Q_n + \frac{\sqrt{n}}{1-p}\left[\int_{q_n}^{\infty}\{1 - F_n(x)\}dx - \int_q^{\infty}\{1 - F(x)\}dx\right] \\
&= Q_n + \frac{\sqrt{n}}{1-p}\left[\int_{q_n}^{\infty}\{1 - F_n(x)\}dx - \int_{q_n}^{\infty}\{1 - F(x)\}dx\right] \\
&\quad + \frac{\sqrt{n}}{1-p}\int_{q_n}^{q}\{1 - F(x)\}dx \\
&= -\frac{1}{1-p}\int_{q_n}^{\infty}\mathbb{F}_n(x)dx + Q_n + \frac{\sqrt{n}}{1-p}\int_{q_n}^{q}\{1 - F(x)\}dx \\
&= -\frac{1}{1-p}\int_{q_n}^{\infty}\mathbb{F}_n(x)dx + \frac{\sqrt{n}}{1-p}\int_{q_n}^{q}\{F(q) - F(x)\}dx \\
&= -\frac{1}{1-p}\int_{q}^{\infty}\mathbb{F}_n(x)dx - \frac{1}{1-p}\int_{q_n}^{q}\mathbb{F}_n(x)dx \\
&\quad + \frac{\sqrt{n}}{1-p}\int_{q_n}^{q}\{F(q) - F(x)\}dx.
\end{aligned}$$

Because of the convergence of $\mathbb{F}_n$, the continuity of F and the tightness of Q_n, it follows that $\frac{1}{1-p}\int_{q_n}^{q} \mathbb{F}_n(x)dx$ and $\frac{\sqrt{n}}{1-p}\int_{q_n}^{q} \{F(q)-F(x)\}dx$ converge in probability to 0. Finally, one can write

$$-\int_q^{\infty} \mathbb{F}_n(x)dx = \frac{1}{\sqrt{n}}\sum_{i=1}^{n}\left[(X_i-q)^+ - E\left\{(X-q)^+\right\}\right] \rightsquigarrow Z,$$

where $Z \sim N\left(0, \sigma^2_{ES(p)}\right)$, and $\sigma^2_{ES(p)}$ is the variance of $(X-q)^+$, by the central limit theorem (Theorem B.4.1).

■

4.I.5 Proof of Proposition 4.3.1

Set $W = A^\top Z$. Note that W is Gaussian since it is a linear transformation of a Gaussian random vector. It remains to find its mean and covariance matrix. For the mean, we have

$$E(W) = E\left(A^\top Z\right) = A^\top E(Z) = 0.$$

As for the covariance matrix, we have

$$E\left(WW^\top\right) = E\left(A^\top ZZ^\top A\right) = A^\top E\left(ZZ^\top\right)A = A^\top A = I$$

where the last equality is due to $A^{-1} = A^\top$. As a result, $W \sim N_m(0, I)$, so W has the same distribution as Z. Using W instead of Z makes it possible to write Q has a sum of independent standard Gaussian variables, since

$$Q = \beta^\top AW + \frac{1}{2}W^\top \Lambda W = \sum_{j=1}^{m}\left(h_j W_j + \frac{1}{2}\lambda_j W_j^2\right), \tag{4.39}$$

using the fact that $A^{-1} = A^\top$. In particular, we deduce from (4.39) that

$$M_Q(u) = \prod_{j=1}^{m} E\left\{e^{u\left(h_j W_j + \frac{1}{2}\lambda_j W_j^2\right)}\right\}.$$

To obtain formula (4.18), it suffices to compute $M_j(u) = E\left\{e^{u\left(h_j W_j + \frac{1}{2}\lambda_j W_j^2\right)}\right\}$ for all $j \in \{1,\ldots,m\}$. Recalling that $W_j \sim N(0,1)$, if $|u||\lambda_j| < 1$, we have

$$\begin{aligned}
M_j(u) &= \int_{-\infty}^{+\infty} \frac{e^{-\frac{w^2}{2}}}{\sqrt{2\pi}} e^{u\left(h_j w + \frac{1}{2}\lambda_j w^2\right)} dw = \int_{-\infty}^{+\infty} \frac{e^{uh_j w - \frac{w^2(1-u\lambda_j)}{2}}}{\sqrt{2\pi}} dw \\
&= \frac{1}{\sqrt{1-u\lambda_j}} \int_{-\infty}^{+\infty} \frac{e^{\frac{uh_j}{\sqrt{1-u\lambda_j}}x - \frac{x^2}{2}}}{\sqrt{2\pi}} dx \\
&= \frac{e^{\frac{1}{2}\frac{u^2 h_j^2}{1-u\lambda_j}}}{\sqrt{1-u\lambda_j}} \int_{-\infty}^{+\infty} \frac{e^{-\frac{1}{2}\left(x - \frac{uh_j}{\sqrt{1-u\lambda_j}}\right)^2}}{\sqrt{2\pi}} dx = \frac{e^{\frac{1}{2}\frac{u^2 h_j^2}{1-u\lambda_j}}}{\sqrt{1-u\lambda_j}}.
\end{aligned}$$

Hence, if $|u| < \dfrac{1}{\max_{1\le j\le m}|\lambda_j|}$, we have $|u||\lambda_j| < 1$, and

$$\begin{aligned}
\ln\{M_Q(u)\} &= \ln\left\{E\left(e^{uQ}\right)\right\} = \frac{1}{2}\sum_{j=1}^{m}\frac{u^2h_j^2}{(1-u\lambda_j)} - \frac{1}{2}\sum_{j=1}^{m}\ln(1-u\lambda_j) \\
&= \frac{1}{2}\sum_{j=1}^{m}u^2h_j^2\left(\sum_{k=0}^{\infty}u^k\lambda_j^k\right) + \frac{1}{2}\sum_{j=1}^{m}\sum_{n=1}^{\infty}\frac{u^n\lambda_j^n}{n} \\
&= \frac{1}{2}u\sum_{j=1}^{m}\lambda_j + \frac{1}{2}\sum_{n=2}^{\infty}\frac{u^n}{n!}\left[\sum_{j=1}^{m}\left\{(n-1)!\lambda_j^n + h_j^2 n!\lambda_j^{n-2}\right\}\right].
\end{aligned}$$

It follows from the definition of cumulants in Section A.5.1, that for all $|u| < \dfrac{1}{\max_{1\le j\le m}|\lambda_j|}$, $\ln\{M_Q(u)\} = \sum_{n=1}^{\infty}\kappa_n\frac{u^n}{n!}$. As a result, we have

$$\begin{aligned}
\kappa_1 &= \frac{1}{2}\sum_{j=1}^{m}\lambda_j = \frac{1}{2}Trace(\Lambda) \\
\kappa_n &= \frac{(n-1)!}{2}\sum_{j=1}^{m}\lambda_j^{n-2}\left(\lambda_j^2 + nh_j^2\right), \quad n \ge 2.
\end{aligned}$$

■

4.I.6 Proof of Proposition 4.4.1

Since$\sqrt{n}\left(\hat{\mathcal{S}}_{n,1} - \mathcal{S}_1\right) = \left\{\sqrt{n}(\hat{\mu}_n - \mu) - \mathcal{S}_1\sqrt{n}(\hat{\sigma}_n - \sigma)\right\}/\hat{\sigma}_n$, one can apply Proposition B.4.1 to obtain that

$$\sigma^2_{\mathcal{S}_1} = \frac{1}{\sigma^2}\left\{V_{11} - 2\mathcal{S}_1V_{12} + \mathcal{S}_1^2V_{22}\right\} = 1 + \mathcal{S}_1^2(\gamma_2 - 1)/4 - \mathcal{S}_1\gamma_1.$$

The convergence in probability of the variance of the pseudo-observations W_n follows from the law of large numbers.

■

4.I.7 Proof of Proposition 4.4.2

With the same notations as in Section 4.4.4.4 with $t = r$, we have

$$\sqrt{n}(\hat{\mathcal{S}}_{n,2} - \mathcal{S}_2) = \sqrt{n}(\hat{\mu}_n - \mu)/\hat{D}_n - \mathcal{S}_2\sqrt{n}(\hat{D}_n^2 - D^2)/\{\hat{D}_n(\hat{D}_n + D)\}.$$

The proof is completed by applying the central limit theorem (Theorem B.4.1) and by noting that the convergence in probability of the variance of the pseudo-observations W_n follows from the law of large numbers.

■

4.I.8 Proof of Proposition 4.4.4

Set $\mathbb{Y}_n = \sqrt{n}(\bar{Y}_n - \mu_Y)$ and $\mathbb{Z}_n = \sqrt{n}(\bar{Z}_n - \mu_Z)$. Then, $\text{Cov}(\mathbb{Y}_n, \mathbb{Z}_n) = -\mu_Y \mu_Z$, and using the central limit theorem (Theorem B.4.1), we obtain that $(\mathbb{Y}_n, \mathbb{Z}_n)^\top \rightsquigarrow N_2(0, C)$, where $C = \begin{pmatrix} \sigma_Y^2 & -\mu_Y\mu_Z \\ -\mu_Y\mu_Z & \sigma_Z^2 \end{pmatrix}$, with $\sigma_Y^2 = \text{var}(Y_i)$ and $\sigma_Z^2 = \text{var}(Z_i)$. These two variances can be estimated by their empirical counterparts $s_Y^2 = \frac{1}{n-1}\sum_{i=1}(Y_i - \bar{Y}_n)^2$ and $s_Z^2 = \frac{1}{n-1}\sum_{i=1}(Z_i - \bar{Z}_n)^2$ respectively. Finally, using the delta method (Theorem B.3.1), we see that the asymptotic variances of $\hat{\Omega}_n(t)$ and $\hat{\bar{G}}_n(t)$ can be estimated by (4.30) and (4.31).

■

Bibliography

B. Abdous and B. Rémillard. Relating quantiles and expectiles under weighted-symmetry. *Ann. Inst. Statist. Math.*, 47(2):371–384, 1995.

M. Abramowitz and I. E. Stegun. *Handbook of Mathematical Functions with Formulas, Graphs, and Mathematical Tables*, volume 55 of *Applied Mathematics Series*. National Bureau of Standards, tenth edition, 1972.

Y. Aït-Sahalia and J. Yu. Saddlepoint approximations for continuous-time Markov processes. *Journal of Econometrics*, 134(2):507– 551, 2006.

P. Artzner, F. Delbaen, J.-M. Eber, and D. Heath. Coherent measures of risk. *Mathematical Finance*, 9:203–228, 1999.

R. W. Butler. *Saddlepoint Approximations with Applications.* Cambridge Series in Statistical and Probabilistic Mathematics. Cambridge University Press, Cambridge, 2007.

P. Carr and D. Madan. Option pricing and the fast Fourier transform. *Journal of Computational Finance*, 2:61–73, 1999.

P. Carr and D. Madan. Saddlepoint methods for option pricing. *The Journal of Computational Finance*, 13:49–61, 2009.

A. Cherny and D. Madan. New measures for performance evaluation. *Review of Financial Studies*, 22(7):2571–2606, 2009.

H. E. Daniels. Saddlepoint approximations in statistics. *Ann. Math. Statist.*, 25:631–650, 1954.

R. B. Davies. Numerical inversion of a characteristic function. *Biometrika*, 60:415–417, 1973.

B. Efron. Bootstrap methods: another look at the jackknife. *Ann. Statist.*, 7 (1):1–26, 1979.

P. Glasserman, P. Heidelberger, and P. Shahabuddin. Variance reduction techniques for estimating Value-at-Risk. *Management Science*, 46:1349–1364, 2000.

S. Heston. A closed-form solution for options with stochastic volatility with application to bond and currency options. *Review of Financial Studies*, 6 (2):327–343, 1993.

R. Jarrow and A. Rudd. Approximate option valuation for arbitrary stochastic processes. *Journal of Financial Economics*, 10:347–369, 1982.

S. R. Jaschke. The Cornish-Fisher expansion in the context of Delta-Gamma-normal approximations. *Journal of Risk*, 4(4):33–52, 2002.

C. Keating and W. F. Shadwick. A universal performance measure. Technical report, The Finance Development Centre, London, 2002.

D. Kuonen. Saddlepoint approximations for distributions of quadratic forms in normal variables. *Biometrika*, 86(4):929–935, 1999.

P. Laroche and B. Rémillard. A statistical test for the Omega measure of performance. Technical report, Innocap, 2008.

E. L. Lehmann. Consistency and unbiasedness of certain nonparametric tests. *Ann. Math. Statistics*, 22:165–179, 1951.

R. Lugannani and S. Rice. Saddle point approximation for the distribution of the sum of independent random variables. *Adv. in Appl. Probab.*, 12(2), 1980.

J. Mina and A. Ulmer. Delta-gamma four ways. Technical report, RiskMetrics Group, 1999.

W. K. Newey and J. L. Powell. Asymmetric least squares estimation and testing. *Econometrica*, 55(4):819–847, 1987.

W. Sharpe. Mutual fund performance. *Journal of Business*, 39:119–138, 1966.

G. R. Shorack and J. A. Wellner. Limit theorems and inequalities for the uniform empirical process indexed by intervals. *Ann. Probab.*, 10(3):639–652, 1982.

D. Tasche. Expected shortfall and beyond. *Journal of Banking & Finance*, 26(7):1519 – 1533, 2002.

Y. Yamai and T. Yoshiba. Comparative analyses of expected shortfall and value-at-risk (2): expected utility maximization and tail risk. *Monetary and Economic Studies*, April:95–115, 2002.

Chapter 5

Modeling Interest Rates

In this chapter, we will present estimation methods for the parameters of stochastic interest rate models. First, we present the main results of Vasicek on the relationship between the instantaneous interest rate and the value of zero-coupon bonds, introducing also the concept of market risk premium. Then we present a first model, the so-called Vasicek model, where the interest rates are modeled by a Ornstein-Uhlenbeck process. We will study some of its properties: distribution, measurement scale, interpretation of parameters, and long-term behavior. Then we compute the value of a zero-coupon bond, when the process remains a Ornstein-Uhlenbeck process under an equivalent martingale measure. We will also show how to estimate the parameters, using a methodology proposed by Duan [1994, 2000]. We give some examples and illustrate the pitfalls to avoid. We then look at another popular model introduced by Cox et al. [1985] (CIR hereafter), where the interest rates are modeled by a Feller process. Contrary to the Ornstein-Uhlenbeck process, the Feller process has the benefit of being non-negative. Again we study some of its properties: distribution, measurement scale, interpretation of parameters, and long-term behavior. We compute the value of a zero-coupon bond under the CIR model, when the process remains a Feller process under an equivalent martingale measure. We will also show how to estimate the parameters. Finally, in an informal section, we will present more general processes for the interest rates.

5.1 Introduction

Until now, we assumed that the short-term interest rate was constant, or at worst deterministic. Now we will consider stochastic models for the instantaneous (spot) interest rate $\mathfrak{r}$, following the exposition of Vasicek [1977].

5.1.1 Vasicek Result

Vasicek [1977] poses the following problem: Given a spot interest rate $\mathfrak{r}$ satisfying the stochastic differential equation

$$d\mathfrak{r}(t) = \mu\{t, \mathfrak{r}(t)\}dt + \gamma\{t, \mathfrak{r}(t)\}dW(t), \quad \mathfrak{r}(0) = r,$$

what is the value $P(t,T)$, at time t, of a zero-coupon bond with face value \$1 at maturity T? Here W is a Brownian motion.

Another way of characterizing the process is through the associated martingale problem: The distribution of $\mathfrak{r}$ is such that for any infinitely differentiable function f, and for any $r \in [0,\infty)$ and for $t \geq 0$,

$$f\{\mathfrak{r}(t)\} - f(r) - \int_0^t \mu\{u,\mathfrak{r}(u)\} f'\{\mathfrak{r}(u)\} du - \frac{1}{2}\int_0^t \gamma^2\{u,\mathfrak{r}(u)\} f''\{\mathfrak{r}(u)\} du$$

is a martingale starting at 0. For more details on martingale problems in financial engineering, see, e.g., Labbé et al. [2012].

Using non-arbitrage arguments and building a portfolio of two contracts with different maturities, Vasicek showed that there exists a deterministic function $q(t,r)$, called the market price of risk, independent of any contract, and such that the value of a contract with payoff $\Phi\{\mathfrak{r}(T)\}$ at maturity T is $C\{t,\mathfrak{r}(t)\}$, where C is the solution to the partial differential equation

$$\frac{\partial}{\partial t}C + (\mu - q\gamma)\frac{\partial}{\partial r}C + \frac{1}{2}\gamma^2 \frac{\partial^2}{\partial r^2}C = rC, \quad t \in (0,T), \tag{5.1}$$

with the boundary condition $C(T,r) = \Phi(r)$. Vasicek also proved that C can be written as an expectation, viz.

$$C(t,r) = E\left[\Phi\{\tilde{\mathfrak{r}}(T)\} e^{-\int_t^T \tilde{\mathfrak{r}}(u)du}\right], \ 0 \leq t \leq T, \tag{5.2}$$

where $\tilde{\mathfrak{r}}$ satisfies $\tilde{\mathfrak{r}}(t) = r$, and

$$d\tilde{\mathfrak{r}}(u) = [\mu\{u,\tilde{\mathfrak{r}}(u)\} - q\{s,\tilde{\mathfrak{r}}(u)\}\gamma\{u,\tilde{\mathfrak{r}}(u)\}]\, du + \gamma\{u,\tilde{\mathfrak{r}}(u)\} d\tilde{W}(u), \tag{5.3}$$

$u \in (t,T)$, for some Brownian motion $\tilde{W}$. In particular, the value of a zero-coupon bond with face value \$1 at maturity T is given by

$$P(t,T) = E\left\{e^{-\int_t^T \tilde{\mathfrak{r}}(u)du}\right\}, \quad 0 \leq t \leq T, \tag{5.4}$$

which is the solution to the partial differential equation (5.1) with boundary condition $P(T,T) = 1$.

Note that $P(t,T)$ depends only on $\mathfrak{r}(t)$ and the parameters of the model; however, the spot rate $\mathfrak{r}(t)$ is not observable. Fortunately, we can observe values of bonds with different maturities. This will be quite useful when estimating parameters of that type of model.

5.2 Vasicek Model

In his article, Vasicek gives an example of a model for the spot interest rate that he attributes to Merton. This stochastic process, known for a long time in Probability under the name Ornstein-Uhlenbeck process [Uhlenbeck and Ornstein, 1930], is a continuous Gaussian process. Therefore, it has the dubious property that the spot rate could take negative values! Taking into account all other advantages of the model, the non-positiveness is considered a negligible problem.

5.2.1 Ornstein-Uhlenbeck Processes

The Ornstein-Uhlenbeck process can be defined by

$$d\mathfrak{r}(t) = \alpha\{\beta - \mathfrak{r}(t)\}dt + \sigma dW(t), \tag{5.5}$$

where W is a Brownian motion and $\sigma > 0$. The solution of (5.5) is

$$\mathfrak{r}(t) = \beta + e^{-\alpha t}\{\mathfrak{r}(0) - \beta\} + \sigma \int_0^t e^{-\alpha(t-u)} dW(u), \quad t \geq 0. \tag{5.6}$$

For properties of stochastic integrals of the form $\int_0^t f(u)dW(u)$, see Appendix 5.A. The Ornstein-Uhlenbeck process thus depends on three parameters: α, β, and σ. In order to be able to interpret and estimate these parameters, we will study some of the model's properties. But first, we need to address the problem of measurement scale for interest rates.

5.2.2 Change of Measurement and Time Scales

The choice of the measurement and time scales for the Ornstein-Uhlenbeck process are quite important. In fact, let λ be a measurement factor and m be the frequency at which $\mathfrak{r}(t)$ is observed, and set $\mathfrak{r}^{(\lambda,m)}(t) = \lambda\mathfrak{r}(mt)$. Then, $\mathfrak{r}^{(\lambda,m)}$ is also a Ornstein-Uhlenbeck process, with parameters $\alpha^{(\lambda,m)} = m\alpha$, $\beta^{(\lambda,m)} = \lambda\beta$, and $\sigma^{(\lambda,m)} = \sqrt{m}\lambda\sigma$. Using this property, one can easily transform the spot rate expressed on a daily basis into annual percentage rates which are easier to manipulate. To do that we have to decide on the number of days in a year.

Convention on Day Count

Depending on the country and the interest rate instrument, the number of days between two dates and the number of days in the year can be computed differently. See, e.g., Hull [2006, Chapter 6]. For example, in the USA, for corporate bonds, the 30/360 convention is used, meaning there are 30 days in the month and 360 days in the year. For other instruments but Treasury bonds, the convention is actual/360, actual meaning the exact number of days between two dates. Note that for most countries of the G8, the convention is actual/365.

For example, under the 30/360 convention, an annual interest rate of 2% corresponds to a spot rate of $r = .02/360 = 5.5556 \times 10^{-5}$, if the time scale is in days. It would mean that working on this measurement/time scale, our data would be of the same order and it could create computational problems when estimating parameters. Note that the new parameters would be $\alpha/360$, $2.78 \times 10^{-5}\beta$, and $5.27\sigma \times 10^{-4}$.

In this chapter, in order to minimize rounding errors and to facilitate interpretation, we will always use interest rates expressed as numbers or percentages, on a yearly basis.

5.2.3 Properties of Ornstein-Uhlenbeck Processes

Proposition 5.2.1 *For $s > t \geq 0$, the conditional distribution of $\mathfrak{r}(s)$ given $\mathfrak{r}(t) = r$, is Gaussian, with mean $\beta + e^{-\alpha(s-t)}(r - \beta)$ and variance $\sigma^2 \left\{\frac{1-e^{-2\alpha(s-t)}}{2\alpha}\right\}$.*

In particular, for any given $h > 0$, setting $X_k = \mathfrak{r}(kh)$, we have

$$X_k - \beta = \phi(X_{k-1} - \beta) + \varepsilon_k, \quad k = 1, \ldots, n, \tag{5.7}$$

where $\phi = e^{-\alpha h}$, the random variables ε_k are independent, ε_k is independent of $X_1, \ldots, X_{k-1}$, and $\varepsilon_k \sim N\left\{0, \sigma^2\left(\frac{1-\phi^2}{2\alpha}\right)\right\}$.

The proof is given in Appendix 5.D.1. Note that times series with representation (5.7) are called AR(1) processes or autoregressive processes of order 1.

5.2.3.1 Moments of the Ornstein-Uhlenbeck Process

Using representation (5.6) and the fact that stochastic integrals with respect to Brownian motion have mean 0, conditionally to $\mathfrak{r}(0) = r$, we have

$$E\{\mathfrak{r}(t)\} = \beta + e^{-\alpha t}(r - \beta), \tag{5.8}$$

$$\mathrm{Var}\{\mathfrak{r}(t)\} = \sigma^2 \frac{\left(1 - e^{-2\alpha t}\right)}{2\alpha}, \tag{5.9}$$

$$\mathrm{Cov}\{\mathfrak{r}(t), \mathfrak{r}(s)\} = \sigma^2 \left\{\frac{e^{-\alpha|t-s|} - e^{-\alpha(t+s)}}{2\alpha}\right\}. \tag{5.10}$$

In particular, if $\alpha > 0$, then $E\{\mathfrak{r}(t)\} \stackrel{t\to\infty}{\longrightarrow} \beta$, $\mathrm{Var}\{\mathfrak{r}(t)\} \stackrel{t\to\infty}{\longrightarrow} \frac{\sigma^2}{2\alpha}$, and $\mathrm{Cov}\{\mathfrak{r}(t), \mathfrak{r}(s)\} \stackrel{t\to\infty}{\longrightarrow} 0$, for any given $s \geq 0$. As a result, the parameter β can be interpreted as the long-term mean. Since the speed of convergence of $E\{\mathfrak{r}(t)\}$ to β depends on α, this parameter is often called the mean-reversion coefficient.

5.2.3.2 Stationary Distribution of the Ornstein-Uhlenbeck Process

When $\alpha > 0$, it is easy to check that if $\mathfrak{r}(0) \sim N\left(\beta, \frac{\sigma^2}{2\alpha}\right)$, then $\mathfrak{r}(t) \sim N\left(\beta, \frac{\sigma^2}{2\alpha}\right)$ for any $t \geq 0$. In that case, the process is stationary, its distribution at time t being the same for all t. Moreover, under this stationary distribution,

$$\text{Cov}\{\mathfrak{r}(t), \mathfrak{r}(s)\} = \frac{\sigma^2}{2\alpha} e^{-\alpha|t-s|}, \text{ and } \text{Cor}\{\mathfrak{r}(t), \mathfrak{r}(s)\} = e^{-\alpha|t-s|}.$$

In fact, whatever the distribution of $\mathfrak{r}(0)$, if $\alpha > 0$, then the distribution of $\mathfrak{r}(t)$ converges in law to the stationary distribution $N\left(\beta, \frac{\sigma^2}{2\alpha}\right)$. Note that the limiting case $\alpha = 0$ corresponds to $\mathfrak{r}(t) = \mathfrak{r}(0) + \sigma W(t)$, which is not a stationary process.

The MATLAB function *SimVasicek* can be used to generate a sample path of a Ornstein-Uhlenbeck process and verify its properties for different values of the parameters.

Mean-Reversion Interpretation

The so-called mean-reversion property of the process (when $\alpha > 0$) is often wrongly interpreted. It does not imply that the process $\mathfrak{r}(t)$ converges to a constant value. It rather be interpreted as the property that the distribution of $\mathfrak{r}(t)$ converges to a stationary distribution. In that sense, speaking of "mean-reversion" is misleading.

5.2.4 Value of Zero-Coupon Bonds under a Vasicek Model

For simplicity, suppose that the face value of the bond at maturity T is \$1 and that the spot rate is expressed as a number. According to Vasicek [1977], the value of a zero-coupon bond is given by (5.4), i.e.,

$$P(t,T) = E\left\{e^{-\int_t^T \tilde{\mathfrak{r}}(s)ds}\right\},\ 0 \leq t \leq T,$$

where $\tilde{\mathfrak{r}}(s)$ satisfies $\tilde{\mathfrak{r}}(t) = r$, and

$$d\tilde{\mathfrak{r}}(s) = [\alpha\{\beta - \tilde{\mathfrak{r}}(s)\} - \sigma q\{s, \tilde{\mathfrak{r}}(s)\}]\, ds + \sigma d\tilde{W}(s),\ t < s \leq T.$$

In practice, the market price of risk is unknown and there is no way to be able to find it. However, if we assume that the process $\tilde{\mathfrak{r}}$ is also a Ornstein-Uhlenbeck process, then we can find the general form of q. It is easy to check that $\tilde{\mathfrak{r}}$ is a Ornstein-Uhlenbeck process if and only if $q(s,r) = q_1 + q_2 r$, for some parameters q_1 and q_2. In this case,

$$d\tilde{\mathfrak{r}}(s) = a\{b - \tilde{\mathfrak{r}}(s)\}ds + \sigma d\tilde{W}(s),\ t < s \leq T, \tag{5.11}$$

where $a = \alpha + q_2\sigma$ and $b = \frac{\alpha\beta - q_1\sigma}{a}$. Note that in Vasicek [1977], the author proposed to use $q(s,r) = q_1$, i.e., he assumed implicitly that $q_2 = 0$.

Remark 5.2.1 *Under a change of measurement and time scales by factors λ and m respectively, that is $\tilde{\mathfrak{r}}^{(\lambda,m)}(s) = \lambda\tilde{\mathfrak{r}}(ms)$, we have $a^{(\lambda,m)} = ma$, $b^{(\lambda,m)} = \lambda b$, $\sigma^{(\lambda,m)} = \sqrt{m}\lambda\sigma$, $q_1^{(\lambda,m)} = \sqrt{m}q_1$ and $q_2^{(\lambda,m)} = \sqrt{m}q_2/\lambda$.*

Under our choice of market price of risk q, and according to (5.6), the explicit solution of (5.11) is

$$\tilde{\mathfrak{r}}(s) = b + e^{-a(s-t)}(r - b) + \sigma \int_t^s e^{-a(s-u)} d\tilde{W}(u), \; 0 \le t \le s \le T.$$

Since $\tilde{\mathfrak{r}}$ is a continuous Gaussian process, it follows from Proposition 5.B.1, $R = \int_t^T \tilde{\mathfrak{r}}(s)ds$ is Gaussian. As a result,

$$P(t,T) = E\left\{e^{-\int_t^T \tilde{\mathfrak{r}}(s)ds}\right\} = E\left(e^{-R}\right) = e^{-E(R)+\frac{1}{2}\mathrm{Var}(R)}.$$

It suffices now to find the mean and variance of R. This is done in the next proposition, proven in Appendix 5.D.2.

5.2.4.1 Vasicek Formula for the Value of a Bond

Proposition 5.2.2 *Suppose that the spot interest rate $\mathfrak{r}$ (expressed as a number) is a Ornstein-Uhlenbeck process with parameters α, β, σ given by (5.6). Then, for a market price of risk of the form $q(r) = q_1 + q_2 r$, we have*

$$P(t,T) = e^{A_\tau - \mathfrak{r}(t)B_\tau}, \tag{5.12}$$

where $\tau = T - t \ge 0$, $a = \alpha + q_2\sigma$, $b = \frac{\alpha\beta - q_1\sigma}{a}$, $B_\tau = \frac{1-e^{-a\tau}}{a}$, and $A_\tau = -\left(b - \frac{\sigma^2}{2a^2}\right)(\tau - B_\tau) - \frac{\sigma^2}{4a}B_\tau^2$.

Remark 5.2.2 *If $a > 0$, then $0 < B_\tau < \tau$ for any $\tau > 0$. In this case, A_τ is an increasing function of q_1. This can be useful when finding adequate starting values for the estimation of the parameters.*

5.2.4.2 Annualized Bond Yields

In practice, instead of the bond value, we often observe the corresponding annual yield $R(t,T)$ expressed in percentage. Also, the most common maturities are 1, 3, 6, and 12 months. For the 30/360 convention, these maturities correspond to 1/12, 1/4, 1/2, and 1 year respectively.

Since $P(t,T) = e^{-(T-t)R(t,T)/100}$, the corresponding spot rate in percentage for a maturity τ (expressed in years) is

$$\mathfrak{r}(t) = r_\theta(t) = \frac{\tau R(t,t+\tau) + 100 \times A_\tau(\tilde{\theta})}{B_\tau(\tilde{\theta})}, \tag{5.13}$$

where $\theta = (\alpha, \beta, \sigma, q_1, q_2)^\top$ are the parameters under the percentage scale, while $\tilde{\theta} = (\alpha, \beta/100, \sigma/100, q_1, 100q_2)^\top$ are the parameters when numbers are used, i.e., under which A_τ and B_τ must be computed. Also,

$$R(t, t+\tau) = \frac{\mathfrak{r}(t) B_\tau(\tilde{\theta}) - 100 \times A_\tau(\tilde{\theta})}{\tau}. \tag{5.14}$$

Long-Term Behavior

Since $B_\tau/\tau \to 0$ and $A_\tau/\tau \to -\left(\tilde{b} - \frac{\tilde{\sigma}^2}{2a^2}\right)$ as $\tau \to \infty$, it follows that

$$R(t, \infty) = \lim_{\tau \to \infty} R(t, t+\tau) = b - \frac{\sigma^2}{200a^2}. \tag{5.15}$$

Consequently, the long-term rate does not depend on the spot rate, a property that annoys some practitioners. On the positive side, one can use formula (5.15) to help finding realistic parameters for the model. For example, if $R(t, \infty) = 35\%$, the choice of b would be obviously incorrect!

The MATLAB function *BondVasicek* can be used to generate an Ornstein-Uhlenbeck process and compute the corresponding values of zero-coupon bonds. As one can check by varying the market price of risk parameters, their effect is not very significant. This is an indication that their estimation might be difficult.

5.2.5 Estimation of the Parameters of the Vasicek Model Using Zero-Coupon Bonds

The estimation of the parameters of the spot rate model, including the parameters of the market price of risk, is complicated since $\mathfrak{r}$ is not observable. However we can observe the bond yields. For the rest of the section, we assume that $\alpha > 0$ and $\beta > 0$, so that the process $\mathfrak{r}$ has a unique stationary distribution. First, a remark is in order on the estimation of the parameters for a Ornstein-Uhlenbeck process.

Remark 5.2.3 (Estimation of Ornstein-Uhlenbeck Processes) *Even if the spot rates $\mathfrak{r}$ were observable, there would be estimation problems for the mean-reversion coefficient α. Suppose for the moment that we use 360 days per year. Since the rates are observed daily, we would get a AR(1) model as given by (5.7), with autocorrelation coefficient $\phi = e^{-\alpha h}$, with $h = 1/360$. It is known, e.g., Example B.7.1, that as $n \to \infty$, $\sqrt{n}\left(\hat{\phi}_n - \phi\right) \rightsquigarrow N(0, 1-\phi^2)$. It then follows from the delta method (Theorem B.3.1) that $\sqrt{n}\,(\hat{\alpha}_n - \alpha)$ is asymptotically Gaussian with mean 0 and variance $\frac{1-\phi^2}{h^2\phi^2}$. Therefore, using k* *years of data, $n = k/h$, and the estimation error corresponding to a 95% confidence level would be approximately*

$$1.96 \times \sqrt{\frac{e^{2\alpha h} - 1}{nh^2}} \approx 1.96 \times \sqrt{\frac{2\alpha}{k} + \frac{2\alpha^2}{n}}.$$

It follows that for 1 year of observations, we would have an error larger that 1.96, assuming for example that $\alpha = 0.5$. Therefore, unless we have many years of data available, we can expect that the error of estimation on α for the Vasicek model will be also large in general. However, the estimation error on $\phi = e^{-\alpha h}$ should be small. Note that the maximum likelihood estimation of the parameters of a Ornstein-Uhlenbeck process is implemented in the MATLAB function EstOU.

5.2.5.1 Measurement and Time Scales

Suppose that we have, for n consecutive days, the annual yields $R_1, \ldots, R_n$ (expressed in percent) of zero-coupon bonds with maturities $\tau_1, \ldots, \tau_n$ expressed in years.

Maturities

It is crucial to have different maturities for the bond yields, for if they all have the same maturity, one could not estimate all parameters. See Remark 5.2.4 later on.

It follows from (5.13) that if the spot rate $\mathfrak{r}$ is modeled by an Ornstein-Uhlenbeck process with parameters $\alpha > 0$, $\beta > 0$, $\sigma > 0$, and if the market price of risk q is $q(t,r) = q_1 + q_2 r$, then setting $r_k = \mathfrak{r}(kh)$, we have

$$r_k = r_k^{(\theta)} = \frac{\tau_k R_k + 100 A_{\tau_k}\left(\tilde{\theta}\right)}{B_{\tau_k}\left(\tilde{\theta}\right)},$$

where $\tilde{\theta} = (\alpha, \beta/100, \sigma/100, q_1, 100q_2)^\top$, with A_τ and B_τ as given in Proposition 5.2.2. For a given choice of θ, $r_k^{(\theta)}$ is called the implied rate. We now described a methodology of estimation proposed by Duan [1994], using the implied rates.

5.2.5.2 Duan Approach for the Estimation of Non Observable Data

Since one cannot observe $r_1, \ldots, r_n$, we might have a problem. Also, we know the joint law of $r_1, \ldots, r_n$, but we do not know the law of the observations $R_1, \ldots, R_n$.

In such cases, Duan [1994, 2000] proposed an ingenious and simple solution. It suffices that there exists an invertible and differentiable mapping[1] of $(r_1, \ldots, r_n)$ to $(R_1, \ldots, R_n)$. This way, we can compute the joint density of $R_1, \ldots, R_n$ if we know the joint density of $r_1, \ldots, r_n$. Then one can use the maximum likelihood principle with the observations $R_1, \ldots, R_n$.

A similar approach was proposed in Chen and Scott [1993] and Fisher and Gilles [1996]. In the latter case, they proposed using maximum likelihood, but they ended up using only a quasi-likelihood method.

To apply Duan methodology, we have to find first the joint density of $r_1, \ldots, r_n$. Since the distribution of r_1 is not known, we will find instead the joint density of $r_2, \ldots, r_n$, given r_1.

5.2.5.3 Joint Conditional Density of the Implied Rates

The following proposition gives us the log-likelihood of an Ornstein-Uhlenbeck process observed at times $h, \ldots, nh$.

Proposition 5.2.3 *If we assume that there are $1/h$ days in a year, the joint conditional density of $r_2, \ldots, r_n$ given r_1, is given by*

$$f(r_2, \ldots, r_n | r_1) = \prod_{k=2}^{n} \frac{e^{-\frac{1}{2}\frac{\{r_k - \beta - \phi(r_{k-1} - \beta)\}^2}{\gamma^2}}}{\gamma\sqrt{2\pi}},$$

where $\phi = e^{-\alpha h}$ and $\gamma^2 = \sigma^2\left(\frac{1-\phi^2}{2\alpha}\right)$.

In particular,

$$-\ln(f) = (n-1)\ln\left(\gamma\sqrt{2\pi}\right) + \frac{1}{2}\sum_{k=2}^{n} \frac{\{r_k - \beta - \phi(r_{k-1} - \beta)\}^2}{\gamma^2}.$$

PROOF. According to Proposition 5.2.1, the conditional distribution of r_k given $r_1, \ldots, r_{k-1}$ is Gaussian, with mean $\beta + \phi(r_{k-1} - \beta)$ and variance $\gamma^2 = \sigma^2\left(\frac{1-\phi^2}{2\alpha}\right)$. Hence the conditional density of r_k given $r_1, \ldots, r_{k-1}$ is

$$(\gamma\sqrt{2\pi})^{-1} e^{-\frac{1}{2}\frac{\{r_k - \beta - \phi(r_{k-1} - \beta)\}^2}{\gamma^2}},$$

which only depends on r_{k-1}, by Proposition 5.2.1. The result is then a simple application of the multiplication formula (A.20) stating that the joint density is the product of these conditional densities.

■

[1] Often called a diffeomorphism in Mathematics.

5.2.5.4 Change of Variables Formula

The other ingredient in Duan methodology is the change of variables formula in Proposition A.8.1, stating that the density of $Y = T(X)$ at point $y = T(x) \in \mathbb{R}^n$, is given by

$$g_Y(y) = \frac{f_X(x)}{|J(x)|},$$

where J is the Jacobian determinant of the mapping T, defined by $J(x) = \det\{H(x)\}$, and H is the Jacobian matrix defined by

$$H_{jk}(x) = \frac{\partial T_j(x)}{\partial_{x_k}}, \quad j,k \in \{1\ldots,n\}.$$

The main idea in Duan [1994, 2000] is to apply the change of variables formula in the following case:

- We have observations $y_1,\ldots,y_n$ of $Y = (Y_1,\ldots,Y_n)$ but the density of Y is not given.
- The density f_θ of X is known (up to parameter θ), but X cannot be observed.
- There exists a nice mapping (invertible and differentiable) of the form $Y = T_\theta(X)$ between Y and X.
- Set $x = T_\theta^{-1}(y)$. Then the density g_θ of Y at point $y = (y_1,\ldots,y_n)$ is given by $g_\theta(y) = \frac{f_\theta(x)}{|J_\theta(x)|}$.

  In particular, if $T_\theta = (T_{\theta,1},\ldots,T_{\theta,n})$, with $T_{\theta,k}$ only depending on x_k, and if $\frac{\partial}{\partial x_k}T_{\theta,k}(x_k) = T'_{\theta,k}(x_k) \neq 0$, then

$$\ln\{|J_\theta(x)|\} = \sum_{k=1}^n \ln\left\{\left|T'_{\theta,k}(x_k)\right|\right\}.$$

5.2.5.5 Application of the Change of Variable Formula to the Vasicek Model

In our setting, the mapping is given by

$$r_k = \frac{\tau_k R_k + 100 A_{\tau_k}(\tilde{\theta})}{B_{\tau_k}(\tilde{\theta})}, \quad k \in \{1,\ldots,n\}.$$

Applying Duan approach to $X = (r_1,\ldots,r_n)$ and $Y = (R_1,\ldots,R_n)$ yields

$$\frac{dR_k}{dr_k} = \frac{B_{\tau_k}(\tilde{\theta})}{\tau_k} = \frac{1 - e^{-a\tau_k}}{a\tau_k} \neq 0,$$

from which we can compute the Jacobian determinant of the transformation.

Before we give the joint density of $R_1, \ldots, R_n$, one needs to talk again about rounding errors.

Rounding Errors

Even if the expression $\frac{1-e^{-a\tau}}{a}$ makes sense for any $a \in \mathbb{R}$, with limiting value τ as $a \to 0$, MATLAB or any other package will not calculate the correct value when a is very small. As a result, when computing such expressions, it is better to define it by part, one part for values near 0 and the other part for other values. For example, to compute $B(a) = \frac{1-e^{-a\tau}}{a}$ correctly for small values of a, one could replace B by $\tilde{B}$, where

$$\tilde{B}(a) = \begin{cases} \tau + ac\tau^2/2, & \text{if } a \in [0, a_0/\tau], \\ B(a), & \text{if } a \geq a_0/\tau, \end{cases}$$

where $c = 2\left(\frac{1-a_0-e^{-a_0}}{a_0^2}\right)$. Here one could take $a_0 = 10^{-10}$.

Using Proposition 5.2.3 and Proposition A.8.1 from Appendix A.8, we obtain the following proposition.

Proposition 5.2.4 *Set $\theta = (\alpha, \beta, \sigma, q_1, q_2)^\top$. The conditional log-likelihood $L(\theta)$ of $R_2, \ldots, R_n$ given R_1, is*

$$\begin{aligned} -L(\theta) &= -L(\theta; R_1, \ldots, R_n) \\ &= \frac{1}{2\gamma^2} \sum_{k=2}^{n} \left\{ r_k^{(\theta)} - \beta - \phi\left(r_{k-1}^{(\theta)} - \beta\right)\right\}^2 \\ &\quad + \sum_{k=2}^{n} \ln\left\{\frac{B_{\tau_k}(\tilde{\theta})}{\tau_k}\right\} + (n-1)\ln(\gamma\sqrt{2\pi}), \end{aligned}$$

where $\phi = e^{-\alpha h}$ and $\gamma = \sigma\sqrt{\frac{1-\phi^2}{2\alpha}}$.

Remark 5.2.4 *Setting $v_k = \gamma\frac{B_{\tau_k}(\tilde{\theta})}{\tau_k}$, and $\mu_k = \frac{\beta B_{\tau_k}(\tilde{\theta}) - A_{\tau_k}(\tilde{\theta})}{\tau_k}$, the log-likelihood can be expressed as*

$$-L(\theta) = \frac{1}{2}\sum_{k=2}^{n} \frac{\{R_k - \mu_k - \phi(R_{k-1} - \mu_k)\}^2}{v_k^2} + \sum_{k=2}^{n} \ln\left(v_k\sqrt{2\pi}\right).$$

In particular, if we had chosen the same maturity for all bonds, that is $\tau_k = \tau$, then $\mu_k = \mu = \frac{(\beta B_\tau - A)}{\tau}$ and $v_k = v = \gamma\frac{|B_\tau|}{\tau}$, $k \in \{2, \ldots, n\}$. As a result, using the maximum likelihood, we would have been able to estimate $\alpha = -\frac{1}{h}\ln(\phi)$, μ and v, and there would have been an infinite number of solutions for β, σ,

q_1 and q_2 since we have two equations $\beta B - A = \tau\mu$ and $\gamma B = \tau v$, with four unknowns. To identify all parameters, it is thus necessary to have different maturities. Just to be sure, we should choose 3 maturities. For example, we could take the first third of the returns with the same maturity τ_1, the second third with a maturity $\tau_2 \neq \tau_1$, and the last third with maturity $\tau_3 \neq \tau_1, \tau_2$. In practice, we often have 4 maturities: 1,3,6, and 12 months so it would make sense to use all of them, dividing roughly the sample into four parts.

Finally, to estimate $\theta = (\alpha, \beta, \sigma, q_1, q_2)^\top$, we just have to minimize $-L(\theta)$ as given in Proposition 5.2.4.

Positiveness

Recall that you have to take into account the positiveness of the parameters α, β, and σ.

5.2.5.6 Precision of the Estimation

The results of Berndt et al. [1974] can also be applied to Markov processes, as in the present case. Consequently, we obtain the following result for the asymptotic behavior of the estimation error.

Proposition 5.2.5 *If $\hat{\theta}_n$ is the estimation of θ using the maximum likelihood principle, then*

$$\sqrt{n}(\hat{\theta}_n - \theta) \rightsquigarrow N_5(0, \hat{V}),$$

where $\hat{V}$ is the Moore-Penrose inverse of the estimation $\mathcal{I}$ of the Fisher information.

As discussed in Appendix B.5.1, $\mathcal{I}$ can be estimated using the Hessian matrix H, in which case $\hat{\mathcal{I}} = H/n$, or it can be estimated as the covariance matrix of the numerical gradients.

Implied Spot Rate

For the Black-Scholes model, the implied volatilities for different strike prices or maturities are different in general. A similar phenomenon is observed here. For different maturities, the implied spot rate will be different, when computed from market values on a given day. To avoid these inconsistencies, it is therefore recommended to use only one yield per calendar date.

Example 5.2.1 *To illustrate the methodology, consider the simulated data in the MATLAB file* DataVasicek *representing annualized rates expressed* *in percentage. They have been generated with the parameters* $\theta = (0.5, 2.55, 0.365, 0.3, 0)^\top$. *Here* $h = 1/360$, *and the sample consists of 4 years of daily prices for bonds with maturities of 1, 3, 6, and 12 months.*

Table 5.1 contains the confidence intervals for the estimation of the parameters, using the MATLAB function EstVasicek. *The starting parameters were chosen to correspond to those of a Ornstein-Uhlenbeck process for* R, *i.e.,* $\alpha_0 = -\frac{\ln(\phi_0)}{h}$, $\beta_0 = \bar{R}$, $\sigma_0 = \gamma_0\sqrt{2\alpha_0/(1-\phi_0^2)}$, *where* ϕ_0 *is the first order autocorrelation and* γ_0 *is the standard deviation of* R. *Almost all intervals cover the true value of each parameter. If we would have used only one year of data, the results would not have been so good. Note that even for large samples, the precision of the estimation of the parameters of the market price of risk is not that good. As predicted, the estimation of* α *is not as precise as the estimation of* β *and* σ.

TABLE 5.1: 95% confidence intervals for the estimation of the parameters of the Vasicek model. Here $\phi = e^{-\alpha h}$, with $h = 1/360$.

Parameter	True Value	Confidence Interval
α	0.5	1.6606 ± 1.1593
β	2.55	2.7107 ± 0.2484
σ	0.365	0.3636 ± 0.0219
q_1	0.3	9.6920 ± 9.5092
q_2	0	-3.2835 ± 3.2782
ϕ	0.9986	0.9954 ± 0.0032

Since we generated the data, we can compare the real spot rates with the implied ones. This is illustrated in Figure 5.1. Both series are very close to each other.

One can verify that for other starting points yielding realistic values for the implied rates series, the estimations of the parameters α, β, σ *are quite stable. For example, taking* $\alpha_0 = 2$, $\beta_0 = \bar{R}$, *and* $\sigma_0 = 1$ *yields almost the same results.*

Remark 5.2.5 *Parameters* q_1 *and* q_2 *seem to be difficult to estimate. However, as mentioned before, the effect of* q_2 *on the value of the bonds seems negligible. It would be preferable to ignore it when its confidence interval contains* 0, *and then re-estimate the remaining four parameters.*

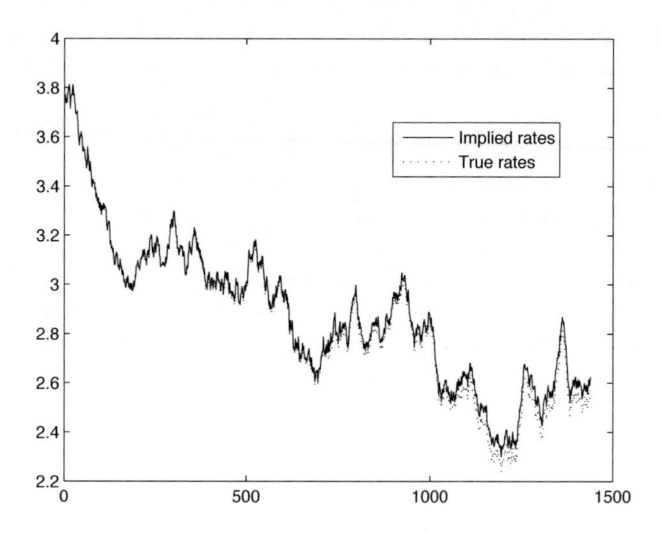

FIGURE 5.1: Implied and true spot rates for the Vasicek model with $\alpha = 0.5$, $\beta = 2.55$, $\sigma = 0.365$, $q_1 = 0.3$ and $q_2 = 0$.

5.3 Cox-Ingersoll-Ross (CIR) Model

The CIR model, introduced in Cox et al. [1985] for modeling the spot rate, is defined by the stochastic differential equation

$$d\mathfrak{r}(t) = \alpha\left\{\beta - \mathfrak{r}(t)\right\}dt + \sigma\sqrt{\mathfrak{r}(t)}dW(t), \quad \mathfrak{r}(0) = r > 0, \tag{5.16}$$

where $\alpha > 0$, $\beta > 0$, and $\sigma > 0$. This stochastic process, known as the Feller process in the Probability literature, was first studied in Feller [1951]. The condition $\alpha\beta > 0$ ensures that the process is non-negative.

5.3.1 Representation of the Feller Process

Contrary to the Ornstein-Uhlenbeck process, there is no real explicit solution to (5.16). See Remark 5.3.1. However, for some values of β, one can construct a stochastic process having the same law and solving a similar equation. In fact, for $i \in \{1, \ldots, n\}$, set

$$X_i(t) = \frac{\sigma}{2}\int_0^t e^{-\frac{\alpha}{2}(t-s)}dW^{(i)}(s) + e^{-\frac{\alpha}{2}t}\sqrt{\frac{r}{n}},$$

where $W^{(1)}, \ldots, W^{(n)}$ are independent Brownian motions. Using representation (5.6), the processes $X_1, \ldots, X_n$ are independent Ornstein-Uhlenbeck processes satisfying $X_i(0) = \sqrt{\frac{r}{n}}$, and

$$dX_i(t) = -\frac{\alpha}{2}X_i(t)dt + \frac{\sigma}{2}dW^{(i)}(t), \quad 1 \le i \le n.$$

Then, set

$$\mathfrak{r}(t) = \sum_{i=1}^{n} X_i^2(t), \quad t \ge 0. \tag{5.17}$$

One can show that $\mathfrak{r}(0) = r$, and

$$d\mathfrak{r}(t) = \alpha \left\{ n\frac{\sigma^2}{4\alpha} - \mathfrak{r}(t) \right\} dt + \sigma\sqrt{\mathfrak{r}(t)}d\hat{W}(t),$$

where $\hat{W}$ is a Brownian motion defined by

$$\hat{W}(t) = \sum_{i=1}^{n} \int_0^t \frac{X_i(s)}{\mathfrak{r}(s)} dW^{(i)}(s).$$

Therefore (5.17) is an explicit representation of the solution to (5.16) when $\beta = n\frac{\sigma^2}{4\alpha}$, i.e., when the parameter ν, defined by $\nu = 4\frac{\alpha\beta}{\sigma^2}$ is equal to n.

Remark 5.3.1 *The process* $\mathfrak{r}$ *defined by* (5.17) *is not really the solution to* (5.16), *since the solution is not expressed in terms of the given Brownian motion W but instead, it is expressed in terms of a new Brownian motion $\hat{W}$. Nevertheless, it means that the distribution of both processes are the same. In Kouritzin and Rémillard [2002], it is shown that a solution can be constructed on a random time interval under the condition $\nu = 4\frac{\alpha\beta}{\sigma^2} = 1$.*

According to equation (5.17), we see that for a given t, $\mathfrak{r}(t)$ is a sum of the squares of n independent Gaussian random variables with mean $e^{-\frac{\alpha}{2}t}\sqrt{\frac{\mathfrak{r}(0)}{n}}$ and variance $\omega_t = \sigma^2\left(\frac{1-e^{-\alpha t}}{4\alpha}\right)$. Hence, using equation (A.11) in Section A.6.12, we obtain that the conditional distribution of $\mathfrak{r}(t)/\omega_t$, given $\mathfrak{r}(0) = r$, is a non-central chi-square variable with n degrees of freedom, and with non-centrality parameter $D_t = \frac{1}{\omega_t}\sum_{i=1}^{n}\left(e^{-\frac{\alpha}{2}t}\sqrt{\frac{r}{n}}\right)^2 = \frac{e^{-\alpha t}}{\omega_t}r$. It is then natural to guess what is the conditional distribution of $\mathfrak{r}(t)/\omega_t$ given $\mathfrak{r}(0) = r$ in the general case, i.e., $\nu \neq n$. The next proposition shows that the guess is correct.

Proposition 5.3.1 *The conditional distribution of $\mathfrak{r}(t)/\omega_t$ given $\mathfrak{r}(0) = r$, is a non-central chi-square distribution with $\nu = 4\frac{\alpha\beta}{\sigma^2}$ degrees of freedom, and non-centrality parameter $D_t = \frac{e^{-\alpha t}}{\omega_t}r$, where $\omega_t = \sigma^2\left(\frac{1-e^{-\alpha t}}{4\alpha}\right)$.*

Moreover, the moment generating function $M(u)$ of $\mathfrak{r}(t)$, given $\mathfrak{r}(0) = r$, exists for any $u < \frac{1}{2\omega_t}$, and is given by

$$M(u) = E\left\{ e^{u\mathfrak{r}(t)} \middle| \mathfrak{r}(0) = r \right\} = \frac{e^{\frac{u\omega_t D_t}{1-2u\omega_t}}}{(1-2u\omega_t)^{\nu/2}}. \tag{5.18}$$

The proof is given in Appendix 5.D.3.

Remark 5.3.2 *Since the coefficients of the stochastic differential equation (5.16) are time-independent, it follows that the Feller process is a homogeneous Markov process, i.e., the conditional law of r_{t+s} given $\mathfrak{r}(s) = r$ and all the information on the trajectory $\mathfrak{r}(u)$, $u \in [0, s]$, depends only on r and t. Moreover, it is the same law as the law of $\mathfrak{r}(t)$ given $\mathfrak{r}(0) = r$. These properties are stated in the next proposition.*

5.3.1.1 Properties of the Feller Process

Proposition 5.3.2 *For $s, t > 0$, the conditional distribution of r_{t+s}/ω_t, given the trajectory $\mathfrak{r}(u)$, $u \in [0, s]$, is a non-central chi-square distribution with parameters $\nu = 4\frac{\alpha\beta}{\sigma^2}$ and $D_t = \mathfrak{r}(s)\frac{e^{-\alpha t}}{\omega_t}$, where $\omega_t = \sigma^2\left(\frac{1-e^{-\alpha t}}{4\alpha}\right)$. Moreover, the cumulants of $\mathfrak{r}(t)$ given $\mathfrak{r}(0) = r$, are*

$$\kappa_n = (2\omega_t)^{n-1}(n-1)!\left\{\beta\left(1 - e^{-\alpha t}\right) + nre^{-\alpha t}\right\}, \quad n \geq 1.$$

In particular, $\kappa_1 = E\{\mathfrak{r}(t)|\mathfrak{r}(0) = r\} = \beta + e^{-\alpha t}(r - \beta)$, and $\kappa_2 = \mathrm{Var}\{\mathfrak{r}(t)|\mathfrak{r}(0) = r\} = \frac{\sigma^2\beta}{2\alpha}\left(1 - e^{-\alpha t}\right)^2 + \frac{\sigma^2 r}{\alpha}e^{-\alpha t}\left(1 - e^{-\alpha t}\right)$.

The proof is given in Appendix 5.D.4.

Remark 5.3.3 *It follows from Proposition 5.3.2 and the Markov property of $\mathfrak{r}$ that for any $s, t \geq 0$,*

$$\begin{aligned}
E\{\mathfrak{r}(t)\mathfrak{r}(t+s)\} &= E[\mathfrak{r}(t)E\{\mathfrak{r}(t+s)|\mathfrak{r}(t)\}] \\
&= E\left[\mathfrak{r}(t)\left\{\beta(1 - e^{-\alpha s}) + e^{-\alpha s}\mathfrak{r}(t)\right\}\right] \\
&= \beta(1 - e^{-\alpha s})E\{\mathfrak{r}(t)\} + e^{-\alpha s}E\left\{\mathfrak{r}(t)^2\right\}.
\end{aligned}$$

In addition,

$$\begin{aligned}
E\{\mathfrak{r}(t)\}E\{\mathfrak{r}(t+s)\} &= E\{\mathfrak{r}(t)\}E[E\{\mathfrak{r}(t+s)|\mathfrak{r}(t)\}] \\
&= E\{\mathfrak{r}(t)\}E\left\{\beta(1 - e^{-\alpha s}) + e^{-\alpha s}\mathfrak{r}(t)\right\} \\
&= \beta(1 - e^{-\alpha s})E\{\mathfrak{r}(t)\} + e^{-\alpha s}E^2\{\mathfrak{r}(t)\}.
\end{aligned}$$

Hence

$$\mathrm{Cov}\{\mathfrak{r}(t), \mathfrak{r}(t+s)\} = e^{-\alpha s}\,\mathrm{Var}\{\mathfrak{r}(t)\}.$$

In particular, if $\alpha > 0$, then $E\{\mathfrak{r}(t)\} \stackrel{t\to\infty}{\longrightarrow} \beta$, $\mathrm{Var}\{\mathfrak{r}(t)\} \stackrel{t\to\infty}{\longrightarrow} \frac{\sigma^2\beta}{2\alpha}$, and $\mathrm{Cov}\{\mathfrak{r}(t), \mathfrak{r}(s)\} \stackrel{t\to\infty}{\longrightarrow} 0$, for any fixed $s \geq 0$. Therefore, as it was the case for the Ornstein-Uhlenbeck process, β can be interpreted as the long-run mean, and α is the mean-reversion coefficient, determining the speed of convergence of the mean to β.

Moreover, since $\omega_t \to \omega_\infty = \frac{\sigma^2}{4\alpha}$ *and* $D_t \to 0$, *as* $t \to \infty$, *we deduce from* (5.18) *that for any* $u < \frac{1}{2}$,

$$\lim_{t\to\infty} E\left\{ e^{u\frac{\mathfrak{r}(t)}{\omega_t}} \middle| \mathfrak{r}(0) = r \right\} = \frac{1}{(1-2u)^{\nu/2}},$$

which is the moment generating function of a chi-square random variable with ν *degrees of freedom. Using* (5.18) *and assuming that* $\mathfrak{r}(0)/\omega_\infty \sim \chi^2(\nu)$, *we see that* $\mathfrak{r}(t)/\omega_\infty \sim \chi^2(\nu)$, *for every* $t \geq 0$. *It shows that the* Gamma $\left(\frac{\nu}{2}, \frac{\sigma^2}{2\alpha}\right)$ *distribution is the unique stationary distribution of the process.*

The MATLAB function *SimCIR* can be used to generate a Feller process and verify some of its properties by varying the parameters.

5.3.1.2 Measurement and Time Scales

Again, we have to be careful with the measurement and time scales. In fact, setting $\mathfrak{r}^{(\lambda,m)}(t) = \lambda\mathfrak{r}(mt)$, it follows that $\mathfrak{r}^{(\lambda,m)}$ is also a Feller process, with parameters $\alpha^{(\lambda,m)} = m\alpha$, $\beta^{(\lambda,m)} = \lambda\beta$, and $\sigma^{(\lambda,m)} = \sqrt{\lambda m}\,\sigma$.

5.3.2 Value of Zero-Coupon Bonds under a CIR Model

For simplicity, assume that the face value of the bond is \$1. According to Vasicek [1977], the value of a zero-coupon bond is given by (5.4), i.e.,

$$P(t,T) = E\left\{ e^{-\int_t^T \tilde{\mathfrak{r}}(s)ds} \right\}, \; 0 \leq t \leq T,$$

where $\tilde{\mathfrak{r}}$ satisfies $\tilde{\mathfrak{r}}(t) = r$, and

$$d\tilde{\mathfrak{r}}(s) = \left[\alpha\{\beta - \tilde{\mathfrak{r}}(s)\} - \sigma\sqrt{\tilde{\mathfrak{r}}(s)}\, q\{s, \tilde{\mathfrak{r}}(s)\}\right] ds + \sigma\sqrt{\tilde{\mathfrak{r}}(s)} d\tilde{W}(s), \; t < s \leq T,$$

for some market price of risk q.

Again, in order to be able to do computations, we suppose that q is such that $\tilde{\mathfrak{r}}$ remains a Feller process under the equivalent martingale measure. It thus follows that $q(s,r) = \frac{q_1}{\sqrt{r}} + q_2\sqrt{r}$, for some parameters q_1, q_2. In this case, we have

$$d\tilde{\mathfrak{r}}(s) = a\{b - \tilde{\mathfrak{r}}(s)\}ds + \sigma\sqrt{\tilde{\mathfrak{r}}(s)} d\tilde{W}(s), \; t < s \leq T,$$

where $a = \alpha + q_2\sigma > 0$ and $b = \frac{\alpha\beta - q_1\sigma}{a} > 0$.

Note also that under a change of measurement scale, i.e., if $\tilde{\mathfrak{r}}^{(\lambda)}(t) = \lambda\tilde{\mathfrak{r}}(t)$, then $\tilde{\mathfrak{r}}^{(\lambda)}$ is a Feller process with parameters $a^{(\lambda)} = a$, $b^{(\lambda)} = \lambda b$ and $\sigma^{(\lambda)} = \sqrt{\lambda}\sigma$. This shows that $q_1^{(\lambda)} = \sqrt{\lambda} q_1$ and $q_2^{(\lambda)} = q_2/\sqrt{\lambda}$. This is equivalent to $q^{(\lambda)}(s, \lambda r) = q(s,r)$ for every $s, r \geq 0$.

To compute the value of the bond, we need to solve (5.26) with $V(r) = -r$ and $\Phi \equiv 1$.

5.3.2.1 Formula for the Value of a Zero-Coupon Bond under the CIR Model

Proposition 5.3.3 *Suppose that the spot interest rate* $\mathfrak{r}$ *(expressed as a number) is a Feller process with parameters* α, β, σ *given by (5.16). Then, for a market price of risk of the form* $q(r) = \frac{q_1}{\sqrt{r}} + q_2\sqrt{r}$*, we have*

$$P(t,T) = e^{A_\tau - \mathfrak{r}(t)B_\tau}, \tag{5.19}$$

where $\tau = T - t \geq 0$*,* $a = \alpha + q_2\sigma$*,* $b = \frac{\alpha\beta - q_1\sigma}{a}$*,* $\gamma = \sqrt{a^2 + 2\sigma^2}$*,* $B_\tau = \frac{2(1-e^{-\gamma\tau})}{(\gamma+a)(1-e^{-\gamma\tau})+2\gamma e^{-\gamma\tau}}$*, and*

$$A_\tau = \frac{2ab}{\sigma^2}\ln\left\{\frac{2\gamma e^{-\frac{(\gamma-a)}{2}\tau}}{(\gamma+a)(1-e^{-\gamma\tau})+2\gamma e^{-\gamma\tau}}\right\}.$$

The proof is given in Appendix 5.D.5.

By working a little bit harder, we can get a more general result. The proof is based again on the Feynmann-Kac formula (5.26); see Lamberton and Lapeyre [2008] for details.

Proposition 5.3.4 *If* $\tilde{\mathfrak{r}}(t) = r$*, then for any* $u \geq 0$ *and* $\tau = T - t \geq 0$*,*

$$E\left\{e^{-u\tilde{\mathfrak{r}}(T) - \int_t^T \tilde{\mathfrak{r}}(s)ds}\right\} = e^{\phi_{u,\tau} - r\psi_{u,\tau}}, \tag{5.20}$$

where

$$\phi_{u,\tau} = \frac{2ab}{\sigma^2}\ln\left\{\frac{2\gamma e^{-(\gamma-a)\tau/2}}{u\sigma^2(1-e^{-\gamma\tau}) + (\gamma+a)(1-e^{-\gamma\tau}) + 2\gamma e^{-\gamma\tau}}\right\}$$

and

$$\psi_{u,\tau} = \frac{u\{(\gamma-a)(1-e^{-\gamma\tau}) + 2\gamma e^{-\gamma\tau}\} + 2(1-e^{-\gamma\tau})}{u\sigma^2(1-e^{-\gamma\tau}) + (\gamma+a)(1-e^{-\gamma\tau}) + 2\gamma e^{-\gamma\tau}}.$$

In particular, $A_\tau = \phi_{0,\tau}$ *and* $B_\tau = \psi_{0,\tau}$.

Remark 5.3.4 *For any* $\tau > 0$*, we have* $A_\tau < 0$*. In fact, in the limiting case* $r = 0$*,* $e^{A_\tau} = P(0,\tau) < 1$*. Also,* A_τ *is an increasing function of* q_1.

Measurement Scale

As usual, one must be careful with the scale of measurement. The computation of A and B only makes sense if the process $\mathfrak{r}$ is a number, with the same time scale as the maturities.

5.3.2.2 Annualized Bond Yields

As mentioned for the Vasicek model, in practice, instead of the bond value, we often work with the corresponding annualized yield $R(t,T)$ expressed in percentage.

Then, $P(t,T) = e^{-(T-t)R(t,T)/100}$, so that we have basically the same relationship as for the Vasicek model, with different functions A and B, i.e., the corresponding spot rate in percentage for a maturity τ (expressed in years) is

$$\mathfrak{r}(t) = \mathfrak{r}_\theta(t) = \frac{\tau R(t, t+\tau) + 100 \times A_\tau(\tilde{\theta})}{B_\tau(\tilde{\theta})}, \tag{5.21}$$

where $\theta = (\alpha, \beta, \sigma, q_1, q_2)^\top$ are the parameters under the percentage scale, while $\tilde{\theta} = (\alpha, \beta/100, \sigma/10, q_1/10, 10q_2)^\top$ are the parameters when numbers are used, i.e., under which A_τ and B_τ must be computed. In addition,

$$R(t, t+\tau) = \frac{\mathfrak{r}(t)B_\tau(\tilde{\theta}) - 100A_\tau(\tilde{\theta})}{\tau}. \tag{5.22}$$

Positiveness Constraints

In the CIR model, the rates $\mathfrak{r}(t)$ are positive, so not all parameters $\theta = (\alpha, \beta, \sigma, q_1, q_2)^\top$ are admissible.

Since $B_\tau/\tau \to 0$ and $\frac{A_\tau}{\tau} \to -\frac{\tilde{a}\tilde{b}(\tilde{\gamma}-\tilde{a})}{\tilde{\sigma}^2}$, as $\tau \to \infty$, where $\tilde{\gamma} = \sqrt{a^2 + 2\sigma^2/100}$, it follows that

$$R(t,\infty) = \lim_{\tau\to\infty} R(t, t+\tau) = 100\frac{ab(\tilde{\gamma} - a)}{\sigma^2} = \frac{2ab}{a + \tilde{\gamma}}. \tag{5.23}$$

As a result, the long-term rate does not depend on the actual spot rate, as it was observed in the Vasicek model.

The MATLAB function *BondCIR* can be used to generate a Feller process and compute the values of zero-coupon bonds. As one can check by varying the market price of risk parameters, they are not very significant. This indicates that their estimation might be difficult, as in the case of the Vasicek model.

5.3.2.3 Value of a Call Option on a Zero-Coupon Bond

Cox et al. [1985] showed that the value at time t of a European or American call option with maturity T_e, on a zero-coupon bond with maturity $T \geq T_e$, is

$$C(r,t,T_e,T,K)$$
$$= P(r,t,T)F\left\{2r^*(\tilde{\phi}+\tilde{\psi}+B_{T-T_e});\nu,\frac{2\tilde{\phi}^2 re^{\gamma(T_e-t)}}{\tilde{\phi}+\tilde{\psi}+B_{T-T_e}}\right\}$$
$$- KP(r,t,T_e)F\left\{2r^*(\tilde{\phi}+\tilde{\psi});\nu,\frac{2\tilde{\phi}^2 re^{\gamma(T_e-t)}}{\tilde{\phi}+\tilde{\psi}}\right\},$$

where K is the strike price, $\mathfrak{r}(t)=r$, $r^*=\frac{A_{T-T_e}-\ln(K)}{B_{T-T_e}}$, that is $P(r^*,T_e,T)=K$, $\tilde{\phi}=\frac{2\gamma}{\sigma^2\{e^{\gamma(T_e-t)}-1\}}$, $\tilde{\psi}=\frac{\gamma+a}{\sigma^2}$, and $F(x;\nu,D)$ is the distribution function of a non-central chi-square random variable with parameters ν and D.[2]

5.3.2.4 Put-Call Parity

Let $V(r,t,T_e,T,K)$ be the value at time t, given $\mathfrak{r}(t)=r$, of a European put option with maturity T_e, on the same bond as for the call option. Then we have the following put-call parity result.

Proposition 5.3.5

$$C(r,t,T_e,T,K)-V(r,t,T_e,T,K)=e^{A^*_{T-T_e,T_e-t}-rB^*_{T-T_e,T_e-t}}-KP(t,T_e),$$

where $A^*_{u,\tau}=A_u+\phi_{B_u,\tau}$, $B^*_{u,\tau}=\psi_{B_u,\tau}$, *with* $\phi_{u,\tau}$ *and* $\psi_{u,\tau}$ *defined in Proposition 5.3.4.*

PROOF. Put-call parity results are usually based on the identity: $\max(x-K,0)-\max(K-x,0)=x-K$. Applying the latter to $x=P(T_e,T)=e^{A_{T-T_e}-r_{T_e}B_{T-T_e}}$, one obtains that

$$C-V=E\left\{e^{-\int_t^{T_e}\tilde{\mathfrak{r}}(s)ds}\left(e^{A_{T-T_e}-\tilde{\mathfrak{r}}_{T_e}B_{T-T_e}}-K\right)\right\}.$$

To complete the proof, just use formula (5.20).

■

5.3.3 Parameters Estimation of the CIR Model Using Zero-Coupon Bonds

In this section we follow the same approach as in the Vasicek model. For the rest of the section, we assume that $\alpha>0$ and $\beta>0$, so that the process $\mathfrak{r}$ has a unique stationary distribution.

[2]This distribution function is available in the Statistics toolbox of MATLAB under the name *ncx2cdf*.

5.3.3.1 Measurement and Time Scales

Suppose that we have, for n consecutive days, the annualized yields (expressed in percentage) $R_1, \ldots, R_n$ of zero-coupon bonds with maturities $\tau_1, \ldots, \tau_n$ expressed in years. Also define $r_k = \mathfrak{r}(kh)$, $k \in \{1, \ldots, n\}$.

Maturities

Here again it is crucial to have different maturities for the returns, for if they all have the same maturity, one could not estimate all parameters.

It follows from the relationship (5.21) that if the spot rate $\mathfrak{r}$ is modeled by a Feller process with parameters $\alpha > 0$, $\beta > 0$, $\sigma > 0$, and the market price of risk is $\frac{q_1}{\sqrt{t}} + q_2\sqrt{r}$, then

$$r_k = r_k^{(\theta)} = \frac{\tau_k R_k + 100 A_{\tau_k}(\tilde{\theta})}{B_{\tau_k}(\tilde{\theta})}, \quad k \in \{1, \ldots, n\}.$$

with $\tilde{\theta} = (\alpha, \beta/100, \sigma/10, q_1/10, 10q_2)^\top$, where A_{τ_k} and B_{τ_k} are given in Proposition 5.3.3. For the estimation of the parameters, we will also use Duan methodology with the annualized bond yields $R_1, \ldots, R_n$. The first step is to find the conditional density of $r_2, \ldots, r_n$, given r_1.

5.3.3.2 Joint Conditional Density of the Implied Rates

Proposition 5.3.6 *For a Feller process $\mathfrak{r}$ with parameters α, β, σ, if we assume that there are $1/h$ days in a year, the joint conditional density of $r_2, \ldots, r_n$ given r_1, is*

$$f(r_2, \ldots, r_n | r_1) = \prod_{k=2}^{n} \left\{ \frac{f\left(\frac{r_k}{\omega}; \nu, \phi \frac{r_{k-1}}{\omega}\right)}{\omega} \right\},$$

where $f(x; \nu, \mu)$ is the density of a non-central chi-square non-central distribution with ν degrees of freedom and non-centrality parameter μ, $\phi = e^{-\alpha h}$, $\omega = \sigma^2 \left(\frac{1-\phi}{4\alpha}\right)$, and $\nu = \frac{4\alpha\beta}{\sigma^2}$. In particular, the log-likelihood L is given by

$$L = \sum_{k=2}^{n} \ln \left\{ f\left(\frac{r_k}{\omega}; \nu, \phi \frac{r_{k-1}}{\omega}\right) \right\} - (n-1)\ln(\omega).$$

It happens often that the value of the density f is so small that the logarithm cannot be computed with MATLAB or any other package. One trick is to replace f by $\tilde{f} = \max\left(f, 10^{-15}\right)$, for example.

PROOF. The result is an application of the multiplication formula (A.20) together with Proposition 5.3.2. In fact, the conditional density of r_k given $r_1, \ldots, r_{k-1}$ is

$$\frac{f\left(\frac{r_k}{\omega}; \nu, \phi\frac{r_{k-1}}{\omega}\right)}{\omega},$$

so the joint density is the product of these conditional densities.

■

Remark 5.3.5 *It follows from Remark 5.3.3 that for observations $r_1, r_2, \ldots, r_n$, we have, for k large enough,*

$$E(r_k) \approx \beta, \quad \mathrm{Var}(r_k) \approx \sigma^2 \frac{\beta}{2\alpha}, \quad \textit{and } \mathrm{Cor}(r_k, r_{k-1}) \approx \phi = e^{-\alpha h}.$$

These values could serve as starting points for the estimation of the parameters of a Feller process. This approach is implemented in the MATLAB function EstFeller. *The precision of the estimation is usually satisfactory for all parameters but α, even if the estimation error on ϕ is small. Note that we had a similar problem with the Ornstein-Uhlenbeck process.*

5.3.3.3 Application of the Change of Variable Formula for the CIR Model

In our setting, we have

$$r_k = r_k^{(\theta)} = \frac{\tau_k R_k + 100 A_{\tau_k}(\tilde{\theta})}{B_{\tau_k}(\tilde{\theta})}, \quad k \in \{1, \ldots, n\}.$$

We then apply Duan methodology to $X = \left(r_1^{(\theta)}, \ldots, r_n^{(\theta)}\right)$ and $Y = (R_1, \ldots, R_n)$. As before, we find a similar expression as in the Vasicek model, i.e., $\frac{dR_k}{dr_k} = \frac{B_{\tau_k}(\tilde{\theta})}{\tau_k} > 0$. The only difference is that the formula for B is not the same. Using Proposition 5.3.6 and formula (A.21) in Appendix A.8, we obtain the following result.

Proposition 5.3.7 *The conditional log-likelihood of $R_2, \ldots, R_n$ given R_1, is*

$$\begin{aligned} L(\theta) &= L(\theta; R_1, \ldots, R_n) \\ &= \sum_{k=}^{n} \ln\left\{ f\left(\frac{r_k^{(\theta)}}{\omega}; \nu, \phi\frac{r_{k-1}^{(\theta)}}{\omega}\right)\right\} - \sum_{k=2}^{n} \ln\left\{\omega \frac{B_{\tau_k}(\tilde{\theta})}{\tau_k}\right\}, \end{aligned}$$

where $f(x; \nu, \mu)$ is the density of a non-central chi-square distribution with parameters ν and μ, with $\nu = \frac{4\alpha\beta}{\sigma^2}$, $\phi = e^{-\alpha h}$, and $\omega = \sigma^2\left(\frac{1-\phi}{4\alpha}\right)$.

The estimation of $\theta = (\alpha, \beta, \sigma, q_1, q_2)^\top$ is then obtained by minimizing $-L(\theta)$.

Positiveness Constraints

We have to take into account the positiveness constraints on $\alpha, \beta, \sigma, a, b$. Also, recall that $r_1, \ldots, r_n$ are also always positive, which is not necessarily guaranteed by the transformation $r_k = \frac{\tau_k R_k + 100 A_{\tau_k}(\tilde{\theta})}{B_{\tau_k}(\tilde{\theta})}$, $k \in \{1, \ldots, n\}$.

5.3.3.4 Precision of the Estimation

As before, one can apply the results of Berndt et al. [1974] to obtain the following result for the asymptotic behavior of the estimation error.

Proposition 5.3.8 *If $\hat{\theta}_n$ is the estimation of θ using the maximum likelihood principle, then*

$$\sqrt{n}(\hat{\theta}_n - \theta) \rightsquigarrow N_5(0, \hat{V}),$$

where $\hat{V}$ is the Moore-Penrose inverse of the estimation $\mathcal{I}$ of the Fisher information.

Recall that I can be estimated by using the Hessian matrix H, in which case $\hat{\mathcal{I}} = H/n$, or it can be estimated by computing the numerical gradient. See Appendix B.5.1.

Example 5.3.1 *As an illustration of the methodology, consider the simulated data set* DataCIR *representing annualized rates expressed in percentage. They have been generated with the parameters $\theta = (0.5, 2.55, 0.365, 0.3, 0)^\top$. Here $h = 1/360$, and we considered 4 years of daily prices for bonds with maturities of 1, 3, 6, and 12 months.*

Table 5.2 contains the confidence intervals for the estimation of the parameters, using the MATLAB function EstCIR. *The starting parameters were chosen to correspond to those of a Feller process for R, i.e., $\alpha_0 = -\frac{\ln(\phi_0)}{h}$, $\beta_0 = \bar{R}$, $\sigma_0 = \gamma_0\sqrt{2\alpha_0/\beta_0}$, where ϕ_0 is the first order autocorrelation and γ_0 is the standard deviation of R. All intervals cover the true value of each parameter. If we would have used only one year of data, the results would not have been so good. Note that even for large samples, the estimation of the parameters of the market price of risk are not that good. As predicted, the estimation of α is not as precise as the estimation of β and σ.*

One can verify that for other starting points yielding realistic values for $r_1, \ldots, r_n$, the estimation of the parameters is quite stable.

Figure 5.2 presents the estimated spot rates and the real ones for the first year of the sample. Both are very close to each other.

TABLE 5.2: 95% confidence intervals for the estimation of parameters of the CIR model. Here $\phi = e^{-\alpha h}$, with $h = 1/360$.

Parameter	True Value	Confidence Interval
α	0.5	1.0447 ± 1.1472
β	2.55	2.6983 ± 0.5727
σ	0.365	0.3477 ± 0.0205
q_1	0.3	4.9409 ± 9.6459
q_2	0	-1.8208 ± 3.3502
ϕ	0.9986	0.9971 ± 0.0032

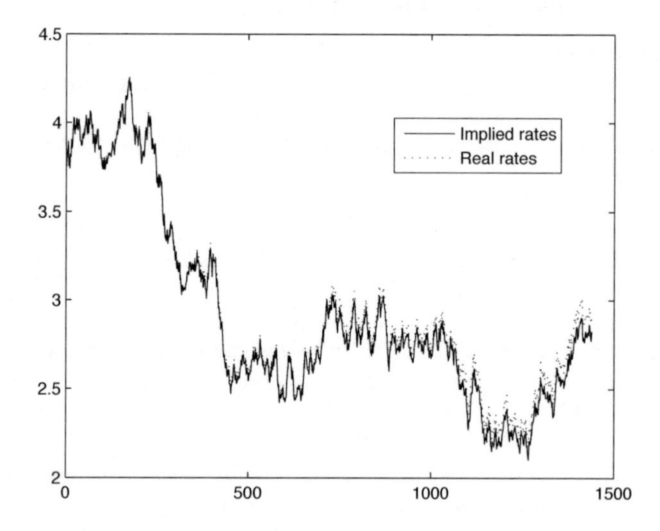

FIGURE 5.2: Implied and true spot rates for the CIR model with $\alpha = 0.5$, $\beta = 2.55$, $\sigma = 0.365$, $q_1 = 0.3$ and $q_2 = 0$.

Remark 5.3.6 *As in the Vasicek model, the parameters of the market price of risk seem to be difficult to estimate.*

5.4 Other Models for the Spot Rates

An interesting property of both the Vasicek and CIR models is that the value of the bond is given by $P(t,T) = e^{A(t,T)-B(t,T)\mathfrak{r}(t)}$, which enables us to implement easily Duan methodology. One can ask if there are other models that share the same property. The answer is yes and they are called affine models.

5.4.1 Affine Models

Suppose that under an equivalent martingale measure, the dynamics of the spot rates are given by

$$d\mathfrak{r}(t) = a(t)\{b(t) - \mathfrak{r}(t)\}dt + \sqrt{\gamma(t)\mathfrak{r}(t) + \delta(t)}\, dW(t), \tag{5.24}$$

$a(t) > 0$, $b(t) > 0$, $\gamma(t) \geq 0$, $\delta(t) \geq 0$, and $\gamma(t) + \delta(t) > 0$ are deterministic functions. Then, the value of a zero-coupon bond is

$$\begin{aligned} P(t,T) &= E\left[e^{-\int_t^T \mathfrak{r}(s)ds} \middle| \mathcal{F}_t \right] \\ &= \exp\left\{ A(t,T) - \mathfrak{r}(t)B(t,T) \right\}, \end{aligned}$$

where B satisfies the Ricatti type ordinary differential equation

$$\frac{dB(t,T)}{dt} - a(t)B(t,T) - \frac{1}{2}\gamma(t)B^2(t,T) = -1, \quad B(T,T) = 0,$$

and

$$\frac{dA(t,T)}{dt} = a(t)b(t)B(t,T) - \frac{1}{2}\delta(t)B^2(t,T), \quad A(T,T) = 0.$$

Note that if the dynamics of $\mathfrak{r}$ satisfy (5.24), in order to obtain a similar model under the equivalent martingale measure, the market price of risk $q(t,r)$ must be given by

$$\frac{q_1(t)}{\sqrt{\gamma(t)\mathfrak{r}(t) + \delta(t)}} + q_2(t)\sqrt{\gamma(t)\mathfrak{r}(t) + \delta(t)},$$

for some continuous deterministic functions q_1 and q_2 depending only on t.

5.5 Suggested Reading

The Vasicek [1977] article is well written and relatively easy to understand. For general results on interest rate models and pricing formulas, see Lamberton and Lapeyre [2008] and Björk [1999, Chapter 17].

Finally, for other applications of Duan estimation methodology, see Duan [1994, 2000].

5.6 Exercises

Exercise 5.1

Suppose that $\tilde{\mathfrak{r}}$ is an Ornstein-Uhlenbeck with parameters a, b, σ under the equivalent martingale measure.

(a) Find
$$E\left\{ e^{u\tilde{\mathfrak{r}}(T)-\int_t^T \tilde{\mathfrak{r}}(s)ds} \middle| \tilde{\mathfrak{r}}(t) = r \right\}, \quad u \in \mathbb{R}. \tag{5.25}$$

(b) For any $x \in \mathbb{R}$ and $u \in \mathbb{R}$, find
$$E\left[e^{u\tilde{\mathfrak{r}}(T)} \mathbb{I}\left\{\tilde{\mathfrak{r}}(T) \le x\right\} e^{-\int_t^T \tilde{\mathfrak{r}}(s)ds} \middle| \tilde{\mathfrak{r}}(t) = r \right].$$

The last expectation can be useful for approximating the value of a Bermudan call or put option on a zero-coupon bond.

Exercise 5.2

Suppose that $\mathfrak{r}$ is the short-term interest rate (expressed in percentage on a yearly basis) and that it is modeled by a Ornstein-Uhlenbeck process with parameters $\alpha = 0.8$, $\beta = 3.25$, $\sigma = 0.9$.

(a) Find the stationary distribution of the process.

(b) Define $X(t) = \frac{1}{36000}\mathfrak{r}(t/360)$, $t \ge 0$. Interpret X and find its distribution.

(c) Suppose that the market price of risk is $q(t,r) = 0.5 + 0.2r$. What is the distribution of $\tilde{\mathfrak{r}}$ under the equivalent martingale measure? Also find the limiting behavior of the annual yield $R(t,T)$ on a zero-coupon bond, as $T \to \infty$.

(d) Suppose that under the equivalent martingale measure, $\tilde{\mathfrak{r}}$ is an Ornstein-Uhlenbeck process with parameters $a = 0.6$ and $b = 1.75$. Find the associated market price of risk.

Exercise 5.3

Consider the Vasicek model for the short-term interest rate $\mathfrak{r}$.

(a) Interpret the so-called mean-reverting property.

(b) Show that $\lim_{t\to\infty} E\{\mathfrak{r}(t)\} = \beta$.

(c) Show that $Var\{\mathfrak{r}(t)\} = \sigma^2 \frac{(1-e^{-2\alpha t})}{2\alpha}$.

(d) Prove that $\lim_{t\to\infty} Var\{\mathfrak{r}(t)\} = \frac{\sigma^2}{2\alpha}$.

(e) Show that $\mathfrak{r}(t)$ converges in law as $t \to \infty$ and find the limiting distribution.

Exercise 5.4

Suppose that $\mathfrak{r}$ is the short-term interest rate (expressed in percentage on a yearly basis) and that it is modeled by a Feller process with parameters $\alpha = 0.6$, $\beta = 2.25$, $\sigma = 0.5$.

(a) Find the stationary distribution of the process.

(b) Define $X(t) = \frac{1}{36000}\mathfrak{r}(t/360)$, $t \geq 0$. Interpret X and find its distribution.

(c) Suppose that the market price of risk is $q(t,r) = 0.25/\sqrt{r} - 0.02\sqrt{r}$. What is the distribution of $\tilde{\mathfrak{r}}$ under the equivalent martingale measure? Also find the limiting behavior of the annual yield $R(t,T)$ on a zero-coupon bond, as $T \to \infty$.

(d) Suppose that under the equivalent martingale measure, $\tilde{\mathfrak{r}}$ is a Feller process with parameters $a = 0.6$ and $b = 2.75$. Find the associated market price of risk.

Exercise 5.5

Consider the CIR model for the short-term interest rate $\mathfrak{r}$.

(a) Interpret the so-called mean-reverting property.

(b) Show that $\lim_{t\to\infty} E\{\mathfrak{r}(t)\} = \beta$.

(c) Prove that $\lim_{t\to\infty} Var\{\mathfrak{r}(t)\} = \frac{\beta\sigma^2}{2\alpha}$.

(d) Show that $\mathfrak{r}(t)$ converges in law as $t \to \infty$ and find the limiting distribution.

(e) For any $x \in \mathbb{R}$ and $u \geq 0$, find

$$E\left[e^{u\tilde{\mathfrak{r}}(T)} \mathbb{I}\left\{\tilde{\mathfrak{r}}(T) \leq x\right\} e^{-\int_t^T \tilde{\mathfrak{r}}(s)ds} \middle| \tilde{\mathfrak{r}}(t) = r \right].$$

The last expectation can be useful for approximating the value of a Bermudan call or put option on a bond.

Exercise 5.6

Suppose that the short-term rate $\mathfrak{r}$ (in percentage on a yearly basis) is modeled by a Ornstein-Uhlenbeck process with parameters $\alpha = 0.5$, $\beta = 2$ and $\sigma = 0.9$. Suppose also that the market price of risk is $q_1 = -0.012$ and $q_2 = 0.01$.

(a) What is the distribution of the process $\tilde{\mathfrak{r}}$ under the equivalent martingale measure?

(b) What is the long term annual yield of a zero-coupon bond?

(c) Today, the annual yield of a 3-month zero-coupon bond is 4.5%. What is the implied spot rate in percentage?

(d) Give the actual value V_0 of a 6-month zero-coupon bond with face value of \$100000.

(e) Give the law of $\mathfrak{r}$ in one month.

(f) Find the value of r such that $P\left\{\mathfrak{r}\left(\frac{1}{12}\right) \geq r\right\} = 0.01$.

(g) Let V be the value in 1 month of a zero-coupon bond expiring in 6 months from now with value \$100000. Find the value v such that $P(V \leq v) = 0.01$.

(g) Find the 1-month VaR of order 99% for the loss $X = V_0 - e^{-r/12}V$ on the investment described in (d). Assume a risk-free rate r of 2%.

Exercise 5.7

What is the effect of increasing the value q_2 of the market price of risk in either the CIR or the Vasicek model? Consider only non-negative values of q_2.

Exercise 5.8

Suppose that the short-term rate $\mathfrak{r}$ (in percentage on a yearly basis) is modeled by a Feller process with parameters $\alpha = 0.9$, $\beta = 2.5$, and $\sigma = 0.95$. Suppose also that the market price of risk is $q_1 = 0.2$ and $q_2 = 1.1$.

(a) What is the distribution of the process $\tilde{\mathfrak{r}}$ under the equivalent martingale measure?

(b) What is the long term annual yield of a zero-coupon bond?

(c) Today, the annual yield of a 1-month zero-coupon bond is 3.5%. What is the implied spot rate in percentage?

(d) Give the actual value V_0 of a 6-month zero-coupon bond with face value of \$100000.

(e) Give the law of $\mathfrak{r}\left(\frac{1}{12}\right)/\omega_{\frac{1}{12}}$.

(f) Find the value of r such that $P\left\{\mathfrak{r}\left(\frac{1}{12}\right) \geq r\right\} = 0.01$.

(g) Let V be the value in 1 month of a zero-coupon bond expiring in 6 months from now with value \$100000. Find the value v such that $P(V \leq v) = 0.01$.

(g) Find the 1-month VaR of order 99% for the loss $X = V_0 - e^{-r/12}V$ on the investment described in (d). Assume a risk-free rate r of 2%.

Exercise 5.9

In (g) of Exercises 5.6 and 5.8, we assumed a risk-free rate of 2%. However the loss should have been written as

$$X = V_0 - e^{A_\tau - \mathfrak{r}(t)*\frac{B_\tau}{100} - \frac{1}{100}\int_0^t \mathfrak{r}(s)ds},$$

with $t = 1/12$ and $\tau = 5/12$. Can you find the law of $B\mathfrak{r}(t) + \int_0^t \mathfrak{r}(s)ds$, for any positive constant B and any $t \geq 0$?

5.7 Assignment Questions

1. Find the value of a call option on a zero-coupon bond when the short-term interest rate is modeled by a Ornstein-Uhlenbeck process. Compute the greeks with respect to α, β, σ, q_1, and q_2.

5.A Interpretation of the Stochastic Integral

When f is a deterministic square integrable function, the stochastic integral

$$M(t) = \int_0^t f(u)dW(u)$$

has a Gaussian distribution with mean 0 and variance $\int_0^t f^2(u)du$. Moreover, M is a martingale and $M(t) - M(s)$ is independent of $\mathcal{F}_s$, for any $0 \leq s \leq t$. More generally, if f is random, left-continuous, square integrable, and $f(s)$ depends only on $\mathcal{F}_s$, then

$$M(t) = \int_0^t f(s)dW(s) \approx \sum_{i=1}^n f(s_{i-1})\{W(s_i) - W(s_{i-1})\}$$

is a mean zero martingale with

$$E\left\{M^2(t)\right\} = \int_0^t E\left\{f^2(s)\right\} ds.$$

5.B Integral of a Gaussian Process

Proposition 5.B.1 *Suppose that the stochastic process X is continuous and Gaussian. Then the distribution of $Y = \int_0^t X(s)ds$ is also Gaussian, with mean $E(Y) = \int_0^t E\{X(s)\}ds$ and variance*

$$\mathrm{Var}(Y) = \int_0^t \int_0^t \mathrm{Cov}\{X(s), X(u)\} ds du.$$

PROOF. If f is a continuous function, then $\int_0^t f(s)ds$ is the limit of the sum

$$\frac{t}{n}\sum_{k=1}^n f\left(\frac{k}{n}t\right) = \int_0^t f\left(\frac{\lfloor un/t \rfloor + 1}{n/t}\right) du.$$

Hence, the distribution of Y can be approximated by the distribution of

$$Y_n = \frac{t}{n}\sum_{k=1}^n X\left(\frac{k}{n}t\right),$$

which is clearly Gaussian, with mean

$$E(Y_n) = \frac{t}{n}\sum_{k=1}^n E\left\{X\left(\frac{k}{n}t\right)\right\} \to \int_0^t E\{X(s)\}ds,$$

and variance

$$\begin{aligned}\mathrm{Var}(Y_n) &= \frac{t^2}{n^2}\sum_{k=1}^n\sum_{j=1}^n \mathrm{Cov}\left\{X\left(\frac{k}{n}t\right), X\left(\frac{j}{n}t\right)\right\} \\ &\to \int_0^t \int_0^t \mathrm{Cov}\{X(s), X(u)\} ds du.\end{aligned}$$

■

5.C Estimation Error for a Ornstein-Uhlenbeck Process

If an Ornstein-Uhlenbeck process is observed at regular times $h, 2h, \ldots$, it follows from Proposition 5.2.1 that we observe indeed a AR(1) process given by

$$r_k = \beta + \phi(r_{k-1} - \beta) + \epsilon_k,$$

with $\phi = e^{-\alpha h}$ and $\epsilon_k \sim N(0,\gamma^2)$, where $\gamma^2 = \sigma^2\left(\frac{1-\phi^2}{2\alpha}\right)$. It is then easy to check that the maximum likelihood estimators of ϕ, β, γ satisfies

$$\begin{aligned}
\hat{\phi}_n &= \frac{\sum_{k=2}^n \left(r_k - \hat{\beta}_n\right)\left(r_{k-1} - \hat{\beta}_n\right)}{\sum_{k=2}^n \left(r_{k-1} - \hat{\beta}_n\right)^2}, \\
\hat{\beta}_n &= \frac{1}{n-1}\sum_{k=2}^n \left(r_k - \hat{\phi}_n r_{k-1}\right) / \left(1 - \hat{\phi}_n\right), \\
\hat{\gamma}_n^2 &= \frac{1}{n-1}\sum_{k=2}^n \left\{r_k - \hat{\beta}_n - \hat{\phi}_n\left(r_{k-1} - \hat{\beta}_n\right)\right\}^2.
\end{aligned}$$

It then follows that

$$\begin{aligned}
\sqrt{n}\left(\hat{\phi}_n - \phi\right) &= \frac{(1-\phi^2)}{\gamma^2}\frac{1}{\sqrt{n}}\sum_{k=2}^n \epsilon_k\left(r_{k-1} - \beta\right) + o_P(1), \\
\sqrt{n}\left(\hat{\beta}_n - \beta\right) &= \frac{1}{n-1}\sum_{k=2}^n \epsilon_k/(1-\phi) + o_P(1), \\
\sqrt{n}\left(\hat{\gamma}_n - \gamma\right) &= \frac{1}{2\gamma\sqrt{n}}\sum_{k=2}^n \left(\epsilon_k^2 - \gamma^2\right) + o_P(1).
\end{aligned}$$

As a result, these estimation errors converge in law to independent centered Gaussian variables with variances $1-\phi^2$, $\gamma^2/(1-\phi)^2$, and $\gamma^2/2$ respectively. Let V be the diagonal matrix with these diagonal elements. To obtain the joint limiting distribution of all these variables plus $\sqrt{n}\,(\alpha_n - \alpha)$ and $\sqrt{n}\,(\sigma_n - \sigma)$, it suffices to compute the Jacobian matrix J of the transformation

$$(\phi, \beta, \gamma)^\top \mapsto \theta = (\phi, \beta, \gamma, \alpha, \sigma),$$

where $\alpha = -\ln(\phi)/h$ and $\sigma = \gamma\sqrt{\frac{2\alpha}{1-\phi^2}}$. It is then easy to check that

$$J = \begin{pmatrix} 1 & 0 & 0 \\ 0 & 1 & 0 \\ 0 & 0 & 1 \\ -\frac{1}{\phi h} & 0 & 0 \\ \zeta & 0 & \frac{\sigma}{\gamma} \end{pmatrix}, \quad \text{with } \zeta = \frac{\sigma^2}{\gamma}\left\{\frac{2\alpha\phi^2 h - (1-\phi^2)}{\phi\left(1-\phi^2\right)^2 h}\right\}.$$

It then follows from the delta method (Theorem B.3.1) than $\sqrt{n}\left(\hat{\theta}_n - \theta\right) \rightsquigarrow N_5(0,\Sigma)$, with $\Sigma = JVJ^\top$. In particular, $\sqrt{n}\,(\hat{\sigma}_n - \sigma)$ has asymptotic variance $\zeta^2\left(1-\phi^2\right) + \frac{\sigma^2}{2}$.

5.D Proofs of the Results

5.D.1 Proof of Proposition 5.2.1

Using representation (5.6), we have

$$\mathfrak{r}(s) = \beta + e^{-\alpha(s-t)}\{\mathfrak{r}(t) - \beta\} + \sigma \int_t^s e^{-\alpha(s-u)} dW(u), \; s \geq t \geq 0.$$

To complete the proof, we have to use the properties of stochastic integrals stated in Appendix 5.A. First, $\sigma \int_t^s e^{-\alpha(s-u)} dW(u) = \sigma e^{-\alpha s} \int_t^s e^{\alpha u} dW(u)$ has a Gaussian distribution with mean 0 and variance

$$\left(\sigma e^{-\alpha s}\right)^2 \int_t^s e^{2\alpha u} du = \sigma^2 \left\{ \frac{1 - e^{-2\alpha(s-t)}}{2\alpha} \right\}.$$

Also, $\sigma \int_t^s e^{-\alpha(s-u)} dW(u)$ is independent of $W(v)$, for all $v \in [0,t]$. As a result, $\sigma \int_t^s e^{-\alpha(s-u)} dW(u)$ is independent of $\mathfrak{r}(u)$, for all $u \in [0,t]$, since

$$\mathfrak{r}(u) = \beta + e^{-\alpha u}\{\mathfrak{r}(0) - \beta\} + \sigma \int_0^u e^{-\alpha(u-v)} dW(v)$$

depends uniquely on $W(v)$, for $v \in [0,u]$. Because of the independence, we may conclude that the conditional distribution of $\mathfrak{r}(s)$ given $\mathfrak{r}(t) = r$, is Gaussian, with mean $\beta + e^{-\alpha(s-t)}(r - \beta)$ and variance $\sigma^2 \left(\frac{1-e^{-2\alpha(s-t)}}{2\alpha} \right)$. In particular, setting $X_k = \mathfrak{r}(kh)$, we have

$$\varepsilon_k = X_k - \beta - e^{-\alpha h}(X_{k-1} - \beta) = \sigma \int_{(k-1)h}^{kh} e^{-\alpha(kh-u)} dW(u),$$

so the ε_k are independent and have a centered Gaussian distribution with variance $\sigma^2 \left(\frac{1-e^{-2\alpha h}}{2\alpha} \right)$. In addition, X_k is independent of $\varepsilon_1, \ldots, \varepsilon_k$.

■

5.D.2 Proof of Proposition 5.2.2

It follows from formulas (5.8)–(5.10) that

$$E\{\tilde{\mathfrak{r}}(s)\} = b + e^{-a(s-t)}(r - b),$$

and

$$\begin{aligned} \mathrm{Cov}\{\tilde{\mathfrak{r}}(s), \tilde{\mathfrak{r}}(u)\} &= \mathrm{Cov}\left\{ \int_t^s e^{-a(s-v)} d\tilde{W}(v), \int_t^u e^{-a(u-v)} d\tilde{W}(v) \right\} \\ &= \int_t^{\min(s,u)} e^{-a(s-v)} e^{-a(u-v)} dv = e^{-a(s+u)} \left. \frac{e^{2av}}{2a} \right|_t^{\min(s,u)} \\ &= \frac{e^{-a|s-u|} - e^{-a(s+u-2t)}}{2a}. \end{aligned}$$

As a result, if $R = \sigma \int_t^T \tilde{\mathfrak{r}}(s)ds$, and $\tau = T - t \geq 0$, then it follows from Proposition 5.B.1 that

$$
\begin{aligned}
E(R) &= E\left\{\int_t^T \tilde{\mathfrak{r}}(s)ds\right\} = \int_t^T E\{\tilde{\mathfrak{r}}(s)\}ds = \int_t^T \left\{b + e^{-a(s-t)}(r-b)\right\} ds \\
&= \int_0^\tau \{b + e^{-au}(r-b)\}\, du = b\tau + (r-b)\left(\frac{1-e^{-a\tau}}{a}\right) \\
&= rB_\tau + b(\tau - B_\tau).
\end{aligned}
$$

Finally, from Proposition 5.B.1, we obtain

$$
\begin{aligned}
\mathrm{Var}(R) &= \sigma^2 \int_t^T \int_t^T \mathrm{Cov}\{\tilde{\mathfrak{r}}(s), \tilde{\mathfrak{r}}(u)\} duds \\
&= \sigma^2 \int_t^T \int_t^T \frac{e^{-a|s-u|} - e^{-a(s+u-2t)}}{2a} duds \\
&= \sigma^2 \int_0^\tau \int_0^\tau \frac{e^{-a|s'-u'|} - e^{-a(s'+u')}}{2a} du'ds' \\
&= \frac{\sigma^2}{a^2} \int_0^\tau e^{-as'} \left\{ \left(e^{au'} + e^{-au'}\right)\Big|_0^{s'} \right\} ds' \\
&= \frac{\sigma^2}{a^2} \int_0^\tau \left(1 - 2e^{-as'} + e^{-2as'}\right) ds' \\
&= \frac{\sigma^2}{2a^3}\left(2\tau a - e^{-2a\tau} + 4e^{-a\tau} - 3\right).
\end{aligned}
$$

Hence $\mathrm{Var}(R) = \frac{\sigma^2}{a^2}(\tau - B_\tau) - \frac{\sigma^2}{2a}B_\tau^2$. We then deduce that

$$
A_\tau = -\left(b - \frac{\sigma^2}{2a^2}\right)(\tau - B_\tau) - \frac{\sigma^2}{4a}B_\tau^2.
$$

■

5.D.3 Proof of Proposition 5.3.1

The proof uses the famous Feynman-Kăc formula stating that for a Feller process $\mathfrak{r}(t)$ with parameters α, β, σ,

$$
f(t,r) = E\left[\Phi\{\mathfrak{r}(t)\}e^{\int_0^t V\{\mathfrak{r}(s)\}ds}\Big|\, \mathfrak{r}(0) = r\right]
$$

is the solution to the partial differential equation

$$
\partial_t f = \alpha(\beta - r)\partial_r f + \frac{1}{2}\sigma^2 r \partial_r^2 f + Vf, \quad t \geq 0, r > 0, \tag{5.26}
$$

with the boundary condition $f(0,r) = \Phi(r)$. In particular, if $\Phi(r) = e^{ur}$ and $V = 0$, $f(t,r)$ is the moment generating function of $\mathfrak{r}(t)$, given $\mathfrak{r}(0) = r$ at point u. It is then easy to check that indeed (5.18) satisfies (5.26), so by the uniqueness of the solution, f is given by (5.18). Next, using the results in Section A.6.12, when X has a non-central chi-square distribution with parameters ν and D, its moment generating function is given by

$$M(u) = E\left(e^{uX}\right) = \frac{e^{\frac{uD}{1-2u}}}{(1-2u)^{\nu/2}}, \quad u < \frac{1}{2}.$$

Therefore

$$E\left\{e^{u\frac{\mathfrak{r}(t)}{\omega_t}} \middle| \mathfrak{r}(0) = r\right\} = f\left(t, r, \frac{u}{\omega_t}\right) = \frac{e^{\frac{uD_t}{1-2u}}}{(1-2u)^{\nu/2}}, \quad u < \frac{1}{2},$$

which completes the proof.

■

5.D.4 Proof of Proposition 5.3.2

The proof of the conditional distribution follows from the Markov property stated in Remark 5.3.2. The rest of the proof follows for the Taylor expansion of the moment generating function, as in Section A.6.12. In fact, if u is small, then $(1-2uc)^{-1} = \sum_{n=0}^{\infty}(2uc)^n$ and $-\ln(1-2uc) = \sum_{n=1}^{\infty}\frac{(2uc)^n}{n}$. Setting $c = \omega_t$, we obtain

$$\ln\{M(u)\} = \frac{ure^{-\alpha t}}{1-2uc} - \frac{\nu}{2}\ln(1-2uc) = \frac{re^{-\alpha t}}{2c}\sum_{n=1}^{\infty}(2uc)^n + \frac{\nu}{2}\sum_{n=1}^{\infty}\frac{(2uc)^n}{n}.$$

Hence, $\kappa_n = (2c)^{n-1}\left(re^{-\alpha t}n! + \nu c(n-1)!\right)$ and $\nu c = \beta\left(1-e^{-\alpha t}\right)$.

■

5.D.5 Proof of Proposition 5.3.3

Since $\tilde{\mathfrak{r}}$ is also a Feller process, we have, using the Feynman-Kac formula (5.26) that $P(t,T) = f(\tau, r)$, where $f(0,r) = 1$, and

$$\partial_t f = a(b-r)\partial_t f + \frac{1}{2}\sigma^2 r \partial_r^2 f - rf, \quad t \geq 0.$$

It is then easy to check that $f(t,r) = e^{A_t - rB_t}$ satisfies the partial differential equation with boundary conditions $A_0 = B_0 = 0$.

■

Bibliography

E. K. Berndt, B. H. Hall, R. E. Hall, and J. A. Hausman. Estimation and inference in nonlinear structural models. *Annals of Economics and Social Measurement*, pages 653–665, 1974.

T. Björk. *Arbitrage Theory in Continuous Time.* Oxford University Press, 1999.

R.-R. Chen and L. Scott. Maximum likelihood estimation for a multifactor equilibrium model of the term structure of interest rates. *Journal of Fixed Income*, pages 14–31, 1993.

J. C. Cox, J. E. Ingersoll, and S.A. Ross. An intertemporal general equilibrium model of asset prices. *Econometrica*, 53:363–384, 1985.

J.-C. Duan. Maximum likelihood estimation using price data of the derivative contract. *Math. Finance*, 4:155–167, 1994.

J.-C. Duan. Correction: Maximum likelihood estimation using price data of the derivative contract. *Math. Finance*, 10:461–462, 2000.

W. Feller. Two singular diffusion problems. *Ann. of Math. (2)*, 54:173–182, 1951.

M. Fisher and C. Gilles. Estimating exponential-affine models of the term structure. Technical report, Federal Reserve Board, 1996.

J. C. Hull. *Options, Futures, and Other Derivatives.* Prentice-Hall, sixth edition, 2006.

M. A. Kouritzin and B. Rémillard. Explicit strong solutions of multidimensional stochastic differential equations. Technical report, Laboratory for Research in Probability and Statistics, University of Ottawa-Carleton University, 2002.

C. Labbé, B. Rémillard, and J.-F. Renaud. A Simple Discretization Scheme for Nonnegative Diffusion Processes, with Applications to Option Pricing. *Journal of Computational Finance*, 15:3–35, 2012.

D. Lamberton and B. Lapeyre. *Introduction to Stochastic Calculus Applied to Finance.* Chapman & Hall/CRC Financial Mathematics Series. Chapman & Hall/CRC, Boca Raton, FL, second edition, 2008.

G. E. Uhlenbeck and L. S. Ornstein. On the theory of Brownian motion. *Physical Review*, 36:823–841, 1930.

O. Vasicek. An equilibrium characterization of the term structure. *Journal of Financial Economics*, 5:177–188, 1977.

Chapter 6

Lévy Models

In this chapter, we cover a class of models for assets' returns for which the associated markets are incomplete, the so-called Lévy processes. We start by introducing the diffusion-jump models proposed by Merton [1976] and Kou [2002]. Then we begin the study of general Lévy processes, covering the topics of equivalent martingale measures and parameter estimation.

6.1 Complete Models

The market associated with an asset S (possibly multidimensional) and filtration $\mathbb{F} = \{\mathcal{F}_t; t \geq 0\}$ is said to be complete if there exists a unique positive martingale $\Lambda(t)$ such that $E\{\Lambda(t)\} = 1$ and $e^{-rt}S(t)\Lambda(t)$ is also a martingale. Here for simplicity, we assume that the risk-free rate is constant.

Recall that a process X is a martingale with respect to the filtration $\mathbb{F}$ is for any $0 \leq s \leq t$, $X(t)$ is $\mathcal{F}_t$-measurable (also noted $X(t) \in \mathcal{F}_t$), $X(t)$ is integrable and $E\{X(t)|\mathcal{F}_s\} = X(s)$.

The probability measure Q defined by

$$\left.\frac{dQ}{dP}\right|_{\mathcal{F}_t} = \Lambda(t)$$

is said to be an equivalent martingale measure, since the discounted prices $e^{-rt}S(t)$ form a Q-martingale. It follows from the properties of conditional expectations that if $Y \in \mathcal{F}_T$, then

$$E_Q(Y|\mathcal{F}_t) = \frac{E\{X\Lambda(T)|\mathcal{F}_t\}}{\Lambda(t)}. \tag{6.1}$$

This formula is quite useful for computing conditional expectations relative to a change of measure Q.

Example 6.1.1 *For the one-dimensional Black-Scholes model, one can show that the unique martingale $\Lambda(t)$ is given by*

$$\ln\{\Lambda(t)\} = bW(t) - t\frac{b^2}{2}, \tag{6.2}$$

where $b = \frac{r-\mu}{\sigma}$. *In addition, under the measure* Q, $\tilde{W}(t) = W(t) - bt$ *is a Brownian motion. In fact, from* (6.1), *we have, for any* $0 \le s \le t$, *and any* $u \in \mathbb{R}$,

$$
\begin{aligned}
E_Q\left[e^{u\{\tilde{W}(t)-\tilde{W}(s)\}} \middle| \mathcal{F}_s \right] &= E\left[\frac{\Lambda(t)}{\Lambda(s)} e^{u\{\tilde{W}(t)-\tilde{W}(s)\}} \middle| \mathcal{F}_s \right] \\
&= e^{-(t-s)b^2/2 - ub(t-s)} E\left[e^{(u+b)\{W(t)-W(s)\}} \middle| \mathcal{F}_s \right] \\
&= e^{-(t-s)b^2/2 - ub(t-s) + (t-s)(u+b)^2/2} = e^{(t-s)u^2/2}.
\end{aligned}
$$

Thus the increment $\tilde{W}(t) - \tilde{W}(s)$ *is independent of* $\mathcal{F}_s$ *and has a Gaussian distribution with mean* 0 *and variance* $t - s$. *From Definition 1.2.1, we know that* $\tilde{W}$ *is a Brownian motion with respect to* Q.

Furthermore, since the price S *satisfies* $S(t) = S_0 e^{(\mu-\sigma^2/2)t+\sigma W(t)}$, *it follows that under* Q,

$$
S(t) = S_0 e^{(\mu-\sigma^2/2)t+\sigma(\tilde{W}(t)+bt)} = S_0 e^{(r-\sigma^2/2)t+\sigma\tilde{W}(t)},
$$

proving that $e^{-rt}S(t) = S_0 e^{-t\sigma^2/2+\sigma\tilde{W}(t)}$ *is a* Q*-martingale.*

We will now introduce models for which the market is incomplete. The incompleteness property will be discussed in Section 6.5.5.

6.2 Stochastic Processes with Jumps

Before introducing the models proposed by Merton [1976], we need to define a Poisson process.

Definition 6.2.1 *A stochastic process* $N(t)$ *is said to be a Poisson process of intensity* $\lambda > 0$ *if*

- $N(0) = 0$;
- N *is a process with independent increments, that is for any* $0 \le t_1 \cdots \le t_m$, *the increments* $N(t_1) - N(0), N(t_2) - N(t_1), \ldots, N(t_m) - N(t_{m-1})$ *are independent;*
- *If* $0 \le s \le t$, *then* $N(t) - N(s)$ *has a Poisson distribution with parameter* $(t - s)\lambda$.

Note that by definition, a Poisson process N is a non-decreasing process taking values in $\{0, 1, 2, \ldots\}$ and the size of its jumps is always 1.

If τ_j represents the time of the j-th jump of the Poisson process, and $\tau_0 = 0$, then the increments $\tau_j - \tau_{j-1}$, $j \ge 1$ are independent and identically distributed with an exponential distribution of parameter λ.

Note that for a Poisson process N,

$$P\{N(t) = k\} = e^{-\lambda t}\frac{(\lambda t)^k}{k!}, \qquad k = 0, 1, \ldots,$$

$E\{N(t)\} = \text{Var}\{N(t)\} = \lambda t$, and for any $u \in \mathbb{R}$,

$$E\left\{e^{uN(t)}\right\} = e^{\lambda t(e^u - 1)}. \tag{6.3}$$

6.2.1 Simulation of a Poisson Process over a Fixed Time Interval

To generate a Poisson process N with intensity λ over $[0, T]$, we can use the following algorithm:

Algorithm 6.2.1 *Generate* $M \sim \text{Poisson}(\lambda T)$.

- *If* $M = 0$, *stop. Otherwise, simulate* $V_1, \ldots, V_{M+1} \sim \text{Exp}(1)$.
- *For* $j \in \{1, \ldots, M\}$, *set* $\tau_j = TS_j/S_{M+1}$, *where* $S_j = V_1 + \cdots + V_j$. *These are the jumping times. Note that* $\tau_0 = 0$.
- *Then, set* $N(t) = j - 1$ *on* $[\tau_{j-1}, \tau_j)$, $j \in \{1, \ldots, M\}$, *and* $N(t) = M$ *for* $t \in [\tau_M, T]$.

6.2.2 Jump-Diffusion Models

The so-called jump-diffusion processes, introduced in finance by Merton [1976], are of the form $S(t) = S_0 e^{X(t)}$, with

$$X(t) = \left(\mu - \lambda\kappa - \frac{\sigma^2}{2}\right)t + \sigma W(t) + \sum_{j=1}^{N(t)} \xi_j,$$

where N is a Poisson process of intensity λ, N is independent of the Brownian motion W, and both processes are independent of the jumps $U_j = e^{\xi_j} - 1 > -1$ which are independent and identically distributed with a finite expectation $\kappa = E(U_j) = E\left(e^{\xi_j}\right) - 1$, $j \geq 1$. Here, we adopt the convention that $\sum_{j=1}^{0} \xi_j = 0$. The jump part $\sum_{j=1}^{N(t)} \xi_j$ is called a compound Poisson process.

Using (6.3), we have $E\left\{e^{iuX(t)}\right\} = e^{t\psi(u)}$, where for all $u \in \mathbb{R}$,

$$\psi(u) = iu\left(\mu - \lambda\kappa - \frac{\sigma^2}{2}\right) - u^2\frac{\sigma^2}{2} + \lambda E\left(e^{iu\xi_1} - 1\right), \tag{6.4}$$

while for all u in some open interval $I \supset [0, 1]$, the cumulant generating function is given by $tc(u)$, where

$$c(u) = u\left(\mu - \lambda\kappa - \frac{\sigma^2}{2}\right) + u^2\frac{\sigma^2}{2} + \lambda E\left(e^{u\xi_1} - 1\right). \tag{6.5}$$

It then follows that $E\{S(t)\} = S_0 e^{tc(1)} = S_0 e^{\mu t}$, and μ can be interpreted has the instantaneous mean return.

Note also that if τ_j represents the time of the j-th jump of the Poisson process, and $\tau_0 = 0$, then for any $t \in [\tau_{j-1}, \tau_j)$,

$$S(t) = S(\tau_{j-1}) e^{\left(\mu - \lambda\kappa - \frac{\sigma^2}{2}\right)(t-\tau_{j-1}) + \sigma\{W(t) - W(\tau_{j-1})\}},$$

and the jumps in S happen at times τ_j, i.e.,

$$\Delta S(\tau_j) = S(\tau_j) - S(\tau_j-) = U_j S(\tau_j-)$$

where τ_j- denotes the instant just before τ_j.

An interesting property of jump-diffusion models is that they have stationary independent returns, which facilitates their simulation.

Proposition 6.2.1 *For $j \in \{1, \ldots, n\}$, set $Y_j = \ln\left[\frac{S(jh)}{S\{(j-1)h)\}}\right]$. Then $Y_1, \ldots, Y_n$ are independent and identically distributed, with the same distribution as*

$$Y = h\left(\mu - \lambda\kappa - \frac{\sigma^2}{2}\right) + \sigma\sqrt{h}Z + \sum_{j=1}^{M} \xi_j,$$

where $M \sim$ Poisson(λh) is independent of $Z \sim N(0,1)$, both variables being independent of the jumps ξ_j, $j \geq 1$.

We now introduce some important cases of jump-diffusion models often used in the literature.

6.2.3 Merton Model

In Merton [1976], Merton studied the case of Gaussian jumps, i.e., $\xi_j \sim N(\gamma, \delta^2)$. Hence $\kappa = e^{\gamma+\delta^2/2} - 1$. Then, conditionally on $N(t) = k \geq 0$, we have

$$X(t) \sim N(at + k\gamma, \sigma^2 t + k\delta^2),$$

where $a = \mu - \lambda\kappa - \sigma^2/2$. It follows that the density of $X(t)$ is given by

$$f_{X(t)}(x) = \sum_{k=0}^{\infty} e^{-\lambda t} \frac{(\lambda t)^k}{k!} \frac{e^{-\frac{1}{2}\frac{(x-at-k\gamma)^2}{\sigma^2 t + k\delta^2}}}{\sqrt{2\pi(\sigma^2 t + k\delta^2)}}, \quad x \in \mathbb{R}.$$

The MATLAB function *SimJumpDiffMerton* can be used to generate paths for this model.

6.2.4 Kou Jump-Diffusion Model

In Kou [2002], the author proposed to model the jumps by a distribution that he called the asymmetric double exponential distribution with density

$$f(x) = p\eta_1 e^{-\eta_1 x}\mathbb{I}(x > 0) + (1-p)\eta_2 e^{\eta_2 x}\mathbb{I}(x \le 0), \quad x \in \mathbb{R}, \tag{6.6}$$

where $\eta_1 > 1$, $\eta_2 > 0$, and $p \in (0,1)$. It is easy to check that $\kappa = \frac{1+p\eta_2-(1-p)\eta_1}{(\eta_1-1)(\eta_2+1)}$. The density can also be written as an infinite sum. See Kou [2002] for details.

The MATLAB function *SimJumpDiffKou* can be used to generate paths for this model. Note that the asymmetric double exponential distribution is a particular case of a weighted-symmetric distribution which is defined next.

6.2.5 Weighted-Symmetric Models for the Jumps

In Abdous and Rémillard [1995] and Abdous et al. [2003], the authors studied properties of weighted symmetric distributions, of which the models of Merton and Kou for the distribution of the jumps are special cases.

Definition 6.2.2 *A random variable ξ is (ω, p)-symmetric with respect to θ if there exists a non-negative random variable ζ such that*

$$\xi = \begin{cases} \theta - \zeta & \text{with probability } 1-p, \\ \theta + \omega\zeta & \text{with probability } p. \end{cases}$$

For example, one could take $\zeta = \delta|Z|$, with $Z \sim N(0,1)$. Then, with $\theta = \gamma$, $p = 1/2$, and $\omega = 1$, we have $\xi \sim N(\gamma, \delta^2)$. Such variables are said to be split Gaussian or to have a weighted-symmetric Gaussian distribution. Some densities of weighted-symmetric Gaussian distributions are displayed in Figure 6.1.

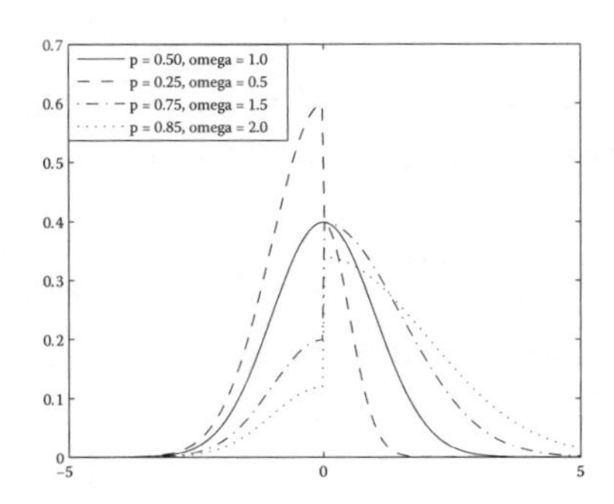

FIGURE 6.1: Densities of weighted-symmetric Gaussian distributions.

Note that for jumps with a weighted-symmetric Gaussian distribution, we have

$$\kappa = 2(1-p)e^{\theta+\delta^2/2}\mathcal{N}(-\delta) + 2pe^{\theta+\delta^2\omega^2/2}\mathcal{N}(\delta\omega) - 1. \tag{6.7}$$

Merton model with weighted-symmetric Gaussian jumps is implemented in the MATLAB function *SimJumpDiffWSG*.

Taking $\zeta = \delta Y$, with $Y \sim \text{Exp}(1)$, $\delta = 1/\eta_2$, $\theta = 0$, and $\omega = \eta_2/\eta_1$, we recover the asymmetric double exponential distribution of Kou. Note that in the MATLAB function *SimJumpDiffKou*, the parameters are μ, σ, λ, p, $\omega = \eta_2/\eta_1$, θ and $\delta = 1/\eta_2$. Further remark that one can also write $Y = |\tilde{Y}|$, with $\tilde{Y} = \sqrt{2}Z\sqrt{V}$, where $Z \sim N(0,1)$ is independent of $V \sim \text{Exp}(1)$. Then $\tilde{Y}$ has a double exponential distribution or Laplace distribution with density $f(y) = \frac{e^{-|y|}}{2}$, $y \in \mathbb{R}$.

More generally, we could consider non-negative random variables $\zeta = \zeta_\alpha = \delta|Z|\sqrt{V_\alpha/\alpha}$, with $Z \sim N(0,1)$, independent of $V_\alpha \sim \text{Gamma}(\alpha, 1)$. This family, indexed by $\alpha > 0$, contains the Kou model ($\alpha = 1$) as well as Merton model ($\alpha \to \infty$). It defines a family of jump-diffusion models with 7 parameters (8 if θ is included) which have fatter tails than the Gaussian distribution. The expression for the density of the symmetric random variable $Y = Z\sqrt{V_\alpha/\alpha} \stackrel{Law}{=} W(V_\alpha/\alpha)$, where W is a Brownian motion, is given in Proposition 6.4.1. It is a called a Linnik distribution [Jacques et al., 1999]. It seems to have been studied first by Pearson et al. [1929]. It was also proposed by McLeish [1982] as an alternative to the Gaussian distribution.

6.3 Lévy Processes

Jump-diffusion processes are a special case of more general processes called Lévy processes which can be decomposed in the following way:

$$X(t) = \ln\{S(t)\} - \ln\{S(0)\} = \alpha t + \sigma W(t) + L(t),$$

where $L(0) = 0$, L is independent of W and it is a càdlàg[1] jump process with independent and stationary increments. This means that for $0 \le t_0 \le t_1 \cdots \le t_n$, the increments

$$L(t_1) - L(t_0), \ldots, L(t_n) - L(t_{n-1})$$

are independent and the distribution of $L(t) - L(s)$ is the same as the distribution of $L(t-s)$, $0 \le s \le t$.

6.3.1 Random Walk Representation

Because a Brownian motion also has independent and stationary increments, it follows that a Lévy process has independent and stationary increments as well. In particular, at times $h, 2h, \ldots, nh$, one can write X as a

[1] From the French *continu à droite avec limite à gauche*, meaning right-continuous with left limits.

random walk, i.e.,

$$X(kh) = \sum_{i=1}^{k} Y_i, \quad k \in \{1, \ldots, n\},$$

where $Y_1, \ldots, Y_n$ are independent and identically distributed observations of $X(h)$. Therefore, if $X(h)$ can be simulated for a given h, then we can generate $X(kh)$ for any integer k.

6.3.2 Characteristics

If the mean of L exists, then $E\{X(t)\} = tE\{X(1)\}$, and if the moment of order 2 exists, then $\text{Var}\{X(t)\} = t\text{Var}\{X(1)\}$. Furthermore the characteristic function of a Lévy process X is given by $E\left\{e^{iuX(t)}\right\} = e^{t\psi(u)}$ for all $u \in \mathbb{R}$, where

$$\psi(u) = iua - u^2\frac{\sigma^2}{2} + \int \left\{e^{iux} - 1 - iux\mathbb{I}(|x| < 1)\right\} \nu(dx), \qquad (6.8)$$

for a measure ν called the Lévy measure[2]. Formula (6.8) is called the Lévy-Khinchin formula. It follows that the law of a Lévy process is completely determined by its characteristics $\{a, \sigma^2, \nu\}$ through the Lévy-Khinchin formula. In fact the law of the pure jump part L is determined by the Lévy measure ν, since $E\left\{e^{iuL(t)}\right\} = e^{t\psi_L(u)}$, with

$$\psi_L(u) = \int \left\{e^{iux} - 1 - iux\mathbb{I}(|x| < 1)\right\} \nu(dx). \qquad (6.9)$$

In addition, if the moment generating function exists for $|u| < u_0$, then $E\left\{e^{uX(t)}\right\} = e^{tc(u)}$, where

$$c(u) = ua + u^2\frac{\sigma^2}{2} + \int \left\{e^{ux} - 1 - ux\mathbb{I}(|x| < 1)\right\} \nu(dx). \qquad (6.10)$$

Therefore, if it exists, the cumulant generating function of X_t is linear in t. It is therefore true also for all existing cumulants. If the latter exists, we have that the cumulants of $X(1)$ are

$$\begin{aligned}
\kappa_1 &= a + \int_{|x|>1} x\nu(dx), \\
\kappa_2 &= \sigma^2 + \int x^2\nu(dx), \\
\kappa_n &= \int x^n\nu(dx), \quad n \geq 3.
\end{aligned}$$

[2] A Lévy measure is a measure defined on $\mathbb{R}$ which satisfies $\nu(\{0\}) = 0$ and $\int_{\mathbb{R}\setminus 0} \min(x^2, 1)\nu(dx) < \infty$. In particular $\nu(A) < \infty$ for any interval $A \subset \mathbb{R} \setminus 0$.

As a result of (6.10), we have that for any $|u| < u_0$, $e^{uX(t)-tc(u)}$ is a positive martingale with mean 1.

Example 6.3.1 (Jump-Diffusion Process) *Suppose that X is a jump-diffusion process of the form*

$$X(t) = at + \sigma W(t) + \sum_{j=1}^{N(t)} \xi_j.$$

Then X is a Lévy process for which

$$\psi(u) = iua - u^2\frac{\sigma^2}{2} + \lambda \int \left(e^{iux} - 1\right) \eta(dx), \tag{6.11}$$

where η is the common distribution of the jumps ξ_j. As a result, its Lévy measure is $\lambda\eta$. Its first characteristic is $a + \lambda E(\xi_1 \mathbb{I}(|\xi_1| \leq 1)$. Moreover, if c exists on an open interval containing $[0,1]$, then e^X is a positive martingale with mean 1 if and only if $c(1) = 0$.

Remark 6.3.1 *The characteristic σ^2 related to the Brownian component of a Lévy process can be easily recovered from the general expression of ψ. In fact, since*

$$\tilde{\psi}(u) = \frac{\psi(u) + \psi(-u)}{2} = -u^2\frac{\sigma^2}{2} - \int \{1 - \cos(ux)\}\nu(dx),$$

it is easy to check that

$$\lim_{u\to\infty} \tilde{\psi}(u)/u^2 = -\frac{\sigma^2}{2}. \tag{6.12}$$

The latter result could be useful in case we cannot compute the Lévy measure to determine whether or not the Lévy process has a Brownian part.

6.3.3 Infinitely Divisible Distributions

A random variable Y is said to be *infinitely divisible* if for any $n \geq 1$, there exists i.i.d. random variables $Y_{1,n}, \ldots, Y_{n,n}$ such that $Y = Y_{1,n} + \cdots + Y_{n,n}$. It can be shown that the characteristic function of an infinitely divisible distribution is necessarily of the form e^ψ, with ψ satisfying the Lévy-Khinchin formula (6.8). In fact, there exists a Lévy process X such that $X(1) = Y$, and the characteristic function of $X(t)$ is $e^{t\psi}$. Conversely, for a Lévy process X, $X(t)$ is infinitely divisible for any $t > 0$. Examples of infinitely divisible distributions include the Gaussian, gamma, and Poisson distributions. Also, the Student distribution is infinitely divisible [Cont and Tankov, 2004].

6.3.4 Sample Path Properties

Most Lévy processes are not continuous. In fact, the only continuous Lévy processes X are of the form $X(t) = at+\sigma W(t)$, where W is a Brownian motion. Other interesting path properties for the pure jump part L are determined by its Lévy measure.

6.3.4.1 Number of Jumps of a Lévy Process

In a finite time interval, the number of jumps of a Lévy process with characteristics $\{a, \sigma^2, \nu\}$ can be finite or infinite, according as $\nu(\mathbb{R}) < \infty$ or $\nu(\mathbb{R}) = \infty$. This follows from the fact that for a given interval A with $\bar{A} \subset \mathbb{R}\backslash 0$,

$$N_A(t) = \sum_{0<s\leq t} \mathbb{I}(\Delta X(s) \in A)$$

is a Poisson process with intensity $\nu(A)$. Here, $\bar{A}$ is the smallest closed interval containing A. As a result,

$$E\{N_A(t)\} = t\nu(A) = E\left[\sum_{0<s\leq t} \mathbb{I}\{\Delta X(s) \in A\}\right].$$

For example, for the jump-diffusion processes introduced before, we have

$$N_A(t) = \sum_{k=1}^{N(t)} \mathbb{I}(\xi_k \in A),$$

which is clearly a Poisson process with intensity $\lambda\eta(A) = \nu(A)$. Note that $\nu(\mathbb{R}) < \infty$ if and only if X is a jump-diffusion process. In this case $\lambda = \nu(\mathbb{R})$, and the distribution of the jumps ξ_j is $\eta = \nu/\lambda$, which is a probability measure.

6.3.4.2 Finite Variation

Let L be a pure jump process. Then L is the sum of its jumps and has finite variation[3] if and only if $\int_{-1}^{1} |x|\nu(dx) < \infty$. In this case,

$$\psi(u) = ibu + \int \left(e^{iux} - 1\right) \nu(dx). \tag{6.13}$$

Furthermore, L is a non-decreasing process (implying it is of finite variation) if $\nu((-\infty, 0)) = 0$ and $\int_0^1 x\nu(dx) < \infty$. In such a case, L is called a *subordinator* and there exists $b \geq 0$, such that

$$\psi_L(u) = ibu + \int_0^\infty \left(e^{iux} - 1\right) \nu(dx). \tag{6.14}$$

Note that the Brownian motion is not a process with finite variation.

[3]A process L has finite variation if and only if for any T and any $0 = t_0 \leq t_1 \leq \cdots \leq t_n = T$, $\sum_{k=1}^{n} |L(t_k) - L(t_{k-1})| \leq V_T$, for some increasing process V.

6.4 Examples of Lévy Processes

We now give some examples of Lévy processes other than the compound Poisson processes and the jump-diffusion models. For more details, see Schoutens [2003].

In view of applications, we restrict ourselves to Lévy processes with a cumulant generating function c existing in an open interval containing $[0,1]$. In fact, since we want to model the returns of assets prices by Lévy processes, we need $E\{S(t)\} = E\left\{S_0^{X(t)}\right\} = S_0 e^{tc(1)}$ to be finite.

Natural Characteristics

Suppose that the pure jump part of a Lévy process has finite variation. The triplet $\{a, \sigma^2, \nu\}$ is called the natural characteristics if the Lévy Khincin formula satisfies

$$\psi(u) = iau - \frac{\sigma^2}{2}u^2 + \int \left(e^{iux} - 1\right)\nu(dx). \quad (6.15)$$

Therefore, the cumulant generating function c has the following representation, for all u is some open interval:

$$c(u) = au + \frac{\sigma^2}{2}u^2 + \int \left(e^{ux} - 1\right)\nu(dx). \quad (6.16)$$

In particular, $a = E\{X(1)\} + \int x\nu(dx)$ and $\text{Var}\{X(1)\} = \sigma^2 + \int x^2\nu(dx)$. It follows that the characteristics are $\left\{a + \int_{-1}^{1} x\nu(dx), \sigma^2, \nu\right\}$.

6.4.1 Gamma Process

A gamma process is a non-decreasing Lévy process, i.e., a subordinator, such that the law at time 1 is Gamma(α, β). It follows that the law at time t is Gamma$(\alpha t, \beta)$. In fact, that for all $u < 1/\beta$, we have

$$tc(u) = -\alpha t \ln(1 - \beta u),$$

which is the cumulant generating function of a Gamma$(\alpha t, \beta)$. Its Lévy measure ν has density[4] $\alpha \frac{e^{-x/\beta}}{x}\mathbb{I}(x>0)$, since

$$\begin{aligned}\int_0^\infty (e^{ux}-1)\frac{e^{-x/\beta}}{x}dx &= \int_0^\infty \int_0^{ux} e^y \frac{e^{-x/\beta}}{x}dydx \\ &= \int_0^u e^{-x(1-\beta z)/\beta}dxdz = -\ln(1-\beta u).\end{aligned}$$

It follows that the natural characteristics are $\{0,0,\nu\}$. Finally, since $\nu((0,\infty)) = \infty$, the gamma process has infinite activity.

6.4.2 Inverse Gaussian Process

For $\alpha, \beta > 0$, let $T_{\alpha,\beta}$ be the first hitting time of point α for the process $\beta t + W(t)$, where W is a Brownian motion. Then the law of $T_{\alpha,\beta}$ is called the *Inverse Gaussian* distribution, denoted $T_{\alpha,\beta} \sim \text{IG}(\alpha,\beta)$. This law is infinitely divisible, with characteristic function

$$\phi_{\alpha,\beta}(u) = E\left[e^{iuT_{\alpha,\beta}}\right] = e^{\alpha\left(\beta - \sqrt{\beta^2-2iu}\right)} = e^{\psi_{\alpha,\beta}(u)}.$$

Remark that $T_{\alpha,\beta} \stackrel{Law}{=} T_{\alpha\beta,1}/\beta^2$, since $\phi_{\alpha,\beta}(\beta^2 u) = \phi_{\alpha\beta,1}(u)$. Similarly, $T_{\alpha,\beta} \stackrel{Law}{=} \alpha^2 T_{1,\alpha\beta}$. Since $t\psi_{\alpha,\beta}(u) = \psi_{\alpha t,\beta}(u)$ for all $u \in \mathbb{R}$, it follows that one can set $X(t) = T_{\alpha t,\beta} \sim \text{IG}(\alpha t, \beta)$, which is clearly a non-decreasing process, so X is a subordinator. Next, the density of $X(1) = T_{\alpha,\beta}$ is given by

$$f(x) = \frac{\alpha e^{\alpha\beta - \frac{1}{2}(\alpha^2 x^{-1} + \beta^2 x)}}{\sqrt{2\pi x^3}}\mathbb{I}_{\{x>0\}} = \frac{\alpha e^{-\frac{(\beta x-\alpha)^2}{2x}}}{\sqrt{2\pi x^3}}\mathbb{I}_{\{x>0\}}, \quad x \in \mathbb{R},$$

and its Lévy measure ν has density

$$\frac{\alpha e^{-\frac{1}{2}\beta^2 x}}{\sqrt{2\pi x^3}}\mathbb{I}_{\{x>0\}}, \quad x \in \mathbb{R}. \tag{6.17}$$

The process has also infinite activity, and its natural characteristics are $\{0,0,\nu\}$. Finally, the mean and variance of $X(1)$ are respectively α/β and α/β^3.

6.4.2.1 Simulation of $T_{\alpha,\beta}$

Here we follow Devroye [1986]. The algorithm is due to Michael et al. [1976]. Since $T_{\alpha,\beta} \stackrel{Law}{=} T_{\alpha\beta,1}/\beta^2$, we only need to know how to generate $T_{\alpha,1}$.

First, generate $Z \sim N(0,1)$, then set $Y = Z^2$, and $X_1 = \alpha + Y - \sqrt{\alpha Y + Y^2/4}$. Next, generate $U \sim \text{Unif}(0,1)$. Then, if $U \le \frac{\alpha}{\alpha + X_1}$, set $T_{\alpha,1} = X_1$; otherwise, set $T_{\alpha,1} = \alpha^2/X_1$.

[4]The term "density" here simply means that $\nu(A) = \int_A f(x)dx$.

6.4.3 Generalized Inverse Gaussian Distribution

One of the most studied family of positive infinitely divisible distribution is the Generalized Inverse Gaussian family (GIG for short). If $X(1) \sim \text{GIG}(\lambda, \alpha, \beta)$, then its density is

$$f_{\lambda,\alpha,\beta}(x) = \frac{1}{2(\alpha/\beta)^\lambda K_\lambda(\beta\alpha)} x^{\lambda-1} e^{-\frac{1}{2}\left(\beta^2 x + \frac{\alpha^2}{x}\right)} \mathbb{I}(x > 0), \quad x \in \mathbb{R},$$

where $\beta \geq 0, \alpha \geq 0$, $\lambda \in \mathbb{R}$. If $\beta = 0$, then $\lambda < 0$; if $\alpha = 0$, then $\lambda > 0$, corresponding to the gamma distribution with parameters $(\lambda, 2/\beta^2)$. Its characteristic function is

$$\phi(u) = \left(1 - 2iu/\beta^2\right)^{-\lambda/2} \frac{K_\lambda\left(\alpha\beta\sqrt{1 - 2iu/\beta^2}\right)}{K_\lambda(\alpha\beta)}, \quad u \in \mathbb{R}.$$

This family contains the gamma, Inverse gamma, Inverse Gaussian, and Hyperbolic distributions to name a few. Here K_λ[5] is the modified Bessel function of the second kind[6] of order λ (see Appendix 6.A). Since $K_{\pm 1/2}(x) = \sqrt{\frac{\pi}{2x}} e^{-x}$, we see that the GIG$(1/2, \alpha, \beta)$ corresponds to the IG(α, β). Unfortunately, the law of $X(t)$ is not known in general.

6.4.4 Variance Gamma Process

The Variance Gamma process, introduced by Madan et al. [1998], can be defined by a random time change of a Brownian motion W viz. $X(t) = \mu V(t) + \sigma W\{V(t)\}$, where V is a gamma process of parameters $(\alpha, 1/\alpha)$, independent of W. The corresponding law for $X(1)$ is denoted $X(1) \sim \text{VG}(\mu, \sigma, \alpha)$. Note that Madan et al. [1998] used a different parametrization.

It is easy to check that $\psi(u) = -\alpha \ln\left\{1 + \frac{1}{\alpha}\left(iu\mu - u^2\sigma^2/2\right)\right\}$, and the cumulant generating function c exists for all u satisfying $(u\sigma + \mu/\sigma)^2 < 2\alpha + \mu^2/\sigma^2$, and

$$c(u) = -\alpha \ln\left\{1 - \frac{1}{\alpha}\left(u\mu + u^2\sigma^2/2\right)\right\}.$$

It then follows that X does not have a Brownian component since $\tilde{\psi}(u)/u^2 \to 0$ as $u \to \infty$.

Set $\beta_1 = \frac{\mu + \sqrt{2\alpha\sigma^2 + \mu^2}}{2\alpha} > 0$ and $\beta_2 = \frac{-\mu + \sqrt{2\alpha\sigma^2 + \mu^2}}{2\alpha} > 0$.

Since $\beta_1\beta_2 = \frac{\sigma^2}{2\alpha}$ and $\beta_1 - \beta_2 = \frac{\mu}{\alpha}$, we obtain that

$$\psi(u) = -\alpha \ln(1 - i\beta_1 u) - \alpha \ln(1 + i\beta_2 u) = \psi_1(u) + \psi_2(-u),$$

where $\psi_i(u) = -\alpha \ln(1 - i\beta_i u)$ is the cumulant generating function of a Gamma(α, β_i), $i \in \{1, 2\}$. This proves that X is the difference between

[5] In MATLAB, the corresponding function is *besselk*.

[6] Also called sometimes modified Bessel function of the third kind!

two independent gamma processes Y_1, Y_2, i.e., $X(t) = Y_1(t) - Y_2(t)$, with $Y_i(t) \sim \text{Gamma}(\alpha t, \beta_i)$, $i \in \{1,2\}$. Therefore X is a process with finite variation, being the difference of two non-decreasing functions. Using the results for gamma processes, one may conclude that the Lévy measure ν of X has density

$$\alpha \frac{e^{-x/\beta_1}}{|x|} \mathbb{I}(x > 0) + \alpha \frac{e^{x/\beta_2}}{|x|} \mathbb{I}(x < 0), \quad x \in \mathbb{R}, \tag{6.18}$$

from which we deduce that the process has infinite activity. Note also that the cumulant of order n is $\alpha(n-1)!\{\beta_1^n + (-1)^n \beta_2^n\}$, $n \geq 1$. We can also find the law of $X(t)$. In fact, $X(t) \sim \text{VG}(\mu t, \sigma\sqrt{t}, \alpha t)$ and the density of $X(1)$ can be deduced from the next proposition by setting $\beta = \sigma^2/\alpha$ and $\theta = \mu/\sigma^2$.

Proposition 6.4.1 *Suppose that $Z \sim N(0,1)$ is independent of $V \sim \text{Gamma}(\alpha, \beta)$. Then the density of $Y = Z\sqrt{V} + \theta V$ is given by*

$$f(y) = 2 \frac{|y|^{\alpha - \frac{1}{2}} e^{\theta y}}{\sqrt{2\pi}\beta^\alpha \Gamma(\alpha)(\theta^2 + \frac{2}{\beta})^{\frac{\alpha}{2} - \frac{1}{4}}} K_{\alpha - \frac{1}{2}}\left(|y|\sqrt{\theta^2 + \frac{2}{\beta}}\right), \quad y \in \mathbb{R}, \tag{6.19}$$

where $K_\lambda(z)$ is the modified Bessel function of the second kind of order λ. In particular, the density of $Y = Z\sqrt{V}$ is

$$f(y) = \frac{|y|^{\alpha - \frac{1}{2}}}{\sqrt{\pi}\beta^{\frac{\alpha}{2} + \frac{1}{4}} 2^{\frac{\alpha}{2} - \frac{3}{4}} \Gamma(\alpha)} K_{\alpha - \frac{1}{2}}\left(|y|\sqrt{2/\beta}\right), \quad y \in \mathbb{R}, \tag{6.20}$$

and

$$\lim_{y \to \infty} \frac{1}{y} \ln\{P(Y > y)\} = -\sqrt{2/\beta}.$$

The Brownian motion with drift μ and volatility σ is a limiting case of the Variance Gamma process when $\alpha \to \infty$, as it can be seen from the limit of its cumulant generating function.

6.4.5 Lévy Subordinators

As seen for the Variance Gamma process, one way of constructing new processes is to look at a Brownian motion W evaluated at a random time given by a subordinator V. One can also replace the Brownian motion by another Lévy process. The question is: Do we always get a Lévy process?

The answer is yes! In fact, suppose that X is a Lévy process with natural characteristics $\{a_X, \sigma_X^2, \nu_X\}$, and V is a subordinator with natural characteristics $\{b, 0, \nu\}$, i.e., $c(u) = ub + \int_0^\infty (e^{ux} - 1)\,\nu(dx)$, $b \geq 0$, with V and X independent. It is assumed that the cumulant generating function c exists for all $u < u_0$, for some $u_0 > 0$, and the cumulant generating function of X exists for all u in some open interval $\mathbb{I}_X$.

Then $Y(t) = X\{V(t)\}$ is a Lévy process, called a Lévy subordinator, with cumulant generating function c_Y given by $c_Y(u) = c\{c_X(u)\}$, for all $u \in I_X$ with $c_X(u) < u_0$. In particular, $E\{Y(1)\} = E\{V(1)\}E\{X(1)\}$ and

$$\text{Var}\{Y(1)\} = \text{Var}\{V(1)\}E^2\{X(1)\} + E\{V(1)\}\text{Var}\{X(1)\}.$$

Moreover its natural characteristics are $\left\{\alpha_Y, \sigma_Y^2, \nu_Y\right\}$, where

$$\begin{aligned} a_Y &= b a_X, \\ \sigma_Y^2 &= b\sigma_X^2, \\ \nu_Y(A) &= b\nu_X(A) + \int_0^\infty P\{X(s) \in A\}\nu(ds), \end{aligned} \tag{6.21}$$

for every interval A, with $\bar{A} \subset \mathbb{R} \setminus \{0\}$. It follows that there is no Brownian component if $b = 0$. Basically, (6.21) means that for nice functions f with $f(x)/x$ bounded about 0,

$$\int f(x)\nu_Y(dx) = b\int f(x)\nu_X(dx) + \int_0^\infty E\left[f\{X(t)\}\right]\nu(dt).$$

Example 6.4.1 (Variance Gamma Process) *Here $X(t) = \mu t + \sigma W(t)$, with characteristics $\left\{\mu, \sigma^2, 0\right\}$ and V is a gamma process with parameters $\left(\alpha, \sigma^2/\alpha\right)$.*

Example 6.4.2 (Negative Binomial Process) *Taking X to be the Poisson process, and V the gamma process, we obtain a negative binomial process, because the law of $Y(1)$ is called the negative binomial distribution. See, e.g., Exercise A.2.*

Example 6.4.3 (Normal Inverse Gaussian Process) *Another quite interesting example is the Normal Inverse Gaussian distribution, obtained as a Lévy subordinator with respect to the Brownian motion, where the subordinator is the Inverse Gaussian process. More precisely, $X(1) \sim \text{NIG}(\alpha, \beta, \delta)$, $|\beta| < \alpha$, $\delta > 0$, if*

$$\psi(u) = \delta\left\{\sqrt{\alpha^2 - \beta^2} - \sqrt{\alpha^2 - (\beta + iu)^2}\right\},$$

so the process has no Brownian component. Its cumulant generating function exists for all $u \in (-\alpha - \beta, \alpha - \beta)$ and is given by

$$c(u) = \delta\left\{\sqrt{\alpha^2 - \beta^2} - \sqrt{\alpha^2 - (\beta + u)^2}\right\}.$$

It follows that $X(t) \sim \text{NIG}(\alpha, \beta, \delta t)$. The Lévy measure of the process has density

$$\frac{\alpha\delta}{\pi}\frac{e^{\beta x}K_1(\alpha|x|)}{|x|}, \quad x \in \mathbb{R}, \tag{6.22}$$

while the density of $X(1)$ *is*

$$\frac{\alpha\delta}{\pi} e^{\delta\sqrt{\alpha^2-\beta^2}+\beta x} \frac{K_1(\alpha\sqrt{\delta^2+x^2})}{\sqrt{\delta^2+x^2}}, \quad x \in \mathbb{R}.$$

Finally, to be able to simulate a NIG process, it suffices to write it as a Lévy subordinator, viz.

$$X(t) = \theta V(t) + \sigma W\{V(t)\},$$

where $\theta = \beta\sigma^2$, *and* V *is an Inverse Gaussian process with parameters* δ/σ *and* $\sigma\sqrt{\alpha^2-\beta^2}$, *i.e.,* $V(1) \sim IG\left(\delta/\sigma, \sigma\sqrt{\alpha^2-\beta^2}\right)$.

Taking again the Brownian motion with the GIG subordinator, we obtain the so-called Generalized Hyperbolic family. See Schoutens [2003] for details.

6.5 Change of Distribution

In this section we define change of measures preserving the Lévy property. We start by defining the so-called Radon-Nikodym density based on the Esscher transform of the process X. The proof of the following result is given in Appendix 6.C.1.

6.5.1 Esscher Transforms

Lemma 6.5.1 *Suppose that* X *is a Lévy process with characteristics* $\{a, \sigma^2, \nu\}$ *such that* $c(u)$ *defined by* (6.10) *exists for all* $|u| < u_0$. *For a given* $\zeta \in (-u_0, u_0)$, *define the Esscher transform* $\Lambda(t) = e^{\zeta X(t) - c(\zeta)t}$.

Then $\Lambda(t)$ *is a positive martingale with mean* 1. *In addition, if the probability measure* $\tilde{P}$ *is defined by the density* $\Lambda(t)$ *on* $\mathcal{F}_t$, *i.e.,* $\left.\frac{d\tilde{P}}{dP}\right|_{\mathcal{F}_t} = \Lambda(t)$, *then under* $\tilde{P}$, X *is a Lévy process with characteristics* $\{\tilde{a}, \sigma^2, \tilde{\nu}\}$, *where the Lévy measure* $\tilde{\nu}$ *has density* $\frac{d\tilde{\nu}}{d\nu}(x) = e^{\zeta x}$ *with respect to* ν, *and*

$$\tilde{a} = a + \zeta\sigma^2 + \int_{-1}^{1} x\left(e^{\zeta x} - 1\right)\nu(dx).$$

In addition, if $|u| < u_0 - |\zeta|$, *then* $E_{\tilde{P}}\left\{e^{uX(t)}\right\} = e^{t\tilde{c}(u)}$, *where*

$$\tilde{c}(u) = u\tilde{a} + u^2\frac{\sigma^2}{2} + \int \{e^{ux} - 1 - ux\mathbb{I}(|x| \le 1)\}\,\tilde{\nu}(dx).$$

In particular, if the Lévy measure ν *has a density* $f(x)$, *then the density of* $\tilde{\nu}$

is $e^{\zeta x}f(x)$. Finally, if the pure jump part of X has finite variation, then the relationship between the first natural characteristics $\tilde{a}^{(nat)}$ and $a^{(nat)}$ is

$$\tilde{a}^{(nat)} = a^{(nat)} + \zeta\sigma^2.$$

In the particular case of jump-diffusion processes, we have the following interesting corollary.

Corollary 6.5.1 *Under the same conditions as in Lemma 6.5.1, if X is a jump-diffusion process with representation $X(t) = at + \sigma W(t) + \sum_{j=1}^{N(t)} \xi_j$, then under measure $\tilde{P}$, X is also a jump-diffusion process with representation*

$$X(t) = \tilde{a}t + \sigma\tilde{W}(t) + \sum_{j=1}^{\tilde{N}(t)} \tilde{\xi}_j,$$

where $\tilde{a} = a+\zeta\sigma^2$, $\tilde{W}$ is a Brownian motion independent of the Poisson process $\tilde{N}$ with intensity $\tilde{\lambda} = \lambda E\left(e^{\zeta\xi_j}\right)$ and the independent jumps $\tilde{\xi}_j$ have distribution $\tilde{\eta}$, with $\tilde{\eta}$ having density $e^{\zeta x}/E\left(e^{\zeta\xi_j}\right)$ with respect to η. In particular, if η has density f, then $\tilde{\eta}$ has density $\frac{e^{\zeta x}f(x)}{\int e^{\zeta y}f(y)dy}$.

PROOF. From Lemma 6.5.1 and (6.11), we have

$$\begin{aligned}
\tilde{c}(u) &= ua + u\zeta\sigma^2 + u^2\frac{\sigma^2}{2} + \lambda\int e^{\zeta x}\left(e^{ux}-1\right)\eta(dx) \\
&= u\tilde{a} + u^2\frac{\sigma^2}{2} + \tilde{\lambda}\int \left(e^{ux}-1\right)\tilde{\eta}(dx).
\end{aligned}$$

Hence, the conclusion follows from (6.11) and the uniqueness of the decomposition of Lévy processes [Protter, 2004].

■

6.5.2 Examples of Application

We are now in a position to look at the effects of the proposed change of distribution for the main models we have defined so far.

6.5.2.1 Merton Model

Applying Corollary 6.5.1 for the Merton model described in Section 6.2.3, the density $\tilde{f}(x)$ of $\tilde{\nu}$ is

$$\tilde{f}(x) = \lambda e^{\gamma\zeta+\zeta^2\delta^2/2} e^{-\frac{(x-\gamma-\zeta\delta^2)^2}{2\delta^2}} \Big/ (\sqrt{2\pi}\,\delta), \quad x \in \mathbb{R}.$$

As a result, under $\tilde{P}$, X is also a Merton jump-diffusion process, where the intensity of the Poisson process is $\tilde{\lambda} = \lambda e^{\zeta\gamma+\zeta^2\delta^2/2}$, and the distribution of the jumps is Gaussian, with mean $\gamma + \zeta\delta^2$ and variance δ^2.

6.5.2.2 Kou Model

Applying Corollary 6.5.1 for the model of Kou described in Section 6.2.4, for any $\zeta \in (-\eta_2, \eta_1)$, the density $\tilde{f}$ of $\tilde{\nu}$ is

$$\tilde{f}(x) = \lambda p \eta_1 e^{-(\eta_1-\zeta)x}\mathbb{I}(x>0) + \lambda(1-p)\eta_2 e^{(\eta_2+\zeta)x}\mathbb{I}(x \le 0), \quad x \in \mathbb{R}.$$

It follows that under $\tilde{P}$, X is also a Kou jump-diffusion with parameters $\tilde{\lambda} = \lambda\left\{\frac{p\eta_1}{\eta_1-\zeta} + \frac{(1-p)\eta_2}{\eta_2+\zeta}\right\}$, $\tilde{\eta}_1 = \eta_1 - \zeta$, $\tilde{\eta}_2 = \eta_2 + \zeta$, and $\tilde{p} = \frac{p\eta_1(\eta_2+\zeta)}{p\eta_1(\eta_2+\zeta)+(1-p)\eta_2(\eta_1-\zeta)}$.

6.5.2.3 Variance Gamma Process

Applying Lemma 6.5.1 for the Variance Gamma process described in Section 6.4.4 and applying formula (6.18) giving the density of the Lévy measure of a Variance Gamma process with parameters (α, β, θ), we see that for any $\zeta \in (-1/\beta_2, 1/\beta_1)$, the density $\tilde{f}$ of the Lévy measure $\tilde{\nu}$ is

$$\begin{aligned}\tilde{f}(x) &= e^{\zeta x} f(x) = \alpha \frac{e^{-x(1-\zeta\beta_1)/\beta_1}}{x}\mathbb{I}(x>0) + \alpha\frac{e^{x(1+\zeta\beta_2)/\beta_2}}{|x|}\mathbb{I}(x<0) \\ &= \alpha\frac{e^{-x/\tilde{\beta}_1}}{x}\mathbb{I}(x>0) + \alpha\frac{e^{x/\tilde{\beta}_2}}{|x|}\mathbb{I}(x<0), \quad x \in \mathbb{R}, \end{aligned} \tag{6.23}$$

where $\tilde{\beta}_1 = \beta_1/(1-\zeta\beta_1)$ and $\tilde{\beta}_2 = \beta_2/(1+\zeta\beta_2)$. It follows that under $\tilde{P}$, X is also a Variance Gamma process with parameters $(\tilde{\mu}, \tilde{\sigma}, \alpha)$, with $\tilde{\mu} = \alpha\left(\tilde{\beta}_1 - \tilde{\beta}_2\right)$ and $\tilde{\sigma} = \sqrt{2\alpha\tilde{\beta}_1\tilde{\beta}_2}$.

6.5.2.4 Normal Inverse Gaussian Process

Applying Lemma 6.5.1 for the Normal Inverse Gaussian process described in Example 6.4.3 and applying formula (6.22) giving the density of the Lévy measure of a Normal Inverse Gaussian process with parameters (α, β, δ), we see that for any $|\beta+\zeta| < \alpha$, the density $\tilde{f}$ of the Lévy measure $\tilde{\nu}$ is

$$\tilde{f}(x) = e^{\zeta x} f(x) = \frac{\alpha\delta}{\pi}\frac{e^{(\beta+\zeta)x}K_1(\alpha|x|)}{|x|}, \quad x \in \mathbb{R}. \tag{6.24}$$

It follows that under $\tilde{P}$, X is also a Normal Inverse Gaussian process with parameters $(\alpha, \beta+\zeta, \delta)$.

6.5.3 Application to Option Pricing

Here is another interesting consequence of Lemma 6.5.1 for pricing purposes. Recall that for a Lévy process X with cumulant generating function c, $e^{-rt+X(t)}$ is a positive martingale with expectation 1 if and only if $c(1) = r$.

Corollary 6.5.2 *Suppose that under an equivalent martingale measure Q,*

$X(t) = \ln\{S(t)/s\}$ is a Lévy process with characteristics $\{a, \sigma^2, \nu\}$ with $c(1) = r$ and with survival function

$$\bar{F}_{a,\sigma,\nu,t}(x) = P(X(t) > x) = \bar{F}_{0,\sigma,\nu,t}(x - at).$$

Further set

$$\tilde{a} = a + \sigma^2 + \int x\,(e^x - 1)\,\mathbb{I}(|x| \le 1)\nu(dx)$$

and define the Lévy measure $\tilde{\nu}$ by $\frac{d\tilde{\nu}}{d\nu} = e^x$.

Then the value at time t of a European call option with strike price K and maturity $t + \tau$, is given by

$$C(\tau, s) = s\bar{F}_{0,\sigma,\tilde{\nu},\tau}(E_1) - Ke^{-r\tau}\bar{F}_{0,\sigma,\nu,\tau}(E_2).$$

where $E_1 = \ln(K/s) - \tau\tilde{a}$ and $E_2 = E_1 + \tau(\tilde{a} - a) = \ln(K/s) - \tau a$.

The proof is given in Appendix 6.C.2. Note that with $\nu \equiv 0$, we retrieve the Black-Scholes formula.

6.5.4 General Change of Measure

Suppose that X is a Lévy process with characteristics $\{a, \sigma^2, \nu\}$. For a function ϕ satisfying the integrability condition

$$\int \left\{e^{\phi(x)/2} - 1\right\}^2 \nu(dx) < \infty, \tag{6.25}$$

set $U_{b,\phi}(t) = bW(t) - t\frac{b^2}{2} + L_\phi(t)$, where L_ϕ is a Lévy process defined as the almost sure uniform limit (on any time interval $[0, T]$) of the sequence of centered compound Poisson processes

$$L_\phi^{(\epsilon)}(t) = \sum_{0<s\le t} \phi\{\Delta X(s)\}\mathbb{I}\{|\Delta X(s)| > \epsilon\} - t\int_{|x|>\epsilon} \left\{e^{\phi(x)} - 1\right\}\nu(dx),$$

as $\epsilon \to 0$. The following result is proven in Sato [1999] but its formulation is taken from Cont and Tankov [2004].

Theorem 6.5.1 *Suppose X is a Lévy process with characteristics $\{a, \sigma^2, \nu\}$. Then X is a Lévy process with characteristics $\{\tilde{a}, \tilde{\sigma}^2, \tilde{\nu}\}$ under an equivalent measure $\tilde{P}$ if and only if there exist $b \in \mathbb{R}$ and ϕ satisfying (6.25) such that $\left.\frac{d\tilde{P}}{dP}\right|_{\mathcal{F}_t} = e^{U_{b,\phi}(t)}$, with $U_{b,\phi}$ defined by (6.5.4). In that case, one must have*

$$\tilde{a} = a + b\sigma + \int_{-1}^{1} x\left\{e^{\phi(x)} - 1\right\}\nu(dx), \tag{6.26}$$

$\tilde{\sigma}^2 = \sigma^2$ and the Lévy measure $\tilde{\nu}$ has density e^ϕ with respect to ν.

Finally, if the jump part of X has finite variation, then the relationship between the first natural characteristics $\tilde{a}^{(nat)}$ and $a^{(nat)}$ is

$$\tilde{a}^{(nat)} = a^{(nat)} + b\sigma.$$

Note that Lemma 6.5.1 is a particular case of Theorem 6.5.1 corresponding to $\phi(x) = \zeta x$ and $b = \zeta\sigma$.

Remark 6.5.1 *Under the assumption that $\int \left|e^{\phi(x)} - 1\right| \nu(dx) < \infty$, ϕ satisfies (6.25) and*

$$L_\phi(t) = \sum_{0<\leq s\leq t} \phi\{\Delta X(s)\} - t \int \left\{e^{\phi(x)} - 1\right\} \nu(dx).$$

6.5.5 Incompleteness

From now on, assume that X is a Lévy process with characteristics $\{a, \sigma^2, \nu\}$, such that its cumulant generating function c exists in an open interval $I \supset [0, 1]$.

For a given ϕ satisfying the integrability condition (6.25), it follows from Theorem 6.5.1 that $e^{-rt}S(t) = S_0 e^{-rt+X(t)}$ is a martingale under $\tilde{P} = \tilde{P}_{b,\phi}$ if and only if

$$\begin{aligned} r &= \tilde{c}_{b,\phi}(1) = a + b\sigma + \frac{\sigma^2}{2} + \int_{-1}^{1} x\left\{e^{\phi(x)} - 1\right\} \nu(dx) \\ &\quad + \int \{e^x - 1 - x\mathbb{I}(|x| \leq 1)\}\, \tilde{\nu}(dx). \end{aligned} \tag{6.27}$$

Note that if the jump part of X has finite variation and natural characteristics $\{a, \sigma^2, \nu\}$, then (6.27) reduces to

$$r = a + b\sigma + \frac{\sigma^2}{2} + \int (e^x - 1)\, \tilde{\nu}(dx). \tag{6.28}$$

As a result, if X has a Brownian motion part, i.e., $\sigma > 0$, there exists an infinite number of equivalent martingale measures, since for any $\zeta \in I$, one can take $\phi(x) = \zeta x$ and b satisfying (6.27).

On the other hand, if $\sigma = 0$, there is a unique $\zeta \in I$ with $r = a + \int e^{\zeta x} (e^x - 1)\, \nu(dx)$. However, we could also take $\phi_{\zeta_1,\zeta_2}(x) = \zeta_1 x\mathbb{I}(x > 0) + \zeta_2 x\mathbb{I}(x \leq 0)$, with $\zeta_1, \zeta_2 \in I$. Then for a given $\zeta_1 \in I$, there is a unique $\zeta_2 \in I$ so that

$$\begin{aligned} r &= a + \int_{-1}^{1} x\left\{e^{\phi_{\zeta_1,\zeta_2}(x)} - 1\right\} \nu(dx) \\ &\quad + \int \{e^x - 1 - x\mathbb{I}(|x| \leq 1)\}\, e^{\phi_{\zeta_1,\zeta_2}(x)} \nu(dx), \end{aligned}$$

proving that in this case too, there exists infinitely many equivalent martingale measures. Finally, note that for Lévy processes with finite activity, i.e., for jump-diffusion processes, then one could take $\phi(x) = \zeta_0 + \zeta_1 x$. This choice of ϕ is the most general one preserving the Merton family, i.e., under $\tilde{P}$, X a jump-diffusion with Gaussian jumps. Taking $\phi(x) = (\zeta_{01} + \zeta_{11}x)\mathbb{I}(x > 0) + (\zeta_{02} + \zeta_{12}x)\mathbb{I}(x \le 0)$ preserves both Kou model and the weighted-symmetric Merton model.

Remark 6.5.2 *For Lévy processes with characteristics $(a, 0, \nu)$, the choice $\phi(x) = \zeta_1 x\mathbb{I}(x > 0) + \zeta_2 x\mathbb{I}(x \le 0)$ often yields processes within the same family. For example, in the Variance Gamma family, according to (6.18), ν has density*

$$f(x) = \alpha \frac{e^{-x/\beta_1}}{|x|}\mathbb{I}(x > 0) + \alpha \frac{e^{x/\beta_2}}{|x|}\mathbb{I}(x < 0), \quad x \in \mathbb{R},$$

so under the change of measure, according to Theorem 6.5.1, $\tilde{\nu}$ has density

$$\begin{aligned} \tilde{f}(x) &= e^{\phi(x)} f(x) = \alpha \frac{e^{-x(1-\zeta_1\beta_1)/\beta_1}}{x}\mathbb{I}(x > 0) + \alpha \frac{e^{x(1+\zeta_2\beta_2)/\beta_2}}{|x|}\mathbb{I}(x < 0) \\ &= \alpha \frac{e^{-x/\tilde{\beta}_1}}{x}\mathbb{I}(x > 0) + \alpha \frac{e^{x/\tilde{\beta}_2}}{|x|}\mathbb{I}(x < 0), \quad x \in \mathbb{R}, \end{aligned} \tag{6.29}$$

where $\tilde{\beta}_1 = \beta_1/(1 - \zeta_1\beta_1)$ and $\tilde{\beta}_2 = \beta_2/(1 + \zeta_2\beta_2)$.

It follows that under $\tilde{P}$, X is also a Variance Gamma process with parameters $(\tilde{\mu}, \tilde{\sigma}, \alpha)$, with $\tilde{\mu} = \alpha\left(\tilde{\beta}_1 - \tilde{\beta}_2\right)$ and $\tilde{\sigma} = \sqrt{2\alpha\tilde{\beta}_1\tilde{\beta}_2}$.

Specializing Theorem 6.5.1 to jump-diffusion processes, we obtain the following corollary.

Corollary 6.5.3 *Suppose that X is a jump-diffusion process with representation $X(t) = at + \sigma W(t) + \sum_{j=1}^{N(t)} \xi_j$, and ϕ is such that $K_\phi = E\left\{e^{\phi(\xi_1)}\right\} < \infty$. Define*

$$U_{b,\phi}(t) = bW(t) - t\frac{b^2}{2} + \sum_{j=1}^{N(t)} \phi(\xi_j) - t\lambda K_\phi.$$

Then under measure $\tilde{P}$, with Radon-Nikodym derivative $e^{U_{b,\phi}}$ with respect to P, X is also a jump-diffusion process with representation

$$X(t) = \tilde{a}t + \sigma\tilde{W}(t) + \sum_{j=1}^{\tilde{N}(t)} \tilde{\xi}_j,$$

where $\tilde{a} = a + b\sigma$, $\tilde{W}$ is a Brownian motion independent of the Poisson process $\tilde{N}$ with intensity $\tilde{\lambda} = \lambda K_\phi$ and the independent jumps $\tilde{\xi}_j$ have distribution $\tilde{\eta}$, with $\tilde{\eta}$ having density $e^{\phi(x)}/K_\phi$ with respect to η. In particular, if η has density f, then $\tilde{\eta}$ has density $f(x)e^{\phi(x)}/K_\phi$.

Note that by choosing $\phi(x) = \zeta_0 + \zeta_1 x$ for a jump-diffusion process with natural characteristics $\{a, \sigma^2, \nu\}$, we have $\tilde{\lambda} = \lambda e^{\zeta_0} E\left(e^{\zeta_1 \xi_1}\right)$ and the law η of the jumps under $\tilde{P}$ has density $e^{\zeta_1 x}/E\left(e^{\zeta_1 \xi_1}\right)$ with respect to original law η. Moreover, $\tilde{P}$ is an equivalent martingale measure if and only if

$$r = a + b\sigma + \frac{\sigma^2}{2} + \lambda e^{\zeta_0} E\left\{e^{(1+\zeta_1)\xi_1} - e^{\zeta_1 \xi_1}\right\}.$$

In particular, if $\zeta_1 = 0$, then the last condition reduces to

$$r = a + b\sigma + \frac{\sigma^2}{2} + \lambda e^{\zeta_0} \kappa,$$

which has an infinite number of solutions (b, ζ_0) provided $\sigma\lambda\kappa \neq 0$. Note that if $\zeta_1 = 0$ then under $\tilde{P}$, the distribution of the jumps remains the same and their intensity is simply $\tilde{\lambda} = \lambda e^{\zeta_0}$.

In Merton [1976], the author chose $\zeta_0 = \zeta_1 = 0$, so $b = \frac{r-\mu}{\sigma}$, which is the same as in the Black-Scholes model.

Remark 6.5.3 *For a family of Lévy processes with parametric characteristics* $\left\{a_\theta, \sigma^2, \nu_\theta\right\}_{\theta \in \Theta}$, *the class of equivalent martingale measures is often restricted to those under which the process is still a Lévy process within the same family, i.e,* $\tilde{\nu} = \nu_{\tilde{\theta}}$ *for some* $\tilde{\theta} \in \Theta$.

In fact, for jump-diffusions models, a popular choice is $\phi \equiv e^{\zeta_0}$, *so that the law of the jumps stay the same and the average number of jumps is* $\tilde{\lambda} = \lambda e^{\zeta_0}$. *The unknown parameter* ζ_0 *is then estimated by calibration, using option prices.*

6.6 Model Implementation and Estimation of Parameters

To estimate the parameters, the simplest method would be to use the moment-matching method with the cumulants. However, in many cases one has to use cumulants of order 6 which are more difficult to estimate precisely. As a result, using this method, we would incur large errors.

Another method of estimation can be based on the maximum likelihood principle, if the density of the returns is available. However, in most applications of interest, the density is an infinite sum so it must be truncated at some point, inducing errors.

These two methods are discussed next. But first we have to look at some particularities of Lévy processes.

6.6.1 Distributional Properties

We assume that the asset prices S are modeled as the exponential a Lévy process X belonging to some parametric family P_θ, with $\theta \in \Theta$, i.e., $S(t) = S_0 e^{X(t)}$, $t \geq 0$.

In practice, the value of the asset is observed at regular intervals of length h, so the observations available take the form

$$R_i = \ln \left[\frac{S(ih)}{S\{(i-1)h\}} \right] = X(ih) - X\{(i-1)h\}, \quad i \in \{1, \ldots, n\},$$

which are independent and identically distributed observations with the same distribution as $X(h)$.

6.6.1.1 Serial Independence

Since the returns are i.i.d., one of the first steps in implementing a Lévy model is to test for serial independence. To do so, one needs a nonparametric test, such as the ones developed in Section 3.1.1. If the null hypothesis of serial independence is not rejected, one can proceed with the estimation of parameters of the model. If the null hypothesis is rejected, a Lévy model is not appropriate for modeling the returns.

6.6.1.2 Lévy Process vs Brownian Motion

Before studying estimation techniques, one has to be aware of another particularity of Lévy processes: they can look like a Brownian motion with drift. This phenomenon is well illustrated in the MATLAB script *DemoCompoundPoisson*, where the paths of compound Poisson processes with jumps $\pm\delta$ are plotted, for several values of the average number of jumps λ. As λ increases, the paths look more and more like Brownian paths. This phenomenon also occurs for processes with infinite activity, such as the Variance Gamma process. See the MATLAB script *DemoVG*. These examples are not coincidental. The next proposition, proven in Appendix 6.C.3, shows why Lévy process can look like a Brownian motion.

Proposition 6.6.1 *If the variance of the Lévy process X exists, then*

$$Y_\lambda(t) = \{X(\lambda t) - \lambda \mu t\}/\sqrt{\lambda} \rightsquigarrow \sigma \tilde{W}(t),$$

as $\lambda \to \infty$, where $\tilde{W}$ is a Brownian motion, the mean of $X(1)$ is μ and its variance is σ^2.

For example, for a compound Poisson process $X_\lambda(t)$, with intensity λ, if the second moment of the jumps ξ_j exists, then $(X_\lambda(t) - \lambda\mu t)/\sqrt{\lambda}$ is approximately a Brownian motion. For a gamma process $X_{\alpha,\beta}$ with parameters (α, β), $(X_{\alpha,\beta}(t) - \alpha\beta t)/\sqrt{\alpha} \rightsquigarrow \beta\tilde{W}(t)$ when $\alpha \to \infty$. See the MATLAB script *DemoLevyGammaStd* for an illustration.

Remark 6.6.1 *If we want to model $S(t) = S_0 e^{X(t)}$ where X is a Lévy process, Proposition 6.6.1 tells us that we could experience difficulties. Indeed, X may look like a Brownian motion, and it could be difficult to distinguish between W and the pure jump process L in $X = at + \sigma W(t) + L(t)$. As a result, the parameter σ could be quite difficult to estimate if the number of jumps is large.*

We are now in a position to discuss the two estimation methods. Since the cumulants are easy to compute, this is the first method we address.

6.6.2 Estimation Based on the Cumulants

Suppose X is a Lévy process with characteristics $\{a, \sigma^2, \nu\}$. Since we assume that the moment generating function exists in a neighborhood of zero, it follows that the cumulants of the returns exist for any integer n, and

$$\kappa_1 = ha + h \int_{|x|>1} x\nu(dx), \tag{6.30}$$

$$\kappa_2 = h\sigma^2 + h \int x^2 \nu(dx), \tag{6.31}$$

$$\kappa_n = h \int x^n \nu(dx), \quad n \geq 3. \tag{6.32}$$

For example, in the case of a jump-diffusion process X, we have

$$R_i = ah + \sigma_h Z_i + \varepsilon_i$$

where $\sigma_h = \sigma\sqrt{h}$, the Z_is are independent standard Gaussian variables, and the ε_is are independent and identically distributed with the same law as $\varepsilon = \sum_{j=1}^N \xi_j$, where N has a Poisson distribution with parameter λh, the jumps ξ_j are independent and identically distributed, and $\kappa = E\left(e^{\xi}\right) - 1$. Then, $\kappa_1 = ah + \lambda h E(\xi)$, $\kappa_2 = \sigma^2 h + \lambda h E\left(\xi^2\right)$ and $\kappa_n = \lambda h E(\xi^n)$, $n \geq 3$.

Example 6.6.1 (Merton Model) *The law of the returns has 5 parameters, and the first 6 cumulants are $\kappa_1 = ah + \lambda h\gamma$, $\kappa_2 = \sigma^2 h + \lambda h(\gamma^2 + \delta^2)$, $\kappa_3 = \lambda h\left(3\gamma\delta^2 + \gamma^3\right)$, $\kappa_4 = \lambda h\left(3\delta^4 + 6\gamma^2\delta^2 + \gamma^4\right)$, $\kappa_5 = \lambda h\left(15\gamma\delta^4 + 10\gamma^3\delta^2 + \gamma^5\right)$, and $\kappa_6 = \lambda h(15\delta^6 + 45\delta^4\gamma^2 + 15\delta^2\gamma^4 + \gamma^6)$. Note that if $\gamma = 0$, then $\kappa_3 = \kappa_5 = 0$, so the sixth cumulant is necessary for the estimation of the remaining 4 parameters.*

Example 6.6.2 (Kou Model) *The law of the returns depends on 6 parameters. The first 6 cumulants are given by $\kappa_1 = a + \lambda h\left(\frac{p}{\eta_1} - \frac{1-p}{\eta_2}\right)$, $\kappa_2 = \sigma^2 h + 2\lambda h\left(\frac{p}{\eta_1^2} + \frac{1-p}{\eta_2^2}\right)$, $\kappa_3 = 6\lambda h\left(\frac{p}{\eta_1^3} - \frac{1-p}{\eta_2^3}\right)$, $\kappa_4 = 24\lambda h\left(\frac{p}{\eta_1^4} + \frac{1-p}{\eta_2^4}\right)$, $\kappa_5 = 120\lambda h\left(\frac{p}{\eta_1^5} - \frac{1-p}{\eta_2^5}\right)$, and $\kappa_6 = 720\lambda h\left(\frac{p}{\eta_1^6} + \frac{1-p}{\eta_2^6}\right)$, since $E\left(\xi^n\right) = n!\left\{\frac{p}{\eta_1^n} + \frac{(1-p)(-1)^n}{\eta_2^n}\right\}$, $n \geq 1$.*

Moreover, for any $\eta_1 > \theta > -\eta_2$,

$$E\left(e^{\theta\xi}\right) = \frac{p}{1-\theta/\eta_1} + \frac{1-p}{1+\theta/\eta_2}.$$

In particular, $\kappa = \frac{1+p\eta_2-(1-p)\eta_1}{(\eta_1-1)(\eta_2+1)}$.

6.6.2.1 Estimation of the Cumulants

Recall that from Section A.5.1, the first six cumulants of a random variable R are given by

$$\begin{aligned}
\kappa_1 &= \mu = E(R), \\
\kappa_2 &= \sigma^2 = \mathrm{Var}(R), \\
\kappa_3 &= E\left\{(R-\mu)^3\right\}, \\
\kappa_4 &= E\left\{(R-\mu)^4\right\} - 3\sigma^4, \\
\kappa_5 &= E\left\{(R-\mu)^5\right\} - 10\sigma^2\kappa_3, \\
\kappa_6 &= E\left\{(R-\mu)^6\right\} - 15\sigma^2\kappa_4 - 10\kappa_3^2 - 15\sigma^6.
\end{aligned}$$

Therefore, using observations $R_1, \ldots, R_n$ of R, we can estimate these cumulants by

$$\begin{aligned}
\hat\kappa_1 &= \bar R = \frac{1}{n}\sum_{i=1}^n R_i, \\
\hat\kappa_2 &= \frac{1}{n}\sum_{i=1}^n (R_i - \bar R)^2, \\
\hat\kappa_3 &= \frac{1}{n}\sum_{i=1}^n (R_i - \bar R)^3, \\
\hat\kappa_4 &= \frac{1}{n}\sum_{i=1}^n (R_i - \bar R)^4 - 3\hat\kappa_2^2, \\
\hat\kappa_5 &= \frac{1}{n}\sum_{i=1}^n (R_i - \bar R)^5 - 10\hat\kappa_2\hat\kappa_3, \\
\hat\kappa_6 &= \frac{1}{n}\sum_{i=1}^n (R_i - \bar R)^6 - 15\hat\kappa_2\hat\kappa_4 - 10\hat\kappa_3^2 - 15\hat\kappa_2^3.
\end{aligned}$$

The estimation error of the cumulants is given by the following result.

Proposition 6.6.2 *Suppose that the observations* $R_1, \ldots, R_n$ *are independent and identically distributed and that the moment of order 12 exists. Then*

$$\sqrt{n}\left(\hat\kappa_1 - \kappa_1, \ldots, \hat\kappa_6 - \kappa_6\right) \rightsquigarrow W = (W_1, \ldots, W_6) \sim N_6(0, V),$$

where

$$\begin{aligned}
W_1 &= Z_1, \\
W_2 &= Z_2, \\
W_3 &= Z_3 - 3\kappa_2 Z_1, \\
W_4 &= Z_4 - 4\kappa_3 Z_1 - 6\kappa_2 Z_2, \\
W_5 &= Z_5 - 5(\kappa_4 - 3\kappa_2^2) Z_1 - 10\kappa_3 Z_2 - 10\kappa_2 Z_3, \\
W_6 &= Z_6 - 6(\kappa_5 - 10\kappa_2\kappa_3) Z_1 - 15(\kappa_4 - 3\kappa_2^2) Z_2 - 20\kappa_3 Z_3 - 15\kappa_2 Z_4,
\end{aligned}$$

with $\mu_j = E\{(R-\mu)^j\}$, $j \in \{1,\ldots,12\}$, $\frac{1}{\sqrt{n}}\sum_{i=1}^n \left\{(R-\mu)^j - \mu_j\right\} \rightsquigarrow Z_j \sim N(0, \mu_{2j} - \mu_j^2)$, *and* $\mathrm{Cov}(Z_j, Z_k) = \mu_{j+k} - \mu_j\mu_k$, $j,k \in \{1,\ldots,6\}$.

The covariance matrix V can be estimated by the covariance matrix of the pseudo-observations $W_i = (W_{i1},\ldots,W_{i6})^\top$, $i \in \{1,\ldots,n\}$, *where* $Z_{ij} = (x_i - \bar{x})^j$, *and*

$$\begin{aligned}
W_{i1} &= Z_{i1}, \\
W_{i2} &= Z_{i2}, \\
W_{i3} &= Z_{i3} - 3\hat\kappa_2 Z_{i1}, \\
W_{i4} &= Z_{i4} - 4\hat\kappa_3 Z_{i1} - 6\hat\kappa_2 Z_{i2}, \\
W_{i5} &= Z_{i5} - 5(\hat\kappa_4 - 3\hat\kappa_2^2) Z_{i1} - 10\hat\kappa_3 Z_{i2} - 10\hat\kappa_2 Z_{i3}, \\
W_{i6} &= Z_{i6} - 6(\hat\kappa_5 - 10\hat\kappa_2\hat\kappa_3) Z_{i1} - 15(\hat\kappa_4 - 3\hat\kappa_2^2) Z_{i2} - 20\hat\kappa_3 Z_{i3} - 15\hat\kappa_2 Z_{i4}.
\end{aligned}$$

In the Gaussian case, V is diagonal and

$$\mathrm{Var}(W_j) = \sigma^{2j} j!, \quad j = 1,\ldots,6.$$

PROOF. The proof is quite similar to the proof of Proposition B.4.1. It suffices to note that $\sqrt{n}\,(\bar{Z}_j - \mu_j) \rightsquigarrow Z_j - j\mu_{j-1}Z_1$ and that $\sqrt{n}\,(\hat\kappa_j\hat\kappa_k - \kappa_j\kappa_k) \rightsquigarrow \kappa_k W_j + \kappa_j W_k$. The representation of V in the Gaussian case follows from Appendix 6.B.

■

6.6.2.2 Application

As an application of Proposition 6.6.2, we can state the following result, whose proof follows from the results in Appendix B.8.

Lemma 6.6.1 *Suppose that the parameters of the Lévy process for returns are expressed by the vector $\theta \in \Theta \subset \mathbb{R}^p$ and assume that there exists an invertible differentiable mapping $T : \Theta \mapsto \mathbb{R}^p$ such that $T_j(\theta) = \kappa_{l_j}$, for $j = 1,\ldots,p$. Let $J(\theta)$ denotes the Jacobian matrix of T, that is $J_{jk}(\theta) = \frac{\partial T_k(\theta)}{\partial \theta_j}$, $j,k \in \{1,\ldots,p\}$.*

If $\hat\theta_n$ is such that $T_j(\hat\theta) = \hat\kappa_{l_j}$, $j = 1,\ldots,p$, then

$$\sqrt{n}(\hat\theta - \theta) \rightsquigarrow N_p(0, \Sigma),$$

where Σ *can be estimated by* $\hat{\Sigma} = \hat{J}^{-1} V \left(\hat{J}^{-1}\right)^\top$, *with* $\hat{J} = J(\hat{\theta})$.

Example 6.6.3 *As an illustration of the methodology, we consider simulated data. The MATLAB file* DataMerton *contains 500 observations simulated from Merton model with parameters* $\mu = 0.08$, $\sigma = 0.22$, $\lambda = 100$, $\gamma = 0.05$, *and* $\delta = 0.005$. *The cumulant-based estimation can be carried out with the MATLAB program* EstMertonCum. *The results are displayed in Table 6.1, while in Figure 6.2, we compare the nonparametric estimation of the density using the Gaussian kernel, with the true density and the density evaluated at the estimated parameters. The estimation of the parameters is not satisfactory in general, and the error of estimation is too large.*

TABLE 6.1: 95% confidence intervals for the parameters of the Merton model using the cumulant matching method, applied to the simulated data set *DataMerton*.

Parameter	True Value	95% C. I.
μ	0.08	0.1704 ± 0.8063
σ	0.22	0.2918 ± 187.8917
λ	100	70.67 ± 220552.7
γ	0.05	0.0479 ± 115.6798
δ	0.005	0.0310 ± 39.9573

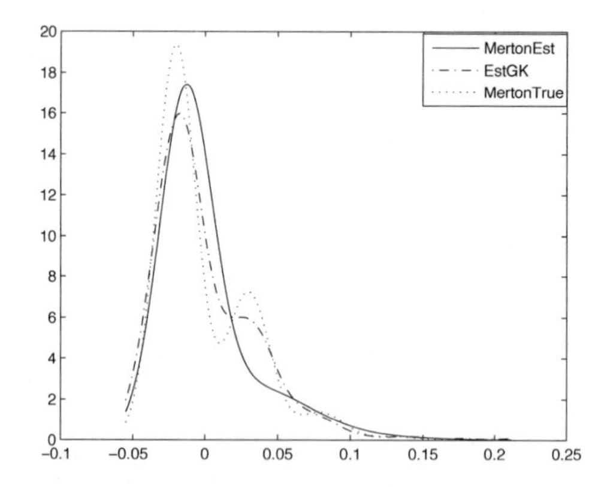

FIGURE 6.2: Nonparametric estimation of the density and Merton densities with true and estimated parameters (cumulants), for the simulated data set *DataMerton*.

6.6.2.3 Discussion

As mentioned previously, the errors of estimation of higher cumulants can be quite large, so the moment matching method using the cumulants can produce extremely poor results, when the solution exists, even in the case of the jump-diffusion model of Merton. This methodology is therefore not recommended if the number of parameters is larger than 4. In fact, it could be recommended if the parameters were a function of the mean, volatility, skewness, and kurtosis. One could then apply the delta method (Theorem B.3.1) together with Proposition B.4.1.

6.6.3 Estimation Based on the Maximum Likelihood Method

To implement the maximum likelihood method, we need an accurate expression for the density. This is the main drawback. However, since the characteristic function and the cumulant generating function are known, we could use the tools developed in Chapter 4 to recover the density by inverting its characteristic function or approximating it by the saddlepoint method.

Also, for jump-diffusion models, because of Proposition 6.6.1, bounds on λ should be set. Because the returns are independent and identically distributed, one can use the results for the maximum likelihood principle for independent observations.

Finally, in Section 3.B, we introduced goodness-of-fit tests. We could apply this methodology here too, using parametric bootstrap, to check if the model makes sense.

Before choosing an estimation method and a parametrization for a given model, one should always simulate data according to the proposed model and estimate the parameters with these data. This way, one might discover potential estimation problems and try to find a solution by using another parametrization or putting constraints on the parameters. For example, one could change the parameters of the model by including the mean of the returns as a parameter. This way, one could estimate it by the sample mean and estimate the remaining parameters by maximum likelihood. However, one must be careful with estimation errors. See Appendix B.9 for details.

We now give study two important cases where we know the densities: Merton model and the Variance Gamma model.

Example 6.6.4 (Merton Model) *Conditionally on $N_i = k \geq 0$, we have*

$$R_i \sim N(a + k\gamma, \sigma^2 h + k\delta^2).$$

Therefore, the density of R_i *is given by*

$$f_{R_i}(r) = \sum_{k=0}^{\infty} e^{-\lambda h} \frac{(\lambda h)^k}{k!} \frac{e^{-\frac{1}{2}\frac{(r-a-k\gamma)^2}{\sigma^2 h + k\delta^2}}}{\sqrt{2\pi(\sigma^2 h + k\delta^2)}}.$$

First we look at simulated data. The MATLAB file DataMerton *contains 500 observations simulated from Merton model with parameters* $\mu = 0.08$, $\sigma = 0.22$, $\lambda = 100$, $\gamma = 0.05$, *and* $\delta = 0.005$. *The estimation can be carried out with the MATLAB program* EstMerton. *The results are displayed in Table 6.2, while in Figure 6.3, we compare the nonparametric estimation of the density using the Gaussian kernel, with the true density and the density evaluated at the estimated parameters. All parameters are satisfactory but* μ. *However it is included in the confidence interval.*

TABLE 6.2: 95% confidence intervals for the parameters of the Merton model using the maximum likelihood method, applied to the simulated data set *DataMerton.*

Parameter	True Value	95% C. I.
μ	0.08	0.1560 ± 0.7433
σ	0.22	0.2256 ± 0.0222
λ	100	96.0595 ± 15.0528
γ	0.05	0.0497 ± 0.0031
δ	0.005	0.0000 ± 0.8240

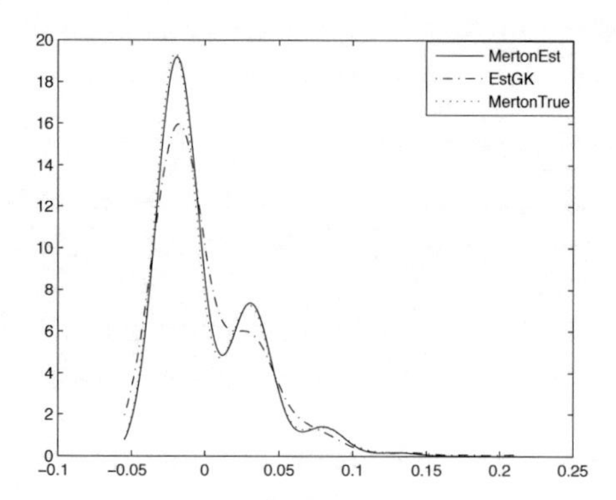

FIGURE 6.3: Nonparametric estimation of the density and Merton densities with true and estimated parameters (mle), for the simulated data set *DataMerton.*

For the next example, we used the prices of Apple, from January 13th, 2009, to January 14th, 2011. We see from Table 6.3 that with the maximum likelihood method, we recover the same estimation of the parameters μ and σ as in the Gaussian case in Section 1.4.1. No surprise then that the estimation of λ is 2×10^{-6}. Here, we could either impose a lower bound on λ or simply decide that the model with no jumps is better than the model with jumps.

As an indication, we also included the estimation of the parameters using the cumulant matching method. The parametric and nonparametric estimations of the density of these returns are displayed in Figures 6.4–6.5 for both the maximum likelihood method and the cumulant matching method.

TABLE 6.3: 95% confidence intervals for the parameters of the Merton model applied to the returns of Apple.

Parameter	95% Confidence Intervals	
	MLE	Cumulants
μ	0.7313 ± 0.4121	0.7314 ± 0.4126
σ	0.2975 ± 0.0145	0.2601 ± 0.1186
λ	0.0000 ± 0.0000	34.18 ± 198.88
γ	1.0720 ± 0.0000	0.0038 ± 0.0062
δ	0.2209 ± 0.0000	0.0244 ± 0.0345

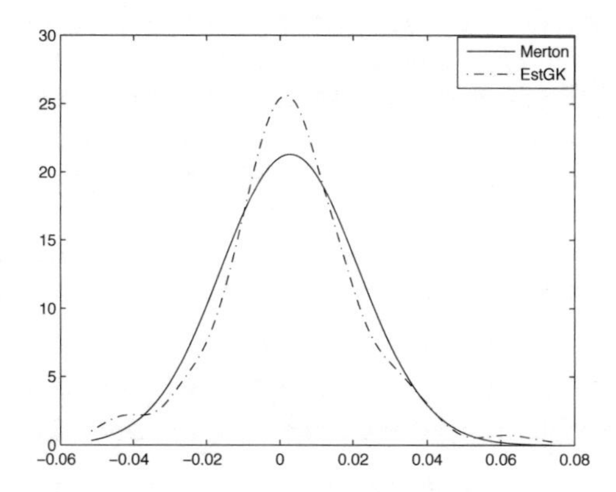

FIGURE 6.4: Nonparametric estimation of the density and Merton density with estimated parameters (mle) for the returns of Apple.

Example 6.6.5 (Variance Gamma Model) *Here again we look at simulated data first. The MATLAB file* DataVG *contains 500 observations simulated from the Variance Gamma model with parameters* $\mu = 0.15$, $\sigma = 0.50$,

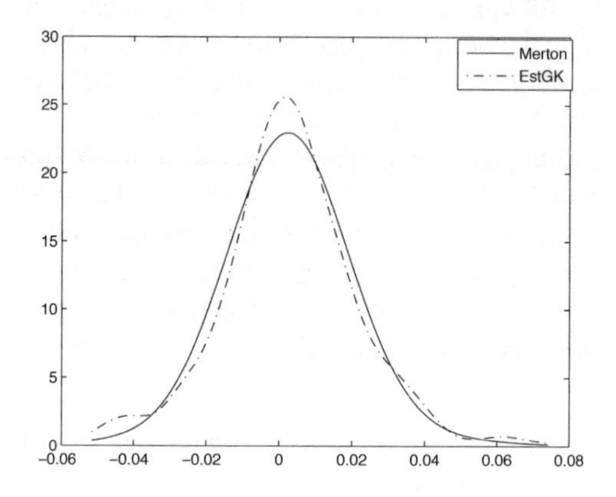

FIGURE 6.5: Nonparametric estimation of the density and Merton density with estimated parameters (cumulants) for the returns of Apple.

and $\alpha = 1000$. *The estimation can be carried out with the MATLAB program* EstVG. *The results are displayed in Table 6.4. The estimations are far off the true values. This could be due to numerical instabilities in the Bessel function.*

In this model, the mean is a parameter, so it could be a good idea to use this to our advantage. Since the maximum likelihood method produced poor results, we could combined the moment estimation of the mean with the maximum likelihood estimation of σ *and* α. *This combined estimation can be carried out with the MATLAB program* EstVGMean. *The results are also displayed in Table 6.4, while Figures 6.6–6.7 compare the nonparametric estimation of the density using the Gaussian kernel, with the true density and the density evaluated at the estimated parameters using both methods. All parameters are satisfactory but* α. *However it is included in the confidence interval. If we look at the daily scale however, the two values are not that different.*

TABLE 6.4: 95% confidence intervals for the parameters of the Variance Gamma model applied to the simulated data set *DataVG*.

Parameter	True Value	95% Confidence Intervals	
		MLE	Combined
μ	0.15	6.1648 ± 0.0018	0.2018 ± 0.6836
σ	0.50	0.0491 ± 0.0027	0.4788 ± 0.0321
α	1000	377.11 ± 0.24	2198.03 ± 3369.43

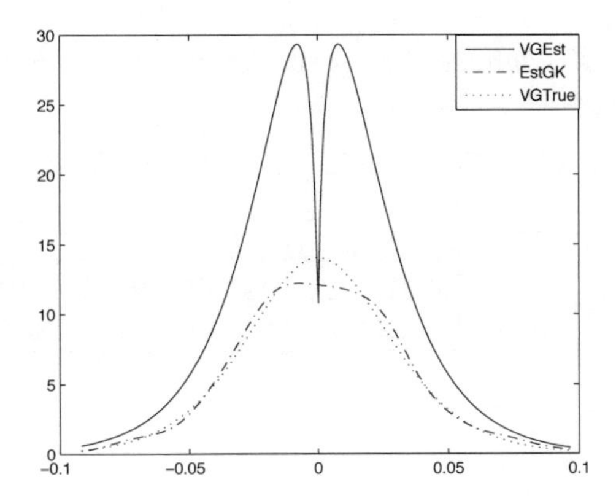

FIGURE 6.6: Nonparametric estimation of the density and Variance Gamma density with the true and estimated parameters (mle) for the simulated data set *DataVG*.

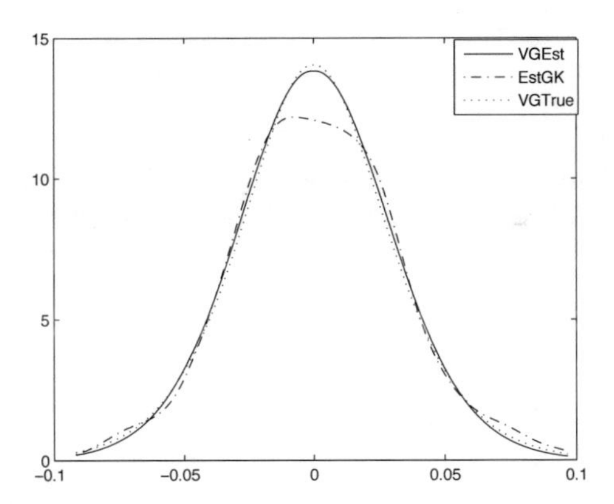

FIGURE 6.7: Nonparametric estimation of the density and Variance Gamma density with the true and estimated parameters (combined) for the simulated data set *DataVG*.

Finally, using the returns of Apple, we can see from the results in Table 6.5 that the maximum likelihood method produces poor estimations compared to the combined method. This is also illustrated in Figures 6.8–6.9.

TABLE 6.5: 95% confidence intervals for the parameters of the Variance Gamma model applied to the returns of Apple.

Parameter	95% Confidence Intervals	
	MLE	Combined
μ	3.5599 ± 0.0014	0.6870 ± 0.4119
σ	0.0195 ± 0.0145	0.2456 ± 0.0210
α	303.97 ± 0.25	469.12 ± 187.96

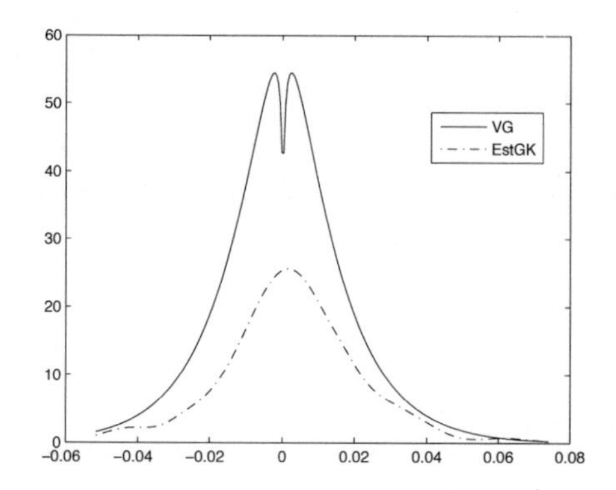

FIGURE 6.8: Nonparametric estimation of the density and Variance Gamma density with estimated parameters (mle) for the returns of Apple.

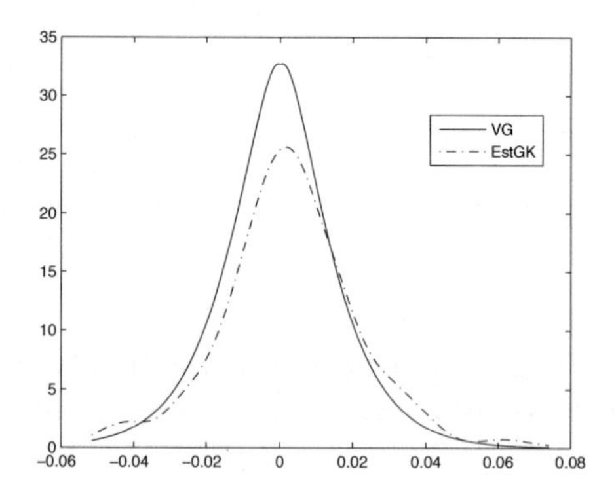

FIGURE 6.9: Nonparametric estimation of the density and Variance Gamma density with estimated parameters (combined) for the returns of Apple.

6.7 Suggested Reading

The main articles on the topics covered in this chapter are Merton [1976] and Kou [2002]. For more details on Lévy processes and financial applications, one should consult Cont and Tankov [2004] and Schoutens [2003].

6.8 Exercises

Exercise 6.1

Consider two jump-diffusion Merton models with parameters μ, σ, λ, γ, δ and $\tilde{\mu}$, σ, $\tilde{\lambda}$, $\tilde{\gamma}$, δ. Find the change of measure to transform the first one into the second one. Can you do it using only an Esscher transform?

Exercise 6.2

Suppose X is a Gamma process with parameters $(\alpha, \beta) = (20, 0.01)$. Find its characteristics.

Exercise 6.3

Consider two jump-diffusion Kou models with parameters μ, σ, λ, p, η_1, η_2 and $\tilde{\mu}$, σ,$\tilde{\lambda}$, $\tilde{p}$, $\tilde{\eta}_1$, $\tilde{\eta}_2$. Find the change of measure to transform the first one into the second one. Can you do it using only an Esscher transform?

Exercise 6.4

Suppose X is an Inverse Gaussian process with parameters $(\alpha, \beta) = (20, 0.01)$. Find its characteristics.

Exercise 6.5

Consider a Normal Inverse Gaussian process with parameters α, β, δ. Using an Esscher transform, what is the law of the process under the new measure?

Exercise 6.6

Let X be a jump-diffusion process of the Merton type with parameters $\mu = .08$, $\sigma = 0.22$, $\lambda = 100$, $\gamma = .05$ and $\delta = .005$. Suppose that the risk-free rate is 2%, and consider the change of measure defined by $U_{b,\phi}$, where $\phi(x) = 2 - 0.5x$ and $b \in \mathbb{R}$.

(a) Find b so that the associated measure is a martingale measure.

(b) Under this change of measure, compute the new coefficients $\tilde{\mu}$, $\tilde{\sigma}$, $\tilde{\lambda}$, $\tilde{\gamma}$, and $\tilde{\delta}$.

(c) Simulate 500 daily observations under this equivalent martingale measure.

Exercise 6.7

Suppose that X is a Lévy process such that $X(1) \sim VG(\mu, \sigma, \alpha)$. Using its cumulant generating function, show that $X(t) \sim VG(\mu t, \sigma\sqrt{t}, \alpha t)$, $t > 0$.

Exercise 6.8

Let X be a jump-diffusion process of the Kou type with parameters $\mu = .08$, $\sigma = 0.22$, $\lambda = 100$, $p = 0.7$, $\eta_1 = 125$, and $\eta_2 = 100$. Suppose that the risk-free rate is 2%, and consider the change of measure defined by $U_{b,\phi}$, where $\phi(x) = (\zeta_{01} + \zeta_{11}x)\mathbb{I}(x > 0) + (\zeta_{02} + \zeta_{12}x)\mathbb{I}(x \le 0)$, and $b \in \mathbb{R}$.

(a) Under this change of measure, the new parameters of the pure-jump part are $\tilde{\lambda} = 500$, $\tilde{p} = 0.9$, $\tilde{\eta}_1 = 250$, and $\tilde{\eta}_2 = 400$. Find ϕ.

(b) Find b so that the associated measure is a martingale measure.

(c) Give the old and new parameters under the weighted-symmetric representation of the jumps, i.e., find ω, $\tilde{\omega}$, δ, and $\tilde{\delta}$.

(d) Simulate 500 daily observations under this equivalent martingale measure.

6.9 Assignment Questions

1. Construct a MATLAB function to compute the new parameters of the Merton model under the change of measure defined by $U_{b,\phi}$, with $\phi(x) = \zeta_0 + \zeta_1 x$ and $b \in \mathbb{R}$. Use the same notations as in Section 6.2.3.

2. Construct a MATLAB function to compute the new parameters of the Kou model under the change of measure defined by $U_{b,\phi}$, with $\phi(x) = (\zeta_{01} + \zeta_{11}x)\mathbb{I}(x > 0) + (\zeta_{02} + \zeta_{12}x)\mathbb{I}(x \le 0)$, and $b \in \mathbb{R}$. Use the same notations as in Section 6.2.4. Note that not all changes are admissible, so the verification of the associated conditions must be taken into account.

3. Construct a MATLAB function for computing the value of a European call or put option for the Merton model.

4. Construct a MATLAB function to estimate the value of a European call or put option, under the Variance Gamma model.

5. Construct a MATLAB function to estimate the parameters of the Variance Gamma model using the cumulants.

6. Construct a MATLAB function to estimate the parameters of the Normal Inverse Gaussian process model using the cumulants.

6.A Modified Bessel Functions of the Second Kind

The modified Bessel function of the second kind of order λ is defined by the differential equation

$$z^2 K_\lambda''(z) + zK_\lambda'(z) = (z^2 + \lambda^2)K_\lambda(z), \qquad z > 0.$$

Here $K_\lambda = K_{-\lambda}$ and when $z \to 0$, we have $K_0(z)/\ln(1/z) \to 1$, and $(z/2)^{|\lambda|}K_\lambda(z) \to \frac{1}{2}\Gamma(|\lambda|)$. Moreover, for any integer n,

$$K_{n-1/2}(z) = \sqrt{\frac{\pi}{2z}}e^{-z}g(1/z),$$

where g is a polynomial of order $n-1$ satisfying

$$z^2 g''(z) + 2(1+z)g'(z) = n(n-1)g(z), \quad g(0) = 1.$$

Hence

$$g(z) = \sum_{k=0}^{n-1} a_{k,n} z^k,$$

where $a_{0,n} = 1$, and for $k \in \{1, \ldots, n-1\}$,

$$a_{k,n} = \frac{(n-k)(n+k-1)}{2k} a_{k-1,n}.$$

Thus for any $k \in \{0, \ldots, n-1\}$

$$a_{k,n} = \frac{1}{2^k k!}\frac{(n+k-1)!}{(n-k-1)!}.$$

For example, if $n = 1$, $g \equiv 1$ and $K_{n-1/2}(z) = \sqrt{\frac{\pi}{2z}}e^{-z}$. If $n = 2$, $g(z) = 1+z$, and $K_{n-1/2}(z) = \sqrt{\frac{\pi}{2z}}e^{-z}(1 + 1/z)$. For more details, see Abramowitz and Stegun [1972].

6.B Asymptotic Behavior of the Cumulants

Suppose that the moment generating function $M(u) = E\left(e^{uX}\right)$ exists for all $|u| < u_0$. Then, for $|u| < u_0$,

$$c(u) = \ln\{M(u)\} = \sum_{j=1}^{\infty} \kappa_j \frac{u^j}{j!}.$$

If $X_1, \ldots, X_n$ are independent observations of X, set

$$M_n(u) = \frac{1}{n}\sum_{i=1}^{n} e^{uX_i}, \qquad c_n(u) = \ln\{M_n(u)\}.$$

Then

$$c_n(u) = \sum_{j=1}^{\infty} \hat{\kappa}_j \frac{u^j}{j!}.$$

We know that for any $j \geq 1$, $n^{1/2}(\hat{\kappa}_j - \kappa_j) \rightsquigarrow W_j$, where W_j has a centered Gaussian distribution. To determine the covariance between W_j and W_k, note by the Central Limit Theorem, for any $u, v \in (u_0/2, u_0/2)$, $n^{1/2}\{M_n(u) - M(u)\}$ and $n^{1/2}\{M_n(v) - M(v)\}$ converge jointly to centered Gaussian variables with covariance $M(u+v) - M(u)M(v)$.

In addition, using the delta method (Theorem B.3.1), we have that $n^{1/2}\{c_n(u) - c(u)\}$ and $n^{1/2}\{c_n(v) - c(v)\}$ converge jointly to centered Gaussian variables with covariance

$$\frac{M(u+v)}{M(u)M(v)} - 1 = e^{c(u+v)-c(u)-c(v)} - 1.$$

On the other hand, using the series representation of c_n and c, we have

$$\frac{M(u+v)}{M(u)M(v)} - 1 = \sum_{j=1}^{\infty}\sum_{k=1}^{\infty} \frac{u^j v^k}{j!k!} \mathrm{Cov}(W_j, W_k).$$

To find the covariance structure, it suffices to expand $\frac{M(u+v)}{M(u)M(v)} - 1$ as a series in v and u and match the coefficients. For example, in the Gaussian case, $M(u) = e^{\mu u + u^2\sigma^2/2}$, so

$$\frac{M(u+v)}{M(u)M(v)} - 1 = e^{uv\sigma^2} - 1 = \sum_{k=1}^{\infty} \sigma^{2k} \frac{u^k v^k}{k!}.$$

As a result, $\mathrm{Cov}(W_j, W_k) = 0$ if $j \neq k$, while $\mathrm{Var}(W_k) = \sigma^{2k} k!$, $k \geq 1$.

6.C Proofs of the Results

6.C.1 Proof of Lemma 6.5.1

If $|u| < u_0 - |\zeta|$, and $0 \le s \le t$, then it follows from (6.1) and the properties of Lévy processes that

$$\begin{aligned} E_{\tilde{P}}\left[e^{u\{X(t)-X(s)\}} \middle| \mathcal{F}_s \right] &= E\left[\Lambda(t) e^{u\{X(t)-X(s)\}} \middle| \mathcal{F}_s \right] / \Lambda(s) \\ &= E\left\{ e^{(u+\zeta)X(t-s)-(t-s)c(\zeta)} \right\} \\ &= e^{(t-s)\{c(u+\zeta)-c(\zeta)\}}. \end{aligned}$$

Next,

$$\begin{aligned} \tilde{c}(u) &= c(u+\zeta) - c(\zeta) \\ &= ua + u\zeta\sigma^2 + u^2\frac{\sigma^2}{2} + u\int x\left(e^{\zeta x} - 1\right)\mathbb{I}(|x| \le 1)\nu(dx) \\ &\quad + \int e^{\zeta x}\left\{e^{ux} - 1 - ux\mathbb{I}(|x| \le 1)\right\}\nu(dx) \\ &= u\tilde{a} + u^2\frac{\sigma^2}{2} + \int \left\{e^{ux} - 1 - ux\mathbb{I}(|x| \le 1)\right\}\tilde{\nu}(dx), \end{aligned}$$

where

$$\tilde{a} = a + \zeta\sigma^2 + \int x\left(e^{\zeta x} - 1\right)\mathbb{I}(|x| \le 1)\nu(dx).$$

Using (6.10), we obtain that under $\tilde{P}$, the increments of X are independent and that X is a Lévy process with characteristics $\left\{\tilde{a}, \sigma^2, \tilde{\nu}\right\}$.

■

6.C.2 Proof of Corollary 6.5.2

Set $\Lambda(t) = e^{X(t)-rt}$. Then according to Lemma 6.5.1, Λ is a positive martingale with mean 1, so under the change of measure $\tilde{Q}$ defined by $\left.\frac{d\tilde{Q}}{dQ}\right| \mathcal{F}_t = \Lambda(t)$, X is a Lévy process with characteristics $\left\{\tilde{a}, \sigma^2, \tilde{\nu}\right\}$, where

$$\tilde{a} = a + \sigma^2 + \int x\left(e^{x} - 1\right)\mathbb{I}(|x| \le 1)\nu(dx)$$

and $\frac{d\tilde{\nu}}{d\nu} = e^x$. As a result,

$$
\begin{aligned}
C(\tau, s) &= e^{-r\tau} E_Q \left\{ \max\left(s e^{X(\tau)} - K, 0 \right) \right\} \\
&= s e^{-r\tau} E_Q \left[e^{X_\tau} \mathbb{I}\{X(\tau) > \ln(K/s)\} \right] \\
&\qquad - K e^{-r\tau} E_Q \left[\mathbb{I}\{X(\tau) > \ln(K/s)\} \right] \\
&= s E_Q \left[\Lambda_\tau \mathbb{I}\{X(\tau) > \ln(K/s)\} \right] \\
&\qquad - K e^{-r\tau} E_Q \left[\mathbb{I}\{X(\tau) > \ln(K/s)\} \right] \\
&= s \tilde{Q} \{X(\tau) > \ln(K/s)\} \\
&\qquad - K e^{-r\tau} Q \{X(\tau) > \ln(K/s)\} \\
&= s \bar{F}_{0,\sigma,\tilde{\nu},\tau} \{\ln(K/s) - \tau \tilde{a}\} \\
&\qquad - K e^{-r\tau} \bar{F}_{0,\sigma,\nu,\tau} \{\ln(K/s) - \tau a\}.
\end{aligned}
$$

■

6.C.3 Proof of Proposition 6.6.1

Let ψ be such that $E\left\{e^{iuX(t)}\right\} = e^{t\psi(u)}$. It follows that $\psi'(0) = i\mu$ and $\psi''(0) = -\sigma^2$. Set $Y_\lambda(t) = \{X_\lambda(t) - \lambda\mu t\}/\sqrt{\lambda}$. Then

$$
\begin{aligned}
E\left\{e^{iuY_\lambda(t)}\right\} &= \exp\left[\lambda t \left\{ \psi(u/\sqrt{\lambda}) - \psi(0) - u\psi'(0)/\sqrt{\lambda} \right\}\right] \\
&\stackrel{\lambda\to\infty}{\longrightarrow} \exp\left\{ \frac{1}{2} t u^2 \psi''(0) \right\} = \exp\left\{ -\frac{1}{2} t u^2 \sigma^2 \right\} \\
&= E\left\{ e^{iu\sigma\tilde{W}(t)} \right\}.
\end{aligned}
$$

Because the increments of the process Y_λ are independent and stationary, it follows that for any integer m, the joint law of $Y_\lambda(t_1), \ldots, Y_\lambda(t_m)$ converges to the joint law of $\sigma W(t_1), \ldots, \sigma W(t_m)$. The tightness of the sequence follows from the following inequality:

$$
E\left[\{Y_\lambda(t) - Y_\lambda(s)\}^2 \{Y_\lambda(s) - Y_\lambda(u)\}^2 \right] = (t-s)(u-s)\sigma^4 \le (t-u)^2\sigma^4,
$$

whenever $0 \le u \le s \le t$.

■

6.C.4 Proof of Proposition 6.4.1

We have

$$
\begin{aligned}
P(Y \le y) &= P\left(Z \le \frac{y - \theta V}{\sqrt{V}} \right) = E\left\{ \mathcal{N}\left(\frac{y - \theta V}{\sqrt{V}} \right) \right\} \\
&= \int_0^\infty \mathcal{N}\left(\frac{y - \theta v}{\sqrt{v}} \right) \frac{v^{\alpha-1} e^{-v/\beta}}{\beta^\alpha \Gamma(\alpha)} dv,
\end{aligned}
$$

where $\mathcal{N}$ is the distribution function of Z. Hence

$$\begin{aligned} f(y) &= \frac{1}{\sqrt{2\pi}} \int_0^\infty e^{-\frac{1}{2}(y-\theta v)^2/v} \frac{v^{\alpha-3/2} e^{-v/\beta}}{\beta^\alpha \Gamma(\alpha)} dv \\ &= \frac{e^{\theta y}}{\sqrt{2\pi}} \int_0^\infty \frac{v^{\alpha-3/2} e^{-v/\beta - v\theta^2/2 - y^2/(2v)}}{\beta^\alpha \Gamma(\alpha)} dv \\ &= \frac{|y|^{\alpha-1/2} e^{\theta y}}{\sqrt{2\pi}\beta^\alpha \Gamma(\alpha)(\theta^2 + 2/\beta)^{\alpha/2-1/4}} \int_0^\infty u^{\alpha-3/2} e^{-\frac{d}{2}\left(u+\frac{1}{u}\right)} du, \end{aligned}$$

where $d = |y|\sqrt{\theta^2 + 2/\beta}$. Now, according to Abramowitz and Stegun [1972, page 376], for any $z > 0$,

$$\int_0^\infty u^{\alpha-3/2} e^{-\frac{z}{2}\left(\frac{1}{u}+u\right)} du = 2K_{\alpha-1/2}(z).$$

Thus the density is given by (6.19). Next, the tail behavior of the distribution function follows from $K_\lambda(z) \sim \sqrt{\frac{\pi}{2z}} e^{-z}$ as $z \to \infty$ [7] [Abramowitz and Stegun, 1972, page 378]. In fact, as $|y| \to \infty$,

$$f(y) \sim \frac{|y|^{\alpha-1} e^{-|y|\sqrt{\frac{2}{\beta}}}}{(2\beta)^{\frac{\alpha}{2}} \Gamma(\alpha)}.$$

■

Bibliography

B. Abdous and B. Rémillard. Relating quantiles and expectiles under weighted-symmetry. *Ann. Inst. Statist. Math.*, 47(2):371–384, 1995.

B. Abdous, K. Ghoudi, and B. Rémillard. Nonparametric weighted symmetry tests. *Canad. J. Statist.*, 31(4):357–381, 2003.

M. Abramowitz and I. E. Stegun. *Handbook of Mathematical Functions with Formulas, Graphs, and Mathematical Tables*, volume 55 of *Applied Mathematics Series*. National Bureau of Standards, tenth edition, 1972.

R. Cont and P. Tankov. *Financial Modelling with Jump Processes*. Chapman & Hall/CRC Financial Mathematics Series. Chapman & Hall/CRC, Boca Raton, FL, 2004.

L. Devroye. *Non-Uniform Random Variate Generation*. Springer-Verlag, New York, 1986.

[7] The notation $f(z) \sim g(z)$ as $z \to \infty$ means that $\lim_{z\to\infty} f(z)/g(z) = 1$.

C. Jacques, B. Rémillard, and R. Theodorescu. Estimation of Linnik law parameters. *Statist. Decisions*, 17(3):213–235, 1999.

S. G. Kou. A jump-diffusion model for option pricing. *Management Science*, 48(8):1086–1101, 2002.

D. B. Madan, P. P. Carr, and E. C. Chang. The variance gamma process and option pricing. *European Finance Review*, 2:79–105, 1998.

D. L. McLeish. A robust alternative to the normal distribution. *Canad. J. Statist.*, 10(2):89–102, 1982.

R. C. Merton. Option pricing when underlying stock returns are discontinuous. *Journal of Financial Economics*, 3:125–144, 1976.

J. R. Michael, W. R. Schucany, and R. W. Haas. Generating random variates using transformations with multiple roots. *The American Statistician*, 30 (2):88–90, 1976.

K. Pearson, G. B. Jeffery, and E. M. Elderton. On the distribution of the first product moment-coefficient, in samples drawn from an indefinitely large normal population. *Biometrika*, 21(1/4):164–201, 1929.

P. E. Protter. *Stochastic Integration and Differential Equations*, volume 21 of *Applications of Mathematics (New York)*. Springer-Verlag, Berlin, second edition, 2004. Stochastic Modelling and Applied Probability.

K.-I. Sato. *Lévy Processes and Infinitely Divisible Distributions*, volume 68 of *Cambridge Studies in Advanced Mathematics*. Cambridge University Press, Cambridge, 1999.

W. Schoutens. *Lévy Processes in Finance : Pricing Financial Derivatives*. Wiley Series in Probability and Statistics. Wiley, 2003.

Chapter 7

Stochastic Volatility Models

In this chapter, we cover first a class of discrete time stochastic volatility models, the ARCH models, introduced in Engle [1982] and later extended to the so-called GARCH models by Bollerslev [1986]. We study some of the properties of these processes, their estimation, including tests of goodness-of-fit, and the pricing of options in a GARCH context, as proposed by Duan [1995]. We also discuss their continuous time approximation by stochastic volatility models of the diffusion type, as studied by Hull and White [1987] and Heston [1993]. Option pricing for these models is also discussed.

7.1 GARCH Models

One way to eliminate the problem of constant volatility inherent in the Black-Scholes model is to assume that the daily log-returns $X_i = \ln(S_i/S_{i-1})$ are modeled by

$$X_i = \mu_i + \sigma_i \varepsilon_i, \quad i > r, \tag{7.1}$$

for some integer $r \geq 1$, where μ_i and σ_i depend on $\mathcal{F}_{i-1} = \sigma\{H_i, X_1, \ldots, X_{i-1}\}$, where H_i is a random vector containing possibly exogenous variables. In addition, $(\varepsilon_i)_{i>r}$ are independent and identically distributed with mean zero and variance one. Moreover, ε_i is independent of $\mathcal{F}_{i-1}$, for all $i > r$. The ε_is are called the innovations.

A particular class of processes having representation (7.1) with non-constant σ_i is the class of Auto-Regressive Conditional Heteroscedasticity models (ARCH for short), introduced by Engle [1982]. In ARCH models, the conditional variance σ_i^2 depends on lagged values of $(X_i - \mu_i)^2$. Then Bollerslev [1986] generalized Engle's ideas to GARCH processes. In these models, σ_i depends also on its own past values, and the conditional distribution of X_i given $\mathcal{F}_{i-1}$ is Gaussian, with mean μ_i and variance σ_i^2, i.e., the innovations are Gaussian. In particular

$$E\left\{(X_i - \mu_i)^2 | \mathcal{F}_{i-1}\right\} = \sigma_i^2.$$

In many cases, μ_i can be modeled by an autoregressive (AR) model of the form

$$\mu_i = \mu + \sum_{j=1}^{P} \phi_j (X_{i-j} - \mu).$$

We could also add a linear combination of exogenous variables. We will come back to AR models in Chapter 10.

Remark 7.1.1 *Models having representation* (7.1) *appear naturally when one approximates diffusion processes.*

Distribution of Innovations

One can also choose other distributions than the Gaussian distribution for the law of the innovations. However, the (standardized) Student is not a good choice for financial engineering applications since its moment generating function does not exist. If the innovations had a Student distribution under an equivalent martingale measure, the value of a call option would then be infinite!

In general, we would like to have conditions on the process σ so that the volatility σ_n forgets its initial values and becomes stationary and that $E\left(\sigma_n^2\right)$ converges to some positive constant as n tends to infinity. These properties are needed for estimation purposes since the process σ is not observable.

We now give some interesting examples of GARCH processes used in practice.

7.1.1 GARCH(1,1)

For this model, $r = 1$, $H = \sigma_1$, $\mu_i = \mu$ and

$$\sigma_i^2 = \omega + \alpha(X_{i-1} - \mu)^2 + \beta\sigma_{i-1}^2 = \omega + \sigma_{i-1}^2\left(\alpha\varepsilon_{i-1}^2 + \beta\right), \quad i \geq 2,$$

where $\omega > 0$, $\alpha \geq 0$, and $\beta \geq 0$. Note that we obtain an ARCH(1) model by setting $\beta = 0$.

When dealing with daily returns, it is often assumed for simplicity that $\mu = 0$. However, estimating μ poses no additional difficulty and omitting it can lead to incorrect inference.

The MATLAB function *DemoGARCH* illustrates the behavior of two GARCH(1,1) trajectories (x, σ) and (y, v) starting from different points (x_1, σ_1) and (y_1, v_1). From Figure 7.1, we see that it does not take long before the effect of the starting points vanishes for the returns (x, y) and the volatilities (σ, v).

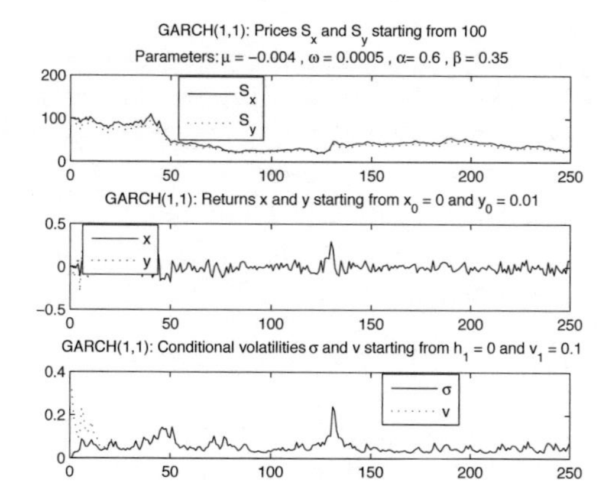

FIGURE 7.1: Graphs of the prices, returns, and volatilities for a GARCH(1,1) model starting from different points. Here $x_1 = 0$, $y_1 = 0.01$, $\sigma_1 = 0$ and $v_1 = 0.1$.

To see under what conditions the expectation of σ_n^2 stabilizes, we compute its expectation:

$$\begin{aligned} E\left(\sigma_n^2\right) &= \omega + \alpha E\left[E\left\{X_{n-2} - \mu)^2 | \mathcal{F}_{n-1}\right\}\right] + \beta E(\sigma_{n-1}^2) \\ &= \omega + (\alpha + \beta) E(\sigma_{n-1}^2) \\ &= \omega \left\{ \frac{1 - (1-\kappa)^{n-1}}{\kappa} \right\} + (1-\kappa)^{n-1} \sigma_1^2, \end{aligned}$$

where $\kappa = 1 - \alpha - \beta$. Hence a necessary and sufficient condition for the expectation $E\left(\sigma_n^2\right)$ to stabilize as $n \to \infty$ is that $\kappa > 0$. In this case, we have

$$\lim_{n\to\infty} E\left(\sigma_n^2\right) = \frac{\omega}{\kappa}.$$

Note also that $(x_i, \sigma_{i+1})_{i \geq 0}$ is a Markov process. Under the condition $\kappa > 0$, the series σ admits a stationary distribution. See Bougerol and Picard [1992] for conditions of stationarity of general Markov chains. In general, the conditions for stationarity are weaker than those for the existence of the limiting moments.

To see under what conditions the volatilities forget their initial starting values, consider two trajectories σ, v starting from σ_1 and v_1 and constructed from the same innovations process ε. It follows that

$$|\sigma_{n+1}^2 - v_{n+1}^2| = |\sigma_n^2 - v_n^2| \left(\alpha \varepsilon_n^2 + \beta\right),$$

so

$$|\sigma_{n+1}^2 - v_{n+1}^2| = |\sigma_1^2 - v_1^2| \prod_{k=1}^{n} \left(\alpha \varepsilon_k^2 + \beta\right).$$

Whenever $E\left\{\ln\left(\alpha\varepsilon_k^2+\beta\right)\right\} < 0$, the righthand side of the last expression converges to 0, meaning that the volatility process forgets its initial value. This follows from the law of large numbers applied to the sequence $\ln\left(\alpha\varepsilon_k^2+\beta\right)$, $k \geq 1$. In fact, by the strong law of large numbers, $\frac{1}{n}\sum_{k=1}^n \ln\left(\alpha\varepsilon_k^2+\beta\right) \to E\left\{\ln\left(\alpha\varepsilon_k^2+\beta\right)\right\} < 0$. This implies that for some $a > 0$, $\sum_{k=1}^n \ln\left(\alpha\varepsilon_k^2+\beta\right) < -na$ for almost all $n \geq 1$, yielding that $\prod_{k=1}^n \left(\alpha\varepsilon_k^2+\beta\right) < e^{-na}$ for almost all $n \geq 1$. By Jensen inequality, since the log function is concave, we have

$$E\left\{\ln\left(\alpha\varepsilon_k^2+\beta\right)\right\} \leq \ln E\left(\alpha\varepsilon_k^2+\beta\right) = \ln(1-\kappa).$$

Hence, the condition $\kappa > 0$, needed for the finiteness of the first moment, is stronger than the stationarity condition $E\left\{\ln\left(\alpha\varepsilon_k^2+\beta\right)\right\} < 0$.

7.1.2 GARCH(p,q)

An immediate extension of the GARCH(1,1) model is to consider more lags for X and σ. In this case, $r = \max(p,q)$, $H = (\sigma_1,\ldots\sigma_r)$, $\mu_i = \mu$ and

$$\sigma_i^2 = \omega + \sum_{k=1}^{p} \beta_k \sigma_{i-k}^2 + \sum_{j=1}^{q} \alpha_j (X_{i-j}-\mu)^2, \quad i > r,$$

where $\omega > 0$ and $\alpha_j, \beta_j \geq 0$, $j \in \{1,\ldots,r\}$. It follows that

$$E\left(\sigma_i^2\right) = \omega + \sum_{j=1}^{r} \lambda_j E\left(\sigma_{i-j}^2\right),$$

where $\lambda_j = \alpha_j + \beta_j$, with $\alpha_j = 0$ if $j > q$, and $\beta_j = 0$ whenever $j > p$. The asymptotic behavior of $E\left(\sigma_n^2\right)$ is given in the next proposition, proven in Appendix 7.B.1. It was shown in part in Bollerslev [1986], where the author also studied conditions on the limit of higher moments of the process σ.

Proposition 7.1.1 *Set* $\kappa = 1 - \sum_{j=1}^{\max(p,q)} \lambda_j$. *Then*

$$\lim_{n\to\infty} E\left(\sigma_n^2\right) = \begin{cases} \omega/\kappa, & \text{if } \kappa > 0; \\ +\infty, & \text{if } \kappa \leq 0. \end{cases}$$

7.1.3 EGARCH

These models, introduced in Nelson [1991], are such that $r = 1$, $H = \sigma_1^2$, and

$$\ln\left(\sigma_i^2\right) = \omega + \alpha\left\{|\varepsilon_{i-1}| - E(|\varepsilon_{i-1}|)\right\} + \gamma\varepsilon_{i-1} + \beta\ln(\sigma_{i-1}^2),$$

$i \geq 2$. One can also define $EGARCH(p,q)$ models. Note that $E\{\ln\left(\sigma_n^2\right)\}$ converges to a finit limite if and only if $|\beta| < 1$, in which case the limit is $\omega/(1-\beta)$.

7.1.4 NGARCH

Introduced in Engle and Ng [1993] to model an asymmetric behavior in the volatility depending on the values of the innovations, these models are such that $r = 1$, $H = \sigma_1$, $\mu_i = \mu + \lambda\sigma_i - \sigma_i^2/2$ and

$$\sigma_i^2 = \omega + \alpha(\varepsilon_{i-1} - \theta)^2\sigma_{i-1}^2 + \beta\sigma_{i-1}^2, \quad i \geq 2,$$

with $\omega > 0$, $\alpha, \beta \geq 0$. It can be shown that $E\left(\sigma_n^2\right)$ converges if and only if $1 - \kappa = \alpha(1+\theta^2) + \beta < 1$. If $\kappa > 0$, then the limit of $E\left(\sigma_n^2\right)$ is ω/κ.

7.1.5 GJR-GARCH

These models, proposed by Glosten et al. [1993] at the same moment as the NGARCH, were motivated almost the same way, i.e., to account for a different behavior of the volatility whenever the innovations are positive or negative. Here $r = 1$, $H = \sigma_1$, and

$$\sigma_i^2 = \omega + \alpha\sigma_{i-1}^2\varepsilon_{i-1}^2 + \beta\sigma_{i-1}^2 + \gamma\sigma_{i-1}^2\{\max(0, -\varepsilon_{i-1})\}^2, \quad i \geq 2.$$

It can be shown that $E\left(\sigma_n^2\right)$ converges if and only if $\kappa = 1 - \alpha - \beta - \gamma/2 > 0$. If $\kappa > 0$, then the limit of $E\left(\sigma_n^2\right)$ is ω/κ.

7.1.6 Augmented GARCH

In Duan [1997], the author introduced a family of models called Augmented GARCH. This family contains all the models described previously. Here we assume that $\alpha_i \geq 0$ for $i \in \{0, \ldots, 5\}$. Note that Duan did no longer assume that $\varepsilon_i \sim N(0,1)$, only that the common distribution of ε_i have mean 0 and variance 1. To describe his model, set $f_\delta(x) = \begin{cases} \frac{x^\delta - 1}{\delta} & \text{if} \quad \delta > 0 \\ \ln(x) & \text{if} \quad \delta = 0 \end{cases}$. Then $f_\delta^{-1}(x) = \begin{cases} (\delta x + 1)^{1/\delta} & \text{if} \quad \delta > 0 \\ e^x & \text{if} \quad \delta = 0 \end{cases}$. The volatility process σ is defined by

$$\sigma_i^2 = f_\lambda^{-1}(\phi_i - 1), \quad i \geq 2,$$

where

$$\begin{aligned} \xi_{1,i} &= \alpha_1 + \alpha_2|\varepsilon_i - c|^\delta + a_3(\max(0, c - \varepsilon_i))^\delta, \\ \xi_{2,i} &= \alpha_4 f_\delta(|\varepsilon_i - c|) + a_5 f_\delta(\max(0, c - \varepsilon_i)), \\ \phi_i &= \alpha_0 + \phi_{i-1}\xi_{1,i-1} + \xi_{2,i-1}. \end{aligned}$$

The series is (strictly) stationary if $E\{\ln(\xi_{1,i})\} < 0$ and $E\{\max(\ln \xi_{2,i}, 0)\} < \infty$. Note that $E(\ln \xi_{1,i}) < 0$ if $E(\xi_{1,i}) < 1$, by Jensen inequality.

7.2 Estimation of Parameters

Suppose that the conditional mean μ_i and the conditional variance σ_i^2 depend on a (multidimensional) parameter θ, First, when the innovations are Gaussian, the conditional log-likelihood L of $X_{r+1}, \ldots, X_n$, given $\mathcal{F}_r$ satisfies

$$-2L = (n-r)\ln(2\pi) + \sum_{i=r+1}^{n} \frac{(x_i - \mu_i)^2}{\sigma_i^2} + \sum_{i=r+1}^{n} \ln\left(\sigma_i^2\right). \tag{7.2}$$

Here we have used the multiplication formula (A.20). Then one can use Proposition B.7.1 to show that the maximum likelihood estimators for θ converge and find the asymptotic errors of estimation.

Often, even if the innovations are not Gaussian, the estimation of the parameters is done by minimizing (7.2). In that case, one talks about quasi-maximum likelihood estimators (QMLE). In most interesting cases, these estimators have also properties similar to those of the maximum likelihood estimators, but their error of estimation is larger in general. See, e.g., Francq and Zakoïan [2010] for its use in a GARCH context. We will come back to that methodology in Section 9.2.5, where the error of estimation is discussed. The MATLAB function *garchfit* in the *Econometrics* Toolbox can be used to estimate parameters using the maximum likelihood methodology for ARMAX models for the conditional mean process μ_i and GARCH, EGARCH, and GJR-GARCH models for the conditional variance, while the innovations are assumed Gaussian or Student. Note that in the case of innovations with density f_θ, the function to minimize is

$$L(\theta) = -\sum_{i=r+1}^{n} \ln\left\{ f_\theta\left(\frac{x_i - \mu_i}{\sigma_i}\right)\right\} + \sum_{i=r+1}^{n} \ln(\sigma_i).$$

Computing the Volatility Process σ

To be able to compute σ_i for $i > r$, we need values for $\sigma_1, \ldots, \sigma_r$. Note that these values should not matter if the stationarity conditions are met since in this case the process forgets its initial values. Often, they are replaced by average values. As a first approximation, the average is approximated by $\sqrt{\frac{1}{n}\sum_{i=1}^{n}(x_i - \hat{\mu}_i)^2}$, while on a second approximation, one uses the relationship between $E\left(\sigma_i^2\right)$ and its previous values. See, e..g., Section 7.2.1.

7.2.1 Application for GARCH(p,q) Models

Consider the model $x_i = \mu + \varepsilon_i \sigma_i$, where

$$\sigma_i^2 = \omega + \sum_{j=1}^{p} \beta_j \sigma_{i-j}^2 + \sum_{j=1}^{q} \alpha_j (x_{i-j} - \mu)^2, \quad i > r = \max(p, q),$$

with $\omega > 0$ and $\alpha_i \geq 0$, $\beta_i \geq 0$ and $\sum_{j=1}^{p} \beta_j + \sum_{j=1}^{q} \alpha_j < 1$. Here $\theta = (\mu, \omega, \beta_1, \ldots, \beta_p, \alpha_1, \ldots, \alpha_q)$. The coefficients $\beta_1, \ldots, \beta_p$ are called the GARCH coefficients, while $\alpha_1, \ldots, \alpha_q$ are the ARCH coefficients. In practice, once the parameters are estimated, one can compute the residuals $e_i = (y_i - \hat{\mu}) / \hat{\sigma}_i$, where

$$\hat{\sigma}_i^2 = \hat{\omega} + \left\{ \sum_{j=1}^{p} \hat{\beta}_j + \sum_{j=1}^{q} \hat{\alpha}_j \right\} s^2, \quad i \in \{1, \ldots, r\}, \tag{7.3}$$

with $s^2 = \frac{1}{n} \sum_{i=1}^{n} (x_i - \hat{\mu})^2$, while

$$\hat{\sigma}_i^2 = \hat{\omega} + \sum_{j=1}^{p} \hat{\beta}_j \hat{\sigma}_{i-j}^2 + \sum_{j=1}^{q} \hat{\alpha}_j (x_{i-j} - \hat{\mu})^2, \quad i > r.$$

Note that the computation of the initial values with equation (7.3) is the approach taken in both MATLAB and R[1].

Remark 7.2.1 *If one wants to implement a more precise numerical method by using the gradient instead of its approximation, it would have to be computed recursively, using the following relations, implemented in the MATLAB function* EstGARCHGaussian*:*

$$\begin{aligned}
\partial_\mu \sigma_i^2 &= -2 \sum_{j=1}^{q} \alpha_j (x_{i-j} - \mu) + \sum_{j=1}^{p} \beta_j \partial_\mu \sigma_{i-j}^2, \\
\partial_\omega \sigma_i^2 &= 1 + \sum_{j=1}^{p} \beta_j \partial_\omega \sigma_{i-j}^2, \\
\partial_{\alpha_k} \sigma_i^2 &= (x_{i-k} - \mu)^2 + \sum_{j=1}^{p} \beta_j \partial_{\alpha_k} \sigma_{i-j}^2, \quad k \in \{1, \ldots, q\}, \\
\partial_{\beta_k} \sigma_i^2 &= \sigma_{i-k}^2 + \sum_{j=1}^{p} \beta_j \partial_{\beta_k} \sigma_{i-j}^2, \quad k \in \{1, \ldots, p\}.
\end{aligned}$$

[1]Be careful in R because the orders of the GARCH are reversed. A real GARCH(2,1) becomes a GARCH(1,2) in R.

7.2.2 Tests

Suppose one wants to test $H_0 : \theta \in \Theta_0$ vs $H_1 : \theta \in \Theta_1$, where Θ_0 is a subset of Θ_1 obtained by setting k parameters to 0. Then, one can use a likelihood ratio test based on the statistic

$$LR = 2L(\hat{\theta}_{n,1}) - 2L(\hat{\theta}_{n,0}),$$

where $\hat{\theta}_{n,i}$ is the estimator obtained under H_i, $i = 0, 1$. Under H_0, the asymptotic distribution of the statistic is a chi-square with k degrees of freedom.

For example, for a GARCH(1,1) model, one could take $\Theta_1 = \{(\omega, \alpha, \beta); \alpha + \beta < 1\}$ and $\Theta_0 = \{(\omega, \alpha, 0); \alpha < 1\}$, so $k = 1$. Then under H_0, the assumed model is a ARCH(1) process. One could also assume that under H_0, the model is a GARCH(1,1), while under H_1, the model is a GARCH(2,2). In this case, $k = 2$.

7.2.3 Goodness-of-Fit and Pseudo-Observations

Let $\hat{\theta}$ be the estimator of θ and set $\hat{\mu}_i = \mu_i\left(\hat{\theta}_n\right)$ and $\hat{\sigma}_i = \sigma_i\left(\hat{\theta}_n\right)$. Then, for each $i \in \{1, \ldots, n\}$, define the pseudo-observation

$$e_i = \frac{x_i - \hat{\mu}_i}{\hat{\sigma}_i},$$

which is an estimation of the non-observable innovation ε_i. These pseudo-observations can then be used to assess the fit of the model.

However, one must be very careful with test statistics based on the pseudo-observations, since they do not behave like independent observations. In fact, the asymptotic behavior of the empirical distribution function constructed from the pseudo-observations is quite complicated. See, e.g., Bai [2003].

Serial Independence

Contrary to the claim in Hull [2006, Chapter 19], the Ljung-Box statistic

$$n(n+2)\sum_{k=1}^{K} \frac{r_k^2}{n-k},$$

where the r_k are the autocorrelation of order k computed from the pseudo-observations $e_{r+1}^2, \ldots, e_n^2$, is not a valid test for checking the hypothesis of serial independence for the innovations, since its limiting distribution is not a chi-square distribution with K degrees of freedom. Note that a valid test of serial independence based on the autocorrelations has been proposed by Berkes et al. [2003].

To check if the innovations are Gaussian, a valid test of goodness-of-fit has been proposed by Bai [2003]. It is based on the Khmaladze transform. It uses also the pseudo-observations and the limiting distribution of the test statistics has the advantage of not depending on the estimated parameters of the conditional mean and variance. In fact, the Khmaladze transform takes the empirical distribution of the pseudo-observations and maps it to a process W that is asymptotically a Brownian motion, under the null hypothesis. The details of the implementation are given in Appendix 7.A. Two test statistics are proposed: the usual Kolmogorov-Smirnov test, and the Cramér-von Mises test. They are both implemented in the MATLAB function *GofGARCHGaussian*. The distribution of these two statistics is well-known. In fact, the distribution function F of the Kolmogorov-Smirnov statistic is

$$F(x) = \frac{4}{\pi}\sum_{k=1}^{\infty} \frac{(-1)^k}{2k+1} e^{-(2k+1)^2\pi^2/(8x^2)}. \tag{7.4}$$

See Shorack and Wellner [1986, p. 36]. In particular, the quantiles of order 90%, 95%, and 99% are respectively 1.96, 2.241, and 2.807. For the Cramér-von Mises statistic, the distribution is the same as the distribution of

$$\frac{4}{\pi^2}\sum_{k=1}^{\infty} \frac{Z_k^2}{(2k-1)^2}, \tag{7.5}$$

where $Z_1, Z_2, \ldots$ are independent and identically distributed standard Gaussian variables. Using Shorack and Wellner [1986, Table 1, p. 748], the quantiles of order 90%, 95%, and 99% are respectively 1.2, 1.657, and 2.8.

Example 7.2.1 *We used the MATLAB function* EstGARCHGaussian *to estimate the parameters of a GARCH model for the data set* DataGARCH, *consisting of* 250 *simulated returns from a GARCH(1,1) model with parameters* $\mu = -0.004$, $\omega = 0.005$, $\beta = 0.3$, *and* $\alpha = 0.6$, *and Gaussian innovations. We see from Table 7.1 that the true values of the parameters are all within the 95% confidence intervals and that the precision of the estimations is satisfactory.*

Using the goodness-of-fit test based on the Khmaladze transform, we obtain that the statistic is 0.7203 *for the Kolmogorov-Smirnov test, yielding a P-value larger than 10% since the 90% quantile is* 1.960, *while the Cramér-von Mises statistic is* 0.1501, *with a P-value larger than 10%, since the 90% quantile is* 1.2. *Therefore, for both tests, we do not reject the null hypothesis that the innovations are Gaussian. The fact that we do not reject the hypothesis of normality for the Kolmogorov-Smirnov test at the 5% level is illustrated in Figure 7.2, since the trajectory of the process W lies entirely with the 95% confidence band about a Brownian path given by* ± 2.241.

Example 7.2.2 *Using the data from Apple (*ApplePrices*) and the MATLAB function* EstGARCHGaussian *for fitting a GARCH(1,1) model on the returns of Apple, both goodness-of-fit tests reject at the 5% the null hypothesis*

TABLE 7.1: 95% confidence intervals for the maximum likelihood estimation of the parameters of a GARCH(1,1) model, using the simulated data set *DataGARCH*.

Parameter	True Value	Confidence Interval
μ	-0.004	-0.0095 ± 0.0162
ω	0.005	0.0044 ± 0.0032
β	0.300	0.3892 ± 0.2271
α	0.600	0.5200 ± 0.2811

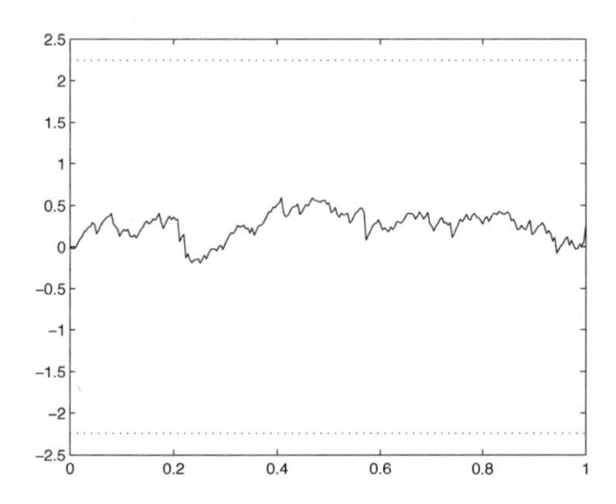

FIGURE 7.2: Trajectory of the process W for the innovations of a GARCH(1,1) model and uniform 95% confidence band for a Brownian motion path.

of the normality for the innovations. In fact, it is rejected for all GARCH(p,q) models with $p, q \in \{1, \ldots, 5\}$. Just as an illustration, the estimated parameters of the GARCH(1,1) model are given in Table 7.2, while in Figure 7.3, we display the trajectory of the process W together with the 95% confidence band for the Brownian motion.

7.2.4 Estimation and Goodness-of-Fit When the Innovations Are Not Gaussian

Consider a GARCH(1,1) model with GED innovations, i.e., $X_i = \mu + \sigma_i \varepsilon_i$, where $\sigma_i^2 = \omega + \alpha \sigma_{i-1}^2 \varepsilon_{i-1}^2 + \beta \sigma_{i-1}^2$, and ε_i have a generalized error distribution

TABLE 7.2: 95% confidence intervals for the maximum likelihood estimation of the parameters of a GARCH(1,1) model, using the data set *ApplePrices*.

Parameter	Confidence Interval
μ	0.0032 ± 0.0015
$10^4\omega$	0.1356 ± 0.1009
β	0.8700 ± 0.0672
α	0.0881 ± 0.0466

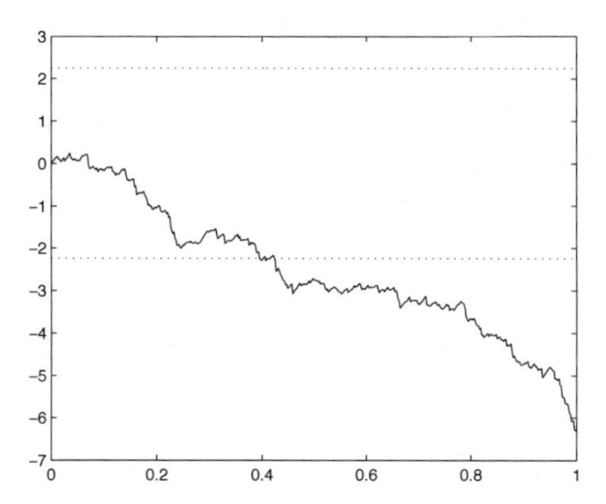

FIGURE 7.3: Trajectory of the process W for the innovations of a GARCH(1,1) model for the returns of Apple and uniform 95% confidence band for a Brownian motion path.

(GED) with parameter $\nu > 0$, i.e., the density is

$$f_\nu(x) = \frac{1}{b_\nu 2^{1+\frac{1}{\nu}}\Gamma(1+\frac{1}{\nu})} e^{-\frac{1}{2}\left(\frac{|x|}{b_\nu}\right)^\nu}, \quad x \in \mathbb{R},$$

where $b_\nu = 2^{-\frac{1}{\nu}}\sqrt{\frac{\Gamma(\frac{1}{\nu})}{\Gamma(\frac{3}{\nu})}}$. Denote the associated distribution function by G_ν. Note that the case $\nu = 2$ corresponds to the Gaussian distribution and that the kurtosis γ_2 is $\frac{\Gamma(\frac{1}{\nu})\Gamma(\frac{5}{\nu})}{\Gamma^2(\frac{3}{\nu})}$. For more details on this distribution, see Appendix A.6.17.

Implementing Khmaladze transform for Gaussian innovations is easy, while implementing it for GED innovations is difficult. However, using the pseudo-observations $u_{n,i} = G_{\hat{\nu}}(e_i)$, $i \in \{1, \ldots, n\}$, the parametric bootstrap approach is always easy to implement for tests based on the empirical distribution function

$$D_n(u) = \frac{1}{n}\sum_{i=1}^{n} \mathbb{I}(u_{n,i} \le u), \quad u \in [0,1].$$

The latter should be close to the uniform distribution function $D(u) = u$, for $u \in [0,1]$, under the null hypothesis that the innovations are GED. Basically, to estimate P-values, one has to implement almost the same approach as the one described in Algorithm 3.B.2, except that we have to generate not only the innovations, but also the associated GARCH process, and estimate its parameters at each iteration. This is time consuming. See, e.g., Rémillard [2011] and Ghoudi and Rémillard [2012].

Example 7.2.3 *For the returns of Apple from January 14th, 2009, to January 14th, 2011, the results of the estimation obtained with the MATLAB function* EstGARCHGED *are given in Table 7.3.*

TABLE 7.3: 95% confidence intervals for the maximum likelihood estimation of the parameters of a GARCH(1,1) model with GED innovations, using the data set *ApplePrices*.

Parameter	Confidence Interval
μ	0.0028 ± 0.0013
$10^4\omega$	0.0714 ± 0.1312
β	0.8968 ± 0.0889
α	0.0817 ± 0.0680
ν	1.3510 ± 0.2387

Using the MATLAB function GofGARCHGED, *we obtain that P-values corresponding to the Kolmogorov-Smirnov test statistic (0.6821) and the Cramér-von Mises statistic (0.0549) are respectively* 19.3% *and* 30.9%. *Here, $N = 1000$ bootstrap samples of a GARCH were simulated to estimate the P-values. Hence the null hypothesis of a GARCH(1,1) model with GED innovations is not rejected. As an illustration, the empirical process D_n is displayed in Figure 7.4 and it lies totally within the uniform 95% band around the uniform distribution function D.*

Remark 7.2.2 *What we have done here for GED innovations can be easily repeated with other models with innovations $\varepsilon_i \sim F_\theta$. For the estimation of the parameters, including θ, one should use maximum likelihood, while goodness-of-fit tests should be based on the empirical distribution function of the pseudo-observations $u_{n,i} = F_{\hat\theta}(e_i)$, $i \in \{1,\ldots,n\}$. To compute P-values, one should use parametric bootstrap. See, e.g. Rémillard [2011].*

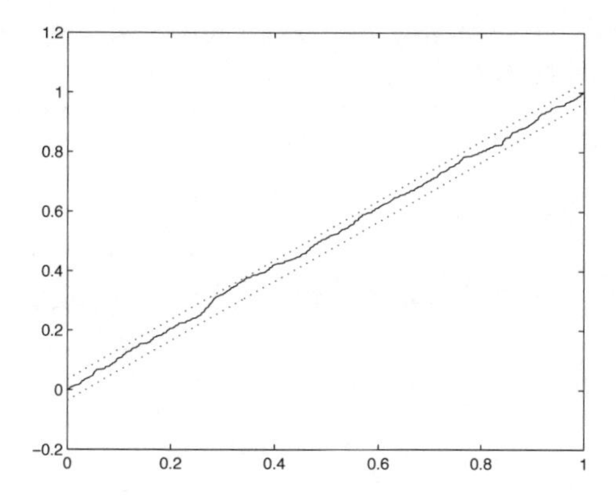

FIGURE 7.4: Empirical distribution D_n and uniform 95% confidence band around D.

7.3 Duan Methodology of Option Pricing

In Duan [1995], the author studied option pricing when the daily returns of the underlying asset are modeled by a special case of NGARCH model, called a GARCH-M. More precisely, he assumed that the returns X_i of the daily prices S_i satisfy

$$X_i = \ln(S_i) - \ln(S_{i-1}) = r_p + \lambda\sigma_i - \frac{1}{2}\sigma_i^2 + \sigma_i\varepsilon_i, \tag{7.6}$$

where r_p is the daily risk-free rate, and $\sigma_i\varepsilon_i \sim GARCH(p,q)$, i.e., the conditional distribution of ε_i given $\mathcal{F}_{i-1}$ is standard Gaussian, and

$$\sigma_i^2 = \alpha_0 + \sum_{j=1}^{q} \alpha_j \sigma_{i-j}^2 \varepsilon_{i-j}^2 + \sum_{j=1}^{p} \beta_j \sigma_{i-j}^2,$$

$\alpha_0 > 0$, $\alpha_i, \beta_j \geq 0$, $\sum_j \alpha_j + \sum_j \beta_j < 1$. The main contribution in Duan [1995] is a criterion to choose an equivalent martingale measure for option pricing. Note that the choice is not necessarily optimal.

7.3.1 LRNVR Criterion

Definition 7.3.1 *A risk neutral measure Q satisfies the LRNVR (locally risk-neutral valuation relationship) criterion if Q is equivalent to the objective measure, if the returns are conditionally Gaussian with the same variance, and if $e^{-r_p i} S_i$ is a martingale under Q.*

In Duan [1995], it is shown that the unique measure Q satisfying the LRNVR criterion is such that

$$X_i = \ln(S_i) - \ln(S_{i-1}) = r_p - \frac{1}{2}\sigma_i^2 + \sigma_i \xi_i,$$

where $\xi_i | \mathcal{F}_{i-1} \sim N(0,1)$ and

$$\sigma_i^2 = \alpha_0 + \sum_{j=1}^{q} \alpha_i \sigma_{i-j}^2 (\xi_{i-j} - \lambda)^2 + \sum_{j=1}^{p} \beta_j \sigma_{i-j}^2.$$

In fact, for a general GARCH model of the form

$$X_i = \mu_i + \sigma_i \varepsilon_i,$$

where the innovations ε_i are independent and standard Gaussian, and

$$\sigma_i = \phi(\sigma_{i-1}, \ldots, \sigma_{i-p}, \varepsilon_{i-1}, \varepsilon_{i-q}), \quad i > r = \max(p,q),$$

for Q to satisfy the LRNVR criterion, it is necessary and sufficient that under Q, $\xi_i = \varepsilon_i - \frac{(r_p - \mu_i - \sigma_i^2/2)}{\sigma_i} \sim N(0,1)$ and ξ_i is independent of $\mathcal{F}_{i-1}$.

One can obtain such a Q through its definition by the densities

$$\left.\frac{dQ}{dP}\right|_{\mathcal{F}_i} = \prod_{j=r+1}^{i} e^{\frac{(r_p - \mu_j - \sigma_j^2/2)}{\sigma_j}\varepsilon_j - \frac{(r_p - \mu_j - \sigma_j^2/2)^2}{2\sigma_j^2}}, \quad i > r.$$

Q is then equivalent to the objective measure since the densities are strictly positive. In this case, for $i > r$, setting $a_i = \frac{r_p - \mu_i - \sigma_i^2/2}{\sigma_i} \in \mathcal{F}_{i-1}$, we have, for any $u \in \mathbb{R}$,

$$E_Q\left(e^{u\xi_i} \middle| \mathcal{F}_{i-1}\right) = E\left\{ e^{(u+a_i)\varepsilon_i - ua_i - \frac{a_i^2}{2}} \middle| \mathcal{F}_{i-1} \right\} = e^{\frac{(u+a_i)^2}{2} - ua_i - \frac{a_i^2}{2}} = e^{\frac{u^2}{2}}.$$

This proves that under Q, $\xi_{r+1}, \ldots,$ are standard Gaussian variables and ξ_i is independent of $\mathcal{F}_{i-1}$. Next note that

$$X_i = \mu_i + \sigma_i \varepsilon_i = \mu_i + \sigma_i(\xi_i + a_i) = r_p - \frac{\sigma_i^2}{2} + \sigma_i \xi_i, \qquad i > r.$$

As a result, the returns are conditionally Gaussian, with the same variance σ_i^2 with respect to $\mathcal{F}_{i-1}$. Finally, since $\xi_i \sim N(0,1)$ is independent of $\mathcal{F}_{i-1}$, we have, for all $i > r$,

$$e^{-r_p} E_Q(S_i | \mathcal{F}_{i-1}) = E\left(S_{i-1} e^{-\frac{\sigma_i^2}{2} + \sigma_i \xi_i} | \mathcal{F}_{i-1} \right) = S_{i-1} e^{-\frac{\sigma_i^2}{2}} e^{\frac{\sigma_i^2}{2}} = S_{i-1}.$$

This proves that $e^{-r_p i} S_i$ is a martingale under Q, so Q satisfies the LRNVR criterion. We also have the following representation for the conditional variance:

$$\sigma_i = \phi(\sigma_{i-1}, \ldots, \sigma_{i-p}, \xi_{i-1} + a_{i-1}, \xi_{i-q} + a_{i-q}), \quad i > r = \max(p,q),$$

Going back to the model (7.6), we have $\mu_i = r_p + \lambda\sigma_i - \sigma_i^2/2$, so $a_i = -\lambda$.

Remark 7.3.1 *In Duan [1995], the author also proposed a way to "hedge" any option based on a risk measure Q satisfying the LRNVR criterion but that methodology was proven to be incorrect in Garcia and Renault [1998]. However, since $(X_i, \ldots, X_{i-q+1}, \sigma_{i+1}, \ldots, \sigma_{i+2-p})$ is a Markov process, one could use the optimal hedging methodology of Chapter 3, without having to define a LRNVR measure Q.*

Once an equivalent martingale measure is chosen, one still has to evaluate the option. One could use Monte Carlo methods, or one could use fast approximation methods, as in Duan et al. [2000].

7.3.2 Continuous Time Limit

Prior to this chapter, we were always dealing with continuous time models. Are the discrete time GARCH models approximations of continuous time models? Duan [1997] answered positively to this question.

For example, take a sequence of simple GARCH(1,1)-M models X_n with parameters $\mu_{n,i} = \frac{\mu}{n} + \frac{\lambda}{\sqrt{n}}\sigma_{n,i} - \frac{\sigma_{n,i}^2}{2}$, $\omega_n = \omega/n^2$, $\alpha_n = \alpha/\sqrt{n}$ and $\beta_n = 1 - \kappa/n - \alpha_n$, $\kappa > 0$, with $\sigma_{n,i} = \sqrt{V_{n,i}/n}$. Then

$$X_{n,i} = \frac{\mu + \lambda\sqrt{V_{n,i}} - V_{n,i}/2}{n} + \sqrt{V_{n,i}}\frac{\varepsilon_i}{\sqrt{n}}, \tag{7.7}$$

$$V_{n,i} = V_{n,i-1} + \frac{\omega - \kappa V_{n,i-1}}{n} + \alpha V_{n,i-1}\frac{(\varepsilon_{i-1}^2 - 1)}{\sqrt{n}}, \tag{7.8}$$

where $\gamma_2 = E\left(\varepsilon_i^4\right)$ is the kurtosis of ε_i. Further set $\ln\{S_n(t)\} = \sum_{i=1}^{\lfloor nt \rfloor} X_{n,i}$, where $\lfloor u \rfloor$ is the largest integer not larger than u. It then follows that if $V_n(t) = V_{n,\lfloor nt \rfloor}$, then the sequence of processes (S_n, V_n) converge in law to a continuous diffusion process (S, V) solving the stochastic differential equation

$$dS(t) = \left\{\mu + \lambda\sqrt{V(t)}\right\} S(t)dt + S(t)\sqrt{V(t)}\, dZ_1(t), \tag{7.9}$$

$$dV(t) = \kappa\left\{\frac{\omega}{\kappa} - V(t)\right\} dt + (\gamma_2 - 1)^{1/2}\alpha V(t)dZ_2(t), \tag{7.10}$$

where Z_1 and Z_2 are correlated Brownian motions with correlation $\rho = \frac{\gamma_1}{\sqrt{\gamma_2 - 1}}$, $\gamma_1 = E\left(\varepsilon_i^3\right)$ being the skewness coefficient of ε_i. In particular, in the case considered in Duan [1995], the innovations ε_i are Gaussian, so $\gamma_1 = 0$, $\gamma_2 = 3$, and $\rho = 0$. Hence

$$dS(t) = \left\{\mu + \lambda\sqrt{V(t)}\right\} S(t)dt + S(t)\sqrt{V(t)}\, dZ_1(t),$$

$$dV(t) = \kappa\left\{\frac{\omega}{\kappa} - V(t)\right\} dt + 2^{1/2}\alpha V(t)dZ_2(t),$$

where Z_1 and Z_2 are independent Brownian motions.

Remark 7.3.2 *Processes satisfying* (7.10) *were proposed in Merton [1974] for modeling the value of a firm. Note that* (7.8) *provides a way to simulate (approximately) this process.*

Note also that under the equivalent martingale measure proposed in Duan [1995], the sequence of processes (S_n, V_n) converge in law to a continuous diffusion process (S, V) solving the stochastic differential equation

$$\begin{aligned} dS(t) &= r_f S(t)dt + S(t)\sqrt{V(t)}\, dZ_1(t), \\ dV(t) &= \kappa\left\{\frac{\omega}{\kappa} - V(t)\right\} dt + 2^{1/2}\alpha\, V(t)dZ_2(t), \end{aligned}$$

where Z_1 and Z_2 are independent Brownian motions, and r_f is the risk-free rate. As we will see in the next section, this is a particular case of the Hull-White model. In Duan [1997], the continuous time limit of the Augmented GARCH models is also studied, leading to a more general class of stochastic volatility models.

7.3.2.1 A New Parametrization

Equations (7.7) to (7.10) suggest the following parametrization for the daily returns modeled by a GARCH(1,1) model:

$$\begin{aligned} X_i &= \mu h + \sigma_i \varepsilon_i, \\ \sigma_i^2 &= \omega h^2 + \left(1 - \kappa h - \alpha\sqrt{h}\right) \sigma_{i-1}^2 + \alpha\sqrt{h}\, \sigma_{i-1}^2 \varepsilon_{i-1}^2, \end{aligned}$$

where $h = 1/252$. With this new parametrization, using Tables 7.2 and 7.3, we would get new parameters, expressed on a annual time scale. These parameters are given in Table 7.4.

TABLE 7.4: Estimation of the parameters of Apple on an annual time scale, using Gaussian and GED innovations.

New Parameter	Estimation	
	Gaussian	GED
μ	0.8177	0.7011
ω	0.8608	0.4531
κ	10.5707	5.4191
α	1.3986	1.2971
ν	2.0000	1.3510

Since $\gamma_2 = 4.1651$ for the GED with $\hat{\nu} = 1.351$, it follows that the value S of Apple, for 2009-2011, could be modeled, on an annual time scale, by a stochastic volatility model of the form

$$\begin{aligned} dS(t) &= .7011 S(t)dt + S(t)\sqrt{V(t)}\, dZ_1(t), \\ dV(t) &= \{.4531 - 5.4191 V(t)\}\, dt + 2.23075 V(t)dZ_2(t), \end{aligned}$$

where Z_1 and Z_2 are independent Brownian motions. It is worth noting that $\frac{\omega}{\kappa}$, which was previously related to the long term average of the conditional variance, is still interpreted as the long term average of V, while κ is the mean reversion rate.

7.4 Stochastic Volatility Model of Hull-White

In Hull and White [1987], the authors proposed to model assets using a stochastic volatility model of the form

$$\begin{aligned} dS(t) &= \phi\{S(t), V(t), t\}S(t)dt + \sqrt{V(t)}S(t)dW(t) \\ dV(t) &= \mu\{V(t), t\}V(t)dt + \xi\{V(t), t\}V(t)dZ(t) \end{aligned}$$

where W and Z are two correlated Brownian motions with correlation ρ. As seen in the previous section, such models arise from the continuous time limit of GARCH processes.

Note that the hypothesis $\rho = 0$ could be justified from the approximation by GARCH models in which the innovations are Gaussian, GED, or have skewness 0.

7.4.1 Market Price of Volatility Risk

As we encountered in the case of short-rate stochastic models, the value of any option based on the underlying S depends on a common deterministic function $\lambda = \lambda(t, s, v)$ called here the market price of volatility risk. In fact, one can show that the value at time t of an option with payoff $\Phi\{S(T)\}$ at maturity T is $C\{t, S(t), V(t)\}$, where C satisfies the following partial differential equation:

$$\begin{aligned} r_f C &= \frac{\partial C}{\partial t} + r_f s\frac{\partial C}{\partial s} + (\mu v - \lambda)\frac{\partial C}{\partial v} \\ &\quad + \frac{1}{2}vs^2\frac{\partial^2 C}{\partial s^2} + \frac{1}{2}\xi^2 v^2\frac{\partial^2 C}{\partial v^2} + \rho\xi s v^{3/2}\frac{\partial^2 C}{\partial v \partial s}, \end{aligned} \tag{7.11}$$

with boundary condition $C(T, s, v) = \Phi(s)$. Here, r_f is the risk-free rate, assumed to be deterministic. Note that the process V is not observable, so even if we have an explicit expression for C, we need a way to predict V. This is the kind of problem that will be studied in Chapter 9. One could also find a GARCH-type approximation and replace V by a function of the stochastic volatility σ.

7.4.2 Expectations vs Partial Differential Equations

For a twice continuously differentiable function f, set

$$L_t f(x) = \sum_{j=1}^{d} b_j(t,x)\partial_{x_j} f(x) + \frac{1}{2}\sum_{j=1}^{d}\sum_{k=1}^{d} A_{jk}(t,x)\partial_{x_j}\partial_{x_k} f(x).$$

The following result, also called the Feynman-Kac formula, is proven in Appendix 7.B.2. It is the essential formula relating the solution of (7.11) to an expectation of a diffusion process under an equivalent martingale measure.

Proposition 7.4.1 (Feynman-Kac Formula) *Under weak conditions, we can show that the solution to the partial differential equation*

$$\partial_t u(t,x) + L_t u(t,x) = r_f(t)u(t,x), \quad t \in (0,T), \quad u(T,x) = \Phi(x),$$

is given by

$$u(t,x) = \frac{B(t)}{B(T)} E[\Phi\{X(T)\}|\mathcal{F}_t, X_t = x],$$

where $B(t) = \exp\left\{\int_0^t r_f(s)ds\right\}$ *and* X_t *is a diffusion "satisfying" the stochastic differential equation*

$$dX_j(t) = b_j\{t,X(t)\}dt + \sum_{k=1}^{d} \sigma_{kj}\{t,X(t)\}dW_k(t),$$

where $A = \sigma^\top \sigma$ *and where* $W_1,\ldots,W_d$ *are independent Brownian motions.*

Remark 7.4.1 *Instead of assuming the strong condition that* X_t *satisfies a stochastic differential equation, it is sufficient to suppose that* X *is a stochastic process so that for any function* f *that is smooth enough,*

$$f\{t,X(t)\} - \int_0^t [\partial_u f\{u,X(u)\} + L_u f\{u,X(u)\}]du$$

is a martingale. In this case, the law of X *is said to be the solution of the martingale problem associated with the operator* L*, called an infinitesimal generator. Note that the operator does not depend on* σ *but only on* $A = \sigma\sigma^\top$*, so there can be in general an infinite number of* σ *yielding the same law. Recall that we discussed briefly the topic of martingale problems in Chapter 5.*

7.4.3 Option Price as an Expectation

It follows from the Feynman-Kac formula that if $C(T,s,v) = \Phi(s)$ and

$$\frac{\partial C}{\partial t} + r_f s\frac{\partial C}{\partial s} + r_v v\frac{\partial C}{\partial v} + \frac{1}{2}vs^2\frac{\partial^2 C}{\partial s^2} + \frac{1}{2}\xi^2 v^2\frac{\partial^2 C}{\partial v^2} = r_f C,$$

with $r_v = \mu - \lambda/v$, then

$$C(t,s,v) = e^{-r_f(T-t)} E_Q \left[\Phi\{S(T)\} | \mathcal{F}_t, S(t) = s, V(t) = v \right], \tag{7.12}$$

where, under the equivalent martingale measure Q, (S, V) now satisfies

$$\begin{aligned} dS &= r_f S dt + \sqrt{V} S d\tilde{W}, \\ dV &= r_v V dt + \xi V d\tilde{Z}, \end{aligned}$$

with $\tilde{W}$ and $\tilde{Z}$ being two Brownian motions with correlation ρ under Q.

In order to be able to do some computations, Hull and White [1987] assumed that ξ and r_v are constant, and that $\rho = 0$. It follows that V is a geometric Brownian motion, i.e.,

$$V(p) = V_t e^{(r_v - \xi^2/2)(s-t) + \xi\{\tilde{Z}(\delta) - \tilde{Z}(t)\}}, \; s \in [t, T].$$

In addition, V is independent of $\tilde{W}$ since it depends only on the Brownian motion $\tilde{Z}$ which is independent $\tilde{W}$. Also, in this case, one can show that

$$S(t) = S_0 e^{r_f t} \exp\left\{ -\frac{1}{2} \int_0^t V(u) du + \int_0^t \sqrt{V(u)} d\tilde{W}(u) \right\},$$

and

$$S(T) = S(t) e^{r_f(T-t)} \exp\left\{ -\frac{1}{2} \int_t^T V(u) du + \int_t^T \sqrt{V(u)} d\tilde{W}(u) \right\}.$$

Given the whole path of V and the past of S up to time t, it follows from the properties of stochastic integrals that the conditional distribution of $\ln\left(\frac{S_T}{S_t}\right)$ is Gaussian with mean $(T-t)(r - \bar{V}_t/2)$ and variance $\bar{V}_t$, where

$$\bar{V}(t) = \frac{1}{T-t} \int_t^T V(u) du.$$

Let $C^{BS}(T-t, s, \sigma^2)$ be the value of the same option under a Black-Scholes model with constant volatility σ. Then, using the independence of V and $\tilde{W}$, if $S(t) = s$ and $V(t) = v$, we obtain that

$$\begin{aligned} C(t,s,v) &= E_Q[C^{BS}\{T-t, s, \bar{V}_t\}] \\ &= \int_0^\infty C^{BS}(T-t, s, y) \psi(y; v) dy, \end{aligned}$$

where $\psi(y; v)$ is the density of $\bar{V}(t)$ at y, given $V(t) = v$. Note that since V is a geometric Brownian motion, the conditional distribution of $\bar{V}_t$ given $V(t) = v$ is the same as the law of $vH(T-t)$, where

$$H(\tau) = \frac{1}{\tau} \int_0^\tau e^{(r_v - \xi^2/2)u + \xi \check{Z}(u)} du,$$

and where $\check{Z}$ is a Brownian motion which is independent $\tilde{W}$. It follows that

$$C(t,s,v) = E\left[C^{BS}\{\tau, s, vH(\tau)\}\right] = \int_0^\infty C^{BS}(\tau, s, vh)\psi_\tau(h)dh, \tag{7.13}$$

where $\tau = T - t$ and ψ_τ is the density of $H(\tau)$. Note that the process H appears in the expression of Asian option in a Black-Scholes model; however, its density is quite complicated [Dufresne, 2000], so it could be better to approximate the expectation in (7.13). This is the topic of the next section.

7.4.4 Approximation of Expectations

As shown in (7.13), the evaluation of $C(t,s,v) = E\left[C^{BS}\{\tau, s, vH(\tau)\}\right]$ is just a computation of an expectation of the form $c = E\{F(X)\}$. There exists many ways to estimate it:

- Monte Carlo method;
- Taylor expansion of F;
- Edgeworth approximation of the density of X;
- Approximate distribution.

The first two methods were considered in Hull and White [1987].

7.4.4.1 Monte Carlo Methods

Here the idea is to generate a large number N of X_i having the same distribution, or a distribution close to the one of X, and then compute

$$\hat{c} = \frac{1}{N}\sum_{i=1}^N F(X_i).$$

In Hull and White [1987], the authors suggested to approximate the distribution of $H(\tau)$ by the mean of the variables Y_{ni}, where

$$Y_{n,i} = Y_{n,i-1} e^{\xi Z_i \sqrt{\tau/n} + (r_v - \xi^2/2)\tau/n},$$

and $Z_1, \ldots, Z_n$ are independent standard Gaussian random variables.

7.4.4.2 Taylor Series Expansion

Suppose that F is differentiable and that we approximate $F(X)$ by a Taylor polynomial of order k around $\mu = E(X)$, i.e.,

$$F(X) \approx F(\mu) + \sum_{j=1}^k F^{(j)}(\mu)\frac{(X-\mu)^j}{j!},$$

where $F^{(j)}(\mu)$ is the derivative of order j or F. Then

$$c = E\{F(X)\} \approx F(\mu) + \sum_{j=1}^{k} \frac{F^{(j)}(\mu)}{j!} E\{(X-\mu)^j\}.$$

Then it suffices to be able to compute the centered moments of X. This methodology was implemented in Hull and White [1987] where they used $k = 3$ and assumed that $r_v = 0$. In this case, $H(\tau)$ has the same law as $\tilde{H}(\xi^2\tau)$, where $\tilde{H}(\tau) = \frac{1}{\tau}\int_0^\tau e^{B_u - u/2} du$, and where B is a Brownian motion. Then

$$\begin{aligned} E\left\{\tilde{H}(\tau)\right\} &= 1, \\ E\left\{\tilde{H}^2(\tau)\right\} &= \frac{2\left(e^\tau - 1 - \tau\right)}{\tau^2}, \end{aligned}$$

and

$$E\left\{\tilde{H}^3(\tau)\right\} = \frac{e^{3\tau} - 9e^\tau + 6\tau + 8}{6\tau^3}.$$

7.4.4.3 Edgeworth and Gram-Charlier Expansions

Instead of approximating F, one could approximate the density of X using properties of characteristic functions and cumulants. This approach was popularized by Jarrow and Rudd [1982]. The proof of the following proposition is given in Appendix 7.B.3.

Proposition 7.4.2 *Suppose that g is the density of Y and f is the density of X, and assume that both variables have moments of order n. Denote by $\kappa_j(X)$ and $\kappa_j(Y)$ the cumulants of order $j \in \{1, \ldots, n\}$ of X and Y respectively. Then if $E_1, \ldots, E_n$ are defined by the Taylor expansion of order n of $\exp\left\{\sum_{j=1}^n \frac{u^j}{j!}\{\kappa_j(Y) - \kappa_j(X)\}\right\}$ about $u = 0$, i.e.,*

$$\exp\left\{\sum_{j=1}^{n} \frac{u^j}{j!}\{\kappa_j(Y) - \kappa_j(X)\}\right\} = 1 + \sum_{j=1}^{n} \frac{u^j}{j!} E_j + O(u^{n+1}), \tag{7.14}$$

then

$$g(x) = f(x) + \sum_{j=1}^{n} \frac{(-1)^j}{j!} E_j f^{(j)}(x) + r_n(x), \tag{7.15}$$

where r_n is a reminder of order x^n.

Of course the precision of this expansion depends on the difference between the cumulants and also the choice of f. If Y has a limiting distribution, then f should be its density, if possible. Note that one can also compute the first n cumulants if we know only the first n moments of X, since for small u,

$$\ln\left\{1 + \sum_{j=1}^{n} \frac{u^j}{j!} E(X^j) + O(u^{n+1})\right\} = \sum_{j=1}^{n} \frac{u^j}{j!} \kappa_j + O(u^{n+1}).$$

This kind of calculation can be done easily with MAPLE. It can also be done with MATLAB, provided the *Symbolic Math* Toolbox is available. For example, with MATLAB, if the first 3 moments are $2, 5, 7$, then using the Taylor expansion of order 4 of $\ln(1 + 2u + 5u^2/2 + 7u^3/6)$, we obtain

```
>> syms u
>> f = log(1+2*u+5*u^2/2 +7*u^3/6);
>> taylor(f,'Order',4)

ans =

- (7*u^3)/6 + u^2/2 + 2*u

>>
```

showing that the first 3 cumulants are 2, 1, and -7. Note that there is no need of the fourth moment if one only wants to compute the first 3 cumulants. In fact, if the fourth moment is 43, then

```
>> syms u
>> f = log(1+2*u+5*u^2/2 +7*u^3/6+43*u^4/24);
>> taylor(f,'Order',4)

ans =

- (7*u^3)/6 + u^2/2 + 2*u
```

giving exactly the same values. Similarly, the values of $E_1, \ldots, E_n$ in (7.14), can also be computed this way. For example, if f is the density of a standard Gaussian distribution, then $(-1)^k \frac{f^{(k)}}{f(x)} = H_k(x)$, where the polynomials H_k are defined in Section 4.3.5. Hence, if $E(Y) = 0$ and $E(Y^2) = 1$, then according to Proposition 7.4.2, $E_1 = E_2 = 0$, and

$$g(x) \approx \frac{e^{-x^2/2}}{\sqrt{2\pi}} \left(1 + \frac{\kappa_3}{3!} H_3(x) + \frac{\kappa_4}{4!} H_4(x) + \frac{\kappa_5}{5!} H_5(x) + \left(\frac{\kappa_6}{6!} + \frac{\kappa_3^2}{72}\right) H_6(x)\right)$$

since the cumulants of X of order $k \geq 3$ are all 0, and

$$\exp\left\{\frac{\kappa_3}{3!}u^3 + \frac{\kappa_4}{4!}u^4 + \frac{\kappa_5}{5!}u^5 + \frac{\kappa_6}{6!}u^6 + O(u^7)\right\}$$

$$= 1 + \frac{\kappa_3}{3!}u^3 + \frac{\kappa_4}{4!}u^4 + \frac{\kappa_5}{5!}u^5 + \left(\frac{\kappa_6}{6!} + \frac{\kappa_3^2}{72}\right) u^6 + O(u^7).$$

The last expression was obtained with the *Symbolic Math* Toolbox of MATLAB:

```
>> syms u kappa3 kappa4 kapp5 kappa6
```

```
g = exp(kappa3*u^3/6+kappa4*u^4/24+kapp5*u^5/120+kappa6*u^6/720);

taylor(g,'Order',7)

ans =

(kappa3^2/72 + kappa6/720)*u^6 + (kapp5*u^5)/120 + (kappa4*u^4)/24 +
(kappa3*u^3)/6 + 1
```

Remark 7.4.2 *The series* (7.15) *is not the Edgeworth expansion of g even if f is the Gaussian density. It is called the Gram-Charlier expansion. The Edgeworth expansion is obtained from the Gram-Charlier expansion by collecting terms of the same power n for variables Y of the form $Y = \frac{(X_1+\ldots+X_n)-n\mu}{\sigma\sqrt{n}}$. The Edgeworth expansion, already treated in Section 4.3.5, is used normally to approximate a density of a sum of variables converging in law to a Gaussian distribution. In that case, f is the Gaussian density with the same mean and variance as the sum.*

7.4.4.4 Approximate Distribution

The main inconvenience of any kind of expansion is that we do not end up with a density in general. In order to estimate the expectation $c = E\{F(X)\}$, one could replace X by its limiting distribution if possible.

In Dufresne [1990] and Milevsky and Posner [1998], it was shown that if $a < \frac{\sigma^2}{2}$ and t is large,

$$I_t = \int_0^t \exp\left\{\left(a - \frac{\sigma^2}{2}\right)u + \sigma W(u)\right\} du \approx \frac{1}{X},$$

where $X \sim \text{Gamma}\left(\alpha = 1 - \frac{a}{\frac{\sigma^2}{2}}, \beta = \frac{\sigma^2}{2}\right)$. Therefore, the density of I_t could be approximated by the density of the inverse gamma distribution, given by

$$f(x) = \frac{x^{-1-\alpha}}{\beta^\alpha \Gamma(\alpha)} e^{-\frac{1}{x\beta}} I_{(0,\infty)}(x).$$

Then one could use Monte Carlo methods. In our case, $a = r_v$ and $\sigma = \xi$, and $H(t) = I_t/t$. Note that the moment of order k exists for that distribution if and only if $\alpha > k$, i.e., if $a < -\frac{k-1}{\frac{\sigma^2}{2}}$.

One could probably do a better approximation by choosing a larger family of distributions containing the inverse gamma.

7.5 Stochastic Volatility Model of Heston

In Heston [1993], the author studied the evaluation of options for the following stochastic volatility model:

$$\begin{aligned} dS(t) &= \phi\{S(t), V(t), t\}S(t)dt + \sqrt{V(t)}S(t)dW(t) \\ dV(t) &= \mu\{V(t), t\}dt + \sigma\{V(t), t\}dZ(t) \end{aligned}$$

where W and Z are two correlated Brownian motions with correlation ρ, $\mu(v,t) = \kappa(\theta - v)$ and $\sigma(v,t) = \xi\sqrt{v}$. He also used a linear market price of volatility risk, that is $\lambda(s,t,t) = \lambda v$. In the general case, under the equivalent martingale measure Q, the dynamics are given by

$$\begin{aligned} dS(t) &= r_f S(t)dt + \sqrt{V(t)}S(t)dW(t) \\ dV(t) &= [\mu\{V(t), t\} - \lambda\{S(t), V(t), t\}]\, dt + \sigma\{V(t), t\}dZ(t) \end{aligned}$$

where W and Z are two correlated Brownian motions with correlation ρ. Setting $S(T) = S(t)e^{X_{t,T}}$, it is then easy to show that the value at time t of a call option of S with strike price K at maturity T is given by

$$\begin{aligned} C(s,v,t,T) &= e^{-r_f(T-t)} E_Q\left[\max\{0, S(T) - K\} | \mathcal{F}_t, S(t) = s, V(t) = v\right] \\ &= sE_Q\left[e^{X_{t,T} - r_f(T-t)} \mathbb{I}\{X_{t,T} > \ln(K/s)\} | \mathcal{F}_t\right] \\ &\quad - Ke^{-r_f(T-t)} Q\{X_{t,T} > \ln(K/s)\} \\ &= s\tilde{Q}\{X_{t,T} > \ln(K/s)\} - Ke^{-r_f(T-t)} Q\{X_{t,T} > \ln(K/s)\}, \end{aligned}$$

where $\tilde{Q}$ is the probability distribution of $X_{t,T}$ with density $e^{X_{t,T} - r_f(T-t)}$ with respect to Q.

If $\phi_u(t,x,v)$ is the characteristic function of $X_{t,T}$ given $\mathcal{F}_t$ under Q, then the characteristic function $\tilde{\phi}_u(t,x,v)$ of $X_{t,T}$ under $\tilde{Q}$ would be

$$\tilde{\phi}_u(t,x,v) = \phi_{u-i}(t,x,v)e^{-r_f(T-t)}.$$

Using the results of Section 7.4.2 applied to $E\left[e^{iu\ln\{S(T)\}} | \mathcal{F}_t\right]$, it follows that $\phi_u(t,x,v)$ satisfies the partial differential equation

$$\begin{aligned} 0 = &\left\{iu(r_f - v/2) - vu^2/2\right\}\phi_u + \partial_t\phi_u \\ &+ (r_f - v/2 + iuv)\partial_x\phi_u + (\mu - \lambda + iu\rho\sigma\sqrt{v})\partial_v\phi_u \\ &+ \frac{1}{2}\left\{v\partial_x^2\phi_u + 2\rho\sigma\sqrt{v}\partial_x\partial_v\phi_u + \sigma^2\partial_v^2\phi_u\right\}, \end{aligned}$$

with the boundary condition $\phi_u(T,x,v) \equiv 1$.

As a result, ϕ_{u-i} satisfies $\phi_{u-i}(T,x,v) \equiv 1$ and

$$\begin{aligned} 0 &= \left\{iu(r_f + v/2) - (vu^2/2 - r_f)\right\}\phi_{u-i} + \partial_t \phi_{u-i} \\ &\quad +(r_f + v/2 + iuv)\partial_x \phi_{u-i} \\ &\quad\quad +(\mu - \lambda + \rho\sigma\sqrt{v} + iu\rho\sigma\sqrt{v})\partial_v \phi_{u-i} \\ &\quad\quad +\frac{1}{2}\left\{v\partial_x^2 \phi_{u-i} + 2\rho\sigma\sqrt{v}\partial_x\partial_v \phi_{u-i} + \sigma^2\partial_v^2 \phi_{u-i}\right\}. \end{aligned}$$

If we can solve those two equations, then using the Gil-Pelaez formula (4.24), we can compute the two probabilities. In Heston [1993], the author solved it for his particular model and he was able to compute the associated characteristic functions. See Heston [1993] for more details. See also Carr and Madan [1999] for a review of the use of characteristic functions in option pricing.

7.6 Suggested Reading

The main articles on the topics covered in the chapter are Duan [1995, 1997], Hull and White [1987], Heston [1993], and Carr and Madan [1999]. For a review on Fourier methods in financial computations, I also suggest Cherubini et al. [2010].

7.7 Exercises

Exercise 7.1

For the EGARCH, NGARCH, and GJR-GARCH models, find sufficient conditions for the lack of memory property, i.e., for two processes with the same innovations but starting from different points, the difference of their respective conditional variances $\sigma_n^2 - v_n^2$ converge to 0, as $n \to \infty$.

Exercise 7.2

Consider the random variable

$$X = \frac{4}{\pi^2}\sum_{k=1}^{\infty} \frac{Z_k^2}{(2k-1)^2},$$

where $Z_1, Z_2, \ldots$ are independent and identically distributed standard Gaussian variables. It is known than the quantiles of order 90%, 95%, and 99% are respectively 1.2, 1.657, and 2.8.

(a) Find the first six cumulants of X. Note that the famous zeta[2] function is defined by

$$\zeta(s) = \sum_{n=1}^{\infty} \frac{1}{n^s}, \quad s > 1.$$

(b) Use the Edgeworth expansion to try to approximate the distribution function of X.

(c) Estimate the quantiles of order 90%, 95%, and 99% using the Cornish-Fisher expansion. Also plot the graph of the quantile function determined by the Cornish-Fisher expansion.

Is the approximation better than the one obtained by the Edgeworth expansion?

(d) Generate 100000 approximations of X by truncating the infinite sum to a sum over $k \leq 100$, i.e., generate

$$\tilde{X} = \frac{4}{\pi^2} \sum_{k=1}^{100} \frac{Z_k^2}{(2k-1)^2}.$$

(e) Using the simulated values of $\tilde{X}$, estimate its distribution function and compare it to the one obtained in (b). Is the Edgeworth expansion precise enough? Also estimate the quantiles of order 90%, 95%, and 99%. Compare them with the theoretical values. Finally, compare the empirical quantiles curve with the Cornish-Fisher expansion.

Exercise 7.3

For the GARCH(1,1), EGARCH, NGARCH, and GJR-GARCH models, find sufficient conditions for the existence of the limit $E\left\{\sigma_i^4\right\}$. Can you find the limit?

Exercise 7.4

Suppose that W is a Brownian motion, and that the time scale is in days. Find c so that $P\left\{\sup_{0\leq t\leq 21} |W(t)| \leq c\right\} = 0.95$.

Exercise 7.5

Suppose that the innovations are of the form $\varepsilon_i = (\epsilon_i - \nu)/\sqrt{\nu}$, with $\epsilon_i \sim \mathit{Gamma}(\nu, 1)$, with $\nu > 0$. Find the continuous time approximation of a GARCH(1,1) process with these innovations.

Exercise 7.6

[2]This function is implemented in MATLAB. It can also be computed exactly for any even integer s, using, e.g., the *Symbolic Math* Toolbox.

Suppose that the estimated parameters of a GARCH(1,1) are $\mu = 0.002$, $\omega = 2.5 \times 10^{-6}$, $\beta = .88$ and $\alpha = .105$. Here we assume Gaussian innovations. What are the corresponding parameters of the associated stochastic volatility model? Write also the two stochastic differential equations satisfied by S and V. What is the correlation between the two Brownian motions?

Exercise 7.7

Consider the random variable

$$X = \frac{4}{\pi^2} \sum_{k=1}^{\infty} \frac{Z_k^2}{(2k-1)^2},$$

where $Z_1, Z_2, \ldots$ are independent and identically distributed standard Gaussian variables. Show that the law of X is infinitely divisible. If Y is a Lévy process with $Y(1) = X$, find a representation for $Y(t)$, for any $t > 0$. Does Y have a Brownian component?

Exercise 7.8

Suppose that the estimated parameters of a GARCH(1,1) are $\mu = 0.0005$, $\omega = 1.5 \times 10^{-7}$, $\beta = .81$ and $\alpha = .17$. Here we assume the innovations and of the form $\varepsilon_i = (\epsilon_i - \nu)/\sqrt{\nu}$, with $\epsilon_i \sim \text{Gamma}(\nu, 1)$. Suppose the estimation of ν is 5.2. What are the corresponding parameters of the associated stochastic volatility model? Write also the two stochastic differential equations satisfied by S and V. What is the correlation between the two Brownian motions?

7.8 Assignment Questions

1. Construct a MATLAB function to compute the distribution function

$$F(x) = \frac{4}{\pi} \sum_{k=1}^{\infty} \frac{(-1)^k}{2k+1} e^{-(2k+1)^2\pi^2/(8x^2)}.$$

   Recall that quantiles of order 90%, 95%, and 99% of F are respectively 1.96, 2.241, and 2.807.

2. Construct a MATLAB function to simulate a E-GARCH, a NGARCH, a GARCH-M, and a GJR-GARCH. The innovations can be either Gaussian or GED.

3. Construct a MATLAB function to simulate approximative daily values of the process V satisfying the stochastic differential equation

$$dV(t) = \alpha \{\beta - V(t)\} \, dt + \gamma V(t) dW(t), \quad V(0) = V_0 > 0.$$

   Here, $\alpha, \gamma > 0$. What could happen if $\beta < 0$?

7.A Khmaladze Transform

Let $e_i = \frac{x_i - \hat{\mu}_i}{\hat{\sigma}_i}$, $i \in \{1, \ldots, n\}$, be the pseudo-observations associated with an ARMA-GARCH(p,q) model. Set $u_i = \mathcal{N}(e_i)$, $i \in \{1, \ldots, n\}$, and denote by $v_1, \ldots, v_n$ the associated order statistics. Further set $v_0 = 0$ and $v_{n+1} = 1$.

According to Bai [2003], set $\dot{g}(s) = \left(1, -\mathcal{N}^{-1}(s), 1 - \left\{\mathcal{N}^{-1}(s)\right\}^2\right)^\top$, and define

$$C(s) = \int_s^1 \dot{g}(t)\dot{g}^\top(t)dt, \quad s \in (0,1).$$

Then one can easily see that if $a = \mathcal{N}^{-1}(s)$ and $x = \mathcal{N}'(a) = \frac{e^{-a^2/2}}{\sqrt{2\pi}}$, then, for any $s \in (0,1)$,

$$C(s) = \begin{pmatrix} 1-s & -x & -ax \\ -x & 1-s+ax & x(1+a^2) \\ -ax & x(1+a^2) & 2(1-s)+ax(1+a^2) \end{pmatrix}$$

Setting $V_n(s) = \frac{1}{\sqrt{n}} \sum_{i=1}^n \{\mathbb{I}(v_i \leq s) - s\}$, the Khmaladze transform of V_n is defined by

$$W_n(s) = V_n(s) - \int_0^s \left\{\dot{g}^\top(t)C^{-1}(t) \int_t^1 \dot{g}(\tau)dV_n(\tau)\right\} dt, \quad s \in [0,1).$$

Under the null hypothesis that the innovations ε_i are Gaussian (with mean 0 and variance 1), W_n is approximately a Brownian motion. One can check that if $D_k = \sum_{j=k}^n \dot{g}(v_j)$, then for any $j \in \{1, \ldots, n\}$,

$$W_n(v_j) = \frac{1}{\sqrt{n}} \left[j - \sum_{k=1}^j \left\{ \int_{v_{k-1}}^{v_k} C^{-1}(t)\dot{g}(t)dt \right\}^\top D_k \right]. \tag{7.16}$$

7.A.1 Implementation Issues

In Bai [2003], the author proposed to approximate $C(v_j)$ by

$$\sum_{k=j}^n (v_{k+1} - v_k)\dot{g}(v_k)\dot{g}^\top(v_k). \tag{7.17}$$

This creates numerical instabilities for values of j near n, since C has to be invertible. Luckily, for Gaussian innovations, we can compute C exactly, so there is no need to use (7.17). However, one cannot compute $\int_{v_{k-1}}^{v_k} C^{-1}(t)\dot{g}(t)dt$. Bai [2003] suggested to approximate it by

$$C^{-1}(v_{k-1}) \int_{v_{k-1}}^{v_k} \dot{g}(t)dt = C^{-1}(v_k)\left\{g(v_k) - g(v_{k-1})\right\}.$$

This also creates numerical instability for values of k near n. Therefore, to compute $\int_{v_{k-1}}^{v_k} C^{-1}(t)\dot{g}(t)dt$, we suggest to use a numerical quadrature that can handle moderate singularities at the endpoints, like the Gauss-Kronrod quadrature implemented in MATLAB. Another solution, proposed by Kilani Ghoudi, is to use the midpoint quadrature, i.e.,

$$\int_{v_{k-1}}^{v_k} C^{-1}(t)\dot{g}(t)dt \approx (v_k - v_{k-1})C^{-1}\left(\frac{v_{k-1}+v_k}{2}\right)\dot{g}\left(\frac{v_{k-1}+v_k}{2}\right).$$

This solution, which is much faster than the previous one, is also implemented in the MATLAB function *GofGARCHGaussian*, together with the quadrature method. Once these computations are done, we can use two familiar tests statistics: the Kolmogorov-Smirnov statistic, defined by

$$KS = \max_{j\in\{1,\ldots,n\}} |W_n(v_j)|,$$

and the Cramér-von Mises statistic, defined by

$$CVM = \frac{1}{n}\sum_{j=1}^{n} W_n^2(v_j)(v_{j+1} - v_j).$$

According to Shorack and Wellner [1986][pp. 34,748], the limiting critical values for a 95% level are 2.2241 for the Kolmogorov-Smirnov statistic, and 1.657 for the Cramér-von Mises statistic. For 90% level, they are respectively 1.96 and 1.2, while for the 99% level, they are 2.807 and 2.8.

For a comparison of goodness-of-fit test for GARCH models, see, e.g., Ghoudi and Rémillard [2012].

7.B Proofs of the Results

7.B.1 Proof of Proposition 7.1.1

Set $K(z) = z^r - \sum_{j=1}^r \lambda_j z^{r-j}$. We want to show that $\kappa = K(1) > 0$ implies $\lim_{n\to\infty} E\left(\sigma_n^2\right) = \frac{\omega}{K(1)}$, while $\kappa \le 0$ implies $\lim_{n\to\infty} E\left(\sigma_n^2\right) = +\infty$. Note that the condition $K(1) > 0$ is equivalent to the condition that all the roots of the polynomial $K(z)$ are inside the unit ball in the complex plane. Furthermore, these roots are the eigenvalues of the matrix

$$A = \begin{pmatrix} \lambda_1 & \lambda_2 & \cdots & \lambda_r \\ 1 & 0 & \cdots & 0 \\ 0 & 1 & \cdots & 0 \\ \vdots & \vdots & \vdots & 0 \\ 0 & \cdots & 1 & 0 \end{pmatrix},$$

so $K(1) > 0$ is equivalent to saying that the spectral radius[3] of A is smaller than 1. Obviously, if the roots of K are inside the unit ball, then $K(1) > 0$. For if $K(1) = 0$ then 1 is a root not in the unit ball, which is a contradiction, and if $K(1) < 0$, then $K(z) \to \infty$ as $z \to \infty$, so by continuity, there would be a real root of K greater than 1. Suppose now that $K(1) > 0$. Then there exists $z_0 \in (0,1)$ so that $K(z) > 0$ for all $z \in [z_0, 1]$. Fix such a z and set

$$v_i = \begin{pmatrix} E\left(\sigma_i^2\right) \\ zE(\sigma_{i-1}^2) \\ \vdots \\ z^{r-1}E(\sigma_{i-r+1}^2) \end{pmatrix} \text{ and } \tilde{\omega} = \begin{pmatrix} \omega \\ 0 \\ \vdots \\ 0 \end{pmatrix}. \text{ Then}$$

$$E\left(\sigma_i^2\right) = \omega + \sum_{j=1}^{r} \lambda_j E(\sigma_{i-j}^2), \quad i \geq r+1,$$

is equivalent to $v_i = \tilde{\omega} + A_z v_{i-1}$ for $i \geq r+1$, where

$$A_z = \begin{pmatrix} \lambda_1 & \lambda_2/z & \cdots & \lambda_r/z^{r-1} \\ z & 0 & \cdots & 0 \\ 0 & z & \cdots & 0 \\ \vdots & \vdots & \vdots & 0 \\ 0 & \cdots & z & 0 \end{pmatrix}.$$

Defining the norm $\|B\| = \max_{1 \leq i \leq r} \left(\sum_{j=1}^{r} |B_{ij}| \right)$, we obtain that $\|A_z\| = z < 1$, since $K(z) > 0$. Next, it follows that

$$v_n = \left(I + A_z + \cdots + A_z^{n-r-1}\right)\tilde{\omega} + A_z^{n-r} v_r.$$

Set $|x| = \max_{1 \leq j \leq r} |x_j|$. Then it is easy to check that for any matrix B, $|Bx| \leq \|B\||x|$, so $|A_z^{n-r} v_r| \leq \|A_z^{n-r}\||v_r| \leq z^{n-r}|v_r| \to 0$ as $n \to \infty$. Also, because $\|A_z\| < 1$, one may conclude that $I + A_z + \cdots + A_z^{n-r-1} \to (I - A_z)^{-1}$, as $n \to \infty$, so v_n converges to $(I - A_z)^{-1}\tilde{\omega} = \omega y$, where $(I - A_z)y = e_1 = (1, 0, \ldots, 0)^\top$. Hence $y_j = z^{j-1} y_1$, and $1 = y_1 - \sum_{j=1}^{r} \lambda_j z^{-j+1} y_j = y_1 K(1)$. As a result $y_1 = 1/K(1)$, and $E\left(\sigma_n^2\right)$ converges to $\omega/K(1)$.

It remains to show that $K(1) \leq 0$ implies that $E\left(\sigma_n^2\right) \to \infty$, as $n \to \infty$.

Case (i): Suppose that $K(1) = 0$. Then A is the transition matrix of an irreducible Markov chain so that $A^n \to B$, with $B_{ij} = \pi_j = \frac{\sum_{k=j}^{r} \lambda_k}{\sum_{k=1}^{r} k\lambda_k}$, $i, j \in \{1, \ldots, r\}$. As a result, $E\left(\sigma_n^2\right)/n \to \omega\pi_1 = \frac{\omega}{\sum_{k=1}^{r} k\lambda_k}$. It follows that

[3] If $\|\cdot\|$ is any norm of the set of $r \times r$ matrices, then $\lim_{n\to\infty} \|A^n\|^{1/n} = \rho$, and ρ is called the spectral radius of A.

if $\sum_{k=1}^{r} \lambda_k = 1$, then $E\left(\sigma_n^2\right) \to \infty$, as $n \to \infty$.

Case (ii): Assume that $K(1) < 0$, so $a = \sum_{j=1}^{r} \lambda_j > 1$. It follows that for $n > r$,

$$E\left(\sigma_n^2\right) > \frac{E\left(\sigma_n^2\right)}{a} = \frac{\omega}{a} + \sum_{j=1}^{r} \frac{\lambda_j}{a} E(\sigma_{n-j}^2).$$

Let u_n be the solution of

$$u_n = \frac{\omega}{a} + \sum_{j=1}^{r} \frac{\lambda_j}{a} u_{n-j}, \quad n > r,$$

with $u_j = E(\sigma_j^2)$, $j \in \{1, \ldots, r\}$. Then by induction, $E\left(\sigma_n^2\right) - u_n > 0$ for every $n > r$. Since u_n/n converges to a positive number (using the conclusion of case (1) since $\sum_{j=1}^{r} \frac{\lambda_j}{a} = 1$), it follows that $E\left(\sigma_n^2\right) \to \infty$, as $n \to \infty$.

■

7.B.2 Proof of Proposition 7.4.1

Set $u(t,x) = \frac{B(t)}{B(T)} g(t,x)$, where

$$g(t,x) = E\left[\Phi\{X(T)\} | \mathcal{F}_t, X_t = x\right].$$

By construction, $u(T,x) = g(T,x) = \Phi(x)$, and

$$g\{t, X(t)\} = E\left[\phi\{X(T)\} | \mathcal{F}_t\right]$$

is a martingale. Moreover, the law of X is the solution of the martingale problem for L, so

$$g(t, X(t)) - \int_0^t \left[\partial_s g\{s, X(s)\} + L_s g\{s, X(s)\}\right] ds$$

is a martingale and consequently

$$V(t) = \int_0^t \left[\partial_s g\{s, X(s)\} + L_s g\{s, X(s)\}\right] ds$$

is also a martingale. Since V is differentiable with respect to t, it is necessarily constant and so it is 0 since $V_0 = 0$. Hence $\partial_t g + L_t g = 0$. Finally,

$$\begin{aligned}
\partial_t u(t,x) + L_t u(t,x) &= \frac{g(t,x)}{B(T)} \partial_t B(t) + \frac{B(t)}{B(T)} \partial_t g(t,x) \\
&\quad + \frac{B(t)}{B(T)} L_t g(t,x) \\
&= \frac{r_f(t) B(t)}{B(T)} g(t,x) + \frac{B(t)}{B(T)} \times 0 \\
&= r_f(t) u(t,x).
\end{aligned}$$

■

7.B.3 Proof of Proposition 7.4.2

If $\psi(u)$ is the characteristic function of Y and if $\int_{-\infty}^{\infty} |\psi(u)|du < \infty$, then Y has a density g and

$$g(y) = \frac{1}{2\pi} \int_{-\infty}^{\infty} e^{-iuy} \psi(u)du.$$

Moreover, if $\int_{-\infty}^{\infty} |u|^n |\psi(u)du < \infty$, then the derivatives of order $k \le n$ of g exist and are given by

$$g^{(k)}(y) = \frac{(-1)^k}{2\pi} \int_{-\infty}^{\infty} (iu)^k e^{-iuy} \psi(u)du, \quad k \in \{0, \ldots, n\}.$$

Suppose now that X is another variable such that its characteristic function $\phi(u)$ has the same integrability properties as ψ. Let f be its density. Then we have

$$\ln \psi(u) = \ln \phi(u) + \sum_{k=1}^{n} \frac{(iu)^k}{k!} d_k + O(u^{n+1}),$$

where $d_k = \kappa_k(Y) - \kappa_k(X)$, $k \in \{1, \ldots, n\}$. As a result,

$$\psi(u) = \phi(u) \left\{ \sum_{j=0}^{n} \frac{(iu)^j}{j!} E_j + O(u^{n+1}) \right\},$$

where $E_0 = 1$ and

$$\exp\left\{ \sum_{j=1}^{n} \frac{u^j}{j!} d_j + O(u^{n+1}) \right\} = \sum_{j=0}^{n} \frac{u^j}{j!} E_j + O(u^{n+1}).$$

Therefore,

$$g(x) = \sum_{j=0}^{n} \frac{(-1)^j}{j!} E_j f^{(j)}(x) + r_n(x).$$

■

Bibliography

J. Bai. Testing parametric conditional distributions of dynamic models. *The Review of Economics and Statistics*, 85(3):531–549, 2003.

I. Berkes, L. Horvth, and P. Kokoszka. Asymptotics for GARCH squared residual correlations. *Econometric Theory*, 19(4):515–540, 2003.

T. Bollerslev. Generalized autoregressive conditional heteroskedasticity. *Journal of Econometrics*, 31:307–327, 1986.

P. Bougerol and N. Picard. Stationarity of GARCH processes and of some nonnegative time series. *Journal of Econometrics*, 52:115–127, 1992.

P. Carr and D. Madan. Option pricing and the fast Fourier transform. *Journal of Computational Finance*, 2:61–73, 1999.

U. Cherubini, G. Della Lunga, S. Mulinacci, and P. Rossi. *Fourier Transform Methods in Finance*. Wiley Finance. Wiley, New York, 2010.

J.-C. Duan. The GARCH option pricing model. *Math. Finance*, 5(1):13–32, 1995.

J.-C. Duan. Augmented GARCH(p, q) process and its diffusion limit. *J. Econometrics*, 79(1):97–127, 1997.

J.-C. Duan, G. Gauthier, and J.-G. Simonato. An analytical approximation for the GARCH option pricing model. *Journal of Computational Finance*, 2:75–116, 2000.

D. Dufresne. The distribution of a perpetuity, with applications to risk theory and pension funding. *Scand. Actuar. J.*, 9:39–79, 1990.

D. Dufresne. Laguerre series for Asian and other options. *Mathematical Finance*, 10:407–428, 2000.

R. F. Engle. Autoregressive conditional heteroskedasticity with estimates of the variance of united kingdom inflation. *Econometrica*, 50:987–1007, 1982.

R. F. Engle and V. K. Ng. Measuring and testing the impact of news on volatility. *The Journal of Finance*, 48(5):1749–1778, 1993.

C. Francq and J.-M. Zakoïan. *GARCH Models: Structure, Statistical Inference and Financial Applications*. John Wiley & Sons, 2010.

R. Garcia and Renault. A note on hedging in ARCH and stochastic volatility option pricing models. *Math. Finance*, 8(2):153–161, 1998.

K. Ghoudi and B. Rémillard. Comparison of specification tests for GARCH models. Technical report, SSRN Working Paper Series No. 2046072, 2012.

L. R. Glosten, R. Jagannathan, and D. E. Runkle. On the relation between the expected value and the volatility of the nominal excess return on stocks. *The Journal of Finance*, 48(5):1779–1801, 1993.

S. Heston. A closed-form solution for options with stochastic volatility with application to bond and currency options. *Review of Financial Studies*, 6 (2):327–343, 1993.

J. C. Hull. *Options, Futures, and Other Derivatives.* Prentice-Hall, sixth edition, 2006.

J. C. Hull and A. White. The pricing of options on assets with stochastic volatilities. *The Journal of Finance*, XLII, No.2:281–300, 1987.

R. Jarrow and A. Rudd. Approximate option valuation for arbitrary stochastic processes. *Journal of Financial Economics*, 10:347–369, 1982.

R. C. Merton. On the pricing of corporate debt: The risk structure of interest rates. *Journal of Finance*, 29:449–470, 1974.

M. A. Milevsky and S. E. Posner. Asian options, the sum of lognormals, and the reciprocal gamma distribution. *Journal of Financial and Quantitative Analysis*, 33:409–422, 1998.

D. B. Nelson. Conditional heteroskedasticity in asset returns: A new approach. *Econometrica*, 59(2):347–370, 1991.

B. Rémillard. Validity of the parametric bootstrap for goodness-of-fit testing in dynamic models. Technical report, SSRN Working Paper Series No. 1966476, 2011.

G. R. Shorack and J. A. Wellner. *Empirical Processes with Applications to Statistics.* Wiley Series in Probability and Mathematical Statistics: Probability and Mathematical Statistics. John Wiley & Sons Inc., New York, 1986.

Chapter 8

Copulas and Applications

In this chapter, we start by looking at two financial applications where the modeling of dependence between assets is very important. Then we will see why correlation should not be considered as a good measure of dependence outside the family of Gaussian distributions. Next, we will define copulas and study the properties of several measures of dependence. Then, we define several multivariate copula families. We will also see how to: simulate observations from a copula, estimate parameters, and perform goodness-of-fit tests. We will also go through all the steps needed to implement a copula model using real data.

8.1 Weak Replication of Hedge Funds

Given the structure of management fees in hedge funds, Kat and Palaro [2005] proposed to replicate the statistical properties of a hedge fund (hence the name weak replication), including its dependence with a portfolio, using another portfolio.

More precisely, Kat and Palaro [2005] showed that with two risky assets S_1 and S_2, one can always reproduce the distribution of a third asset S_3, together with its dependence with asset S_1, in the sense that there exists, for a given time T, a function g such that the joint distribution of S_{T1} and $g\left(S_{T1}, S_{T2}\right)$ is the same as the joint distribution of S_{T1} and S_{T3}.

Strong vs Weak Replication

Note that the values of S_{T3} are not reproduced, only the statistical properties like the mean, volatility, skewness, kurtosis, and the correlation with S_1. This is totally different from the so-called (strong) replication of hedge funds that will be covered in Chapter 10.

Once g is defined, we have to find a way to generate a payoff with return $g\left(S_{T1}, S_{T2}\right)$ without investing in S_3, thereby avoiding paying management fees. Weak replication can be seen as replicating an option whose payoff is given by $G\left(S_{T1}, S_{T2}\right) = 100e^{g(S_{T1}, S_{T2})}$, where the initial value of the portfolio is 100. As such, Kat and Palaro [2005] proposed to replicate the value $G(S_{T1}, S_{T2})$ with a portfolio composed of a non-risky asset S_0 and the two assets risky assets S_1 and S_2, traded at periods $t = 0, \ldots, T-1$. In practice, they used a maturity of one month and daily trading.

As seen in Section 3.3, what is needed in that case is a discrete time hedging strategy $(V_0, \vec{\varphi})$ so that the value of the portfolio $V_T(\vec{\varphi})$ at maturity T is as close as possible to $G(S_{T1}, S_{T2})$. Moreover, because we are trading in discrete time, there will be a hedging error so we have to be careful about the choice of the strategy. Kat and Palaro [2005] simply proposed to use delta hedging, as if the Black-Scholes model applies, which is not the case in any of their examples. As seen before, we often can do much better by using the optimal hedging described in Section 3.4.

Coming back to hedge fund replication, the initial value V_0 is related the so-called alpha of the manager, since the theoretical return of the portfolio is

$$\ln\left\{\frac{G(S_{T1}, S_{T2})}{V_0}\right\} = \ln\left(\frac{100}{V_0}\right) + g(S_{T1}, S_{T2}). \tag{8.1}$$

Note that all statistics based on centered moments are not affected by the value of $\alpha = \ln\left(\frac{V_0}{100}\right)$. However, the average return of the portfolio would be the average target minus α. If $\alpha < 0$, then one could do better by replicating the funds than investing in it. If $\alpha > 0$, the portfolio will have a smaller return on average, so it might not be worth trying to replicate it. Fortunately, since we are interested in statistical properties, we could use Monte Carlo simulations and check more realistically the performance of the replicating portfolio. One could even add transaction fees to be more realistic.

In short, the approach of Kat and Palaro [2005] for the replication of S_{T3} follows these steps:

1. Model the distribution of all single assets at time T.
2. Model the dependence between S_{T1} and S_{T2}, and between S_{T1} and S_{T3}.
3. Compute the function g.
4. Choose the portfolio strategy.
5. Validate, using Monte Carlo methods.

8.1.1 Computation of g

Since we want the joint distribution of S_{T1} and $g(S_{T1}, S_{T2})$ to be the same as the joint distribution of S_{T1} and S_{T3}, we must have, for any $x, y > 0$,

$$P(S_{T1} \le x, S_{T3} \le y) = P\left\{S_{T1} \le x, g(S_{T1}, S_{T2}) \le y\right\}.$$

Assuming that g is increasing with respect to the second variable, we then have

$$P(S_{T1} \le x, S_{T3} \le y) = P\left\{S_{T1} \le x, S_{T2} \le g^{-1}(y|S_{T1})\right\},$$

where

$$P\{S_{T3} \leq g(x,y)|S_{T1} = x\} = P(S_{T2} \leq y|S_{T1} = x).$$

As a result,

$$g(x,y) = q\{x, P(S_{T2} \leq y|S_{T1} = x)\}, \tag{8.2}$$

where $q(x,u)$ is the quantile function of the conditional distribution of S_{T3} given $S_{T1} = x$, i.e., for all $u \in (0,1)$,

$$P\{S_{T3} \leq q(x,u)|S_{T1} = x\} = u.$$

Since for a given $x > 0$, $q(x,u)$ is non-decreasing in u, and $P(S_{T2} \leq y|S_{T1} = x)$ is non-decreasing in y, $g(x,y)$ is also non-decreasing in y. We will see how to compute these conditional distributions in Section 8.4.2.

Here is another example where modeling dependence is important.

8.2 Default Risk

Suppose we want to compute the monthly premium to pay, for at most one year, to insure a debt of face value F against the default of a firm. If there is a default at period τ, with $\tau \leq 12$, we receive a (random) fraction RR of the debt F and the contract is terminated. Under an equivalent martingale measure Q, the monthly premium is $P_m \times F$, with P_m determined by the equation

$$E_Q\left\{P_m \sum_{i=1}^{12} \beta_i \mathbb{I}(\tau > i) - \beta_\tau(1 - RR)\mathbb{I}(\tau \leq 12)\right\} = 0,$$

where β_k is the discounting factor at period k. Hence

$$P_m = \frac{E_Q\{\beta_\tau(1 - RR)\mathbb{I}(\tau \leq 12)\}}{E_Q\left\{\sum_{i=1}^{12} \beta_i \mathbb{I}(\tau > i)\right\}}.$$

8.2.1 n-th to Default Swap

To cover the risk of default of several firms, there exist "n-th to default swap" contracts. Such a contract pays a fixed predetermined amount if there are at least n defaults amongst a basket of d firms, with $d \geq n$. For these contracts, the premium to pay cannot be determined explicitly, so we will typically have to rely on Monte Carlo methods.

The are two major steps for the implementation of the pricing of n-th to default swaps: the modeling of default times of each firms, and the modeling of the dependence between the default times. The latter is quite important in view of contagion effects. If the risk of contagion is high, then the premium should be higher. To understand the problems inherent to the modeling of the default times, we consider first the following simplified model. We basically follow the treatment in Li [2000].

8.2.2 Simple Model for Default Time

Consider a dynamical model with two states:

$$X_i = \begin{cases} 1 = \text{'non default'}, \\ 2 = \text{'default'}, \end{cases}$$

where X_i is the state of the firm at period $i \geq 0$. To make things simpler, assume that X is a homogeneous Markov chain, the state "default" being an absorbing state. As illustrated in Table 8.1, the dynamics only depends on parameter p_D, the probability of default one step ahead.

TABLE 8.1: Transition probabilities from the state at period n to the state at period $n+1$.

	X_{n+1}	
X_n	1	2
1	$1-p_D$	p_D
2	0	1

Table 8.1 can be summarized by the transition matrix

$$P = \begin{pmatrix} 1-p_D & p_D \\ 0 & 1 \end{pmatrix}.$$

It is then easy to check that the default time τ has a geometric distribution with parameter p_D. To simulate the Markov chain X, generate independent uniform[1] random variables $U_1, U_2, \ldots$, and set

$$X_i = \begin{cases} 2 & \text{if } X_{i-1} = 2 \\ 1 & \text{if } X_{i-1} = 1 \text{ and } U_i \leq 1 - p_D, \\ 2 & \text{if } X_{i-1} = 1 \text{ and } U_i > 1 - p_D. \end{cases}$$

This can also be stated as follows: If $j \in \{1,2\}$ and $X_{i-1} = j$, then $X_i = 1$ if $U_i \leq P_{j1}$, and $X_i = 2$ otherwise.

If Y_i is another Markov chain with the same state space (representing the state of another firm), with transition matrix

$$Q = \begin{pmatrix} 1-q_D & q_D \\ 0 & 1 \end{pmatrix},$$

[1]Throughout this chapter, uniform means uniform on $(0,1)$.

one can also simulate Y by generating another sequence of uniform $V_1, V_2, \ldots$, the transition rule being: If $j \in \{1,2\}$ and $Y_{i-1} = j$, then $Y_i = 1$ if $V_i \le Q_{j1}$, and $Y_i = 2$ otherwise.

8.2.3 Joint Dynamics of X_i and Y_i

The question now is how do the two chains evolve simultaneously? In fact, assuming that the joint dynamics is still a homogeneous Markov chain, the transition matrix of the pair (X, Y) is given in Table 8.2.

TABLE 8.2: Transition probabilities from the states at period n to the states at period $n+1$.

	(X_{n+1}, Y_{n+1})			
(X_n, Y_n)	(1,1)	(1,2)	(2,1)	(2,2)
(1,1)	p_{11}	$p_D - p_{22}$	$q_D - p_{22}$	p_{22}
(1,2)	0	$1 - p_D$	0	p_D
(2,1)	0	0	$1 - q_D$	q_D
(2,2)	0	0	0	1

Note that the following constraints hold: $p_{22} \le p_D$, $p_{22} \le q_D$, and $p_{11} = 1 + p_{22} - p_D - q_D = P(U_i \le 1 - p_D, V_i \le 1 - q_D) \ge 0$. One can do the same for d firms but then it becomes exponentially tedious to write the transition matrix. For example, with $d = 10$ firms, there are $2^{10} = 1024$ joint states and the transition matrix has $2^{20} = 1048576$ entries. There must be a better way to define the joint dynamics. The answer is simple: It suffices to determine the joint distribution (U_i, V_i) used in the simulation of (X_i, Y_i). As we shall see later, the so-called "copula" is precisely the distribution function C of the pair (U_i, V_i).

For example, suppose that the transition rules are still the same, i.e., for $j \in \{1,2\}$, if $X_{i-1} = j$, then $X_i = 1$ if $U_i \le P_{j1}$, and $X_i = 2$ otherwise, and if $Y_{i-1} = j$, then $Y_i = 1$ if $V_i \le Q_{j1}$, and $Y_i = 2$ otherwise. One can then write, if necessary, the transition matrix displayed in Table 8.2 in terms of joint probabilities involving (U_i, V_i). For example,

$$\begin{aligned} p_{22} &= P(U_i > 1 - p_D, V_i > 1 - q_D), \\ p_{11} &= P(U_i \le 1 - p_D, V_i \le 1 - q_D) = C(1 - p_D, 1 - q_D). \end{aligned} \tag{8.3}$$

It follows from (8.3) that there exist an infinite number of copulas so that the transition matrix of the Markov chain (X_i, Y_i) generated from the uniform variables U_i, V_i is given by Table 8.2.

8.2.4 Simultaneous Evolution of Several Markov Chains

Suppose now that we have a more "complicated" model for the dynamics of the states of the firms. More precisely, we assume that $X_i = (X_{i1}, \ldots, X_{id})$ is a homogeneous Markov chain with state space $\{1, \ldots, m+1\}^d$, where $m+1$ represents the state of default and where the components of X are also Markov chains with the transition matrices $P^{(1)}, \ldots, P^{(d)}$ respectively. Using the individual transition matrices, we can determine the joint dynamics of X by using independent random vectors $U_1, U_2, \ldots$, where $U_i = (U_{i1}, \ldots, U_{id})$, with $U_{ij} \sim \text{Unif}(0,1)$ for all $j \in \{1, \ldots, d\}$, and with joint distribution function C. C is then called a "copula." Setting $F_{j,0}^{(\alpha)} = 0$, and $F_{j,k}^{(\alpha)} = \sum_{l=1}^{k} P_{jl}^{(\alpha)}$, $j, k \in \{1, \ldots, m+1\}$, and $\alpha \in \{1, \ldots, d\}$, the transition $X_{i-1} \to X_i$ is simply defined by following rule:

$$\text{If } X_{(i-1)\alpha} = j, \text{ then } X_{i\alpha} = k, \text{ if } F_{j,k-1}^{(\alpha)} < U_{k\alpha} \le F_{j,k}^{(\alpha)},$$

$j, k \in \{1, \ldots, m+1\}$, $\alpha \in \{1, \ldots, d\}$.

8.2.4.1 CreditMetrics

As stated in Li [2000] and in the CreditMetrics documentation, the Markov chain approach coupled with a copula is basically what is used by CreditMetrics for modeling the migration of credit ratings of several firms. Here, all Markov chains have the transition matrix P, and the copula that is used is the so-called Gaussian copula. More precisely, one generates random Gaussian vectors $Z_i = (Z_{i1}, \ldots, Z_{id}) \sim N_d(0, \rho)$, with correlation matrix ρ, and one sets $U_{i\alpha} = \mathcal{N}(Z_{i\alpha})$, $\alpha \in \{1, \ldots, d\}$, where $\mathcal{N}$ is the distribution function of a standard Gaussian variable. In the CreditMetrics documentation, it is implicitly assumed that the Markov chains have the same distribution under the equivalent martingale measure. The common transition matrix can be based on one-year transition probabilities, as illustrated in Table 8.3.[2]

To illustrate the proposed methodology, we use the MATLAB function *DemoBondPortfolio* to create a portfolio of 28 zero-coupon bonds (four for each rating category), each bond having a face value of \$100 at maturity in

[2]Source: Standard & Poors CreditWeek (15 April '96). Standard & Poor's Financial Services LLC (S&P) does not guarantee the accuracy, completeness, timeliness or availability of any information, including ratings, and is not responsible for any errors or omissions (negligent or otherwise), reganlless of the cause, or for the results obtained trom the use of ratings. S&P GIVES NO EXPRESS OR IMPLIED WARRANTIES, INCLUDING, BUT NOT LIMITED TO, ANY WARRANTIES OF MERCHANTABILITY OR FITNESS FORA PARTICULAR PURPOSE OR USE. S&P SHALL NOT BE LIABLE FOR ANY DIRECT, INDIRECT, INCIDENTAL, EXEMPLAR Y, COMPENSATORY, PUNITIVE, SPECIAL OR CONSEQUENTIAL DAMAGES, COSTS, EXPENSES, LEGAL FEES, or LOSSES (INCLUDING LOST INCOME OR PROFITS AND OPPORTUNITY COSTS) IN CONNECTION WITH ANY USE OF RATINGS. S&P's ratings are statements of opinions and are not statements of fact or recommendations to purchase, hold or sell securities. They do not address the market value of securities or the suitability of securities for investment purposes, and should not be relied on as investment advice.

TABLE 8.3: Example of a transition matrix.

Rating	Rating at the End of the Year (%)							
	AAA	AA	A	BBB	BB	B	CCC	Default
AAA	90.81	8.33	0.68	0.06	0.12	0	0	0
AA	0.70	90.65	7.79	0.64	0.06	0.14	0.02	0
A	0.09	2.27	91.05	5.52	0.74	0.26	0.01	0.06
BBB	0.02	0.33	5.95	86.93	5.30	1.17	0.12	0.18
BB	0.03	0.14	0.67	7.73	80.53	8.84	1.00	1.06
B	0	0.11	0.24	0.43	6.48	83.46	4.07	5.20
CCC	0.22	0	0.22	1.30	2.38	11.24	64.86	19.79

four years. We assume, as in CreditMetrics, that the recovery rates depend on the bond rating prior to default. These values are respectively 53.8%, 52.465%, 51.13%, 44.825%, 38.52%, 27.805%, 17.09%, and 0. It follows that the value $B_{i,n}$ of a defaultable bond at period i, with state j, maturity n, recovery rates R_j and face value 1 is

$$\begin{aligned} B_{i,n} &= \mathcal{B}_{i,n}\left\{P_j(\tau > n|\tau > i) + \sum_{l=1}^{m}\sum_{k=i+1}^{n} P_j(\tau = k, X_{k-1} = l)R_l\right\} \\ &= \mathcal{B}_{i,n}\left\{1 - P_{j,m+1}^{(n-i)} + \sum_{l=1}^{m}\sum_{k=0}^{n-i-1} \left(P^k\right)_{j,l} P_{l,m+1}R_l\right\}, \end{aligned}$$

where $\mathcal{B}_{k,n}$ is the corresponding value of a non-defaultable bond. We then use $N = 10^5$ Monte Carlo simulations to obtain the distribution of losses on that portfolio. Table 8.4 provides examples of VaR estimations for several models of dependence which will be described later in the chapter. Kendall tau, which is a measure of dependence, is equal to 0.9 for all models but independence.

TABLE 8.4: VaR for the loss of the portfolio, based on 10^5 simulations.

	Copula Family					
VaR	Ind.	Gaussian	Student	Clayton	Gumbel	Fréchet
95	157.60	457.87	457.87	437.71	457.87	334.18
99	222.96	822.24	880.12	686.49	887.25	691.75
99.9	309.52	1345.06	1358.84	877.21	1358.84	1358.84

The distribution of losses on each bond does not depend at all on the copula family, but the distribution of portfolio losses is quite sensitive to it. Note the difference in VaR for the independence model compared to the other models. Assuming independence can lead to bankruptcy.

8.2.5 Continuous Time Model

A more realistic version of the previous example is to consider that the credit rating X evolves in continuous time (see Appendix 8.A) on the same state space $\{1, \ldots, m, m+1\}$. This was first proposed by Jarrow et al. [1997] in the case of a single firm. The rest of the section is inspired by Berrada et al. [2006].

For a discrete state space Markov process X in continuous time, the distribution of the process is determined by its (infinitesimal) generator Λ, where

$$\lim_{t \to 0} \frac{P_j\{X(t) = k\}}{t} = \Lambda_{jk}, \quad k \neq j,$$

and $\lim_{t \to 0} \frac{P_j\{X(t)=j\}-1}{t} = \Lambda_{jj}$. Note that $\Lambda_{jj} = -\sum_{k \neq j} \Lambda_{jk}$ and if $P(t)$ is the transition probability matrix at time t defined by $P_{jk}(t) = P_j\{X(t) = k\}$, then P satisfies the differential equation

$$\frac{dP(t)}{dt} = \Lambda P(t) = P(t)\Lambda, \quad P(0) = I.$$

The unique solution of this equation is

$$P(t) = e^{\Lambda t} = \sum_{n=0}^{\infty} \Lambda^n \frac{t^n}{n!},$$

where Λ^0 is the identity matrix. With MATLAB, the function *expm* can be used to compute $P(t)$. The main advantage of the continuous time model is that instead of being able to compute the probability of default every year, we can compute it for any time.

We assume that the state $m+1$ is the default state, so it is an absorbent state. The time of default τ is defined by

$$\tau = \inf\{t > 0; X(t) = m+1\}.$$

The distribution of the default time, given $X(0) = j$, can be found explicitly since

$$P_j(\tau \leq t) = P\{X(t) = m+1 | X(0) = j\} = P_{j,m+1}(t).$$

In fact, if Λ can be diagonalized, i.e., factored as $\Lambda = M\Delta M^{-1}$, where Δ is a diagonal matrix, then

$$P(t) = e^{t\Lambda} = M e^{t\Delta} M^{-1},$$

where $e^{t\Delta}$ is the diagonal matrix with entries $\left(e^{t\Delta}\right)_{jj} = e^{t\Delta_{jj}}$, $j \in \{1, 1, \ldots, m+1\}$. Note that the diagonal values are the eigenvalues of the matrix Λ and M can be constructed using the corresponding eigenvectors as column vectors. More precisely, one can write $\Lambda = \sum_{j=1}^{m+1} \Delta_{jj} x_j y_j^\top$, where $x_j^\top y_j = 1$ and $x_j^\top y_k = 0$ if $j \neq k$. In this case,

$$P(t) = \sum_{l=1}^{m+1} e^{t\Delta_{ll}} x_l y_l^\top \text{ and } P_j(\tau \le t) = \sum_{l=1}^{m+1} e^{t\Delta_{ll}} x_{lj} y_{l,m+1}.$$

These computations, done with the MATLAB script *DemoBondMarkovChain*, are illustrated in Figure 8.1. The generator Λ is obtained from the transition matrix Q, described in Table 8.3, through the equation $Q = P(1) = e^{\Lambda}$. Note that such a solution exists if and only if all the eigenvalues of Q are positive. This is the case here.

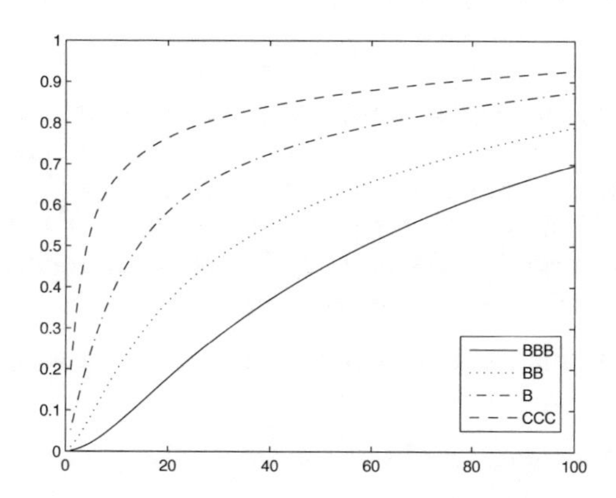

FIGURE 8.1: Default probabilities on a 100-year horizon.

We also have

$$P\{\tau \le T, X(\tau-) = k | X(t) = j\} = \Lambda_{k,m+1} \int_0^{T-t} P_{jk}(s) ds.$$

and

$$P\{\tau > T | X(t) = j\} = \sum_{k=1}^{m} P_{jk}(T-t) = 1 - P_{j,m+1}(T-t).$$

Suppose that $Z_j(t)$ is the recovery rate when the state before default is j and let $B(t,T)$ be the value of a non-defaultable zero-coupon bond, with the same face value. Then the value $V_j(t,T)$ of a defaultable bond, given $X(t) = j < m+1$, is

$$\begin{aligned} \frac{V_j(t,T)}{B(t,T)} &= E_j\left[\left\{\mathbb{I}(\tau > T) + \mathbb{I}(\tau \le T) Z_{X_{\tau-}}(\tau-)\right\} | X(t) = j\right] \\ &= \sum_{k=1}^{m} P_{jk}(T-t) + \sum_{k=1}^{m} \Lambda_{k,m+1} \int_0^{T-t} P_{jk}(u) Z_k(t+u) du. \end{aligned}$$

For the joint modeling of credit ratings, we have to allow simultaneous transitions otherwise the Markov chains $X_1, \ldots, X_d$ will be independent. This is proven in Berrada et al. [2006].

8.2.5.1 Modeling the Default Time of a Firm

The model is not perfect. For example, two firms with the same credit rating have the same distribution of default. Even if that model is simplistic, it is the one used by CreditMetrics. However, it can be justified, see, e.g., Berrada et al. [2006], that for a given firm, the real distribution function of the default time F and the distribution function G under the equivalent martingale measure are related by

$$G(t) = F(\theta t), \tag{8.4}$$

for a given constant θ to be determined, depending on the firm. This way, the value of the bonds of two firms with the same rating may differ, provided they have different values of θ. Note that in our setting, θ is a time scale parameter. If $\theta = 1/10$ then defaults will occur 10 times faster in the risk neutral world than in the real world. For the pricing of credit risk products, Berrada et al. [2006], proposed the following methodology:

- Estimate the distribution of defaults F using the credit ratings, as shown previously.
- Estimate θ for each firm by calibration, using historical values of simple contracts like bonds or CDS.

The main advantage of that methodology is that we have explicit formulas for the default probabilities.

8.2.6 Modeling Dependence Between Several Default Times

Whatever the approach for modeling the default time of individual firms, one must be able to model the joint behavior of the credit ratings or values of the firms. Introducing dependence between default times may have significant impact on the pricing of multi-assets credit products like n-th to default swaps. Table 8.5, reproduced from table 5 in Berrada et al. [2006], illustrates the variations in the value of contracts caused by selecting different dependence models.

8.3 Modeling Dependence

In their often-cited article, Embrechts et al. [2002] warn risk managers against the use of Pearson correlation as a measure of dependence between two

TABLE 8.5: Premia (in basis points) for the nth to default based on different families of copula

	Order of Default						
Model	1st	2nd	3rd	4th	5th	6th	7th
Clayton	151.1	25.21	3.48	0.40	0.03	0.00	0.00
Frank	149.8	28.97	5.12	0.77	0.12	0.02	0.00
Gumbel	110.8	35.90	15.68	7.49	3.88	2.01	0.73
Gaussian	136.7	32.29	8.57	2.10	0.50	0.11	0.01
Student	109.8	39.26	17.01	7.53	3.15	1.06	0.24

risky assets. Their message is simple: Outside the Gaussian world, (Pearson) correlation

- is inadequate;
- does not always exist (e.g., for pairs of Cauchy observations);
- can be close to 0 even if the dependence is very strong;
- only measures linear relationships.

8.3.1 An Image is Worth a Thousand Words

By looking at the graphs displayed in Figure 8.2 representing the estimated densities of two random variables X_1, X_2, using 10000 pairs of observations, there is no reason to disbelieve that variables X_1 and X_2 are both Gaussian.

In addition, the correlation between X_1 and X_2 is -0.001625 and a test of independence based on the correlation yields a P-value of 24.5%. For somebody who believes in the virtues of correlation, these two variables are independent and Gaussian, therefore jointly Gaussian! However, by looking at the scatter plot of the 10000 pairs of points, in Figure 8.3, the previous conclusion, based on correlation, is totally incorrect. There is a very strong dependence between the two variables.

How it that possible? In fact, the observations of X_1, X_2 were generated the following way: First generate U_1 and U_2, where $U_1 \sim \text{Unif}(0,1)$, and

$$U_2 = T(U_1) = 2\min(U_1, 1-U_1). \tag{8.5}$$

The mapping T is the famous tent map, used in chaos theory. Note that $U_2 \sim \text{Unif}(0,1)$, since for any $u_2 \in [0,1]$,

$$\begin{aligned}
P(U_2 \le u_2) &= P(2U_1 \le u_2, U_1 < 1/2) + P(2(1-U_1) \le u_2, U_1 \ge 1/2) \\
&= P(U_1 \le u_2/2) + P(U_1 \ge 1 - u_2/2) \\
&= u_2/2 + 1 - (1 - u_2/2) = u_2.
\end{aligned}$$

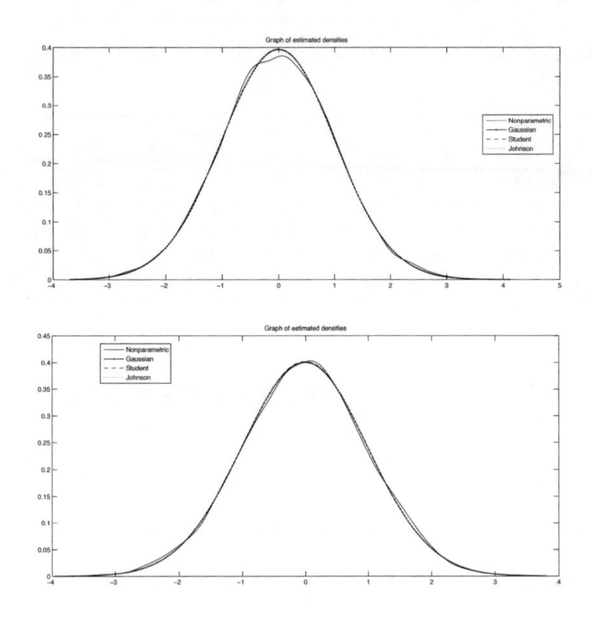

FIGURE 8.2: Estimation of the densities of (X_1, X_2) using 10000 pairs of observations.

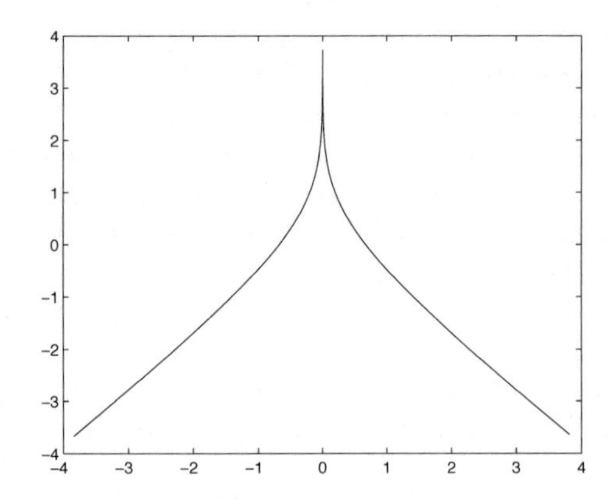

FIGURE 8.3: Graph of 10000 pairs of points (X_{i1}, X_{i2}).

It is also easy to show that $E(U_1U_2) = 1/4$, so $\text{Cor}(X_1, X_2) = 0$. Next, let $\mathcal{N}$ be the distribution function of a standard Gaussian variable, and set

$$X_1 = \mathcal{N}^{-1}(U_1) \quad \text{and} \quad X_2 = \mathcal{N}^{-1}(U_2).$$

Since U_1 and U_2 are uniform, it follows that $X_1, X_2 \sim N(0,1)$. The fact that the correlation is zero follows from Proposition 8.4.1.

Having convinced the reader that no correlation does not mean independence, even if the marginal distributions are Gaussian, we are finally in a position to define the concept of copula.

8.3.2 Joint Distribution, Margins and Copulas

Suppose that $(X_1, X_2) \sim H$ and that $X_1 \sim F_1$, $X_2 \sim F_2$, i.e., H is the (joint) distribution function of (X_1, X_2), while F_1 is the margin of X_1 and F_2 is the margin of X_2. Note that the marginal distributions F_1 and F_2 can be deduced from H, since for any $x_1, x_2 \in \mathbb{R}$,

$$H(x_1, \infty) = F_1(x_1) \quad \text{and} \quad H(\infty, x_2) = F_2(x_2).$$

We will show next that there is still another useful relationship between H and its margins.

8.3.3 Visualizing Dependence

Three set of $n = 1000$ pairs of independent observations (X_{i1}, X_{i2}) with exponential, Gaussian, and Cauchy distributions were generated. For each of these three samples, we computed the associated normalized ranks defined by $\left(\frac{R_{i1}}{n+1}, \frac{R_{i2}}{n+1}\right)$, where R_{i1} is the rank of X_{i1} amongst $X_{11}, \ldots, X_{n1}$ and R_{i2} is the rank of X_{i2} amongst $X_{12}, \ldots, X_{n2}$. By convention, the smallest observation has rank 1. The three graphs are displayed in Figures 8.4–8.6.

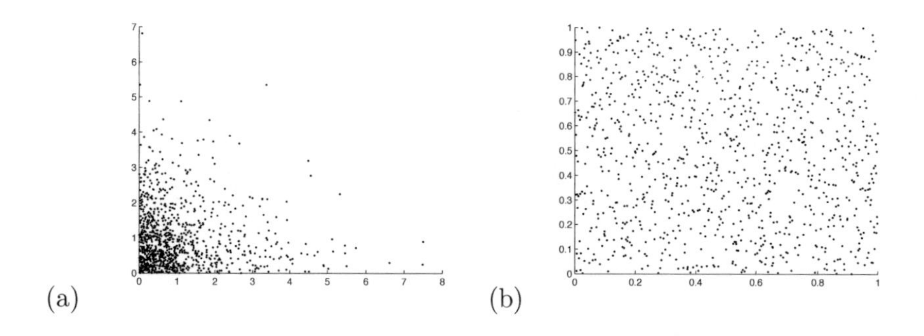

FIGURE 8.4: (a) Graph of 1000 pairs (X_{i1}, X_{i2}) of independent exponential observations; (b) Graph of 1000 pairs of their normalized ranks $\left(\frac{R_{i1}}{1001}, \frac{R_{i2}}{1001}\right)$.

By looking at the raw data, it is difficult to conclude that the observations are independent because in each panel (a) the effect of the margin distribution (which determine the scale of the graphs) interferes with the dependence structure. This problem does not appear in panels (b) since the ranks are

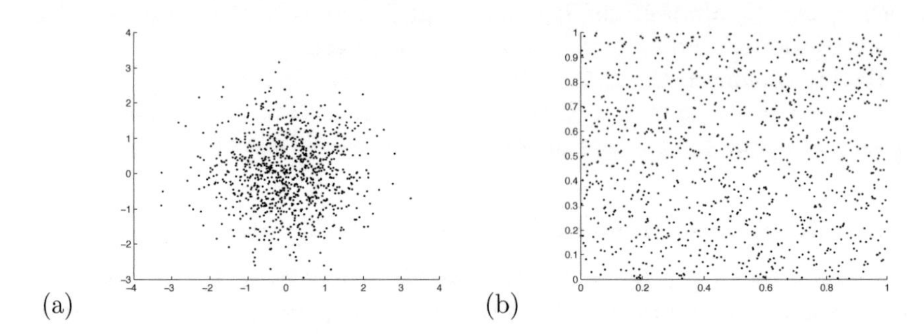

FIGURE 8.5: (a) Graph of 1000 pairs (X_{i1}, X_{i2}) of independent Gaussian observations; (b) Graph of 1000 pairs of their normalized ranks $\left(\frac{R_{i1}}{1001}, \frac{R_{i2}}{1001}\right)$.

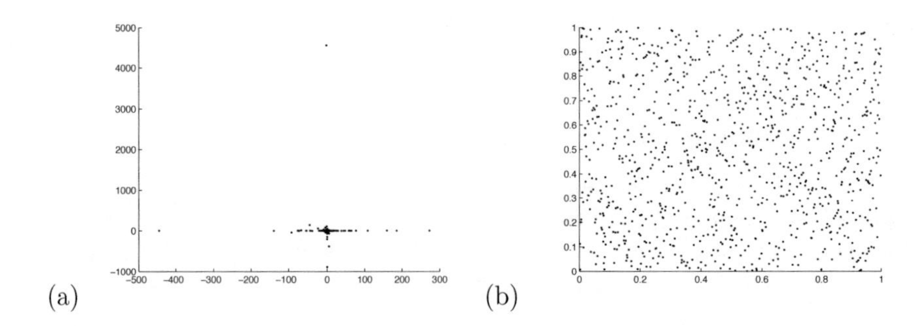

FIGURE 8.6: (a) Graph of 1000 pairs (X_{i1}, X_{i2}) of independent Cauchy observations; (b) Graph of 1000 pairs of their normalized ranks $\left(\frac{R_{i1}}{1001}, \frac{R_{i2}}{1001}\right)$.

invariant to increasing transformations. The marginal distribution does not affect the ranks. Indeed the graphs in panels (b) look like the graphs of independent uniform variables. The reason is quite simple: For $j = 1, 2$, the empirical distribution function F_{nj} defined by

$$F_{nj}(x_j) = \text{Card}\{k; X_{kj} \le x_j\}/(n+1),$$

is a very good estimate of the distribution function F_j of X_j (recall the results of Section 4.2.2.1). Hence, if F_1 and F_2 are continuous, we have

$$\hat{U}_{ij} = \frac{R_{ij}}{n+1} = F_{nj}(X_{ij}) \approx U_{ij} = F_j(X_{ij}) \sim \text{Unif}(0,1),$$

i.e., the pseudo-observations $\hat{U}_{ij}$ are (very) close to the (non-observable) random variables U_{ij}. The conclusion is, therefore, that measuring dependence is best achieved by getting rid of the marginal distributions. If the margins F_1

and F_2 are continuous, then

$$\begin{aligned} X_1 \sim F_1 &\Rightarrow U_1 = F_1(X_1) \sim \text{Unif}(0,1), \\ X_2 \sim F_2 &\Rightarrow U_2 = F_2(X_2) \sim \text{Unif}(0,1), \\ (X_1, X_2) \sim H &\Rightarrow (U_1, U_2) \sim C. \end{aligned}$$

The distribution function C of (U_1, U_2) is the so-called copula. In practice however, the margins are not known. The next best thing to do is to estimate them by their empirical distributions, i.e., to compute the normalized ranks.

In the next section, we define correctly what is a bivariate copula, and we state the famous Sklar theorem.

8.4 Bivariate Copulas

Definition 8.4.1 *A copula $C(u_1, u_2)$ is a distribution function with uniform margins, i.e., there exist two random variables $U_1, U_2 \sim \text{Unif}(0,1)$ such that for all $u_1, u_2 \in [0,1]$,*

$$P(U_1 \le u_1, U_2 \le u_2) = C(u_1, u_2).$$

In particular, for all $u_1, u_2 \in [0,1]$,

$$C(u_1, 1) = u_1, C(u_1, 0) = 0 \quad \text{and} \quad C(1, u_2) = u_2, C(0, u_2) = 0,$$

i.e., every copula has the same values on the boundary of the square $[0,1]^2$.

8.4.1 Examples of Copulas

Example 8.4.1 (Independence Copula) *The independence copula, denoted by $C_\perp$, is defined, for all $u_1, u_2 \in [0,1]$, by*

$$C_\perp(u_1, u_2) = u_1 u_2.$$

Example 8.4.2 (The Tent Map Copula Family) *Suppose $U_1, V, W \sim \text{Unif}(0,1)$ are independent. Let $\theta \in [0,1]$ be fixed. Then, we set*

$$U_2 = \begin{cases} V & \text{if} \quad W > \theta, \\ 2\min(U_1, 1-U_1) & \text{if} \quad W \le \theta. \end{cases}$$

It follows that (U_1, U_2) is a mixture of the independence copula (with probability $1-\theta$, and the tent map transformation, with probability θ. Since $U_2 \sim \text{Unif}(0,1)$, the joint distribution of (U_1, U_2) is indeed a copula C_θ, depending on θ. Moreover, for any $u_1, u_2 \in [0,1]$, we have

$$C_\theta(u_1, u_2) = \theta \left\{ \min\left(u_1, \frac{u_2}{2}\right) + \left(u_1 + \frac{u_2}{2} - 1\right)^+ \right\} + (1-\theta) u_1 u_2. \quad (8.6)$$

In particular, $C_0(u_1,u_2) = u_1u_2$ *and*

$$C_1(u_1,u_2) = \min\left(u_1, \frac{u_2}{2}\right) + \left(u_1 + \frac{u_2}{2} - 1\right)^+.$$

For C_1*, we have the following interesting result, proven in Appendix 8.G.1.*

Proposition 8.4.1 *Suppose that* F_1 *is a symmetric distribution and* F_2 *is a distribution function such that the variance of* X_1 *and* X_2 *exist. Then* $X_1 = F_1^{-1}(U_1) \sim F_1$, $X_2 = F_2^{-1}(U_2) \sim F_2$, *and* $\mathrm{Cor}(X_1,X_2) = 0$*. In particular, the result holds when* F_1 *and* F_2 *are the distribution of the standard Gaussian variable or the uniform distribution.*

As we shall see later, there are many copula-based measures of dependence of the form

$$\rho_C = \mathrm{Cor}\left\{F^{-1}(U_1), F^{-1}(U_2)\right\},$$

where $(U_1,U_2) \sim C$*. Proposition 8.4.1 proves that for these measures,* $\rho_C = 0$.

8.4.2 Sklar Theorem in the Bivariate Case

Theorem 8.4.1 (Sklar [1959]) *Suppose* (X_1,X_2) *has distribution function* H *with margins* F_1 *and* F_2*. Then there exists a copula* C *such that for all* $x_1, x_2 \in \mathbb{R}$,

$$H(x_1,x_2) = P(X_1 \le x_1, X_2 \le x_2) = C\left\{F_1(x_1), F_2(x_2)\right\}. \tag{8.7}$$

The copula C *is unique when restricted to* $\overline{Image(F_1)} \times \overline{Image(F_2)} \subset [0,1]^2$*, where* $\overline{Image(F_j)} = \{0,1\} \cup \{F_j(y), F_j(y-); y \in \mathbb{R}\}$*. In particular, it is unique if* F_1 *and* F_2 *are continuous.*

Example 8.4.3 *If* $X_1 \in \{0,1\}$ *with probabilities* $1/2, 1/2$ *and* $X_2 \in \{0,1\}$ *with probabilities* $1/4$, $3/4$ *respectively, then*

$$\overline{Image(F_1)} = \{0,1/2,1\} \quad \text{and} \quad \overline{Image(F_2)} = \{0,1/4,1\}.$$

Hence the only known value of the copula inside $[0,1]^2$ *is* $C(1/2,1/4) = P(X_1 \le 0, X_2 \le 0)$.

Proposition 8.4.2 *If* C *is a copula, then it is Lipschitz continuous, i.e.,*

$$|C(u_1,u_2) - C(v_1,v_2)| \le |u_1 - v_1| + |u_2 - v_2|, \quad u_1,u_2,v_1,v_2 \in [0,1].$$

Moreover, its partial derivatives of first order exist almost surely, are bounded by 1, and for any $u_1, u_2 \in [0,1]$,

$$C(u_1,u_2) = \int_0^{u_1} \partial_{u_1} C(s,u_2)ds = \int_0^{u_2} \partial_{u_2} C(u_1,s)ds.$$

The proof is given in Appendix 8.G.2.

When the margins are continuous, we have a more interesting result, that enables the identification of the unique copula in the Sklar theorem.

Theorem 8.4.2 *Suppose that $(X_1, X_2) \sim H$ with continuous margins F_1 and F_2. Set $U_1 = F_1(X_1)$ and $U_2 = F_2(X_2)$. Then $U_1 \sim \text{Unif}(0,1)$, $U_2 \sim \text{Unif}(0,1)$ and the unique copula C associated with (X_1, X_2) is the distribution function of (U_1, U_2), i.e.,*

$$C(u_1, u_2) = P(U_1 \le u_1, U_2 \le u_2), \quad u_1, u_2 \in [0,1].$$

Moreover, if ϕ and ψ are increasing functions, then C is also the copula associated with $\{\phi(X_1), \psi(X_2)\}$. The copula is therefore invariant by increasing transformations of the margins.

Proposition 8.4.3 *Suppose that $(X_1, X_2) \sim H$ with margins F_1, F_2 and densities f_1, f_2 respectively. Let C be the unique copula of H. Then H has a density h if and only if C has a density c. In either case, for all $x_1, x_2 \in \mathbb{R}$, we have*

$$h(x_1, x_2) = c\{F_1(x_1), F_2(x_2)\} f_1(x_1) f_2(x_2). \tag{8.8}$$

Furthermore, for all $u_1, u_2 \in (0,1)$,

$$c(u_1, u_2) = \frac{h\left\{F_1^{-1}(u_1), F_2^{-1}(u_2)\right\}}{f_1\left\{F_1^{-1}(u_1)\right\} f_2\left\{F_2^{-1}(u_2)\right\}}. \tag{8.9}$$

As announced at the beginning of the chapter, we will now show the link between conditional distributions and copulas.

Proposition 8.4.4 *Suppose that $(X_1, X_2) \sim H$, with continuous margins F_1, F_2 and copula C. Then, almost surely,*

$$P(U_2 \le u_2 | U_1 = u_1) = \frac{\partial C(u_1, u_2)}{\partial u_1} \tag{8.10}$$

and

$$P(X_2 \le x_2 | X_1 = x_1) = \left. \frac{\partial C(u_1, u_2)}{\partial u_1} \right|_{u_1 = F_1(x_1), u_2 = F_2(x_2)}. \tag{8.11}$$

PROOF. For a fixed u_2, setting $h(u_1) = P(U_2 \le u_2 | U_1 = u_1)$, and using the definition of the conditional expectation, we must have, for any $0 \le b \le 1$,

$$\begin{aligned} E\{h(U_1)\mathbb{I}(a < U_1 \le b)\} &= \int_a^b h(u_1) du_1 \\ &= P(a < U_1 \le b, U_2 \le u_2) \\ &= C(b, u_2) - C(a, u_2). \end{aligned}$$

By Proposition 8.4.2, $\frac{\partial C(u_1, u_2)}{\partial u_1} = h(u_1)$ almost surely. The conclusion then follows from the definition of a conditional expectation.

■

The following corollary shows the relationship between the conditional quantiles of a distribution function H and those of its copula C.

Corollary 8.4.1 *Let $q_C(u_1, p)$ be the conditional quantile of order p of U_2 given $U_1 = u_1$, i.e., for any $p \in (0, 1)$,*

$$P\{U_2 \leq q_C(u_1, p)|U_1 = u_1\} = p.$$

Let $q_H(x_1, p)$ be the conditional quantile of order p of X_2 given $X_1 = x_1$, i.e., for any $p \in (0, 1)$,

$$P\{X_2 \leq q_H(x_1, p)|X_1 = x_1\} = p.$$

Then

$$q_H(x_1, p) = F_2^{-1}\left[q_C\{F_1(x_1), p\}\right]. \tag{8.12}$$

Remark 8.4.1 *Returning to the function g defined by (8.2) in Section 8.1, if C_{12} is the copula between S_1 and S_2, and C_{13} is the copula between S_1 and S_3, we have, using Proposition 8.4.4 and Corollary 8.4.1,*

$$g(x, y) = F_3^{-1} \circ q_{C_{13}}\left[F_1(x), \partial_{u_1} C_{12}\{F_1, (x) F_2(y)\}\right].$$

8.4.3 Applications for Simulation

If we can simulate $(U_1, U_2) \sim C$, then we can generate observations $(X_1, X_2) \sim H$, with H given by

$$H(x_1, x_2) = P(X_1 \leq x_1, X_2 \leq x_2) = C(F_1(x), F_2(x_2)),$$

by setting

$$X_1 = F_1^{-1}(U_1), \quad X_2 = F_2^{-1}(U_2),$$

where we recall that $F_j^{-1}(u) = \inf\{x \in \mathbb{R}; F_j(x) \geq u\}$, $j = 1, 2$. The next step is to show how to generate $(U_1, U_2) \sim C$.

8.4.4 Simulation of $(U_1, U_2) \sim C$

First, for $u_1 \in (0, 1)$ and $u_2 \in (0, 1)$, set

$$F_2(u_2|u_1) = P(U_2 \leq u_2|U = u_1) = \frac{\partial}{\partial u_1} C(u_1, u_2) \ a.s.$$

Then, as we will see later in Theorem 8.6.2, U_1 and $F_2(U_2|U_1)$ are independent. So by inverting the transform $(u_1, u_2) \mapsto \{u_1, F_2(u_2|u_1)\}$, and applying it to $U_1 \sim \text{Unif}(0, 1)$ and $V \sim \text{Unif}(0, 1)$, with V independent of U_1, we get

$$U_2 = F_2^{-1}(V|U_1) = q_C(U_1, V) \quad \text{and } (U_1, U_2) \sim C.$$

Example 8.4.4 *For the tent map copula with $\theta = 1$, we have*

$$C_1(u_1, u_2) = \begin{cases} u_1, & \text{if } u_1 < u_2/2, \\ u_2/2, & \text{if } u_2 \leq 2\min(u_1, 1 - u_1), \\ u_1 + u_2 - 1, & \text{if } u_2/2 > (1 - u_1). \end{cases}$$

As a result,

$$\frac{\partial}{\partial u_1} C(u_1, u_2) = \begin{cases} 0, & \text{if } u_2 < T(u_1), \\ 1, & \text{if } u_2 > T(u_1). \end{cases}$$

Therefore the conditional distribution of U_2 is discrete. It does not come as a surprise since $U_2 = T(U_1)$. For $\theta < 1$, we have

$$F_2(u_2|u_1) = \begin{cases} (1-\theta)u_2, & \text{if } u_2 < T(u_1), \\ (1-\theta)u_2 + \theta, & \text{if } u_2 \geq T(u_1). \end{cases}$$

Hence, setting $b = (1 - \theta)T(u_1)$, we find that

$$F_2^{-1}(u_2|u_1) = \begin{cases} \frac{u_2}{1-\theta}, & \text{if } u_2 \in [0, b), \\ T(u_1), & \text{if } u_2 \in [b, b + \theta), \\ \frac{u_2 - \theta}{1-\theta}, & \text{if } u_2 \in [b + \theta, 1]. \end{cases}$$

This method of generating (U_1, U_2) is different from the one proposed in Example 8.4.2.

8.4.5 Modeling Dependence with Copulas

As seen before, the dependence[3] is

- Margin free;
- Completely determined by the copula.

If we change the copula in the formula

$$H(x_1, x_2) = C\{F_1(x_1), F_2(x_2)\},$$

we change the dependence without modifying the margins, since for any $x_1, x_2 \in \mathbb{R}$,

$$\begin{aligned} H(x_1, \infty) &= C\{F_1(x_1), 1\} = F_1(x_1), \\ H(\infty, x_2) &= C\{1, F_2(x_2)\} = F_2(x_2). \end{aligned}$$

This is what makes copulas interesting for financial applications. We can only change the dependence structure and check the changes in the price of the financial instruments, as in Tables 8.4 and 8.5.

[3]For arguments for and against the usefulness of copulas and the concept of dependence, see Mikosch [2006] and the discussions within, particularly Genest and Rémillard [2006].

8.4.6 Positive Quadrant Dependence (PQD) Order

When modeling dependence, it can be interesting to compare the models and check if the dependence is stonger or weaker. The following partial order, called PQD order, is a natural order for comparing copulas.

Definition 8.4.2 *The PQD order is a partial order on copulas, defined by*

$$C \preceq_{PQD} C^\star \text{ if and only if } C(u) \leq C^\star(u),$$

for all $u = (u_1, u_2) \in [0,1]^2$, *which is equivalent to*

$$P(U_1 > u_1, U_2 > u_2) \leq P(U_1^\star > u_1, U_2^\star > u_2),$$

where $U \sim C$ *and* $U^\star \sim C^\star$. *The dependence induced by* C *is said to be positive (respectively negative) if* $C_\perp \preceq_{PQD} C$ *(respectively* $C \preceq_{PQD} C_\perp$*).*

Definition 8.4.3 (Fréchet-Hoeffding Bounds) *The Fréchet-Hoeffding lower and upper bounds are the copulas defined respectively by*

$$C_-(u_1, u_2) = \max(u_1 + u_2 - 1, 0) = (u_1 + u_2 - 1)^+,$$

which is the copula of $(U_1, 1 - U_1)$, *and by*

$$C_+(u_1, u_2) = \min(u_1, u_2),$$

which is the copula of (U_1, U_1).

The next result shows that with respect to the PQD order, there exist a smallest and a largest copula.

Proposition 8.4.5 *For any copula* C,

$$C_- \preceq_{PQD} C \preceq_{PQD} C_+.$$

PROOF. $C(u_1, u_2) \leq C(u_1, 1) = u_1$ and $C(u_1, u_2) \leq C(1, u_2) = u_2$. Thus $C \preceq_{PQD} C_+$. To complete the proof, note that

$$C(u_1, u_2) = u_1 + u_2 - 1 + P(U_1 > u_1, U_2 > u_2) \geq C_-(u_1, u_2).$$

■

This result will prove very useful for measures of dependence, which is the next topic.

8.5 Measures of Dependence

Suppose that $(X_1, X_2) \sim H$, with continuous margins F_1 and F_2, and (unique) copula C. Set $U_1 = F_1(X_1)$ and $U_2 = F_2(X_2)$. Since the dependence is uniquely determined by the copula, a good measure of dependence should be margin free. In what follows, we will see several measures of dependence that are margin free such as Kendall tau, Spearman rho, and van der Waerden coefficient. We will also show how to estimate these measures and give results on the precision of the estimation. Note that all these measures are based on ranks. First we begin by showing how to estimate the copula, which is the ultimate measure of dependence. Then we define another important function related to dependence, the probability integral transform, also called Kendall function.

Recall that for a data set

$$(X_{11}, X_{12}), \ldots, (X_{n1}, X_{n2}),$$

we associate the pairs of ranks

$$(R_{11}, R_{12}), \ldots, (R_{n1}, R_{n2}),$$

where, for $j = 1, 2$, R_{ij} is the rank of X_{ij} amongst $X_{1j}, \ldots, X_{nj}$. We then define the pseudo-observations

$$\left(\hat{U}_{i1}, \hat{U}_{i2}\right) = \left(\frac{R_{i1}}{n+1}, \frac{R_{i2}}{n+1}\right), \quad i \in \{1, \ldots, n\}.$$

Example 8.5.1 *To illustrate the calculations, the following simple data set will be used throughout the section. The associated ranks and pseudo-observations are given in Table 8.7.*

TABLE 8.6: Data set.

i	1	2	3	4	5
X_{i1}	1.5	-2.4	4.2	-6.1	-8.9
X_{i2}	5.3	2.6	-3.4	-5.6	-9.7

TABLE 8.7: Ranks and pseudo-observations for the data set.

i	1	2	3	4	5
X_{i1}	1.5	-2.4	4.2	-6.1	-8.9
R_{i1}	4	3	5	2	1
$\hat{U}_{i1}$	4/6	3/6	5/6	2/6	1/6
X_{i2}	5.3	2.6	-3.4	-5.6	-9.7
R_{i2}	5	4	3	2	1
$\hat{U}_{i2}$	5/6	4/6	3/6	2/6	1/6

8.5.1 Estimation of a Bivariate Copula

For a sample $(X_{11}, X_{12}), \ldots, (X_{n1}, X_{n2})$ of observations from the bivariate distribution H, with continuous margins F_1, F_2 and copula C, we justified the use of pseudo-observations by the property that for all $i \in \{1, \ldots, n\}$ and $j = 1, 2$,

$$\hat{U}_{ij} = \frac{R_{ij}}{n+1} \approx U_{ij} = F_j(X_{ij}).$$

According to Theorem 8.4.2, the copula C is the distribution function of (U_1, U_2), so it can be estimated by the empirical copula of the pseudo-observations by setting

$$C_n(u_1, u_2) = \frac{1}{n} \sum_{i=1}^{n} \mathbb{I}\left(\frac{R_{i1}}{n+1} \le u_1, \frac{R_{i2}}{n+1} \le u_2 \right), \quad u_1, u_2 \in [0,1]. \qquad (8.13)$$

8.5.1.1 Precision of the Estimation of the Empirical Copula

It can be shown [Gänssler and Stute, 1987, Fermanian et al., 2004] that if the first order partial derivatives of C are continuous, then $\sqrt{n}\,(C_n - C) \rightsquigarrow \mathbb{C}$, where $\mathbb{C}$ is a continuous centered Gaussian process on $[0,1]^2$. Comparing C_n and C can also be a way of choosing the best copula model [Genest and Rémillard, 2008]. The covariance of the process is in general quite complicated, except when C is the independence copula, in which case

$$\operatorname{Cov}\left\{\mathbb{C}(u_1, u_2), \mathbb{C}(v_1, v_2)\right\} = \left\{\min(u_1, v_1) - u_1 v_1\right\}\left\{\min(u_2, v_2) - u_2 v_2\right\},$$

$u_1, u_2, v_1, v_2 \in [0,1]$. In particular the variance of $\mathbb{C}(u_1, u_2)$ is $u_1(1-u_1)u_2(1-u_2)$. Such a process is called a Brownian drum because it is 0 on the boundary of the square $[0,1]^2$. For the data set of Table 8.6, we have plotted the pseudo-observations and the corresponding empirical copula in Figure 8.8, using the MATLAB function *Graph3dCopula*.

8.5.1.2 Tests of Independence Based on the Empirical Copula

Since independence is characterized by the copula $C_\perp$, independence between X_1, X_2 can be based on $\mathbb{C}_n$ or on the related process

$$\begin{aligned} \mathbb{G}_{n,\{1,2\}}(x_1, x_2) &= \frac{1}{\sqrt{n}} \sum_{i=1}^{n} \prod_{j=1}^{2} \{\mathbb{I}(X_{ij} \le x_j) - F_{nj}(x_j)\} \\ &= \frac{1}{\sqrt{n}} \sum_{i=1}^{n} \prod_{j=1}^{2} \left[\mathbb{I}\{U_{ij} \le F_j(x_j)\} - D_{nj}\{F_j(x_j)\}\right], \end{aligned}$$

where $U_{ij} = F_j(X_{ij})$, and

$$F_{nj}(y) = D_{nj}\{F_j(y)\} = \frac{1}{n} \sum_{i=1}^{n} \mathbb{I}(X_{ij} \le y) = \frac{1}{n} \sum_{i=1}^{n} \mathbb{I}(U_{ij} \le F_j(y)\}, \quad y \in \mathbb{R},$$

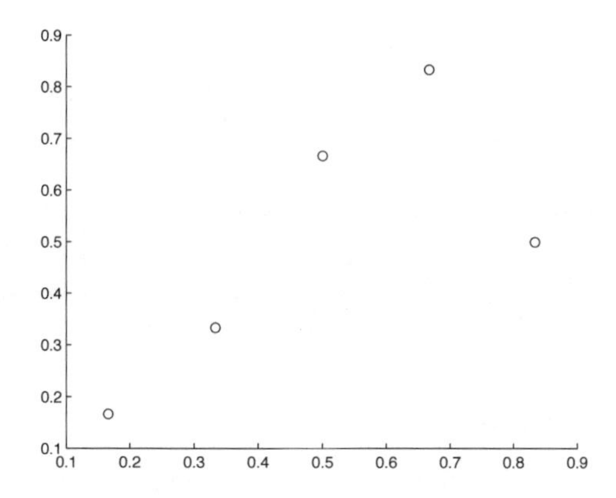

FIGURE 8.7: Graphs of the pseudo-observations for the data of Table 8.6.

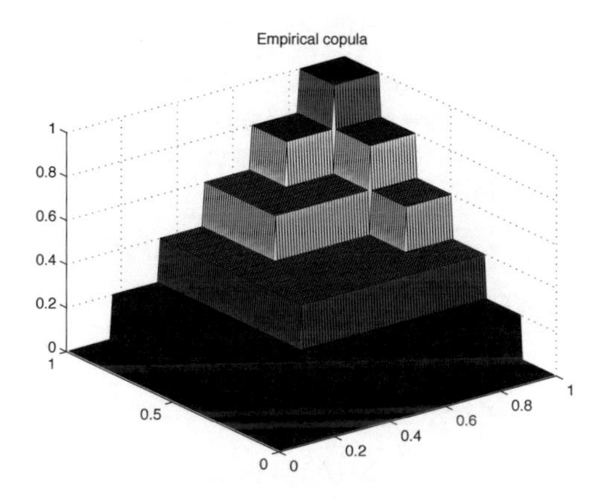

FIGURE 8.8: Graphs of the estimated copula for the data of Table 8.6.

is the empirical distribution function estimating the continuous F_j, $j \in \{1,2\}$. It is shown in Genest and Rémillard [2004] that under the null hypothesis of independence, i.e., $C = C_\perp$, $\mathbb{G}_{n,\{1,2\}}(x_1, x_2) \rightsquigarrow \mathbb{C}\{F_1(x_1), F_2(x_2)\}$, and the Cramér-von Mises statistics

$$T_{n,\{1,2\}} = \frac{36}{n} \sum_{i=1}^{n} \mathbb{G}^2_{n,\{1,2\}}(X_{i1}, X_{i2}) \approx 36 \int_0^1 \int_0^1 \mathbb{C}^2_n(u_1, u_2) du_1 du_2$$

converge in law to

$$T_2 = 36 \int_0^1 \int_0^1 \mathbb{C}^2(u_1, u_2) du_1 du_2 \stackrel{\text{Law}}{=} \frac{36}{\pi^4} \sum_{i=1}^{\infty} \sum_{j=1}^{\infty} \frac{Z_{ij}^2}{i^2 j^2},$$

where the random variables Z_{ij} are i.i.d. $N(0,1)$. Tables for T_2 can be obtained by simulation. For example, using 10^6 simulations of T_2 (truncated at $i, j \leq 100$), one obtains that the 95% quantile is 2.08. One can also estimate P-values using Algorithm 8.10.2. For more details, see Appendix 8.B.

8.5.2 Kendall Function

Recall that $(X_1, X_2) \sim H$, with continuous margins F_1 and F_2, and copula C. Also $U_1 = F_1(X_1)$ and $U_2 = F_2(X_2)$. Set

$$V = H(X_1, X_2) = C\{F_1(X_1), F_2(X_2)\} = C(U_1, U_2) \in [0, 1].$$

Then the law of V does not depend on the margins and its distribution function K, defined by

$$K(t) = P(V \leq t), \quad 0 \leq t \leq 1,$$

is called the Kendall function or the probability integral transform. Its expectation is related to a measure of dependance called Kendall tau, as we shall see later. We will also see that even though K is a univariate function, it characterizes some families of copulas.

Example 8.5.2 *Here are some examples of Kendall functions.*

- *For the independence copula $C_\perp$, we have*

$$K(t) = t - t\ln(t), \quad t \in [0, 1]. \tag{8.14}$$

  Indeed, $V = C(U_1, U_2) = U_1 U_2$, and

$$-\ln(V) = -\ln(U_1) - \ln(U_2) \sim \text{Gamma}(2, 1).$$

- *For the tent map copula C_1 associated with $U_2 = 2\min(U_1, 1 - U_1)$, we have*

$$K(t) = \min(2t, 1) \quad 0 \leq t \leq 1, \tag{8.15}$$

  since

$$V = C(U_1, U_2) = U_2/2 \sim \text{Unif}(0, 1/2).$$

- *For the Fréchet-Hoeffding upper bound C_+, $U_2 = U_1$ and $V = C_+(U_1, U_1) = \min(U_1, U_1) = U_1$, so*

$$K(t) = t, \quad t \in [0, 1]. \tag{8.16}$$

- *For the Fréchet-Hoeffding lower bound C_-, $U_2 = 1 - U_1$ and $V = C_-(U_1, 1 - U_1) = \max(0, U_1 + 1 - U_1 - 1) \equiv 0$, so*

$$K(t) = 1, \quad t \in [0, 1]. \tag{8.17}$$

8.5.2.1 Estimation of Kendall Function

To estimate K, Genest and Rivest [1993] proposed to compute first the pseudo-observations

$$\begin{aligned}
\hat{V}_i &= \text{Card}\{j; X_{j1} < X_{i1} \text{ and } X_{j2} < X_{i2}\}/(n-1) \\
&= \text{Card}\{j; R_{j1} < R_{i1} \text{ and } R_{j2} < R_{i2}\}/(n-1) \\
&= \text{Card}\{j; \hat{U}_{j1} < \hat{U}_{i1} \text{ and } \hat{U}_{j2} < \hat{U}_{i2}\}/(n-1) \\
&= \frac{n}{n-1}\left\{C_n\left(\hat{U}_{i1}, \hat{U}_{i2}\right)\right\} - \frac{1}{n-1} \\
&\approx V_i = C(U_{i1}, U_{i2}), \quad i \in \{1, \ldots, n\}.
\end{aligned}$$

Then the estimation of K is the empirical distribution function of these pseudo-observations, namely,

$$K_n(t) = \text{Card}\{i; \hat{V}_i \le t\}/n, \quad t \in [0,1]. \tag{8.18}$$

Since K and K_n are univariate functions, they can be compared graphically. Comparing K_n and K can also be a way of choosing the best copula model [Genest et al., 2006a]. For the data set of Table 8.6, we then have

$$\hat{V}_1 = 3/4, \ \hat{V}_2 = 2/4, \ \hat{V}_3 = 2/4, \hat{V}_4 = 1/4, \hat{V}_5 = 0.$$

The corresponding graph of K_n is displayed in Figure 8.9. The MATLAB function *Pseudos* can be used to compute the pseudo-observations $\hat{U}_1, \ldots, \hat{U}_n$ and $\hat{V}_1, \ldots, \hat{V}_n$.

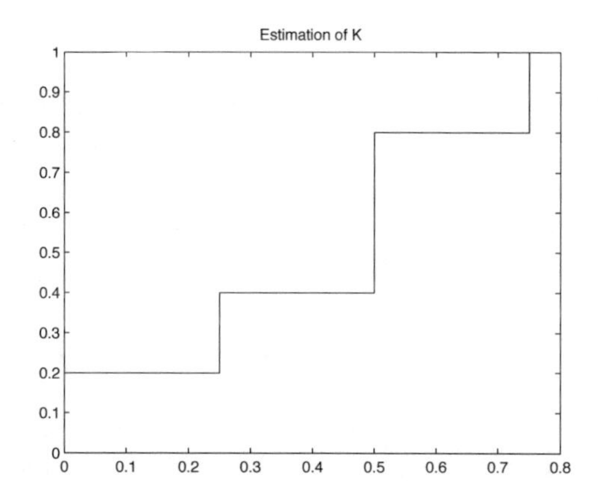

FIGURE 8.9: Estimation of K for the data of Table 8.6.

8.5.2.2 Precision of the Estimation of the Kendall Function

It can be shown [Barbe et al., 1996] that under smoothness conditions on the density of K, $\sqrt{n}\,(K_n - K) \rightsquigarrow \mathbb{K}$, where $\mathbb{K}$ is a continuous centered Gaussian process on $[0,1]$. The covariance of the process is in general quite complicated, except when C is the independence copula, for which

$$\mathrm{Cov}\left\{\mathbb{K}(s), \mathbb{K}(t)\right\} = s\left\{t - 1 - \ln(t)\right\}, \quad 0 \le s \le t \le 1.$$

In particular the variance of $\mathbb{K}(t)$ is $t\left\{t - 1 - \ln(t)\right\}$. According to Figure 8.10, the maximal variance 0.1619 is attained at $t = 0.2032$.

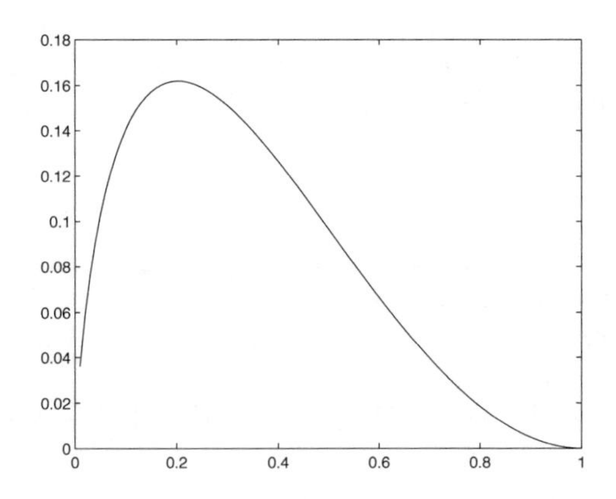

FIGURE 8.10: Variance of $\mathbb{K}$ for the independence copula.

8.5.2.3 Tests of Independence Based on the Empirical Kendall Function

Using the results of the last section, tests of independence between X_1, X_2 can be based on $\mathbb{K}_n$, with $K(t) = K_\perp(t) = t - t\ln(t)$. For example, we can use the Kolmogorov-Smirnov statistic

$$KS_n = \sqrt{n}\max_{1\le i\le n}\left|K_n\left(\hat{V}_i\right) - K_\perp\left(\hat{V}_i\right)\right|. \tag{8.19}$$

By simulating 10^5 samples of different sizes n of pairs of independent uniform variables (U_1, U_2), we obtain a value of 1.12 for the 95% quantile of the statistic and 1.33 for the 99% quantile. A uniform 95% confidence band around $K_\perp$ is then given by $K_\perp \pm 1.12/\sqrt{n}$. The null hypothesis of independence is rejected at the 5% level if K_n is not entirely within the band. P-values can also be obtained by simulation using Algorithm 8.5.1.

Algorithm 8.5.1 *To estimate the P-values of a Kolmogorov-Smirnov statistic KS for testing independence, for N large and for $k \in \{1, \ldots, N\}$, repeat the following steps:*

a) Generate independent $\left(U_{i1}^{(k)}, U_{i2}^{(k)}\right) \sim C_{\perp}$, $i \in \{1, \ldots, n\}$.

b) Compute the associated pseudo-observations $\hat{V}_i^{(k)}$, $i \in \{1, \ldots, n\}$.

c) Compute the Kolmogorov-Smirnov statistic $KS_n^{(k)}$ *according to formula* (8.19).

An approximate P-value is then given by

$$\frac{1}{N} \sum_{k=1}^{N} \mathbb{I}\left(KS_n^{(k)} > KS_n\right).$$

One can also base a test of independence on the Cramér-von Mises statistic

$$CVM_n = \sum_{i=1}^{n} \left\{K_n\left(\hat{V}_i\right) - K_{\perp}\left(\hat{V}_i\right)\right\}^2. \tag{8.20}$$

Using simulations again, we obtain 0.30 for the 95% quantile and 0.48 for the 99% quantile. P-values can be estimated by adapting Algorithm 8.5.1 in an obvious way.

Example 8.5.3 *An instructive example is to compute K_n for a simulated sample of* 500 *pairs of observations (X_1, X_2) with Gaussian margins and tent map copula C_1, which can be found in the MATLAB file* DataTentMapCop.

By looking at Figure 8.11, K_n is a quite good estimator of $K(t) = \min(1, 2t)$, as given in (8.15)*, and it lies outside the 95% confidence band of the independence copula. As expected, the null hypothesis of independence between X_1 and X_2 or between U_1 and U_2 is rejected.*

We can also compute the test statistics for the two tests based on the empirical Kendall function, the test based on the copula, three tests based on the dependence measures that will be introduced in the next sections, and Pearson correlation, using the MATLAB programs IndKendallTest, IndGRTest, *and* EstDep.

The results are displayed in Table 8.8. In the last three cases, the P-values were estimated using $N = 10000$ replications. As expected, the null hypothesis of independence between X_1 and X_2 or between U_1 and U_2 is clearly rejected for both tests based on Kendall function and the test based on the copula. However, the null hypothesis is not rejected for the usual dependence measures, showing the lack of power of these tests.

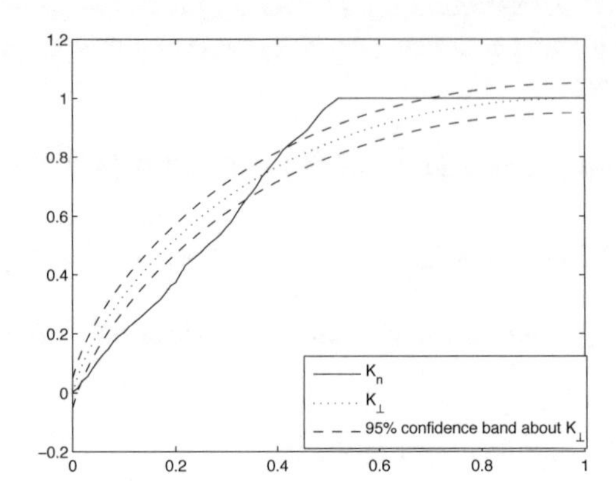

FIGURE 8.11: Graph of K_n and 95% confidence band around Kendall function for the independence copula.

TABLE 8.8: 95% confidence intervals and P-values for tests of independence based on classical statistics (Pearson rho, Kendall tau, Spearman rho, van der Waerden rho), and on Kolmogorov-Smirnov and Cramér-von Mises statistics using the empirical Kendall function and the empirical copula. The quantiles and the P-values were computed using $N = 10000$ iterations.

Statistic	95% C. I.	P-Value (%)
Pearson rho	$0 \in -0.002 \pm 0.088$	96.2
Kendall tau	$0 \in 0.024 \pm 0.101$	41.9
Spearman rho	$0 \in 0.024 \pm 0.117$	59.4
van der Waerden rho	$0 \in 0.011 \pm 0.103$	81.3
Cramér-von Mises (C)	$149.96 \notin [0, 2.08]$	0
Cramér-von Mises (K)	$4.82 \notin [0, 0.30]$	0
Kolmogorov-Smirnov (K)	$3.44 \notin [0, 1.12]$	0

Example 8.5.4 *As another illustration, we can look at the dependence between the returns of Apple and Microsoft, from January 14th, 2009, to January 14th, 2011. The results are reported in Table 8.9, and the estimated Kendall function is displayed in Figure 8.12.*

TABLE 8.9: 95% confidence intervals and P-values for tests of independence based on classical statistics (Pearson rho, Kendall tau, Spearman rho, van der Waerden rho), and on Kolmogorov-Smirnov and Cramér-von Mises statistics for the returns of Apple and Microsoft. The quantiles and the P-values were computed using $N = 10000$ iterations.

Statistic	95% C. I.	P-Value (%)
Kendall tau	$0 \notin 0.35 \pm 0.06$	0
Spearman rho	$0 \notin 0.48 \pm 0.08$	0
van der Waerden rho	$0 \notin 0.51 \pm 0.08$	0
Cramér-von Mises (C)	$4.88 \notin [0, 2.08]$	0
Cramér-von Mises (K)	$4.88 \notin [0, 0.30]$	0
Kolmogorov-Smirnov (K)	$3.63 \notin [0, 1.12]$	0

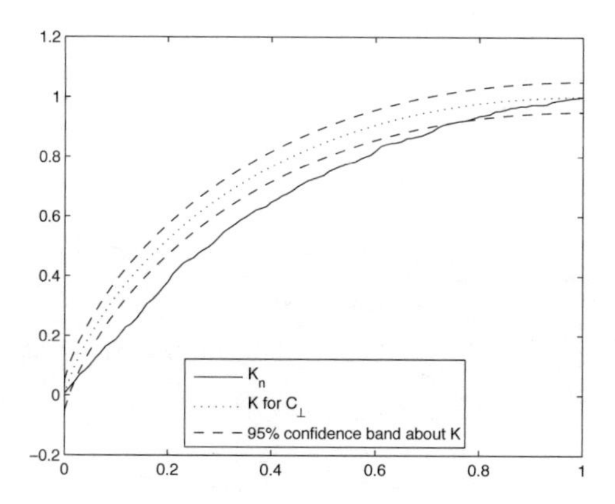

FIGURE 8.12: Graph of K_n and 95% confidence band around Kendall function for the independence copula of returns of Apple and Microsoft.

Copulas and Serial Dependence

Note that since there is serial dependence in the returns of Apple, as seen in Example 3.1.1, one cannot really test independence between the returns of Apple and Microsoft using the tests based on the copula, the Kendall function, or the classical measures of dependence. One can either model dependence and try to work with the residuals, as in Section 8.8.1, or model the interdependence and serial dependence at the same time with copulas, as proposed in Rémillard et al. [2012].

8.5.3 Kendall Tau

Kendall tau is a measure of concordance. Two pairs of points (x_1, x_2) and (y_1, y_2) are concordant if and only if $x_1 \le y_1$ and $x_2 \le y_2$, or $y_1 \le x_1$ and $y_2 \le x_2$. Otherwise they are said to be discordant. Geometrically, being discordant means that the line passing through (x_1, x_2) and (y_1, y_2) has a negative slope. Recall that $(X_1, X_2) \sim H$, with continuous margins F_1, F_2 and copula C. Set $U_1 = F_1(X_1)$, $U_2 = F_2(X_2)$, and $V = C(U_1, U_2)$. Suppose that $(Y_1, Y_2) \sim H$ is independent of (X_1, X_2). Then Kendall tau is defined by

$$
\begin{aligned}
\tau = \tau_C &= P\{(X_1, X_2) \text{ and } (Y_1, Y_2) \text{ are concordant }\} \\
&\quad - P\{(X_1, X_2) \text{ and } (Y_1, Y_2) \text{ are discordant }\} \\
&= 2P\{(X_1, X_2) \text{ and } (Y_1, Y_2) \text{ are concordant }\} - 1 \\
&= 4P(Y_1 \le X_1, Y_2 \le X_2) - 1 = 4E\left[C\{F_1(X_1), F_2(X_2)\}\right] - 1 \\
&= 4E(V) - 1 \\
&= 3 - 4\int_0^1 K(t)dt,
\end{aligned}
$$

where K is the Kendall function. It is easy to check that $\tau = 1$ for the Fréchet-Hoeffding upper bound, $\tau = -1$ for the Fréchet-Hoeffding lower bound, and $\tau = 0$ for the independence copula.

Independence vs Kendall Tau

One can have $\tau_C = 0$ for $C \ne C_\perp$. Recall that by taking the tent map copula C_1, we have $V = C(U_1, U_2) = \min(U_1, 1 - U_2) = U_2/2$. Hence,

$$\tau = 4E(V) - 1 = 2E(U_2) - 1 = 0,$$

without U_1 and U_2 being independent.

The following proposition, proven in Appendix 8.G.3, shows that Kendall tau is monotone with respect to the PQD order.

Proposition 8.5.1 *If $C \preceq_{PQD} C^\star$ then $\tau_C \le \tau_{C^\star}$, with equality if and only if $C = C^\star$. Hence $\tau = 1$ if and only if C is the Fréchet-Hoeffding upper bound, and $\tau = -1$ if and only if C is the Fréchet-Hoeffding lower bound.*

8.5.3.1 Estimation of Kendall Tau

Let P_n and Q_n be the number of concordant and discordant pairs in a sample of size n. Then $P_n + Q_n = \binom{n}{2} = n(n-1)/2$. Following the definition of τ, a natural estimate is

$$\hat{\tau} = \tau_n = \frac{2(P_n - Q_n)}{n(n-1)} = \frac{4P_n}{n(n-1)} - 1 = 1 - \frac{4Q_n}{n(n-1)}. \tag{8.21}$$

Using the pseudo-observations $\hat{V}_1, \ldots, \hat{V}_n$ defined in Section 8.5.2.1, we have

$$\hat{\tau} = \tau_n = -1 + \frac{4}{n}\sum_{i=1}^{n} \hat{V}_i. \tag{8.22}$$

Remark 8.5.1 *$\hat{\tau}$ depends only on the ranks. It is easy to see that $\hat{\tau} = 1$ if and only if $R_{i1} = R_{i2}$ for all $i \in \{1, \ldots, n\}$. Similarly, $\hat{\tau} = -1$ if and only if $R_{i1} = n + 1 - R_{i2}$ for all $i \in \{1, \ldots, n\}$.*

8.5.3.2 Precision of the Estimation of Kendall Tau

The following proposition shows that $\hat{\tau}$ is a good estimator of τ.

Proposition 8.5.2 *For any copula C, we have*

$$\sqrt{n}(\tau_n - \tau)/\sigma_n \rightsquigarrow N(0, 1),$$

where σ_n is the standard deviation of the pseudo-observations

$$4(2\hat{V}_i - \hat{U}_{i1} - \hat{U}_{i2}), \quad i \in \{1, \ldots, n\}.$$

Moreover, under the null hypothesis of independence, i.e., when $C = C_\perp$, we have $\sqrt{n}\tau_n \rightsquigarrow N(0, 4/9)$.

The proof of the result follows from an application of U-statistics or directly from Ghoudi and Rémillard [2004].

Example 8.5.5 *For the data of Table 8.6, using the graphs of the pairs of points as in Figure 8.8, we have 8 pairs of concordant points and 2 pairs of discordant points, so $\hat{\tau} = (8 - 2)/10 = 3/5$. Moreover $\sigma_n = 0.5578$.*

8.5.4 Spearman Rho

Spearman rho is simply defined as the Pearson correlation between $U_1 = F_1(X_1)$ and $U_2 = F_2(X_2)$, i.e.,

$$\rho^{(S)} = \rho_C^{(S)} = \text{Cor}(U_1, U_2) = 12\left\{E(U_1U_2) - 1/4\right\} = 12E(U_1U_2) - 3.$$

Note that $\rho^{(S)} = 1$ for the Fréchet-Hoeffding upper bound, $\rho^{(S)} = -1$ for the Fréchet-Hoeffding lower bound, and $\rho^{(S)} = 0$ for $C = C_\perp$.

Independence vs Spearman Rho

Again we can have $\rho^{(S)} = 0$ with $C \neq C_\perp$. In fact, taking the tent map copula C_1, we may conclude from Proposition 8.4.1 that $\rho^{(S)} = 0$.

Using Hoeffding equality (8.56) in Lemma 8.F.1, we can write $\rho^{(S)}$ directly in terms of the copula C viz.

$$\rho^{(S)} = 12 \int_0^1 \int_0^1 \{C(u_1, u_2) - u_1 u_2\} du_1 du_2.$$

This representation enables us to derive the following monotonicity property of Spearman rho with respect to the PQD order.

Proposition 8.5.3 *If $C \preceq_{PQD} C^\star$ then $\rho_C^{(S)} \le \rho_{C^\star}^{(S)}$, with equality if and only if $C = C^\star$. Hence $\rho^{(S)} = 1$ if and only if C is the Fréchet-Hoeffding upper bound, and $\rho^{(S)} = -1$ if and only if C is the Fréchet-Hoeffding lower bound.*

Remark 8.5.2 *It has been proven by Daniels [1950] that $\left|3\tau - 2\rho^{(S)}\right| \le 1$.*

8.5.4.1 Estimation of Spearman Rho

Because of its definition, Spearman rho $\rho^{(S)}$ is naturally estimated by correlation between the pairs of ranks (R_{11}, R_{12}), ..., (R_{n1}, R_{n2}), i.e.,

$$\begin{aligned}
\hat\rho^{(S)} &= \rho_n^{(S)} = \frac{12}{n(n^2-1)} \sum_{i=1}^n \left(R_{i1} - \frac{n+1}{2}\right)\left(R_{i2} - \frac{n+1}{2}\right) \\
&= \frac{12}{n(n^2-1)} \sum_{i=1}^n R_{i1} R_{i2} - 3\frac{n+1}{n-1} \\
&= 1 - \frac{6}{n(n^2-1)} \sum_{i=1}^n (R_{i1} - R_{i2})^2.
\end{aligned}$$

Remark 8.5.3 *As for the estimation of Kendall tau, $\hat\rho^{(S)}$ depends only on the ranks. It is easy to see that $\hat\rho^{(S)} = 1$ if and only if $R_{i1} = R_{i2}$ for all $i \in \{1, \ldots, n\}$. Similarly, $\hat\rho^{(S)} = -1$ if and only if $R_{i1} = n + 1 - R_{i2}$ for all $i \in \{1, \ldots, n\}$.*

8.5.4.2 Precision of the Estimation of Spearman Rho

The following proposition shows that $\hat\rho^{(S)}$ is a good estimator of $\rho^{(S)}$.

Proposition 8.5.4 *For any copula C, we have*

$$\sqrt{n}\left(\rho_n^{(S)} - \rho^{(S)}\right)/\sigma_n \rightsquigarrow N(0,1),$$

where σ_n is the standard deviation of the pseudo-observations

$$12\left\{(\hat U_{i1} - 1/2)(\hat U_{i2} - 1/2) + \hat W_{i1} + \hat W_{i2}\right\}, \quad i \in \{1, \ldots, n\},$$

with

$$\hat W_{i1} = \frac{1}{n} \sum_{k=1}^n \{\mathbb{I}(\hat U_{i1} \le \hat U_{k1}) - \hat U_{k1}\}(\hat U_{k2} - 1/2)$$

and

$$\hat{W}_{i2} = \frac{1}{n}\sum_{k=1}^{n}\{\mathbb{I}(\hat{U}_{i2} \le \hat{U}_{k2}) - \hat{U}_{k2}\}(\hat{U}_{k1} - 1/2).$$

In addition, if $C = C_\perp$, *then* $\sqrt{n}\,\rho_n^{(S)} \rightsquigarrow N(0,1)$.

The proof of the result follows from an application of U-statistics or directly from Ghoudi and Rémillard [2004].

Example 8.5.6 *For the data of Table 8.6, we have*

$$\sum_{i=1}^{n}(R_{i1} - R_{i2})^2 = 6,$$

so $\hat{\rho}^{(S)} = 7/10$. *Also* $\sigma_n = 0.3211$.

8.5.5 van der Waerden Rho

Recall that $\mathcal{N}$ is the distribution function of the standard Gaussian. The van der Waerden coefficient $\rho^{(W)}$ measures the correlation between the standard Gaussian variables $\mathcal{N}^{-1}\{F_1(X_1)\} = \mathcal{N}^{-1}(U_1)$ and $\mathcal{N}^{-1}\{F_2(X_2)\} = \mathcal{N}^{-1}(U_2)$, i.e.,

$$\rho^{(W)} = \mathrm{Cor}\left\{\mathcal{N}^{-1}(U_1), \mathcal{N}^{-1}(U_2)\right\} = E\left\{\mathcal{N}^{-1}(U_1)\mathcal{N}^{-1}(U_2)\right\}.$$

Note that in general the joint distribution of $\mathcal{N}^{-1}(U_1)$ and $\mathcal{N}^{-1}(U_2)$ is not Gaussian. We have $\rho^{(W)} = 1$ for the Fréchet-Hoeffding upper bound, $\rho^{(W)} = -1$ for the Fréchet-Hoeffding lower bound, and $\rho^{(W)} = 0$ for $C = C_\perp$.

Independence vs van der Waerden Rho

We can have $\rho^{(W)} = 0$ without X_1, X_2 being independent. As before, the counter example comes from the tent map copula C_1. By Proposition 8.4.1, we known that $\rho^{(W)} = 0$.

Using Hoeffding equality (8.56) in Lemma 8.F.1, we can write $\rho^{(W)}$ directly in terms of the copula C viz.

$$\rho^{(W)} = \int_{\mathbb{R}^2}\left[C\left\{\mathcal{N}(x_1), \mathcal{N}(x_2)\right\} - \mathcal{N}(x_1)\mathcal{N}(x_2)\right] dx_1 dx_2.$$

This representation enables us to derive the following monotonicity property of van der Waerden coefficient with respect to the PQD order.

Proposition 8.5.5 *If* $C \preceq_{PQD} C^\star$ *then* $\rho_C^{(W)} \le \rho_{C^\star}^{(W)}$, *with equality if and only if* $C = C^\star$. *Hence* $\rho^{(W)} = 1$ *if and only if* C *is the Fréchet-Hoeffding upper bound, and* $\rho^{(W)} = -1$ *if and only if* C *is the Fréchet-Hoeffding lower bound.*

8.5.5.1 Estimation of van der Waerden Rho

The van der Waerden rho is estimated by the correlation between the pairs $(\hat{Z}_{i1}, \hat{Z}_{i2})$, where $\hat{Z}_{ij} = \mathcal{N}^{-1}(\hat{U}_{ij}) = \mathcal{N}^{-1}\left(\frac{R_{ij}}{n+1}\right)$, $j = 1, 2$, $i \in \{1, \ldots, n\}$. Its expression is given by

$$\hat{\rho}^{(W)} = \rho_n^{(W)} = \frac{\sum_{i=1}^n \hat{Z}_{i1}\hat{Z}_{i2}}{\sum_{i=1}^n \hat{Z}_{i1}^2}, \tag{8.23}$$

since $\sum_{i=1}^n \hat{Z}_{ij} = \sum_{i=1}^n \mathcal{N}^{-1}\left(\frac{i}{n+1}\right) = 0$ and

$$\sum_{i=1}^n \hat{Z}_{ij}^2 = \sum_{i=1}^n \left\{\mathcal{N}^{-1}\left(\frac{i}{n+1}\right)\right\}^2.$$

8.5.5.2 Precision of the Estimation of van der Waerden Rho

The following proposition shows that $\hat{\rho}^{(W)}$ is a good estimator of $\rho^{(W)}$.

Proposition 8.5.6 *For any copula C, we have*

$$\sqrt{n}\left(\rho_n^{(W)} - \rho^{(W)}\right)/\sigma_n \rightsquigarrow N(0,1),$$

where σ_n is the standard deviation of the pseudo-observations

$$\hat{Z}_{i1}\hat{Z}_{i2} + \hat{W}_{i1} + \hat{W}_{i2}, \quad i \in \{1, \ldots, n\},$$

with

$$\begin{aligned}
\hat{W}_{i1} &= \frac{1}{n}\sum_{k=1}^n \{\mathbb{I}(\hat{U}_{i1} \le \hat{U}_{k1}) - \hat{U}_{k1}\}\hat{Z}_{k2}/\phi(\hat{Z}_{k1}), \\
\hat{W}_{i2} &= \frac{1}{n}\sum_{k=1}^n \{\mathbb{I}(\hat{U}_{i2} \le \hat{U}_{k2}) - \hat{U}_{k2}\}\hat{Z}_{k1}/\phi(\hat{Z}_{k2}),
\end{aligned}$$

and ϕ is the standard Gaussian density function. Moreover, if $C = C_\perp$, $\sqrt{n}\,\rho_n^{(W)} \rightsquigarrow N(0,1)$.

The proof of the result follows from an application of U-statistics or from Ghoudi and Rémillard [2004].

Example 8.5.7 *For the data of Table 8.6, we find $\hat{\rho}^{(W)} = 0.6858$ and $\sigma_n =$ 0.3415.*

8.5.6 Other Measures of Dependence

Several other measures of dependence can be written as

$$\rho_C^{(J)} = E\left\{J^{-1}(U_1)J^{-1}(U_2)\right\}, \quad (U_1, U_2) \sim C,$$

where J is a distribution function of a random variable which is symmetric about 0, with variance 1. For these dependence measures, it is possible to find the best test of independence to detect a particular family on alternative hypotheses, e.g., Genest and Verret [2005], Genest et al. [2006b]. Using Hoeffding equality (8.56), we have

$$\rho_C^{(J)} = \int_{\mathbb{R}^2} \left[C\left\{J(x_1), J(x_2)\right\} - J(x_1)J(x_2)\right] dx_1 dx_2.$$

It is then easy to check that $\rho_C^{(J)}$ is monotone with respect to the PQD order, i.e., if $C \preceq_{PQD} C^*$, then $\rho_C^{(J)} \le \rho_{C^*}^{(J)}$, with equality if and only if $C = C^*$, as long as $\text{Image}(J) = [0,1]$. Note that since $J(u)J(1-u) = -J^2(u)$, we have $-1 = -E\left\{J^2(U)\right\} = \rho_{C_-}^{(J)}$, and $\rho_{C_+}^{(J)} = E\left\{J^2(U)\right\} = 1$. Also, for the independence copula $C_\perp$, $\rho_{C_\perp}^{(J)} = 0$.

Independence vs Measures of Dependence

Again $\rho^{(J)} = 0$ does not imply that X_1, X_2 are independent. For a counter example, take the tent map copula C_1 and apply Proposition 8.4.1 to obtain $\rho^{(J)} = 0$.

Both Spearman rho and van der Waerden rho are special cases of $\rho^{(J)}$. For example, for Spearman rho, take $J^{-1}(u) = \sqrt{12}(u - 1/2)$, corresponding to the uniform on $[-\sqrt{3}, \sqrt{3}]$, while for van der Waerden rho, take $J = \mathcal{N}^{-1}$, corresponding to the standard Gaussian distribution.

8.5.6.1 Estimation of $\rho^{(J)}$

$\rho^{(J)}$ can be estimated by the correlation between the pairs $(\hat{Y}_{i1}, \hat{Y}_{i2})$, where $\hat{Y}_{ij} = J^{-1}\left(\frac{R_{ij}}{n+1}\right)$, $j = 1, 2$ and $i \in \{1, \ldots, n\}$, i.e.,

$$\hat{\rho}^{(J)} = \sum_{i=1}^n \hat{Y}_{i1}\hat{Y}_{i2} \Big/ \sum_{i=1}^n \left(\hat{Y}_{i1}\right)^2.$$

Since for all $j = 1, 2$,

$$\frac{1}{n}\sum_{i=1}^n \hat{Y}_{ij} = \sum_{i=1}^n J^{-1}\left(\frac{R_i}{n+1}\right) = \sum_{i=1}^n J^{-1}\left(\frac{i}{n+1}\right) = 0,$$

and

$$\sum_{i=1}^{n}\left(\hat{Y}_{ij}\right)^2 = \sum_{i=1}^{n}\left\{J^{-1}\left(\frac{i}{n+1}\right)\right\}^2,$$

it follows that $\hat{\rho}^{(J)} \in [-1,1]$, with $\hat{\rho}^{(J)} = 1$ if and only if $R_{i1} = R_{i2}$ for all $i \in \{1,\ldots,n\}$, and $\hat{\rho}^{(J)} = -1$ if and only if $R_{i1} = n+1-R_{i2}$ for all $i \in \{1,\ldots,n\}$.

8.5.6.2 Precision of the Estimation of $\rho^{(J)}$

The next proposition shows that $\hat{\rho}^{(J)}$ is a good estimator of $\rho^{(J)}$.

Proposition 8.5.7 *For any copula C, we have*

$$\sqrt{n}\left(\rho_n^{(J)} - \rho^{(J)}\right)/\sigma_n \rightsquigarrow N(0,1),$$

where σ_n is the standard deviation of the pseudo-observations

$$\hat{Y}_{i1}\hat{Y}_{i2} + \hat{W}_{i1} + \hat{W}_{i2}, \quad i \in \{1,\ldots,n\},$$

where $\hat{Y}_{ij} = J^{-1}(\hat{U}_{ij})$,

$$\hat{W}_{i1} = \frac{1}{n}\sum_{k=1}^{n}\{\mathbb{I}(\hat{U}_{i1} \le \hat{U}_{k1}) - \hat{U}_{k1}\}\hat{Y}_{k2}/J'(\hat{Y}_{k1})$$

and

$$\hat{W}_{i2} = \frac{1}{n}\sum_{k=1}^{n}\{\mathbb{I}(\hat{U}_{i2} \le \hat{U}_{k2}) - \hat{U}_{k2}\}\hat{Y}_{k1}/J'(\hat{Y}_{k2}).$$

Moreover, under the null hypothesis of independence, we have $\sqrt{n}\rho_n^{(J)} \rightsquigarrow N(0,1)$.

8.5.7 Serial Dependence

Instead of measuring or modeling the dependence between two variables, one can be interested in the dependence between $(X_i, X_{i+\ell})$ for a time series X. In this case, we talk about serial dependence. Replacing X by its series of ranks $R_1,\ldots,R_n$, we can define

$$\hat{\rho}_\ell^{(J)} = \frac{1}{n}\sum_{i=1}^{n} J^{-1}\left(\frac{R_{i+\ell}}{n+1}\right)J^{-1}\left(\frac{R_i}{n+1}\right),$$

where $R_{n+i} = R_i$, for $1 \in \{1,\ldots,\ell\}$. See Genest and Rémillard [2004] and references therein. Under serial independence, the behavior is exactly the same as in the previous setting. See also Rémillard et al. [2012] for modeling dependence in time series using copulas.

8.6 Multivariate Copulas

In many financial applications, there are often more than two assets to model, so it is worth defining a copula in a multidimensional setting.

Definition 8.6.1 *A copula is a distribution function on $[0,1]^d$ with uniform margins. Hence $C : [0,1]^d \mapsto [0,1]$ is a copula if and only if there exist random variables $U_1, \ldots, U_d \sim \mathrm{Unif}(0,1)$, such that for all $u = (u_1, \ldots, u_d) \in [0,1]^d$,*

$$C(u_1, \ldots, u_d) = P(U_1 \leq u_1, \ldots, U_d \leq u_d).$$

C is the law of the random vector $U = (U_1, \ldots, U_d)$, denoted by $U \sim C$, and we write $C(u) = P(U \leq u)$, $u \in [0,1]^d$.

One can now state Sklar theorem in a multivariate context.

Theorem 8.6.1 (Sklar [1959]) *For any distribution function H of a random vector $X = (X_1, \ldots, X_d)$, with margins $F_1, \ldots, F_d$, there exists a copula C such that*

$$H(x_1, \ldots, x_d) = C\{F_1(x_1), \ldots, F_d(x_d)\}. \tag{8.24}$$

The copula is unique when restricted to

$$\overline{Image(F_1)} \times \cdots \times \overline{Image(F_d)} \subset [0,1]^d.$$

Remark 8.6.1 *If the margins are continuous, then the copula is unique and it is the distribution function of*

$$U = (U_1, \ldots, U_d), \quad U_i = F_i(X_i), \quad i = 1, \ldots, d.$$

In addition, for any $u \in (0,1)^d$, we have

$$C(u_1, \ldots, u_d) = H\left\{F_1^{-1}(u_1), \ldots, F_d^{-1}(u_d)\right\}, \tag{8.25}$$

where

$$F_j^{-1}(u_j) = \inf\{x_j \in \mathbb{R}; F_j(x_j) \geq u_j\}, \quad j \in \{1, \ldots, d\}.$$

Proposition 8.6.1 *Suppose that $X \sim H$ with margins $F_1, \ldots, F_d$ having densities $f_1, \ldots, f_d$ respectively, and copula C. Then H has a density h if and only if C has a density c. In either case, for all $x = (x_1, \ldots, x_d) \in \mathbb{R}^d$, we have*

$$h(x) = c\{F_1(x_1), \ldots, F_d(x_d)\} \prod_{j=1}^{d} f_j(x_j). \tag{8.26}$$

Also, for all $u = (u_1, \ldots, u_d) \in (0,1)^d$,

$$c(u) = \frac{h\left\{F_1^{-1}(u_1), \ldots, F_d^{-1}(u_d)\right\}}{\prod_{j=1}^{d} f_j\left\{F_j^{-1}(u_j)\right\}}. \tag{8.27}$$

One can now generalize the Kendall function to a multivariate setting.

8.6.1 Kendall Function

Suppose that $X \sim H$, with continuous margins $F_1, \ldots, F_d$, and copula C. Set $U = (U_1, \ldots, U_d)$, where $U_j = F_j(X_j)$, $j \in \{1, \ldots, d\}$. Further set

$$V = H(X) = C\{F_1(X_1), \ldots, F_d(X_d)\} = C(U).$$

The Kendall function K is the distribution function of V, i.e.,

$$K(t) = P(V \le t), \quad t \in [0,1].$$

8.6.2 Conditional Distributions

In many applications of copulas we have to compute conditional distributions of the form

$$P_d(u) = P(U_d \le u_d | U_1 = u_1, \ldots, U_{d-1} = u_{d-1}),$$

where $U \sim C$. The following proposition shows how to compute these conditional probabilities.

Proposition 8.6.2 *If $X \sim H$ with continuous margins $F_1, \ldots, F_d$ and copula C with density c, then*

$$\begin{aligned} P_d(u) &= \frac{\partial_{u_1} \cdots \partial_{u_{d-1}} C(u_1, \ldots, u_{d-1}, u_d)}{\partial_{u_1} \cdots \partial_{u_{d-1}} C(u_1, \ldots, u_{d-1}, \ \ 1)} \\ &= \int_0^{u_d} c(u_1, \ldots, u_{d-1}, s) ds \Big/ \int_0^1 c(u_1, \ldots, u_{d-1}, s) ds. \end{aligned}$$

Also,

$$P(X_d \le x_d | X_1 = x_1, \ldots, X_{d-1} = x_{d-1}) = P_d\{F_1(x_1), \ldots, F_d(x_d)\}.$$

Being able to compute conditional distributions, we can now adapt the Rosenblatt theorem (Theorem A.7.1) for copulas.

Theorem 8.6.2 (Rosenblatt) *Set $W_1 = P_1(U_1) = U_1$, $W_2 = P_2(U_1, U_2)$, $\ldots$, $W_d = P_d(U_1, \ldots, U_d)$. Then $W = \psi(U)$ is an invertible mapping called Rosenblatt transform and $W \sim C_\perp$.*

8.6.2.1 Applications of Theorem 8.6.2

As a first application of the Rosenblatt theorem, consider the problem of simulating $U \sim C$. To do so, first simulate i.i.d. uniform variables $W_1, \ldots, W_d$, i.e., $W = (W_1, \ldots, W_d) \sim C_\perp$. Then set $U = \psi^{-1}(W)$. It follows that $U \sim C$.

Another application is for goodness-of-fit. It is sometimes easier to test $H_0 : W \sim C_\perp$ than to test $H_0 : U \sim C$.

Remark 8.6.2 *Note that for almost all families that we will see in Section 8.7, the Rosenblatt transform is usually easy to compute.*

8.6.3 Stochastic Orders for Dependence

There exist several order relations to compare the strength of dependence between components of two random vector X and Y when they have the same margins. This amounts to defining an order on the associated copulas. For example, the PQD order can be extended in a multivariate setting by defining $X \preceq_{PQD} Y$ if for all $u \in [0,1]^d$,

$$C_X(u) \le C_Y(u).$$

The dependence is said to be positive if $C_\perp \preceq_{PQD} C$ and negative if $C \preceq_{PDQ} C_\perp$. Here, $C_\perp$ is the independence copula defined for all $u = (u_1, \ldots, u_d) \in [0,1]^d$ by $C_\perp(u) = \prod_{j=1}^d u_j$. As seen in Section 8.5 for the case $d = 2$, the classical measures of dependence are ordered with respect to PQD order, i.e., if ρ is such a measure, then

$$X \preceq_{PQD} Y \text{ implies } \rho(X) \le \rho(Y).$$

It has been shown recently that in the case $d = 2$, many families of elliptical copulas (copulas defined from elliptical distributions) are ordered with respect to different order relations of interest. This is also true for many families of Archimedean copulas; see, e.g., Example 8.7.1.

A family is said to be positively (respectively negatively) ordered with respect to the PQD order if $C_\theta \preceq_{PQD} C_{\theta'}$ whenever $\theta \le \theta'$ (respectively $\theta' \le \theta$) componentwise.

For a copula family C_θ depending on a one-dimensional parameter θ in some interval $\mathcal{O}$, there is an easy way to find out when the family is ordered, positively or negatively.

Proposition 8.6.3 *Suppose that C_θ is differentiable with respect to θ. Then the family is positively ordered if and only if for every $\theta \in \mathcal{O}$, $\partial_\theta C_\theta(u) \ge 0$ for all $u \in (0,1)^d$. It is negatively ordered if and only if for every $\theta \in \mathcal{O}$, $\partial_\theta C_\theta(u) \le 0$ for all $u \in (0,1)^d$.*

We can now define the Fréchet-Hoeffding bounds.

8.6.3.1 Fréchet-Hoeffding Bounds

Definition 8.6.2 *The Fréchet-Hoeffding upper bound C_+ is the copula defined by*

$$C_+(u) = \min_{j \in \{1,\ldots,d\}} u_j, \quad u = (u_1, \ldots, u_d) \in [0,1]^d.$$

Its corresponds to the distribution function of $U = (U_1, \ldots, U_1)$, with $U_1 \sim$ Unif$(0,1)$. The Fréchet-Hoeffding lower bound C_- is the function defined by

$$C_-(u) = \max(u_1 + \cdots + u_d - d + 1, 0), \quad u = (u_1, \ldots, u_d) \in [0,1]^d.$$

C_- is not a copula unless $d = 2$. However, for any given $u \in [0,1]^d$, there exists a copula C so that $C(u) = C_-(u)$.

The following result extends Proposition 8.4.5.

Proposition 8.6.4 *For any copula C and any $u \in [0,1]^d$,*

$$C_-(u) \le C(u) \le C_+(u).$$

8.6.3.2 Application

Suppose that $X_1, \ldots, X_d$ have continuous margins $F_1, \ldots, F_d$ and let G be the distribution function of $Y = X_1 + \cdots + X_d$. For any $x = (x_1, \ldots, x_d)$, set

$$S_d(x) = S_d(x_1, \ldots, x_d) = F_1(x_1) + \cdots + F_d(x_d).$$

Then it follows from Proposition 8.6.4 that for all $y \in \mathbb{R}$,

$$\begin{aligned} G^-(y) &= \sup_{x; x_1+\cdots+x_d=y} \max\{0, S_d(x) - d + 1\} \\ &\le G(y) \le \inf_{x; x_1+\cdots+x_d=y} \min\{S_d(x), 1\} = G^+(y). \end{aligned}$$

8.6.3.3 Supermodular Order

An important order in applications is the supermodular order.

Definition 8.6.3 $C \preceq_{\mathrm{SM}} C^*$ *if and only if for random vectors $X \sim C$ and $Y \sim C^*$, we have*

$$\mathrm{E}\{\varphi(X)\} \le \mathrm{E}\{\varphi(Y)\}$$

for any supermodular function[4] *for which the expectations exist.*

As an application, consider a random vector $X = (X_1, \ldots, X_d)$ representing losses and let $Y = X_1 + \cdots + X_d$ be the loss of the associated portfolio. The stop-loss premium, a risk measure often used in Actuarial Science, is defined by

$$\pi(Y, x) = \mathrm{E}\{\max(0, Y - x)\}, \quad x \ge 0.$$

Given losses with the same marginal distributions but with different dependence structures C and C^* for X, Müller [1997] showed that stop-loss premia are ordered with respect to the supermodular order, i.e.,

$$\pi_C(Y, x) \le \pi_{C^*}(Y, x) \quad \text{if } C \preceq_{\mathrm{SM}} C^*.$$

[4] φ is supermodular if and only if $\varphi\{\max(x_1, x_2\} + \varphi\{\min(x_1, x_2)\} \ge \varphi(x_1) + \varphi(x_2)$. A function φ with continuous second order derivatives is supermodular if and only if $\partial^2 \varphi(x_1, \ldots, x_p)/\partial x_i \partial x_j \ge 0$ for all $i \ne j$.

8.7 Families of Copulas

There exist many interesting copula families that are already used in financial applications. We start by (re)defining the independence copula in a multivariate setting. Then we will define Archimedean copulas [Genest and MacKay, 1986a] and elliptical copulas [Fang et al., 2002, 2005, Abdous et al., 2005]. Other families will also be introduced.

8.7.1 Independence Copula

The copula $C_\perp$, defined for all $u = (u_1, \ldots, u_n) \in [0,1]^d$ by

$$C_\perp(u) = u_1 \times u_2 \times \cdots \times u_d,$$

is called the independence copula since $X_1, \ldots, X_d$ are independent if and only if, for all $x = (x_1, \ldots, x_d) \in \mathbb{R}^d$,

$$H(x) = C_\perp \{F_1(x_1), \ldots, F_d(x_d)\} = F_1(x_1) \times \cdots \times F_d(x_d).$$

$C_\perp$ is thus the uniform distribution on the hypercube $[0,1]^d$. Its density is

$$c_\perp(u) = 1 \text{ for all } u \in (0,1)^d.$$

Moreover, it is easy to see that the Kendall function K is

$$K(t) = t \sum_{j=0}^{d-1} \frac{\{-\ln(t)\}^j}{j!}, \quad t \in [0,1], \tag{8.28}$$

since $-\ln(V) = -\sum_{j=1}^{d} \ln(U_j) \sim \text{Gamma}(d, 1)$.

8.7.2 Elliptical Copulas

Suppose that Y has an elliptical distribution, i.e., $Y \sim \mathcal{E}(\mathcal{R}, \mu, \Sigma)$, and there exist $\mu \in \mathbb{R}^d$, a positive random variable $\mathcal{R}$ independent of $S \sim \text{Unif}(\mathcal{S}_d)$ such that

$$Y = \mu + \mathcal{R}^{1/2} A^\top S,$$

with $\Sigma = A^\top A$. If $\mathcal{R}$ has a density, then we say that Y has an elliptical distribution with generator g and parameters μ, Σ, denoted $Y \sim \mathcal{E}(g, \mu, \Sigma)$, if the density h of Y is given by

$$h(y) = \frac{1}{|\Sigma|^{1/2}} g\left\{(y-\mu)^\top \Sigma^{-1} (y-\mu)\right\},$$

where $\frac{\pi^{d/2}}{\Gamma(d/2)} r^{(d-2)/2} g(r)$ is the density of $\mathcal{R}$. For more details on elliptical distributions, see Appendix A.6.20.

An elliptical copula is a copula associated with an elliptical distribution. It follows from Proposition A.6.7 that the copula depends only on $\mathcal{R}$ (up to constant) and the correlation matrix ρ derived from Σ, i.e., $\rho_{ij} = \Sigma_{ij}/\sqrt{\Sigma_{ii}\Sigma_{jj}}$, $1 \le i,j \le d$.

8.7.2.1 Estimation of ρ

To estimate ρ, one can rely on the following interesting result.

Proposition 8.7.1 (Fang et al. [2002]) *If $X \sim \mathcal{E}(\mathcal{R}, \mu, \rho)$, then $\tau(X_i, X_j) = \frac{2}{\pi}\arcsin(\rho_{ij})$. In particular, τ does not depend on $\mathcal{R}$, it depends only on ρ.*

One could be tempted to estimate ρ by R, where $R_{jk} = \sin\left(\frac{\pi}{2}\hat{\tau}_{jk}\right)$, $j,k \in \{1,\ldots,d\}$, based on Proposition 8.7.1. According to Proposition 8.5.2 and using the delta method (Theorem B.3.1), if the sample size n is large enough, then R will be positive definite and in fact the joint distribution of the estimation errors will be asymptotically Gaussian. However, for a given sample n, R may not be positive definite. To circumvent this problem, one can proceed in the following way:

- Diagonalize $RR^\top = R^2$, obtaining $R^2 = M\Delta M^\top$.
- Set $B = M\Delta^{1/2}M^\top$. This is possible since the eigenvalues of R^2, i.e., the elements of the diagonal matrix Δ are non negative;
- Finally set $\hat{\rho} = \delta^{-1}B\delta^{-1}$, where δ is the diagonal matrix with entries $\delta_{jj} = \sqrt{B_{jj}}$, $j \in \{1,\ldots,d\}$.

We are now in a position to define some elliptical copulas.

8.7.3 Gaussian Copula

The Gaussian distribution is a particular case of an elliptical distribution for which the generator g is given by $g(r) = \frac{e^{-r/2}}{(2\pi)^{d/2}}$. Note also that $\mathcal{R} \sim \chi^2(d)$ and if $X \sim N_d(0,I)$, then $S = X/\|X\|$ is independent of $\mathcal{R} = \|X\|^2$ and S is uniformly distributed on the sphere $\mathcal{S}_d$. It follows from Proposition A.6.7 that the associated copula, called the Gaussian copula, only depends on ρ. Denote it by C_ρ. Using formula (8.27), the density c_ρ is given by

$$c_\rho(u) = \frac{1}{|\rho|^{1/2}} e^{-\frac{1}{2}\eta^\top(\rho^{-1}-I)\eta} \tag{8.29}$$

for all $u = (u_1,\ldots,u_d) \in (0,1)^d$, where $\eta_j = \mathcal{N}^{-1}(u_j)$, $j \in \{1,\ldots,d\}$, and

$$C_\rho(u) = \mathcal{N}_d\left\{\mathcal{N}^{-1}(u_1),\ldots,\mathcal{N}^{-1}(u_d),\rho\right\}, \tag{8.30}$$

where $\mathcal{N}_d(y,\rho) = P(Z \le y)$, with $Z \sim N(0,\rho)$. For the Gaussian copula C_ρ, it can be shown that

$$\rho^{(S)}(X_i,X_j) = \frac{6}{\pi}\arcsin(\rho_{ij}/2), \quad i,j \in \{1,\ldots,d\}.$$

In addition, since $\mathcal{N}^{-1}(U_i) \sim N(0,1)$, we have

$$\rho_{ij} = \rho_{ij}^{(W)}, \quad i,j \in \{1,\dots,d\}.$$

For the Gaussian copula, ρ can be interpreted as the matrix of van der Waerden coefficients.

8.7.3.1 Simulation of Observations from a Gaussian Copula

To generate $U \sim C_\rho$, first simulate $Z \sim N(0,\rho)$ and set

$$U = (U_1,\dots,U_d) \quad \text{with} \quad U_j = \mathcal{N}(Z_j), \quad j \in \{1,\dots,d\}.$$

Note that it is exactly the approach described in Section 8.2.4.1.

Remark 8.7.1 *When $d=2$, the density of $\mathcal{R}$ is $\frac{1}{2}e^{-r/2}$ and we can simulate $\mathcal{R}$ by setting $\mathcal{R} = -2\ln(V)$, with $V \sim \text{Unif}(0,1)$. If U and V are independent, then*

$$X = \mathcal{R}^{1/2} S = \left(\sqrt{-2\ln(V)}\cos(2\pi U), \sqrt{-2\ln(V)}\sin(2\pi U)\right)$$

has distribution $N_2(0,I)$. This is the well-known Box-Müller method for generating Gaussian observations.

8.7.4 Student Copula

The multivariate Student distribution, defined in Appendix A.6.19, is another particular case of an elliptical distribution. Since its density is given by (A.14), we have that $X \sim \mathcal{E}(g,0,\Sigma)$, where the generator g is given by

$$g(r) = \frac{\Gamma\left((d+\nu)/2\right)}{\Gamma\left(\nu/2\right)} \frac{(1+r/\nu)^{-(d+\nu)/2}}{(\pi\nu)^{d/2}}.$$

If we set $W = \mathcal{R}/(\mathcal{R}+\nu)$, then $W \sim \text{Beta}(d/2,\nu/2)$. Using Proposition A.6.8, we may conclude that the margins are also Student with ν degrees of freedom. It follows from Proposition A.6.7 that the associated copula, called the Student copula and denoted $C_{\nu,\rho}$, depends only on the parameters ν and ρ. In addition, using formulas (8.27) and (A.14), the density $c_{\nu,\rho}$ is given, for all $u = (u_1,\dots,u_d) \in (0,1)^d$, by

$$c_{\nu,\rho}(u) = \frac{k_d}{|\rho|^{1/2}} \frac{\left(1+\eta^\top \rho^{-1}\eta/\nu\right)^{-(\nu+d)/2}}{\prod_{i=1}^d \left(1+\eta_i^2/\nu\right)^{-(\nu+1)/2}}, \tag{8.31}$$

where $\eta_j = T_\nu^{-1}(u_j)$, $j \in \{1,\dots,d\}$, and $k_d = \frac{\Gamma((\nu+d)/2)}{\Gamma(\nu/2)} \left(\frac{\Gamma(\nu/2)}{\Gamma((\nu+1)/2)}\right)^d$.

Remark 8.7.2 *It follows from Proposition A.6.8 that all possible conditional distributions of a multivariate Student distribution are Student, so the associated Rosenblatt transforms are also easy to compute.*

8.7.4.1 Simulation of Observations from a Student Copula

To simulate $U \sim C_{\nu,\rho}$, first simulate X according to formula (A.13), i.e., $X = Z\Big/\sqrt{V/\nu}$, with $Z \sim N_d(0,\rho)$ and $V \sim \chi^2(\nu)$, and set

$$U = (U_1, \ldots, U_d) \quad \text{with} \quad U_j = T_\nu(X_j), \quad j \in \{1, \ldots, d\},$$

where T_ν is the distribution function of a univariate Student distribution with ν degrees of freedom. An illustration of such a simulation is provided in Figure 8.13.

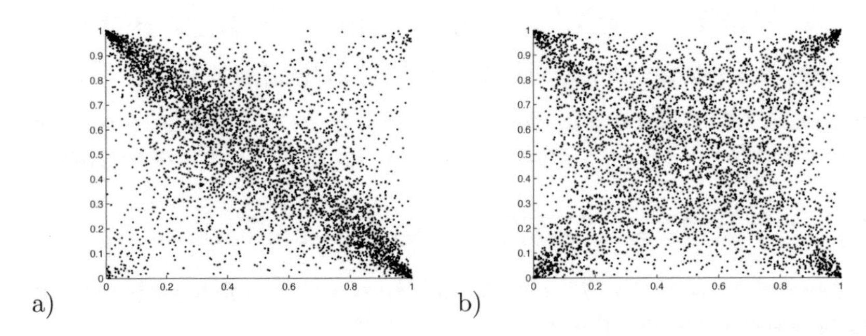

FIGURE 8.13: Graph of 5000 pairs of points from a Student copula with $\nu = 1$ degree of freedom (Cauchy copula) with $\tau = -0.5$ (panel a) and $\tau = 0$ (panel b).

8.7.5 Other Elliptical Copulas

Two other popular elliptical distributions are the Pearson type II and type VII distributions, whose generators are given in Table A.1. Recall that for the Pearson type VII, the case $\alpha = \nu/2$ corresponds to the multivariate Student distribution, while the case $\alpha = 1/2$ and $\nu = 1$ corresponds to the multivariate Cauchy.

Pearson Type VII Copulas

The parameter ν of the Pearson type VII distribution is a scaling parameter. Therefore, all copulas associated with the Pearson type VII distributions are Student copulas.

We now define a very general class of copula families introduced by Genest and MacKay [1986a,b].

8.7.6 Archimedean Copulas

Definition 8.7.1 *ϕ is the generator of a d-dimensional copula if*

(i) $\phi : [0,1] \mapsto [0, \phi(0)]$ is decreasing, with $\phi(1) = 0$, and possibly $\phi(0) = +\infty$;

(ii) for all $0 < s < \phi(0)$ and for all $1 \leq j \leq d$,

$$(-1)^j \frac{\partial^j}{\partial s^j} \phi^{-1}(s) > 0.$$

Set $\mathcal{D}_\phi = \{u = (u_1, \ldots, u_d) \in (0,1]^d; \phi(u_1) + \cdots + \phi(u_d) < \phi(0)\}$.

The Archimedean copula generated by ϕ is defined by

$$C(u) = \begin{cases} \phi^{-1}\{\phi(u_1) + \cdots + \phi(u_d)\}, & \text{if } u \in \mathcal{D}_\phi, \\ 0, & \text{otherwise.} \end{cases}$$

Remark 8.7.3 *Note that the generator associated with a copula is unique up to a multiplying factor since ϕ and $\lambda\phi$ generate the same copula for any $\lambda > 0$.*

8.7.6.1 Financial Modeling

For applications in Finance, Archimedean copulas do not offer much flexibility since each pair (U_i, U_j) has the same distribution. However the type of dependence modeled by Archimedean copulas is quite different from the dependence modeled by elliptical copulas. See, e.g., the MATLAB script *DemoCopula*. Graphical examples in Figure 8.14 display the differences in modeling the dependence, together with the effect of margins. The strength of dependence is measured by Kendall tau. In those figures, we have chosen $\tau \in \{.1, .5, .9\}$. The margins are Gaussian, Laplace (double exponential), and Pareto, with distribution function $F(x) = 1 - x^{-4}$, $x \geq 1$. Even if they lack flexibility to model multivariate assets, the dependence modeled by Archimedean copulas is easy to interpret, being basically a one-factor model. But most of all, computations are relatively easy and since many Archimedean families have only one parameter, estimation is easier.

8.7.6.2 Recursive Formulas

Based on Barbe et al. [1996], set

$$f_j(t) = (-1)^j \left. \frac{d^j}{ds^j} \phi^{-1}(s) \right|_{s=\phi(t)}, \qquad 0 \leq j \leq d. \tag{8.32}$$

Note that by assumption, $f_j(t) > 0$ for all $t \in (0,1]$ and for all $j \in \{0, \ldots, d\}$. Also, $f_0(t) = t$, $f_1(t) = -1/\phi'(t)$, and for $j \in \{1, \ldots, d\}$,

$$f_j(t) = -\frac{1}{\phi'(t)} f'_{j-1}(t) = f_1(t) f'_{j-1}(t), \quad t \in (0,1]. \tag{8.33}$$

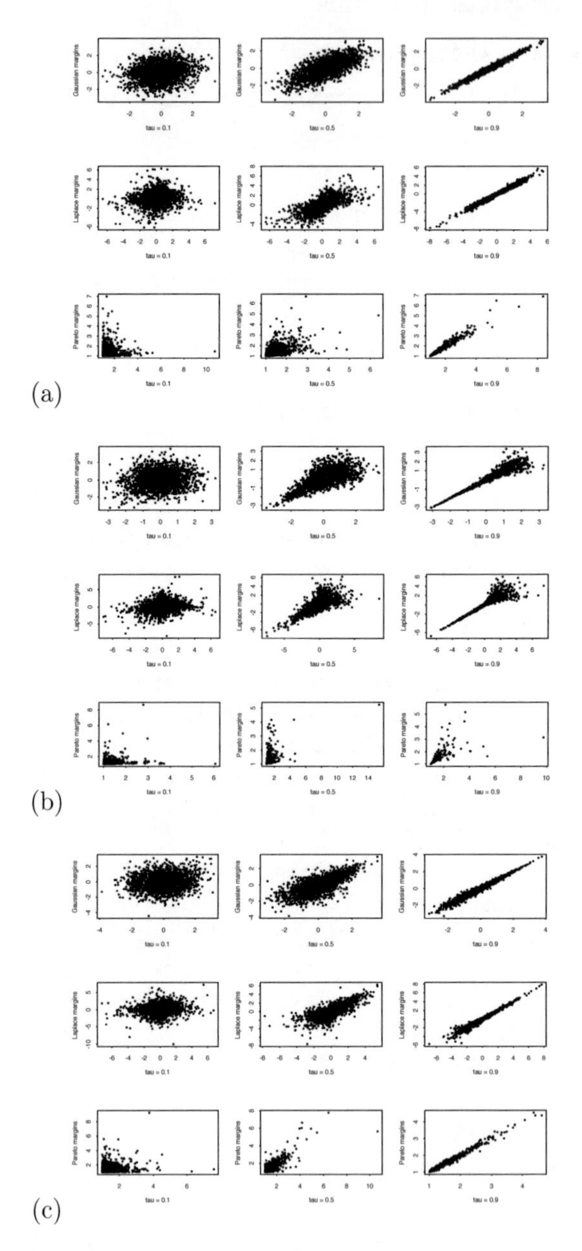

FIGURE 8.14: Gaussian, Laplace, and Pareto margins with (a) Gaussian copula, (b) Clayton copula, (c) Gumbel copula, and $\tau \in \{.1, .5, .9\}$.

It follows that the density of the copula $C = \phi^{-1}\{\phi(u_1) + \cdots + \phi(u_d)\}$ is given by

$$c(u) = f_d\{C(u)\} \prod_{j=1}^{d} \{-\phi'(u_j)\}. \tag{8.34}$$

for all $u = (u_1, \ldots, u_d) \in (0,1)^d$. Similarly, the Kendall function K can be written as

$$K(t) = \sum_{j=0}^{d-1} \frac{\phi^j(t)}{j!} f_j(t), \tag{8.35}$$

and

$$k(t) = K'(t) = -\phi'(t) \frac{\phi^{d-1}(t)}{(d-1)!} f_d(t), \tag{8.36}$$

if $\lim_{t\to 0+} \phi^j(t) f_j(t) = 0$ for all $j \in \{1, \ldots, d-1\}$. Also, the conditional probability P_k can be written as

$$P_k(u_1, \ldots, u_k) = \frac{f_{k-1}\{C_k(u_1, \ldots, u_k)\}}{f_{k-1}\{C_{k-1}(u_1, \ldots, u_{k-1})\}},$$

where $C_j(u_1, \ldots, u_j) = \phi^{-1}\{\phi(u_1) + \cdots + \phi(u_j)\}$ is the copula of dimension j, $j \in \{1, \ldots, d\}$.

8.7.6.3 Conjecture

In the Archimedean case, the Kendall function determine ϕ for $d = 2$. Based on the fact that $y = \phi^{-1}$ satisfies the following differential equation

$$K\{y(s)\} = \sum_{j=0}^{d-1} \frac{(-1)^j}{j!} s^j \frac{\partial^j}{\partial s^j} y(s), \quad s > 0, \quad y(0) = 1,$$

Genest et al. [2006a] conjectured that the result also hold for $d > 2$.

8.7.6.4 Kendall Tau for Archimedean Copulas

In the bivariate case, [Genest and Rivest, 1993] found a way to express the Kendall tau in terms of generator. In fact, they proved that

$$K(t) = t - \frac{\phi(t)}{\phi'(t)}$$

so

$$\tau = 1 + 4 \int_0^1 \frac{\phi(t)}{\phi'(t)} dt.$$

8.7.6.5 Simulation of Observations from an Archimedean Copula

The following proposition follows from the properties of completely monotone functions. See Feller [1971].

Proposition 8.7.2 *Suppose that* $\phi(0) = \infty$. *Then* ϕ *defines a copula for all* $d \geq 2$ *if and only if there exists a positive random variable* S *so that* $E\left(e^{-\lambda S}\right) = \phi^{-1}(\lambda)$ *for all* $\lambda \geq 0$, *i.e.,* ϕ^{-1} *is the Laplace transform of* S.

To simulate such a copula, one can use the following result of Marshall and Olkin [1988].

Proposition 8.7.3 (Marshall and Olkin [1988]) *Suppose that* S *has Laplace transform* ϕ^{-1} *and is independent of* $E_1, \ldots, E_d$, *which are i.i.d, with an exponential distribution with mean* 1, *i.e.,* $P(E_i > t) = e^{-t}$. *Then setting*

$$U_j = \phi^{-1}\left(E_j/S\right), \quad j \in \{1, \ldots, d\}, \tag{8.37}$$

we have $U = (U_1, \ldots, U_d) \sim C$. *Also, the variable* $V = C(U)$ *is generated by* $\phi^{-1}\left(E/S\right)$, *where* $E = \sum_{j=1}^d E_j \sim \text{Gamma}(d, 1)$.

Remark 8.7.4 *For Archimedean families with generator* ϕ *such that* ϕ^{-1} *is a Laplace transform, the dependence in the model is due only to the common factor* S, *as shown by formula* (8.37).

We now look at some important families of Archimedean copulas.

8.7.7 Clayton Family

The Clayton family of copulas has generator $\phi_\theta(t) = \frac{t^{-\theta}-1}{\theta}$, for $\theta \geq -\frac{1}{d-1}$. Then

$$\phi_\theta^{-1}(s) = \begin{cases} (1+\theta s)^{-1/\theta}, & \text{for } s \in [0, -1/\theta], \quad \text{if } -\frac{1}{d-1} \leq \theta < 0, \\ (1+\theta s)^{-1/\theta}, & \text{for } s \geq 0, \quad \text{if } \theta > 0. \end{cases}$$

For $0 > \theta \geq -\frac{1}{d-1}$, we have

$$C_\theta(u) = \left\{\max\left(0, \sum_{i=1}^d u_i^{-\theta} - d + 1\right)\right\}^{-1/\theta},$$

while for $\theta > 0$,

$$C_\theta(u) = \left(\sum_{i=1}^d u_i^{-\theta} - d + 1\right)^{-1/\theta}.$$

Note that $C_0(u) = \lim_{\theta \to 0} C_\theta(u) = C_\perp(u)$, and $C_+(u) = \lim_{\theta \to \infty} C_\theta(u)$, for all $u \in [0,1]^d$. When $d = 2$, C_{-1} is the Fréchet-Hoeffding lower bound. In other words, while $\theta = 0$ yields the independence copula, $\theta = \infty$ returns the upper Fréchet-Hoeffding bound for any $d \geq 2$. Moreover, when $d = 2$, we have $\tau = \frac{\theta}{\theta+2}$, for all $\theta \geq -1$. Finally, using formulas (8.34) and (8.35), we have

$$
\begin{aligned}
f_j(t) &= t^{1+j\theta} \prod_{k=0}^{j-1} (1 + k\theta), \quad j \geq 0, \\
c_d(u) &= \{C(u)\}^{1+d\theta} \left\{ \prod_{j=1}^{d-1} (1 + j\theta) \right\} \left\{ \prod_{j=1}^{d} u_j^{-1-\theta} \right\}, \\
K_d(t) &= t \sum_{k=0}^{d-1} \left(\frac{1 - t^\theta}{\theta} \right)^k \left\{ \prod_{j=1}^{k-1} (1 + j\theta) \right\} \Big/ k!.
\end{aligned}
$$

8.7.7.1 Simulation of Observations from a Clayton Copula

When $\theta > 0$, we can take $\phi(t) = t^{-\theta} - 1$ as the generator of the Clayton copula. In this case, $\phi^{-1}(s) = (1 + s)^{-1/\theta}$ is the Laplace transform of a Gamma$(1/\theta, 1)$. The copula is thus defined for all $d \geq 2$. According to Proposition 8.7.3, to generate $U \sim C_\theta$, we simulate independently $S \sim$ Gamma$(1/\theta, 1)$ and $E_1, \ldots, E_d \sim$ Exp(1). Then we set $U_j = (1 + E_j/S)^{-1/\theta}$, $j \in \{1, \ldots, d\}$. Finally, $0 > \theta \geq -\frac{1}{d-1}$, one has to use the Rosenblatt transform. For example, if $d = 2$, and $-1 < \theta < 0$ or $\theta > 0$, and $(W_1, W_2) \sim C_\perp$, one can set

$$
U_1 = W_1, \quad U_2 = \left(1 + \frac{W_2^{-\theta/(1+\theta)} - 1}{U_1^\theta} \right)^{-1/\theta}.
$$

8.7.8 Gumbel Family

The Gumbel family appears naturally in extreme values models. The generator and its inverse are given by

$$
\phi_\theta(t) = \{-\ln(t)\}^{1/\theta}, \; \phi_\theta^{-1}(s) = e^{-s^\theta}, \quad 0 < \theta \leq 1.
$$

Hence, the associated copula C_θ is given by

$$
C_\theta(u) = \exp\left[- \left\{ \sum_{i=1}^{d} \{-\ln(u_i)\}^{1/\theta} \right\}^\theta \right],
$$

where C_1 is independence copula. Moreover, $C_0(u) = \lim_{\theta \to 0} C_\theta(u) = C_+(u)$ is the Fréchet-Hoeffding upper bound. Note also that $\tau = 1 - \theta$. Finally,

$$
f_j(t) = t \frac{p_{j,\theta}\{-\ln(t)\}}{(-\ln t)^{j/\theta}}, \quad j \geq 0,
$$

where the polynomials $p_{j,\theta}$ are defined by $p_{0,\theta} \equiv 1$ and

$$p_{j+1}(x) = jp_{j,\theta}(x) + \theta x\{p_{j,\theta}(x) - p'_{j,\theta}(x)\}, \qquad j \geq 0.$$

For example, $p_{1,\theta}(x) = \theta x$, $p_{2,\theta}(x) = \theta(1-\theta)x + \theta^2 x^2$, $p_{3,\theta}(x) = \theta(1-\theta)(2-\theta)x + 3\theta^2(1-\theta)x^2 + \theta^3 x^3$, etc. For more details, see Appendix 8.C. Also,

$$K(t) = t\sum_{j=0}^{d-1} \frac{p_{j,\theta}\{-\ln(t)\}}{j!}, \quad t > 0.$$

In the Statistics Toolbox of MATLAB, the Gumbel family is parameterized by $\alpha = 1/\theta$.

8.7.8.1 Simulation of Observations from a Gumbel Copula

To simulate $U \sim C$, we have to simulate S which has a positive stable distribution of index $\theta \in (0,1)$. To do so, simply simulate independently $X \sim \text{Unif}(0,1)$ and $W \sim \text{Exp}(1)$, and set

$$S = \sin(\theta\pi X)\{\sin(\pi X)\}^{-1/\theta} \left\{\frac{\sin((1-\theta)\pi X)}{W}\right\}^{(1-\theta)/\theta}.$$

Then it can be shown [Chambers et al., 1976] that $E\left(e^{-\lambda S}\right) = e^{-\lambda^\theta}$, $\lambda \geq 0$. Next, independently of S, generate independent $E_1, \ldots, E_d \sim \text{Exp}(1)$, and set $U_j = \exp\left\{-\left(E_j/S\right)^\theta\right\}$, $j \in \{1, \ldots, d\}$.

8.7.9 Frank Family

For the Frank family, the generator is given by

$$\phi_\theta(t) = -\ln\left(\frac{1-\theta^t}{1-\theta}\right),$$

with $0 < \theta < \theta_d$, and $\theta_2 = \infty > \theta_3 > \cdots > \theta_\infty = 1$. Hence the copula exists for all dimensions $d \geq 2$ whenever $\theta \leq 1$. The limiting case $\theta = 1$ is the independence copula, while the limiting case $\theta = 0$ corresponds to the Fréchet-Hoeffding upper bound. When $d = 2$, the limiting case $\theta = \infty$ yields the Fréchet-Hoeffding lower bound. In addition

$$\tau(\theta) = \frac{\ln(\theta)^2 + 4\ln(\theta) + 4\text{dilog}(\theta)}{\ln(\theta)^2},$$

where $\text{dilog}(x) = \int_1^x \frac{\ln(t)}{1-t} dt$. Since $\text{dilog}(1/x) = -\text{dilog}(x) - \frac{1}{2}\ln(x)^2$, we have $\tau(1/\theta) = -\tau(\theta)$. If $\theta > 1$,

$$C(u) = \ln\left\{1 + (\theta - 1)\prod_{i=1}^d \left(\frac{\theta^{u_i} - 1}{\theta - 1}\right)\right\} / \ln(\theta),$$

while if $\theta < 1$,

$$C(u) = \ln\left\{1 - (1-\theta)\prod_{i=1}^{d}\left(\frac{1-\theta^{u_i}}{1-\theta}\right)\right\} \Big/ \ln(\theta).$$

In particular, if $d = 2$,

$$C(u) = \ln\left(\frac{\theta + \theta^{u_1+u_2} - \theta^{u_1} - \theta^{u_2}}{\theta - 1}\right) \Big/ \ln(\theta)$$

Next, for $j \geq 1$, we have

$$f_j(t) = p_j\left(\theta^{-t} - 1\right) / \ln(1/\theta),$$

where $p_1(x) = x$ and for all $j \geq 1$,

$$p_{j+1}(x) = x(1+x)p_j'(x).$$

Hence p_j is a polynomial of degree k. For example, $p_2(x) = x(1+x)$, $p_3(x) = x(1+x)(1+2x)$, $p_4(x) = x(1+x)(1+6x+6x^2)$ and $p_5(x) = x(1+x)(1+2x)(1+12x+12x^2)$. If x_d is the largest root of the polynomial $p_d(x)$ in $(-\infty, 0)$, then $\theta_d = 1/(1+x_d)$. For example, $x_2 = -1$ and $\theta_2 = \infty$, $x_3 = -1/2$ and $\theta_3 = 2$, $x_4 = -(3-\sqrt{3})/6$, $\theta_4 = 3 - \sqrt{3}$, $x_5 = -(3-\sqrt{6})/6$, and $\theta_5 = 2(3-\sqrt{6})$. Finally, $\theta_6 \approx 1.0431$. For more details, see Appendix 8.D. The computation of Kendall tau and dilog function can be done with the MATLAB function *tauFrank*.

In the Statistics Toolbox of MATLAB, the Frank family is parameterized by $\alpha = -\ln(\theta)$.

8.7.9.1 Simulation of Observations from a Frank Copula

For $0 < \theta < 1$, $\phi^{-1}(s) = \ln\{1 - (1-\theta)e^{-s}\} / \ln(\theta)$ is the Laplace transform of a discrete random variable S, called the logarithmic series distribution. In fact,

$$\phi^{-1}(s) = \frac{1}{\ln(1/\theta)}\sum_{k=1}^{\infty}(1-\theta)^k \frac{e^{-ks}}{k},$$

so

$$P(S = k) = \frac{1}{\ln(1/\theta)}\frac{(1-\theta)^k}{k}, \qquad k = 1, 2, \ldots.$$

According to Devroye [1986], we can generate S using

$$S = \left\lfloor 1 + \frac{\ln(W_2)}{\ln\left(1 - \theta^{W_1}\right)}\right\rfloor, \tag{8.38}$$

where $W_1, W_2 \sim \text{Unif}(0,1)$ are independent and $\lfloor x \rfloor$ stands for the integer part of x. Then, to simulate $U \sim C$, we simulate independently S according to formula (8.38), and we generate $E_1, \ldots, E_d \sim \text{Exp}(1)$. Finally, we set $U_j = \phi^{-1}(E_j/S) = \ln\left\{1-(1-\theta)e^{-E_j/S}\right\}/\ln(\theta)$, $j \in \{1, \ldots, d\}$.

8.7.10 Ali-Mikhail-Haq Family

The last Archimedean family that we present is the Ali-Mikhail-Haq family, where the generator is given by

$$\phi_\theta(t) = \frac{1}{1-\theta}\ln\left(\frac{1-\theta+\theta t}{t}\right),$$

with parameter $\theta \in [0,1)$. The case $\theta = 0$ corresponds to the independence copula, while the limiting case $\theta = 1$ corresponds to the Clayton copula with parameter 1. In this case, $\tau = 1/3$. Moreover,

$$\tau = \frac{3\theta^2 - 2\theta - 2(1-\theta)^2\ln(1-\theta)}{3\theta^2}.$$

Remark 8.7.5 *This family of copulas suffers from a major limitation since the largest possible value of τ is $1/3$.*

For the Ali-Mikhail-Haq family, we have $f_1(t) = t(1-\theta+\theta t)$ and f_j is a polynomial in t satisfying

$$f_j(t) = t(1-\theta+\theta t)f'_{j-1}(t), \quad j \geq 1, \tag{8.39}$$

according to formula (8.33).

8.7.10.1 Simulation of Observations from an Ali-Mikhail-Haq Copula

For the generator $\phi(t) = \ln\left(\frac{1-\theta+\theta t}{t}\right)$, we find that

$$\phi^{-1}(s) = \frac{1-\theta}{e^s-\theta} = \sum_{k=1}^{\infty}(1-\theta)\theta^{k-1}e^{-sk}$$

which is the Laplace transform of a geometric random variable S with parameter $1-\theta$. That is, $P(S=k) = (1-\theta)\theta^{k-1}$, $k = 1, 2, \ldots$, and S can be generated using

$$S = \left\lfloor 1 + \frac{\ln(W)}{\ln(\theta)} \right\rfloor, \tag{8.40}$$

where $W \sim \text{Unif}(0,1)$. To simulate $U \sim C$, we simulate independently S according to (8.40) and then $E_1, \ldots, E_d \sim \text{Exp}(1)$. Finally we set $U_j = \frac{1-\theta}{e^{E_j/S}-\theta}$, $j \in \{1, \ldots, d\}$.

8.7.11 PQD Order for Archimedean Copula Families

Here, we give a criterion on the generator of an Archimedean family such that the related copulas are ordered with respect to the PQD order.

Theorem 8.7.1 *Let ϕ_θ be the generator of an Archimedean copula family $C_\theta(u) = \phi_\theta^{-1}\{\phi_\theta(u_1) + \cdots + \phi_\theta(u_d)\}$, indexed by $\theta \in \mathcal{O} \subset \mathbb{R}^p$. Then, a sufficient condition for the family to be positively ordered with respect to the PQD order, i.e., $\theta_1 \le \theta_2$ (componentwise) implies $C_{\theta_1} \preceq_{PQD} C_{\theta_2}$, is that the functions*

$$t \mapsto g_{k,\theta}(t) = \partial_{\theta_k} \ln\{-\partial_t \phi_\theta(t)\} = \frac{\partial_{\theta_k}\partial_t \phi_\theta(t)}{\partial_t \phi_\theta(t)} \tag{8.41}$$

are not increasing for every $k \in \{1, \ldots, p\}$ and any θ in the interior of $\mathcal{O}$. The family is negatively ordered with respect to the PQD order if $t \mapsto g_{k,\theta}$ is not decreasing for any $\theta \in \mathcal{O}$ and every $k \in \{1, \ldots, p\}$.

Example 8.7.1 *Consider the Clayton, Frank, Gumbel, and Ali-Mikhail-Haq families. They are all indexed by a one-dimensional parameter, so for simplicity we drop the subscript $k = 1$ from Theorem 8.7.1.*

- *(Clayton) In this case, $\phi_\theta(t) = \frac{t^{-\theta}-1}{\theta}$, so $g_\theta(t) = -\ln(t)$ is decreasing in t. Hence the Clayton family is positively ordered with respect to the PQD order.*
- *(Frank) Here, $\phi_\theta(t) = -\ln\left(\frac{1-\theta^t}{1-\theta}\right)$ so $g_\theta(t) = \frac{t}{\theta(1-\theta^t)} + \frac{1}{\theta\ln(\theta)}$ is increasing in t. Hence the Frank family is negatively ordered with respect to the PQD order.*
- *(Gumbel) In this case, $\phi_\theta(t) = \{-\ln(t)\}^{1/\theta}$ so $g_\theta(t) = -\frac{1}{\theta} - \frac{1}{\theta^2}\ln\{-\ln(t)\}$ is increasing in t. Hence the Gumbel family is also negatively ordered with respect to the PQD order.*
- *(Ali-Mikhail-Haq) Here, $\phi_\theta(t) = \frac{1}{1-\theta}\ln\left(\frac{1-\theta+\theta t}{t}\right)$ so $g_\theta(t) = -\frac{t}{1-\theta+\theta t}$ is decreasing in t. Hence the Ali-Mikhail-Haq family is positively ordered with respect to the PQD order.*

8.7.12 Farlie-Gumbel-Morgenstern Family

This family is neither elliptical nor Archimedean. Its d-dimensional copula is defined for all $u = (u_1, \ldots, u_d) \in [0, 1]^d$ by

$$C_d(u) = \prod_{i=1}^d u_i + \theta \prod_{i=1}^d u_i(1 - u_i), \quad |\theta| \le 1,$$

with density

$$c_d(u) = 1 + \theta \prod_{i=1}^d (1 - 2u_i).$$

We have $C_0 = C_\perp$ and for $d = 2$, we find $\tau = \frac{2}{9}\theta$. Note that when $d > 2$, for any subset of less that d elements, the corresponding variables are independent. In particular, $U_1, \ldots, U_{d-1}$ are independent, even if $U_1, \ldots, U_d$ are dependent when $\theta > 0$. Hence, for a given $d \geq 2$,

$$P(U_d \leq u_d | U_1 = u_1, \ldots, U_{d-1} = u_{d-1}) = u_d \left\{ 1 + \theta(1 - u_d) \prod_{i=1}^{d-1} (1 - 2u_i) \right\},$$

while

$$P(U_j \leq u_j | U_1 = u_1, \ldots, U_{j-1} = u_{j-1}) = u_j, \quad j \in \{1, \ldots, d-1\}.$$

To generate observations from this copula, one generates $W_1, \ldots, W_d \sim \text{Unif}(0,1)$ independently, and one sets $U_1 = W_1, \ldots, U_{d-1} = W_{d-1}$, and

$$U_d = \frac{1 + B - \sqrt{(1+B)^2 - 4BW_d}}{2B} = \frac{2W_d}{1 + B + \sqrt{(1+B)^2 - 4BW_d}},$$

where $B = \theta \prod_{j=1}^{d-1}(1 - 2U_j)$. Note that the expression on the righthand side is useful if $B = 0$, which happens when $\theta = 0$.

8.7.13 Plackett Family

Plackett family is a family of bivariate copulas that has not been extended to higher dimensions. The copula is defined, for any $\theta > 0$, and any $u_1, u_2 \in [0, 1]$, by

$$C_\theta(u_1, u_2) = \frac{2\theta u_1 u_2}{B + \sqrt{\{B^2 - 4u_1u_2\theta(\theta - 1)}} \tag{8.42}$$

$$= \frac{B - \sqrt{\{B^2 - 4u_1u_2\theta(\theta - 1)}}{2(\theta - 1)}, \tag{8.43}$$

where $B = 1 + (\theta - 1)(u_1 + u_2)$. Representation (8.42) shows that $C_1 = C_\perp$, while representation (8.43) shows that $C_0 = C_-$. In addition, as $\theta \to \infty$, $C_\theta \to C_+$. One can also prove that $\rho^{(S)} = \frac{\theta+1}{\theta-1} - \frac{2\theta}{(\theta-1)^2}\ln(\theta)$. There is no explicit expression for τ. Note that using Proposition 8.6.3, the family is positively ordered with respect to PQD order.

8.7.14 Other Copula Families

Between the elliptical copulas which have too many parameters when d is large and many Archimedean copulas that have too few parameters, there exist several other families of copulas. However, to be attractive, these families have to be easy to simulate and easy to estimate.

Recently, new promising families of copulas have been studied, namely the hierarchical families of Archimedean copulas, e.g., McNeil [2008], Hering et al. [2010] and the vine copulas [Kurowicka and Joe, 2011].

8.8 Estimation of the Parameters of Copula Models

In this section, we will consider how to implement a copula model. We will first see three methods to estimate the parameter θ of a copula family $\{C_\theta; \theta \in \mathcal{O} \subset \mathbb{R}^p\}$. In what follows, we assume that we have independent observations $X_1 = (X_{11}, \ldots, X_{1d})$, ..., $X_n = (X_{n1}, \ldots, X_{nd})$ from a distribution function H with continuous margins $F_1, \ldots, F_d$ and copula C_θ, possibly depending on a multidimensional parameter.

8.8.1 Considering Serial Dependence

In practice, we often have to deal with serial dependence in time series. For example, many authors consider GARCH type models of the form

$$X_{ij} = \mu_{ij}(\alpha) + \sigma_{ij}(\alpha)\epsilon_{ij}, \quad j \in \{, \ldots, d\}, \quad i \geq 1, \tag{8.44}$$

where the innovations $\epsilon_i = (\epsilon_{i1}, \ldots, \epsilon_{id})$ are independent, and $\epsilon_i \sim H$, with continuous margins $F_1, \ldots, F_d$ (such as standard Gaussian) and with a copula C that does not depend on the conditional mean and variance parameter α. First, α is estimated by $\hat{\alpha}_n$ and residuals $e_{ij} = \frac{X_{ij} - \mu_{ij}(\hat{\alpha}_n)}{\sigma_{ij}(\hat{\alpha}_n)}$ are computed. For a large class of such models, it has been shown by Chen and Fan [2006] that using these residuals does not affect the estimation of the copula parameters. More recently, it was also proven [Rémillard, 2010] that the nonparametric estimation of the copula was not affected. Neither are the estimation of Kendall tau, Spearman rho, and van der Waerden rho.

It means that for a large class of stochastic volatility models, one can use the residuals and treat them as if they were independent observations with copula C. However, note that this result is valid only for the copula.

Using Residuals

Rémillard [2010] proved that the limiting distribution of the empirical distribution function estimating H does indeed depend on the estimation of α.

These results justify many approaches that were used in dependence modeling in financial applications, for instance Breymann et al. [2003], Panchenko [2005], van den Goorbergh et al. [2005], and Patton [2006].

8.8.2 Estimation of Parameters: The Parametric Approach

Assume that the margins depend on parameters $\alpha_1, \ldots, \alpha_d$, and that the copula C belongs to a parametric family $\{C_\theta; \theta \in \mathcal{O}\}$. We also assume that the density c_θ of C_θ exists and is continuous on $(0,1)^d$. Then

$$c_\theta(u) = \frac{\partial^d}{\partial u_1 \cdots \partial u_d} C_\theta(u_1, \ldots, u_d), \quad u \in (0,1)^d.$$

To implement the parametric approach, we follow the following steps:

1. Estimate the parameters of the margins $\alpha_1, \ldots, \alpha_d$, by $\hat{\alpha}_{n,1}, \ldots, \hat{\alpha}_{n,d}$.

2. Define the pseudo-observations $\hat{U}_{ij} = F_{j,\hat{\alpha}_{n,j}}(X_{ij})$, $i \in \{1, \ldots, n\}$, $j \in \{1, \ldots, d\}$.

3. Estimate θ using the maximum likelihood principle, by maximizing

$$\theta \mapsto \sum_{i=1}^n \ln c_\theta(\hat{U}_i),$$

   or by using another estimation method applied to the pseudo-observations $\hat{U}_1, \ldots, \hat{U}_n$ instead of $U_1, \ldots, U_n$ (which are not available anyway). For example, for one-dimensional parameters θ, one could estimate θ using Kendall tau or Spearman rho, if there is an analytical expression for $\tau = \tau_\theta$ or $\rho = \rho_\theta^{(S)}$.

This methodology, proposed in Xu [1996] is often called the IFM (Inference functions for margin) approach or the two-stage method. See, e.g., Joe [2005] for asymptotic results on the estimation error.

8.8.2.1 Advantages and Disadvantages

On the positive side, this estimation method is relatively easy to implement and intuitive. On the negative side, specifying the margins and the copula mean that we have a full parametric model for H, that is

$$H(x_1, \ldots, x_d) = C_\theta\{F_{1,\alpha_1}(x_1), \ldots, F_{d,\alpha_d}(x_d)\}, \quad x \in \mathbb{R}^d.$$

Hence one could estimate all parameters at the same time which is usually more efficient. Also, errors on the choice of the margins could have significant impact on the estimation of the copula parameter. This is why we strongly suggest the next approach.

8.8.3 Estimation of Parameters: The Semiparametric Approach

Here we do not assume any model for the margins $F_1, \ldots, F_d$. We only assume that they are continuous. We use the pseudo-observations $\hat{U}_1 =$

$(\hat{U}_{11}, \ldots, \hat{U}_{1d})$, ..., $\hat{U}_n = (\hat{U}_{n1}, \ldots, \hat{U}_{nd})$, where $\hat{U}_{ij} = F_{nj}(X_{ij}) = \frac{R_{ij}}{n+1}$, $i \in \{1, \ldots, n\}$, $j \in \{1, \ldots, d\}$, with R_{ij} being the rank of X_{ij} amongst $X_{1j}, \ldots, X_{nj}$. Recall that we want to estimate the parameter θ of C_θ. We suggest to use the so-called maximum pseudo likelihood method [Genest et al., 1995, Shih and Louis, 1995] by maximizing

$$\theta \mapsto \sum_{i=1}^{n} \ln\left\{c_\theta(\hat{U}_i)\right\}.$$

The estimation errors have been shown to be asymptotically Gaussian.[5] One could also use other methods based on the pseudo-observations $\hat{U}_1, \ldots, \hat{U}_n$. For example, if θ is univariate, one could estimate θ by using Kendall tau or Spearman rho, if there is an analytical expression for $\tau = \tau_\theta$ or $\rho = \rho_\theta^{(S)}$. This methodology is often used for Clayton, Frank, and Gumbel families, where τ can be computed explicitly. Here the estimation error is also asymptotically Gaussian, using the results of Section 8.5, together with the delta method.

8.8.3.1 Advantages and Disadvantages

On the positive side, this approach is easier to implement than the two-stage method, avoiding the estimation of additional parameters. Also, we do not make any inference on the margins, hence no error in the estimation thereof.

On the negative side, this approach is usually less efficient than a full parametric method, but the latter cannot be applied here, the margins being unknown. For more on this subject, see, e.g., Genest and Werker [2002].

8.8.4 Estimation of ρ for the Gaussian Copula

For the Gaussian copula C_ρ, one can show that the maximum pseudo likelihood method yields an explicit expression for $\hat{\rho}$, namely that $\hat{\rho}$ is the correlation matrix of the pseudo-observations

$$\hat{Z}_i = \left(\Phi^{-1}(\hat{U}_{i1}), \ldots, \Phi^{-1}(\hat{U}_{id})\right), \quad i \in \{1, \ldots, n\}.$$

Hence, the estimator $\hat{\rho}$ is the matrix of the van der Waerden coefficients between all pairs of variables, and the individual estimation errors can be computed using Proposition 8.5.6.

8.8.5 Estimation of ρ and ν for the Student Copula

To estimate ρ, one can use the methodology described in Section 8.7.2.1.

[5]The MATLAB function *copulafit* in the Statistics Toolbox can be used to estimate the parameters of the following families using the IFM or the maximum pseudo likelihood: Gaussian, Student, Clayton, Frank, and Gumbel. However, be careful: some families have a parametrization different from the one presented here.

One can then estimate ν by maximizing the pseudo likelihood

$$\nu \mapsto \sum_{i=1}^{n} \ln c_{\hat{\rho},\nu}(\hat{U}_i), \tag{8.45}$$

where the density $c_{\rho,\nu}$ is given by formula (8.31). The computation of the estimation error can then be done using the results of Appendix B.9. We can also estimate ρ and ν at the same time by maximizing

$$(\rho,\nu) \mapsto \sum_{i=1}^{n} \ln c_{\rho,\nu}(\hat{U}_i).$$

Here of course, one has to take into account the constraint $\rho \in \mathcal{S}_d^+$. The estimator $\hat{\rho}$ described in Section 8.7.2.1 and the estimation of ν obtained by maximizing (8.45) could then serve as starting points for the optimization algorithm.

8.8.6 Estimation for an Archimedean Copula Family

In the general case, we can estimate θ by the maximum pseudo likelihood method, using formula (8.34) for the density c_θ. If the copula family depends on a one-dimensional parameter θ and the family is ordered with respect to the PQD order, Kendall tau or any dependence measure monotone with respect to the PQD order will have the property that $\tau = \psi(\theta)$ for some monotone function ψ. One can then estimate θ by

$$\hat{\theta}_n = \psi^{-1}(\tau_n).$$

Using the delta method (Theorem B.3.1), one may conclude that

$$\sqrt{n}(\hat{\theta}_n - \theta) \rightsquigarrow N\left(0, \sigma_n^2 / \left\{\psi'(\hat{\theta}_n)\right\}^2\right),$$

where σ_n is defined in Proposition 8.5.2 for Kendall tau. This moment-matching type approach, when applicable, is usually much faster than the pseudo-likelihood method but it is also less efficient in general.

8.8.7 Nonparametric Estimation of a Copula

Suppose we want to estimate C for goodness-of-fit testing or testing for independence. Then the results of Section 8.5.1 extends to the multivariate case. The copula C can be estimated efficiently by the empirical copula C_n based on the pseudo-observations $\hat{U}_1, \ldots, \hat{U}_n$, viz.

$$C_n(u) = \frac{1}{n}\sum_{i=1}^{n} \mathbb{I}\left(\hat{U}_i \le u\right) = \frac{1}{n}\sum_{i=1}^{n}\prod_{j=1}^{d} \mathbb{I}\left(\hat{U}_{ij} \le u_j\right). \tag{8.46}$$

One can show that $\sqrt{n}\,(C_n - C)$ is asymptotically a continuous centered Gaussian process [Gänssler and Stute, 1987, Fermanian et al., 2004], provided the first order derivatives of C are continuous on $(0,1)^d$.

8.8.8 Nonparametric Estimation of Kendall Function

To estimate the Kendall function K, we use the pseudo-observations $\hat{V}_1, \ldots, \hat{V}_n$, where

$$\hat{V}_i = \text{Card}\{k; \hat{U}_{kj} < \hat{U}_{ij} \text{ for all } j = 1, \ldots, d\} \Big/ (n-1), \quad i \in \{1, \ldots, n\}.$$

Here, the pseudo-observations can be obtained either from the parametric approach or the semiparametric approach. Next, K is estimated by

$$K_n(t) = \frac{1}{n} \sum_{i=1}^{n} \mathbb{I}\left(\hat{V}_i \le t\right), \quad t \in [0,1].$$

This estimation is quite efficient since Barbe et al. [1996] proved that $\sqrt{n}\,(K_n - K) \rightsquigarrow \mathbb{K}$, where $\mathbb{K}$ a continuous centered Gaussian process. As seen before, we can base tests of independence on K_n, comparing it with $K_\perp$ given by formula (8.28). One can also use K_n for goodness-of-fit tests, by comparing K_n with $K_{\hat{\theta}_n}$, as in Genest et al. [2006a]. In either cases the parametric bootstrap can be used to estimate p-values. See Genest and Rémillard [2008] for more details.

8.9 Tests of Independence

In Section 8.5.1.2, we proposed a test of independence based on the empirical copula. We will extend the test to the multivariate setting, using the ideas of Genest and Rémillard [2004]. We could also consider tests of independence based on the empirical Kendall function. This is almost word for word the same technique as in Section 8.5.2.3, except for the value of the Kendall function $K_\perp$ associated with the independence copula, which now depends on the dimension d. The Kolmogorov-Smirnov and the Cramér-von Mises tests are implemented in the MATLAB function IndKendallTest. Note that even if the tests are generally powerful, they are not necessarily consistent since one cannot prove that $K_\perp$ characterizes the independence copula, even when restricted to the Archimedean family. The interest of using the Kendall function is that one can assess visually the departure from independence by constructing a uniform confidence band around $K_\perp$ and comparing it to K_n.

8.9.1 Test of Independence Based on the Copula

For any $A \in \mathcal{A}_d = \{B \subset \{1,\ldots,d\}, |B| > 1\}$, one can define Cramér-von Mises statistics $T_{n,A}$ which converge jointly in law to independent random variables T_A, under the null hypothesis of independence. See Appendix 8.B for the precise definition of these statistics. Since the statistics are based on ranks, they are distribution-free, and tables of quantiles and also P-values can be easily computed via Monte Carlo methods, using Algorithm 8.B.1.

As in the case of serial independence discussed in Chapter 3, one can then construct a *dependogram*, i.e., a graph of the P-values $p_{n,A}$ of $T_{n,A}$, for all $A \in \mathcal{A}_d$. P-values below the dashed horizontal line at 5% suggest dependence for the given indices. One can also define a powerful test of independence by combining the P-values of the test statistics $T_{n,A}$, as proposed in Genest and Rémillard [2004]. More precisely, set

$$\mathfrak{F}_n = -2 \sum_{A \subset \mathcal{A}_d} \ln(p_{n,A}).$$

Under the null hypothesis of serial independence, the P-values $p_{n,A}$ converge to independent uniform variables, so $\mathfrak{F}_n \rightsquigarrow \mathfrak{F} \sim \chi^2\left(2^{d+1} - 2d - 2\right)$. This test and the dependogram are implemented in *IndGRTest*.

Remark 8.9.1 *Copula-based tests of independence between innovations of several time series can also be performed using residuals of stochastic volatility models, as discussed in Section 8.8.1. See, e.g., Duchesne et al. [2012].*

8.10 Tests of Goodness-of-Fit

One can use C_n to test the goodness-of-fit of a parametric family like the Gaussian, Student, Clayton, etc. For Archimedean families, one can also use tests based on K_n, as proposed in Genest et al. [2006a].

Suppose that the null hypothesis is $H_0 : C = C_\theta$ for some $\theta \in \mathcal{O}$. We assume that θ is estimated by a function of the pseudo-observations, i.e.,

$$\hat{\theta}_n = T_n(\hat{U}_1, \ldots, \hat{U}_n)$$

and that the goodness-of-fit test is based on the statistic $S_n = \psi_n(\mathbb{C}_n)$, where

$$\mathbb{C}_n = n^{1/2}\left(C_n - C_{\hat{\theta}_n}\right).$$

A good choice for S_n would be the Cramér-von Mises statistic

$$S_n = \sum_{i=1}^{n} \left\{C_n\left(\hat{U}_i\right) - C_{\hat{\theta}_n}\left(\hat{U}_i\right)\right\}^2 \approx \int_{[0,1]^d} \mathbb{C}_n^2(u) dC(u).$$

Another choice would be the (approximate) Kolmogorov-Smirnov statistic

$$\max_{1\leq i\leq n} \sqrt{n}\left|C_n\left(\hat{U}_i\right) - C_{\hat{\theta}_n}\left(\hat{U}_i\right)\right|.$$

However it seems that this statistic lacks power in higher dimensions since one should really be computing

$$\sup_{u\in[0,1]^d} \sqrt{n}\left|C_n(u) - C_{\hat{\theta}_n}(u)\right|,$$

which takes too much time to evaluate.

8.10.1 Computation of P-Values

Finally, assume that we reject the null hypothesis for large values of S_n. To compute the corresponding P-value, we can use the following algorithm based on parametric bootstrap, whose validity for goodness-of-fit for copulas was proven in Genest and Rémillard [2008].

Algorithm 8.10.1 *For N large enough and for all $k \in \{1,\ldots,N\}$, repeat the following steps:*

a) Given $\hat{\theta}_n$, generate n independent observations $U_1^{(k)},\ldots,U_n^{(k)}$ from copula $C_{\hat{\theta}_n}$ and compute the pseudo-observations $\hat{U}_1^{(k)},\ldots,\hat{U}_n^{(k)}$.

b) Estimate θ by $\hat{\theta}_n^{(k)} = T_n\left(\hat{U}_1^{(k)},\ldots,\hat{U}_n^{(k)}\right)$ and calculate the empirical copula of the new pseudo-observations, i.e.,

$$C_n^{(k)}(u) = \frac{1}{n}\sum_{i=1}^{n}\mathbb{I}\left(\hat{U}_i^{(k)} \leq u\right), \quad u\in[0,1]^d.$$

c) Compute the statistic $S_n^{(k)} = \psi_n\left(\mathbb{C}_n^{(k)}\right)$, where

$$\mathbb{C}_n^{(k)} = n^{1/2}\left(C_n^{(k)} - C_{\hat{\theta}_n^{(k)}}\right).$$

An approximative P-value for the test is then given by

$$\frac{1}{N}\sum_{k=1}^{N}\mathbb{I}\left(S_n^{(k)} > S_n\right).$$

Remark 8.10.1 *One problem with this test is that one needs to re-estimate the parameters at each iteration. However there is another way to do it, based on the multipliers methodology, as developed by Rémillard and Scaillet [2009]. See, e.g., Kojadinovic and Yan [2010] and Rémillard [2012b]. The only catch is that we have to work much harder to implement this methodology.*

Another problem is that often one cannot compute explicitly C_θ. In such a case, Genest and Rémillard [2008] proposed the two-stage parametric bootstrap, which is even more time consuming, since one has to use simulations to evaluate C_θ. The best solution to this problem is to use the Rosenblatt transform, which is usually quite easy to compute. This methodology is discussed next.

8.10.2 Using the Rosenblatt Transform for Goodness-of-Fit Tests

In practice, if there is no explicit formula for C_θ, as it is the case for example for the Student copula, one can rely on a two-stage parametric bootstrap proposed in Genest and Rémillard [2008]. However it is very slow. Motivated by the results in Genest et al. [2009], one of the best omnibus test of goodness-of-fit for copulas is based on the Rosenblatt transform described in Section 8.6.2. In fact, to each parametric model C_θ corresponds a parametric Rosenblatt transform Ψ_θ so that, according to Theorem 8.6.2, if $U \sim C_\theta$, then $W = \Psi_\theta(U) \sim C_\perp$. Therefore, instead of computing the empirical copula C_n, we suggest to compute the empirical distribution function

$$D_n(u) = \frac{1}{n} \sum_{i=1}^{n} \mathbb{I}(\hat{W}_i \leq u), \quad u \in [0,1]^d,$$

where the pseudo-observations $\hat{W}_i$ are defined by $\hat{W}_i = \Psi_{\hat{\theta}_n}(\hat{U}_i)$, $i \in \{1,\ldots,n\}$. Under the null hypothesis, $\hat{W}_i \approx W_i \sim C_\perp$, so D_n should be compared to $C_\perp$ and tests of goodness-of fit can be based on statistics of the form $S_n = \psi_n(\mathbb{D}_n)$, where $\mathbb{D}_n = \sqrt{n}(D_n - C_\perp)$. For example, one could take the Cramér-von Mises statistics

$$S_n^{(B)} = \int_{[0,1]^d} \mathbb{D}_n^2(u)du$$

or

$$S_n^{(C)} = \sum_{i=1}^{n} \left\{ D_n\left(\hat{W}_i\right) - C_\perp\left(\hat{W}_i\right) \right\}^2.$$

8.10.2.1 Computation of P-Values

Parametric bootstrap can be used to compute P-values for the test, as described in the following algorithm.

Algorithm 8.10.2 *For N large enough and for all $k \in \{1,\ldots,N\}$, repeat the following steps:*

a) Given $\hat{\theta}_n$, generate n independent observations $U_1^{(k)},\ldots,U_n^{(k)}$ from copula $C_{\hat{\theta}_n}$ and compute the pseudo-observations $\hat{U}_1^{(k)},\ldots,\hat{U}_n^{(k)}$.

b) Estimate θ *by* $\hat{\theta}_n^{(k)} = T_n\left(\hat{U}_1^{(k)}, \ldots, \hat{U}_n^{(k)}\right)$*, compute the new pseudo-observations* $\hat{W}_i^{(k)} = \Psi_{\hat{\theta}_n^{(k)}}\left(\hat{U}_1^{(k)}\right)$*, and calculate the associated empirical distribution function*

$$D_n^{(k)}(u) = \frac{1}{n}\sum_{i=1}^{n} \mathbb{I}\left(\hat{W}_i^{(k)} \le u\right), \quad u \in [0,1]^d.$$

c) Compute the statistic $S_n^{(k)} = \psi_n\left(\mathbb{D}_n^{(k)}\right)$*, where*

$$\mathbb{D}_n^{(k)} = n^{1/2}\left(D_n^{(k)} - C_\perp\right).$$

An approximative P-value for the test is then given by

$$\frac{1}{N}\sum_{k=1}^{N} \mathbb{I}\left(S_n^{(k)} > S_n\right).$$

The goodness-of-fit tests $S_n^{(B)}$ and $S_n^{(C)}$ based on the Rosenblatt transform are implemented in the MATLAB function *GofRosenCopula* for the Clayton, Gaussian, and Student distribution. P-values are calculated using Algorithm 8.10.2.

To illustrate the methodologies introduced in Sections 8.8 through 8.10, we will use the returns of Apple and Microsoft.

8.11 Example of Implementation of a Copula Model

To implement a copula model for a (stationary) multivariate times series one should always start by looking at serial dependence. Note that there exists powerful nonparametric tests to detect the lack of stationarity. These tests are called change point tests. See Appendix 8.E for more details. A univariate change point test is implemented in the MATLAB function *TestChangePoint*, while the change point test for the copula is implemented in the MATLAB function *TestChangePointCopula.*

When the null hypothesis of serial independence is rejected for a series, one must try to remove the serial dependence by modeling it with stochastic volatility model satisfying (8.44), as discussed in Section 8.11. After estimating the parameters, we end up with the residuals. Change point tests should also be applied to the residuals of all series. If the null hypothesis of stationarity is not rejected for the margins and for the copula, then one can proceed with the copula modeling. A commun practice is to estimate the parameters and perform tests of goodness-of-fit on several models.

As an illustration of the implementation, consider the returns of Apple and Microsoft, from January 14th, 2009, to January 14th, 2011.

8.11.1 Change Point Tests

Using the MATLAB function *TestChangePoint* with the returns of Apple, the P-value is 10.5%, while for the returns of Microsoft, the P-value of the test is 3.49%, which could lead to the rejection of the hypothesis of stationarity.

8.11.2 Serial Independence

As seen in Section 3.1.1, the null hypothesis of serial independence was rejected for the returns of Apple. It is also rejected for Microsoft.

8.11.3 Modeling Serial Dependence

We try to fit the simplest model, i.e., a GARCH(1,1) model with GED innovations, since we know from the results of Example 7.3 that the hypothesis of GED innovations was not rejected for the returns of Apple. The estimated parameters for both series are given in Table 8.10. As shown in Figure 8.15, the autocorrelations of the residuals look nice enough.

TABLE 8.10: 95% confidence intervals for the parameters of GARCH(1,1) models with GED innovations, applied to the returns of Apple and Microsoft.

	Series	
Parameter	Apple	Microsoft
μ	0.0028 ± 0.0013	0.0015 ± 0.0016
$10^5\omega$	0.7135 ± 1.3121	0.6264 ± 0.7671
β	0.8968 ± 0.0889	0.9023 ± 0.0678
α	0.0817 ± 0.0680	0.0755 ± 0.0615
ν	1.3510 ± 0.2387	1.1342 ± 0.1816

8.11.3.1 Change Point Tests for the Residuals

This time the P-values are large enough: 67.7% for the residuals of Apple and 11.9% for the residuals of Microsoft. For the change point test on the copula, one obtains a P-value of 95%. Therefore one cannot reject the hypothesis stationarity.

8.11.3.2 Goodness-of-Fit for the Distribution of Innovations

Since the estimation of the parameters were done using GED innovations, it is natural to test that hypothesis, using the MATLAB function *Gof-*

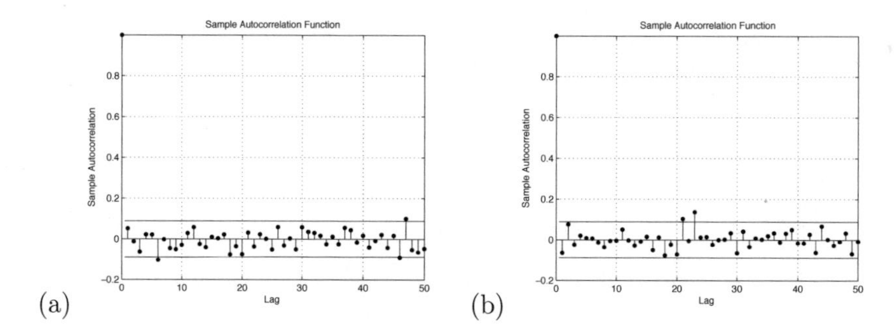

FIGURE 8.15: Autocorrelograms of the residuals of GARCH(1,1) models for Apple (panel a) and Microsoft (panel b).

GARCHGED. The results are displayed in Table 8.11. There is not sufficient evidence to reject the hypothesis of GED innovations. The empirical distribution functions used for the Kolmogorov-Smirnov tests are displayed in Figures 8.16–8.17. They are both within the 95% confidence band of the uniform distribution.

TABLE 8.11: Tests of the hypothesis of GED innovations. The P-values were computed with $N = 1000$ bootstrap samples.

	Apple		Microsoft	
	CVM	KS	CVM	KS
Test	.0549	.6821	.0938	.7991
P-value (%)	29.5	18.0	8.3	7.3

8.11.4 Modeling Dependence Between Innovations

The last step is to try for fit a model of dependence between the innovations of the GARCH processes used for the returns of Apple and Microsoft. The first step should always be to test for independence, i.e., the copula C of the innovations $(\varepsilon_{i1}, \varepsilon_{i2})$ is $C_\perp$. If this hypothesis is rejected, then one can try more complicated models.

8.11.4.1 Test of Independence for the Innovations

Here the null hypothesis is $H_0 : C = C_\perp$. Computing the P-value of the test of independence based on Kendall tau, we find a 0 P-value, and a 95% confidence interval for τ is $\tau = .34 \pm .06$. In fact, the P-values are all four tests computed with *EstDep* is 0. One can also verify if the empirical Kendall function is within the 95% confidence band about $K_\perp$. Figure 8.18 shows

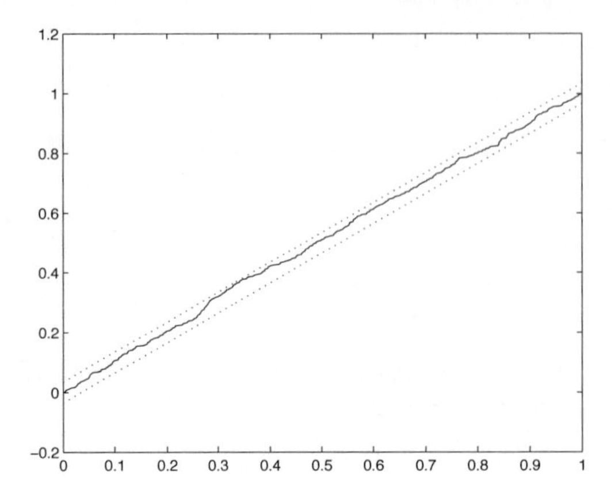

FIGURE 8.16: Empirical distribution function D_n and 95% confidence band about the uniform distribution function for the innovations of Apple.

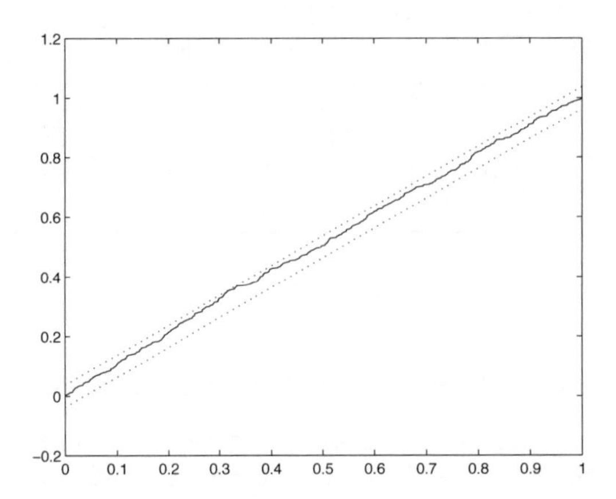

FIGURE 8.17: Empirical distribution function D_n and 95% confidence band about the uniform distribution function for the innovations of Microsoft.

clearly that we reject the null hypothesis of independence. Note that the P-values for the tests of independence based on the copula and the Kendall function are also 0.

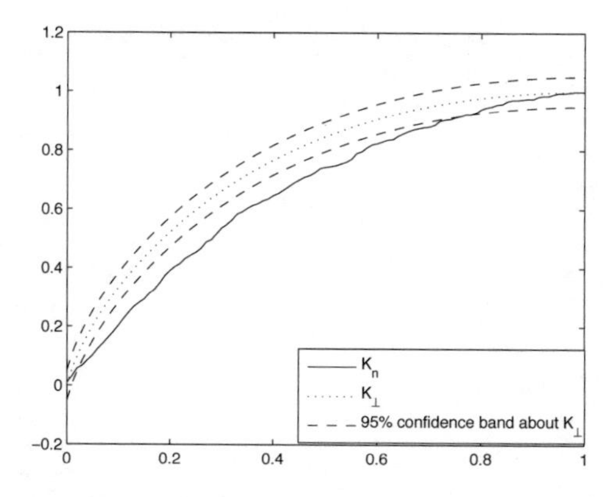

FIGURE 8.18: Graph of the empirical Kendall function and 95% confidence band about $K_\perp$. $N = 1000$ bootstrap samples were generated to compute the P-values and the 95% quantile of the Kolmogorov-Smirnov statistic.

8.11.4.2 Goodness-of-Fit for the Copula of the Innovations

We first use the empirical Kendall function for testing the adequation of the Ali-Mikhail-Haq (AMH), Clayton, Frank, and Gumbel copula families. The results of the tests are given in Table 8.12, where the P-values of the Cramér-von Mises (S_n) and Kolmogorov-Smirnov (T_n) tests based on K_n were computed using $N = 1000$ parametric bootstrap replications. Figures 8.19–8.20 also illustrate why we do not reject the null hypothesis for the Kolmogorov-Smirnov tests for both Clayton and Ali-Mikhail-Haq families.

According to these results, the "best" model would be an Ali-Mikhail-Haq copula, followed by the Clayton copula. Note that the parameter θ of the Ali-Mikhail-Haq copula is almost 1, indicating that it is no significantly different from a Clayton copula with parameter 1, corresponding to a Kendall tau of 1/3. As an indication, a 95% confidence interval for Kendall τ is $\tau = .34 \pm .06$. As mentioned before, the results in Rémillard [2010] make it possible to use the residuals of GARCH models.

Finally, we performed goodness-of-fit tests based on $S_n^{(B)}$ for the Clayton, Gaussian and Student copulas, the other models having been rejected by the test based on the empirical Kendall function. The reason for choosing $S_n^{(B)}$ is that it is one of the best goodness-of-fit tests, being based on the Rosenblatt transform. The results are given in Table 8.13 and the P-values were estimated with $N = 1000$ parametric bootstrap replications. First, note that the best model is the Student copula, with a P-value of 24%. All other models are clearly rejected. The somewhat surprising result is that the Clayton family is rejected with a P-value less than 1%, while it was not for tests based on Kendall function. However, in view of the results on the power of goodness-of-

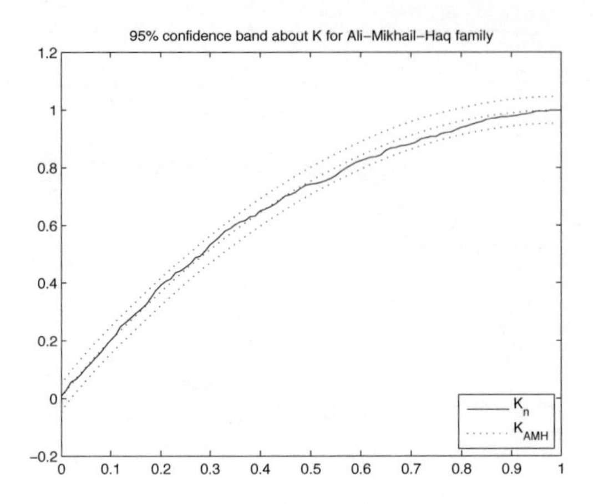

FIGURE 8.19: 95% confidence band about K_{AMH}.

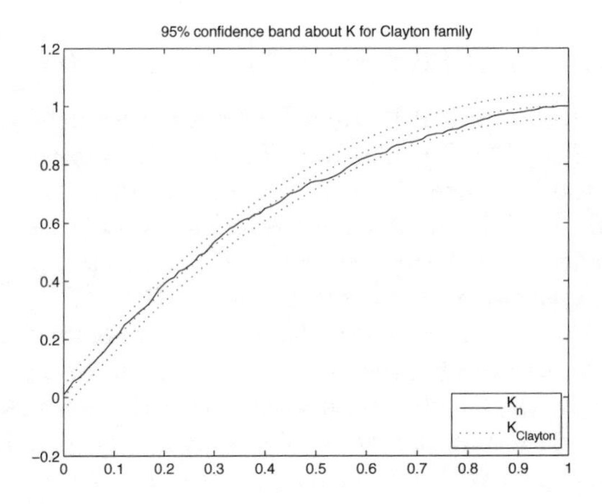

FIGURE 8.20: 95% confidence band about $K_{Clayton}$.

fit tests in Genest et al. [2009], it is not surprising. In fact, it was shown that the tests based on the Kendall function have no power for rejecting the null hypothesis of a Clayton copula, when the true copula is Student, which seems to be the case here. Figure 8.21 illustrates the distribution of the pseudo-observations $\hat{W}_i$, $i \in \{1, \ldots, n\}$ when using the Rosenblatt transform on the Student copula.

Summarizing our findings, both the returns of Apple and Microsoft can be modeled by GARCH(1,1) processes with GED innovations. Moreover, the de-

TABLE 8.12: Tests of goodness-of-fit based on the Cramér–von Mises (S_n) and Kolmogorov–Smirnov (T_n) statistics computed with K_n. $N = 1000$ bootstrap samples were generated.

Model	$\hat{\theta}$	S_n / T_n	95% Quantile	P-value (%)
A-M-H	0.972	0.1066	0.2020	**25.7**
		0.7698	1.0512	**36.0**
Clayton	0.885	0.1142	0.1633	**12.8**
		0.8656	0.9754	**11.8**
Frank	0.030	0.3645	0.1266	0
		1.2940	0.8708	0
Gumbel	0.690	0.5913	0.1478	0
		1.5559	0.9405	0

pendence between the innovations of Apple and Microsoft can be modeled by a Student copula with 3.5 degrees of freedom and correlation parameter 0.51, corresponding to a Kendall tau of 0.34, through the formula $\tau = \frac{2}{\pi}\arcsin(\rho)$ of Proposition 8.7.1. Note that this is exactly the estimation of Kendall tau.

TABLE 8.13: Tests of goodness-of-fit based on $S_n^{(B)}$, using Rosenblatt transform. The P-values were estimated with $N = 1000$ parametric bootstrap replications.

Model	Estimated Parameter(s)	$S_n^{(B)}$	P-value (%)
Clayton	$\hat{\theta} = 0.889$	0.0929	0.2
Gaussian	$\hat{\rho} = 0.476$	0.1255	0.0
Student	$(\hat{\rho}, \hat{\nu}) = (0.51, 3.51)$	0.0420	**23.9**

8.12 Suggested Reading

For financial applications of copulas, I suggest Cherubini et al. [2004], Roncalli [2004], van den Goorbergh et al. [2005], Berrada et al. [2006], Papageorgiou et al. [2008], and Patton [2009]. For copula-based econometric models, see Patton [2006, 2012], Chen and Fan [2006] and Rémillard et al. [2012].

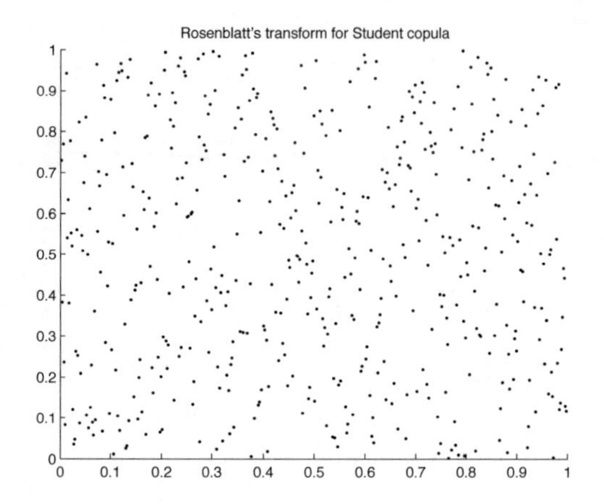

FIGURE 8.21: Rosenblatt transform of the pseudo-observations using the Student copula with parameters $\hat{\rho} = 0.51$ and $\hat{\nu} = 3.51$.

8.13 Exercises

Exercise 8.1

For $\theta \in (-\infty, \infty)$, $\theta \neq 0$, define

$$C_\theta(u_1, u_2) = -\frac{1}{\theta} \log \left\{ \frac{e^{-u_1\theta} + e^{-u_2\theta} - e^{-(u_1+u_2)\theta} - e^{-\theta}}{1 - e^{-\theta}} \right\}, \qquad u_1, u_2 \in [0, 1].$$

(a) Identify the copula family.

(b) Find the value of the following limits:

(i) $\lim_{\theta \to 0} C_\theta(u_1, u_2)$;

(ii) $\lim_{\theta \to \infty} C_\theta(u_1, u_2)$;

(iii) $\lim_{\theta \to -\infty} C_\theta(u_1, u_2)$.

(c) For θ given, find the Rosenblatt transform.

Exercise 8.2

Consider the random variable

$$T_2 = \frac{36}{\pi^4} \sum_{i=1}^{\infty} \sum_{j=1}^{\infty} \frac{Z_{ij}^2}{i^2 j^2}.$$

where the Z_{ij}s are independent and identically distributed standard Gaussian variables.

(a) Find the first six cumulants of X. Note that the famous zeta[6] function is defined by

$$\zeta(s) = \sum_{n=1}^{\infty} \frac{1}{n^s}, \quad s > 1.$$

(b) Use the Edgeworth expansion to try to approximate the distribution function of X.

(c) Estimate the quantiles of order 90%, 95%, and 99% using the Cornish-Fisher expansion. Also plot the graph of the quantile function determined by the Cornish-Fisher expansion. Is the approximation better than the one obtained by the Edgeworth expansion?

(d) Generate 100000 approximations of X by truncating the infinite sum to a sum over $i, j \leq 100$, i.e., generate

$$\tilde{X} = \frac{36}{\pi^4} \sum_{i=1}^{100} \sum_{j=1}^{100} \frac{Z_{ij}^2}{i^2 j^2}.$$

(e) Using the simulated values of $\tilde{X}$, estimate its distribution function and compare it to the one obtained in (b). Is the Edgeworth expansion precise enough? Also estimate the quantiles of order 90%, 95%, and 99%. Finally, compare the empirical quantiles curve with the Cornish-Fisher expansion.

Exercise 8.3

Suppose that the joint distribution of (X, Y) is Student with parameters $\nu > 0$ and $\rho \in (-1, 1)$. Find the distribution of X and the conditional distribution of Y given $X = x$.

Exercise 8.4

Suppose that $\tau = 0.5$. Find the associated parameter for the following models:

(a) Gaussian copula C_ρ.

(b) Student copula $C_{\rho,\nu}$.

(c) Clayton copula C_θ.

---

[6]This function is implemented in MATLAB. It can also be computed exactly for any even integer s, using, e.g., the *Symbolic Math* Toolbox.

(d) Frank copula C_θ.

(e) Gumbel copula C_θ.

Exercise 8.5

For $\theta > 0$, we set

$$C_\theta(u_1, u_2) = \sqrt[\theta]{u_1^\theta + u_2^\theta - 1}, \quad u_1, u_2 \in [0, 1]. \tag{8.47}$$

(a) Identify the copula family.

(b) Find the following limits:

(i) $\lim_{\theta \to 0} C_\theta(u_1, u_2)$;

(ii) $\lim_{\theta \to \infty} C_\theta(u_1, u_2)$;

(iii) $\lim_{\theta \to -\infty} C_\theta(u_1, u_2)$.

(c) For θ fixed, find the Rosenblatt transform.

(d) For θ fixed, find the inverse Rosenblatt transform.

Exercise 8.6

Suppose that for a sample of size n, we have $\tau = 0.3 \pm .08$ at the 95% level of confidence. If the underlying copula is the Clayton, find a 95% confidence interval for θ.

Exercise 8.7

For each of the following cases, identify the (bivariate) copula family and compute the associated Kendall tau.

(i) $C_\theta(u_1, u_2) = \left\{\max\left(0, u_1^{\frac{1}{3}} + u_2^{\frac{1}{3}} - 1\right)\right\}^3$.

(ii) $C_\theta(u_1, u_2) = \exp\left[-\sqrt{\ln^2(u_1) + \ln^2(u_2)}\right]$.

(iii) Archimedean copula with generator $\phi(t) = 4 \ln\left(\frac{1+3t}{4t}\right)$.

Exercise 8.8

Suppose that for a sample of size n, we have $\tau = -0.3 \pm .04$ at the 95% level of confidence. If the underlying copula is the Frank, find a 95% confidence interval for θ.

Exercise 8.9

Explain how you could generate observations from a Plackett copula for a given τ.

Exercise 8.10

Suppose that for a sample of size n, we have $\rho^{(S)} = 0.6 \pm .03$ at the 95% level of confidence. If the underlying copula is the Plackett, find a 95% confidence interval for θ.

Exercise 8.11

Compute the Rosenblatt transform for the bivariate Gumbel copula. Is it invertible explicitly?

Exercise 8.12

One obtains the following output from the MATLAB function *EstDep*.

```
pseudo: [500x3 double]
  stat: [0.3851 0.5533 0.5498 0.5537]
   std: [0.5439 0.7148 0.6723 0.6934]
 error: [0.0477 0.0627 0.0589 0.0608]
pvalue: [0 0 0 0]
 names: {'Kendall' 'Spearman' 'van der Waerden' 'Pearson'}
```

For each of the following families, find the "best" estimation of the related parameter and construct a 95% confidence interval.

(a) Gaussian copula C_ρ.

(b) Student copula $C_{\rho,\nu}$.

(c) Clayton copula C_θ.

(d) Frank copula C_θ.

(e) Gumbel copula C_θ.

(f) Plackett copula C_θ.

Exercise 8.13

Find a way to generate observations from the bivariate Frank copula, for all $\theta > 0$.

Exercise 8.14

Suppose that we have a portfolio composed of 3 risky bonds with face value \$100000, and the time to default is exponentially distributed with a mean of

15. The issuer offers no guaranty, so the loss is total in case of default. As we are interested in evaluating the total risk of the portfolio during the next year, we would like to identify the behavior of the number of defaults in the portfolio. To this end, define $D = \sum_{i=0}^{3}(T_i \leq 1)$.

(a) Suppose that random variables T_1, T_2, T_3 are independent. Compute $P(D = k)$, for $k \in \{0, 1, 2, 3\}$.

(b) Suppose that the copula of T_1, T_2, T_3 is C_+. Compute $P(D = k)$, for $k \in \{0, 1, 2, 3\}$.

(c) Suppose that the copula of T_1, T_2, T_3 is the Clayton copula corresponding to $\tau = 0.7$. Estimate $P(D = k)$, for $k \in \{0, 1, 2, 3\}$.

Exercise 8.15

Show that the Plackett family is positively ordered with respect to the PQD order. This exercise is difficult.

Exercise 8.16

Find a way to generate observations from the Plackett copula, for all $\theta > 0$.

8.14 Assignment Questions

1. Construct a MATLAB function for finding the parameter of a Plackett copula corresponding to a given Kendall tau.

2. Construct a MATLAB function to perform a goodness-of-fit test for the Clayton copula by comparing the empirical Kendall tau with $\frac{\hat{\theta}_n}{\hat{\theta}_n+2}$, where $\hat{\theta}_n$ is the maximum pseudo likelihood estimator of θ.

3. Construct a MATLAB function to compute the Kendall function for the Gumbel, Frank, and Ali-Mikhail-Haq copulas.

4. Construct a MATLAB function to perform a goodness-of-fit test based on the Rosenblatt transform for the Gumbel, Frank, and Ali-Mikhail-Haq copulas.

8.A Continuous Time Markov Chains

Let $S \subset \{1, 2, \ldots\}$ be a countable set. A stochastic process X is a (homogeneous) continuous time Markov chain with state space S and infinitesimal generator Λ if for all $j, k \in S$, we have

$$P\{X(t+h) = k | X(t) = j\} = \begin{cases} h\Lambda_{jk} + o(h), & k \neq j, \\ 1 + h\Lambda_{jj} + o(h), & k = j, \end{cases} \quad (8.48)$$

Here, $f(h) = o(h)$ means that $f(h)/h \to 0$, as $h \to 0$.

$X(t)$ is in fact related to a discrete time Markov chain spending a random time in each state, as shown in the next proposition.

Proposition 8.A.1 *Set $\lambda_i = -\Lambda_{ii}$, $i \in S$. Suppose that $\lambda \geq \max_{i \in S} \lambda_i$, and let N_t be a Poisson process of intensity λ. Then, setting $R = I + \Lambda/\lambda$, R defines a transition matrix of a discrete time Markov chain Y, and*

$$X(t) = Y_{N_t}$$

defines a continuous time Markov chain with infinitesimal generator Λ.

A state j is said to be absorbing if $P\{X(t) = j | X_0 = j\} = 1$ for all $t \geq 0$. Hence j is absorbing if and only if $\Lambda_{jk} = 0$ for all $k \in S$. Furthermore, by definition of the infinitesimal generator, if

$$P_{jk}(t) = P\{X(t) = k | X_0 = j\}, \quad j, k \in S,$$

then $\dot{P}(t) = \frac{dP(t)}{dt} = \Lambda P(t)$, $P(0) = I$. In fact, according to (8.48), and using the Markov property, we have

$$\begin{aligned} P_{jk}(t+h) &= \sum_{l \in S} P_{jl}(t) P_{lk}(h) \\ &= \sum_{l \neq k} P_{jl}(t)\{\Lambda_{lk} h + o(h)\} + P_{jk}(t)\{1 + h\Lambda_{kk} + o(h)\} \\ &= P_{jk}(t) + h\{P(t)\Lambda\}_{jk} + o(h). \end{aligned}$$

As a result,

$$\frac{d}{dt} P(t) = \lim_{h \to 0,\, h \neq 0} \frac{P(t+h) - P(t)}{h} = \lim_{h \to 0,\, h \neq 0} \frac{hP(t)\Lambda + o(h)}{h} = P(t)\Lambda.$$

Finally, one can write

$$P(t) = e^{t\Lambda} = \sum_{n=0}^{\infty} \frac{t^n}{n!} \Lambda^n,$$

where by convention $\Lambda^0 = I$.

8.B Tests of Independence

Suppose $X_1, \ldots, X_n$ are identically distributed d-dimensional random vectors with continuous margins $F_1, \ldots, F_d$. Following Genest and Rémillard [2004], if $d \geq 2$, for any $A \in \mathcal{A}_d = \{B \subset \{1, \ldots, d\}; |B| > 1\}$ and any $x = (x_1, \ldots, x_d) \in \mathbb{R}^d$, set

$$\begin{aligned} \mathbb{G}_{n,A}(x) &= \frac{1}{\sqrt{n}} \sum_{i=1}^{n} \prod_{j \in A} \{\mathbb{I}(X_{ij} \leq x_j) - F_{jn}(x_j)\} \\ &= \frac{1}{\sqrt{n}} \sum_{i=1}^{n} \prod_{j \in A} \left[\mathbb{I}\{U_{ij} \leq F_j(x_j)\} - D_{nj}\{F_j(x_j)\}\right], \end{aligned}$$

where $|A|$ is the cardinality of A, $U_{ij} = F_j(X_{ij})$, and

$$F_{jn}(y) = D_{nj}\{F_j(y)\} = \frac{1}{n} \sum_{i=1}^{n} \mathbb{I}(X_{ij} \leq y) = \frac{1}{n} \sum_{i=1}^{n} \mathbb{I}(U_{ij} \leq F_j(y)\}, \quad y \in \mathbb{R},$$

is the empirical distribution function estimating F_j. It is shown in Genest and Rémillard [2004] that under the null hypothesis of independence, the Cramér-von Mises statistics

$$T_{n,A} = \frac{6^{|A|}}{n} \sum_{i=1}^{n} \mathbb{G}_{n,A}^2(X_i) \tag{8.49}$$

converge jointly in law to T_A and the latter are independent. Moreover, if $|A| = k$, the T_A has the same law as

$$T_k = \frac{6^k}{\pi^{2k}} \sum_{i_1=1}^{\infty} \cdots \sum_{i_k=1}^{\infty} \frac{Z_{i_1,\ldots,i_k}^2}{(i_1 \cdots i_k)^2},$$

where the random variables $Z_{i_1,\ldots,i_k}$ are i.i.d. standard Gaussian. Note that these limiting random variables are distribution-free in the sense that their distribution is independent of the distribution function H. In fact, they are the same as the random variables defined in Chapter 3. Tables for the distribution of T_A can be constructed. One can also compute P-values for the statistics. To do so, one can use the following algorithm.

Algorithm 8.B.1 *For N large (say $N = 1000$), repeat the following steps for each $k \in \{1, \ldots, N\}$:*

- *Generate independent random vectors $U_i^{(k)} = \left(U_{i1}^{(k)}, \ldots, U_{id}^{(k)}\right) \sim C_\perp$, $i \in \{1, \ldots, n\}$;*
- *Compute the associated statistics $T_{n,A}^{(k)}$, $A \in \mathcal{A}_d$, using formula (8.49).*

Then one can approximate the P-value of $T_{n,A}$ by

$$p_{n,A} = \frac{1}{N}\sum_{k=1}^{k} \mathbb{I}\left(T_{n,A}^{(k)} > T_{n,A}\right), \quad A \in \mathcal{A}_d.$$

With these asymptotically independent statistics, one can construct a *dependogram*, i.e., a graph of the P-values $p_{n,A}$, for all $A \in \mathcal{A}_d$. One can also define a powerful test of independence by combining the P-values of the test statistics $T_{n,A}$, as proposed in Genest and Rémillard [2004]. More precisely, set

$$\mathfrak{F}_n = -2 \sum_{A \subset \mathcal{A}_d} \ln(p_{n,A}).$$

Under the null hypothesis of independence, the P-values $p_{n,A}$ converge to independent uniform variables, so $\mathfrak{F}_n \rightsquigarrow \mathfrak{F}$, where $\mathfrak{F}$ has a chi-square distribution with $2^{d+1} - 2d - 2$ degrees of freedom.

8.C Polynomials Related to the Gumbel Copula

Based on the formulas in Barbe et al. [1996], it is easy to check that

$$(-1)^k \frac{d^k}{ds^k} e^{-s^\theta} = \frac{p_{k,\theta}(s^\theta)}{s^k} e^{-s^\theta}, k \geq 0, \tag{8.50}$$

where the sequence of polynomials $p_{k,\theta}$ is defined by $p_{0,\theta} \equiv 1$ and

$$p_{k+1,\theta}(x) = k p_{k,\theta}(x) + \theta x \{p_{k,\theta}(x) - p'_{k,\theta}(x)\}, \qquad k \geq 0. \tag{8.51}$$

For example, $p_{1,\theta}(x) = \theta x$, $p_{2,\theta}(x) = \theta(1-\theta)x + \theta^2 x^2$, $p_{3,\theta}(x) = \theta(1-\theta)(2-\theta)x + 3\theta^2(1-\theta)x^2 + \theta^3 x^3$, etc. Since e^{-s^θ} is completely monotone, and because of representation (8.50), it follows that for all $x > 0$, $p_{k,\theta}(x) > 0$. Furthermore,

$$p_{k,\theta}(x) = \sum_{j=1}^{k} a_{k,j,\theta} x^j,$$

and it follows from (8.51) that for any $k \geq 1$,

$$\begin{cases} a_{k+1,1,\theta} &= (k-\theta) a_{k,1,\theta}, \\ a_{k+1,k+1,\theta} &= \theta a_{k,k,\theta}, \\ a_{k+1,j,\theta} &= (k - j\theta) a_{k,j,\theta} + \theta a_{k,j-1,\theta}, \quad j = 2, \ldots, k. \end{cases} \tag{8.52}$$

Note that (8.52) proves that all coefficients of the polynomials are non-negative and the leading term is positive, reinforcing the observation that $p_{k,\theta}(x) > 0$ for all $x \in (0, 1]$.

8.D Polynomials Related to the Frank Copula

It is easy to check that

$$(-1)^k \frac{d^k}{ds^k} \ln\left\{1-(1-\theta)e^{-s}\right\}/\ln(\theta) = p_k\left(\frac{y}{1-y}\right)/\ln(1/\theta), \tag{8.53}$$

where $y = (1-\theta)e^{-s}$, and the sequence of polynomials p_k is defined by $p_1(x) = x$ and

$$p_{k+1}(x) = x(1+x)p_k'(x), \qquad k \geq 1. \tag{8.54}$$

Thus p_k is a polynomial of degree k. For example, $p_2(x) = x(1+x)$, $p_3(x) = x(1+x)(1+2x)$, $p_4(x) = x(1+x)(1+6x+6x^2)$ and $p_5(x) = x(1+x)(1+2x)(1+12x+12x^2)$. Since $\log\{1-(1-\theta)e^{-s}\}/\log(\theta)$ is completely monotone, and because of representation (8.53), it follows that for all $x > 0$, $p_k(x) > 0$. Furthermore,

$$p_k(x) = \sum_{j=1}^{k} a_{k,j} x^j,$$

and it follows from (8.54) that for any $k \geq 1$,

$$\left\{\begin{array}{lcl} a_{k+1,1} & = & a_{k,1}, \\ a_{k+1,k+1} & = & k a_{k,k}, \\ a_{k+1,j} & = & j a_{k,j} + (j-1)a_{k,j-1}, \quad j = 2,\ldots,k. \end{array}\right. \tag{8.55}$$

Note that (8.55) proves that all coefficients of the polynomials are non-negative and the leading term is positive, reinforcing the observation that $p_k(x) > 0$ for all $x > 0$.

8.E Change Point Tests

For testing stationarity for a univariate time series, using either the series or the residuals estimating the innovations of a dynamic model, one could use the Kolmogorov-Smirnov statistic

$$\mathcal{T}_n = \mathcal{T}_n(e_1,\ldots,e_n) = \frac{1}{\sqrt{n}} \max_{1\leq k\leq n} \max_{1\leq i\leq n} \left| \sum_{j=1}^{k} \mathbb{I}(e_j \leq e_i) - kF_n(e_i) \right|,$$

where $F_n(x) = \frac{1}{n}\sum_{i=1}^{n} \mathbb{I}(e_i \leq x)$. Here $e_1,\ldots,e_n$ are the observations or the residuals.

The null hypothesis is that all the observations (respectively the innovations) have the same distribution, and the alternative hypothesis is that there is at time τ, such that $X_1, \ldots, X_\tau$ have the same distribution G_1, while $X_{\tau+1}, \ldots, X_n$ have distribution $G_2 \neq G_1$. Then one can show that under the null hypothesis, $\mathcal{T}_n$ converges in law to a parameter free distribution. To estimate the P-value associated with $\mathcal{T}_n$, one can use the following algorithm.

Algorithm 8.E.1 *For $k = 1$ to N, N large (say $N = 1000$), do the following:*

- *Generate independent uniformly distributed random variables $U_1^{(k)}, \ldots, U_n^{(k)}$.*
- *Compute $\mathcal{T}_n^{(k)} = \mathcal{T}_n\left(U_1^{(k)}, \ldots, U_n^{(k)}\right)$.*

The P-value can be estimated by

$$\frac{1}{N}\sum_{k=1}^{N} \mathbb{I}\left(\mathcal{T}_n^{(k)} > \mathcal{T}_n\right).$$

This change point test is implemented in the MATLAB function *TestChangePoint*.

8.E.1 Change Point Test for the Copula

According to Rémillard [2012a], for detecting a change point in the copula, one begins first by trying to detect a change point in each univariate series. If the null hypotheses are not rejected, then one can base the change point test for the copula on the Kolmogorov-Smirnov statistic

$$\mathcal{H}_n(e_1, \ldots, e_n) = \frac{1}{\sqrt{n}} \max_{1\le k\le n} \max_{1\le i\le n} \left| \sum_{j=1}^{k} \mathbb{I}(e_j \le e_i) - kH_n(e_i) \right|,$$

where $H_n(x) = \frac{1}{n}\sum_{i=1}^{n} \mathbb{I}(e_i \le x)$. Here $e_1, \ldots, e_n$ are the multivariate observations or residuals. Under the null hypothesis of stationarity, i.e., all observations or innovations have the same continuous distribution H, the statistic $\mathcal{H}_n$ converges in law to a random variable $\mathcal{H}$, whose distribution depends on the unknown copula C of H. The problem of calculating a P-value seems impossibly difficult, but it can be solved with the multipliers methodology, as proposed in Rémillard [2012a]. The algorithm is described next.

Algorithm 8.E.2 *For $k = 1$ to N, N large (say $N = 1000$), do the following:*

- *Generate independent standard Gaussian random variables $\xi_1^{(k)}, \ldots, \xi_n^{(k)}$.*
- *Compute*

$$\mathcal{H}_n^{(k)} = \frac{1}{\sqrt{n}} \max_{1\le k\le n} \max_{1\le i\le n} \left| \sum_{j=1}^{k} \xi_j^{(k)} \left\{\mathbb{I}(e_j \le e_i) - H_n(e_i)\right\} \right|.$$

The P-value can be estimated by

$$\frac{1}{N}\sum_{k=1}^{N} \mathbb{I}\left(\mathcal{H}_n^{(k)} > \mathcal{H}_n\right).$$

This test is implemented in the MATLAB function *TestChangePointCopula*.

8.F Auxiliary Results

The following identity is very useful.

Lemma 8.F.1 (Hoeffding identity) *If $(X_1, X_2) \sim H$, with margins F_1 and F_2, and if $E(X_1^2) < \infty$, $E(X_2^2) < \infty$, then*

$$\mathrm{Cov}(X_1, X_2) = \int_{\mathbb{R}^2} \{H(x_1, x_2) - F_1(x_1)F_2(x_2)\}\, dx_1 dx_2. \tag{8.56}$$

Note that as a particular case, one has

$$\mathrm{Var}(X_1) = 2\int_{x_1<x_2} F_1(x_1)\{1 - F_1(x_2)\}\, dx_1 dx_2.$$

An immediate application of Hoeffding identity is when H is a copula C, leading to the following expression for Spearman rho:

$$\rho^{(S)} = 12\,\mathrm{Cov}(U_1, U_2) = 12\int_0^1\int_0^1 \{C(u_1, u_2) - u_1 u_2\}\, du_1 du_2.$$

Note that one can also apply Hoeffding identity with empirical distribution functions.

8.G Proofs of the Results

8.G.1 Proof of Proposition 8.4.1

Since F_1 is symmetric, it follows that for any $u \in (0,1)$,

$$F_1^{-1}(u) + F_1^{-1}(1-u) = 2E(X_1).$$

Recall that the tent map T is defined by $T(u) = 2\min(u, 1-u)$. Then

$$\begin{aligned}
E(X_1X_2) &= \int_0^1 F_1^{-1}(u)F_2^{-1}\{T(u)\}du \\
&= \int_0^{1/2} F_1^{-1}(u)F_2^{-1}(2u)du + \int_{1/2}^1 F_1^{-1}(u)F_2^{-1}(2-2u)du \\
&= \int_0^{1/2} F_1^{-1}(u)F_2^{-1}(2u)du + \int_0^{1/2} F_1^{-1}(1-u)F_2^{-1}(2u)du \\
&= \int_0^{1/2} \left\{F_1^{-1}(u) + F_1^{-1}(1-u)\right\} F_2^{-1}(2u)du \\
&= 2E(X_1)\int_0^{1/2} F_2^{-1}(2u)du = E(X_1)E(X_2).
\end{aligned}$$

Hence $\text{Cov}(X_1, X_2) = 0$ and $\text{Cor}(X_1, X_2) = 0$.

■

8.G.2 Proof of Proposition 8.4.2

Suppose first that $u_1 \le v_1$ and $u_2 \le v_2$. Then

$$\begin{aligned}
|C(u_1,u_2) - C(v_1,v_2)| &= P(u_1 < U_1 \le v_1, U_2 \le v_2) \\
&\quad + P(U_1 \le u_1, u_2 < U_2 \le v_2) \\
&\le P(u_1 < U_1 \le v_1) + P(u_2 < U_2 \le v_2) \\
&= |u_1 - v_1| + |u_2 - v_2|,
\end{aligned}$$

since U_1 and U_2 are uniformly distributed. The case $v_1 \le u_1$ and $v_2 \le u_2$ is similar. Suppose now that $u_1 \le v_1$ and $v_2 \le u_2$. Then

$$\begin{aligned}
|C(u_1,u_2) - C(v_1,v_2)| &= |P(u_1 < U_1 \le v_1, U_2 \le v_2) \\
&\quad - P(U_1 \le u_1, v_2 < U_2 \le u_2)| \\
&\le P(u_1 < U_1 \le v_1) + P(v_2 < U_2 \le u_2) \\
&= |u_1 - v_1| + |u_2 - v_2|.
\end{aligned}$$

Next, for any $h \ge 0$, $0 \le C(u_1 + h, u_2) - C(u_1, u_2) \le h$, which implies for a fixed u_2 that $C(u_1, u_2)$ is absolutely continuous and by the fundamental theorem of calculus, $\partial_{u_1} C(u_1, u_2)$ exists almost surely, is bounded by 1 and for any $u_1 \in [0, 1]$,

$$C(u_1, u_2) = \int_0^{u_1} \partial_s C(s, u_2) ds.$$

■

8.G.3 Proof of Proposition 8.5.1

Suppose that $(U_1, U_2) \sim C$, $(U_1^\star, U_2^\star) \sim C^\star$, with (U_1, U_2) independent of $(U^\star, U_2^\star)$. Then $C(U_1, U_2) \le C^\star(U_1, U_2)$. Hence

$$\begin{aligned}
\frac{1+\tau_C}{4} &= E\left\{C(U_1,U_2)\right\} \le E\left\{C^\star(U_1,U_2)\right\} \\
&= E\left\{\mathbb{I}(U_1^\star \le U_1)\mathbb{I}(U_2^\star \le U_2)\right\} \\
&= E\left\{1 - \mathbb{I}(U_1 < U_1^\star) - I(U_2 < U_2^\star)\right. \\
&\qquad \left. + I(U_1 < U_1^\star)\mathbb{I}(U_2 < U_2^\star)\right\} \\
&= E\left\{C(U_1^\star, U_2^\star)\right\} \le E\left\{C^\star(U_1^\star, U_2^\star)\right\} = \frac{1+\tau_{C^\star}}{4}.
\end{aligned}$$

On the other hand, if $\tau_{C^\star} = \tau_C$, then $C(U_1^\star, U_2^\star) = C^\star(U_1^\star, U_2^\star)$ and $C(U_1, U_2) = C^\star(U_1, U_2)$. As a result, $C(u_1, u_2) = C^\star(u_1, u_2)$ for all $(u_1, u_2) \in [0,1]^2$.

■

8.G.4 Proof of Theorem 8.7.1

We prove the result only when $t \mapsto g_{k,\theta}$ is not increasing. The other case is similar. For simplicity, suppose that $\mathcal{O}$ is open. First, $\theta \mapsto C_\theta(u)$ is non-decreasing for all $u \in [0,1]^d$ if and only if for every $k \in \{1, \ldots, p\}$, $\partial_{\theta_k} C_\theta(u) \ge 0$ for all $u \in (0,1)^d$. Next, using the differentiation chain rule, we obtain that $\partial_s \phi_\theta^{-1}(s) = 1/\partial_t \phi_\theta\left\{\phi_\theta^{-1}(s)\right\}$ and

$$\partial_{\theta_k} \phi_\theta^{-1}(s) = -\frac{\partial_{\theta_k} \phi_\theta\left\{\phi_\theta^{-1}(s)\right\}}{\partial_t \phi_\theta\left\{\phi_\theta^{-1}(s)\right\}}.$$

As a result,

$$\partial_{\theta_k} C_\theta(u) = -\frac{\partial_\theta \phi_\theta\left\{C_\theta(u)\right\}}{\partial_t \phi_\theta\left\{C_\theta(u)\right\}} + \frac{1}{\partial_t \phi_\theta\left\{C_\theta(u)\right\}} \times \left\{\sum_{j=1}^d \partial_{\theta_k} \phi_\theta(u_j)\right\}.$$

Since $t \mapsto \phi_\theta(t)$ is decreasing, $\theta \mapsto C_\theta(u)$ is not decreasing for all $u \in (0,1)^d$ if and only if

$$\psi_{k,d}(u) = \partial_{\theta_k} \phi_\theta\left\{C_\theta(u)\right\} - \sum_{j=1}^d \partial_{\theta_k} \phi_\theta(u_k) \ge 0$$

for all $u \in (0,1)^d$, all $\theta \in \Theta$ and every $k \in \{1, \ldots, p\}$. Since $\psi_{k,d}(u_1, \ldots, u_{d-1}, 1) = \psi_{k,d-1}(u_1, \ldots, u_{d-1})$, and $\psi_{k,1}(u_1) = 0$, it suffices to show that for all $j \in \{2, \ldots, d\}$ and $u_1, \ldots, u_{j-1} \in (0,1)$, we have

$\partial_{u_j}\psi_{k,j}(\tilde{u}) \le 0$, where $\tilde{u} = (u_1, \ldots, u_j, 1, \ldots, 1)$. Now,

$$\begin{aligned}
\partial_{u_j}\psi_{k,j}(\tilde{u}) &= \partial_{\theta_k}\partial_t\phi_\theta\left\{C_\theta(\tilde{u})\right\}\partial_{u_j}C_\theta(\tilde{u}) - \partial_{\theta_k}\partial_t\phi_\theta(u_j) \\
&= \partial_{\theta_k}\partial_t\phi_\theta\left\{C_\theta(\tilde{u})\right\}\frac{\partial_t\phi_\theta(u_j)}{\partial_t\phi_\theta\left\{C_\theta(\tilde{u})\right\}} - \partial_{\theta_k}\partial_t\phi_\theta(u_j) \\
&= \partial_t\phi_\theta(u_j)\left[g_{k,\theta}\left\{C_\theta(\tilde{u})\right\} - g_{k,\theta}(u_j)\right].
\end{aligned}$$

Since $C_\theta(\tilde{u}) \le u_j$, we have $\partial_{u_j}\psi_j(\tilde{u}) \le 0$ if $g_{k,\theta}$ is not increasing for every $k \in \{1, \ldots, p\}$ and any $\theta \in \mathcal{O}$.

■

Bibliography

B. Abdous, C. Genest, and B. Rémillard. Dependence properties of meta-elliptical distributions. In *Statistical Modeling and Analysis for Complex Data Problems*, volume 1 of *GERAD 25th Anniv. Ser.*, pages 1–15. Springer, New York, 2005.

P. Barbe, C. Genest, K. Ghoudi, and B. Rémillard. On Kendall's process. *J. Multivariate Anal.*, 58(2):197–229, 1996.

T. Berrada, D. J. Dupuis, E. Jacquier, N. Papageorgiou, and B. Rémillard. Credit migration and derivatives pricing using copulas. *J. Comput. Fin.*, 10:43–68, 2006.

W. Breymann, A. Dias, and P. Embrechts. Dependence structures for multivariate high-frequency data in finance. *Quant. Finance*, 3:1–14, 2003.

J. M. Chambers, C. L. Mallows, and B. W. Stuck. A method for simulating stable random variables. *Journal of the American Statistical Association*, 71(354):340–344, 1976.

X. Chen and Y. Fan. Estimation and model selection of semiparametric copula-based multivariate dynamic models under copula misspecification. *Journal of Econometrics*, 135(1-2):125–154, 2006.

U. Cherubini, E. Luciano, and W. Vecchiato. *Copula Methods in Finance.* Wiley Finance. Wiley, New York, 2004.

H. E. Daniels. Rank correlation and population models. *J. Roy. Statist. Soc. Ser. B.*, 12:171–181, 1950.

L. Devroye. *Non-Uniform Random Variate Generation.* Springer-Verlag, New York, 1986.

P. Duchesne, K. Ghoudi, and B. Rémillard. On testing for independence between the innovations of several time series. *Canad. J. Statist.*, 40:447–479, 2012.

P. Embrechts, A. J. McNeil, and D. Straumann. Correlation and dependence in risk management: properties and pitfalls. In *Risk Management: Value at Risk and Beyond (Cambridge, 1998)*, pages 176–223. Cambridge Univ. Press, Cambridge, 2002.

H.-B. Fang, K.-T. Fang, and S. Kotz. The meta-elliptical distributions with given marginals. *J. Multivariate Anal.*, 82(1):1–16, 2002.

H.-B. Fang, K.-T. Fang, and S. Kotz. Corrigendum to: "The meta-elliptical distributions with given marginals" [J. Multivariate Anal. **82** (2002), no. 1, 1–16; mr 1918612]. *J. Multivariate Anal.*, 94(1):222–223, 2005.

W. Feller. *An Introduction to Probability Theory and its Applications*, volume II of *Wiley Series in Probability and Mathematical Statistics.* John Wiley & Sons, second edition, 1971.

J.-D. Fermanian, D. Radulović, and M. H. Wegkamp. Weak convergence of empirical copula processes. *Bernoulli*, 10:847–860, 2004.

P. Gänssler and W. Stute. *Seminar on Empirical Processes*, volume 9 of *DMV Seminar.* Birkhäuser Verlag, Basel, 1987.

C. Genest and J. MacKay. The joy of copulas: bivariate distributions with uniform marginals. *The American Statistician*, 40:280–283, 1986a.

C. Genest and R. J. MacKay. Copules archimédiennes et familles de lois bidimensionnelles dont les marges sont données. *The Canadian Journal of Statistics*, 14(2):145–159, 1986b.

C. Genest and B. Rémillard. Tests of independence or randomness based on the empirical copula process. *Test*, 13:335–369, 2004.

C. Genest and B. Rémillard. Discussion of Copulas: Tales and facts, by Thomas Mikosch. *Extremes*, 9:27–36, 2006.

C. Genest and B. Rémillard. Validity of the parametric bootstrap for goodness-of-fit testing in semiparametric models. *Ann. Inst. H. Poincaré Sect. B*, 44:1096–1127, 2008.

C. Genest and L.-P. Rivest. Statistical inference procedures for bivariate Archimedean copulas. *J. Amer. Statist. Assoc.*, 88(423):1034–1043, 1993.

C. Genest and F. Verret. Locally most powerful rank tests of independence for copula models. *J. Nonparametr. Stat.*, 17(5):521–539, 2005.

C. Genest and B. J. M. Werker. Conditions for the asymptotic semiparametric efficiency of an omnibus estimator of dependence parameters in copula models. In *Distributions with Given Marginals and Statistical Modelling*, pages 103–112. Kluwer Acad. Publ., Dordrecht, 2002.

C. Genest, K. Ghoudi, and L.-P. Rivest. A semiparametric estimation procedure of dependence parameters in multivariate families of distributions. *Biometrika*, 82:543–552, 1995.

C. Genest, J.-F. Quessy, and B. Rémillard. Goodness-of-fit procedures for copula models based on the integral probability transformation. *Scand. J. Statist.*, 33:337–366, 2006a.

C. Genest, J.-F. Quessy, and B. Rémillard. Local efficiency of a Cramér-von Mises test of independence. *J. Multivariate Anal.*, 97:274–294, 2006b.

C. Genest, B. Rémillard, and D. Beaudoin. Omnibus goodness-of-fit tests for copulas: A review and a power study. *Insurance Math. Econom.*, 44: 199–213, 2009.

K. Ghoudi and B. Rémillard. Empirical processes based on pseudo-observations. II. The multivariate case. In *Asymptotic Methods in Stochastics*, volume 44 of *Fields Inst. Commun.*, pages 381–406. Amer. Math. Soc., Providence, RI, 2004.

C. Hering, M. Hofert, J.-F. Mai, and M. Scherer. Constructing hierarchical Archimedean copulas with Lévy subordinators. *J. Multivariate Anal.*, 101 (6):1428–1433, 2010.

R. A. Jarrow, D. Lando, and S. M. Turnbull. A Markov model for the term structure of credit risk spreads. *The Review of Financial Studies*, 10:481–523, 1997.

H. Joe. Asymptotic efficiency of the two-stage estimation method for copula-based models. *J. Multivariate Anal.*, 94:401–419, 2005.

H. M. Kat and H. P. Palaro. Who needs hedge funds? A copula-based approach to hedge fund return replication. Technical report, Cass Business School, City University, 2005.

I. Kojadinovic and J. Yan. A goodness-of-fit test for multivariate multiparameter copulas based on multiplier central limit theorems. *Stat. Comput.*, 21:17–30, 2010.

D. Kurowicka and H. Joe, editors. *Dependence Modeling: Vine Copula Handbook.* World Scientific, 2011.

D.X. Li. On default correlation: a copula function approach. *Journal of Fixed Income*, 9:43–54, 2000.

A. W. Marshall and I. Olkin. Families of multivariate distributions. *Journal of the American Statistical Association*, 83:834–841, 1988.

A. J. McNeil. Sampling nested Archimedean copulas. *Journal of Statistical Computation and Simulation*, 78(6):567–581, 2008.

T. Mikosch. Copulas: Tales and facts. *Extremes*, 9:3–20, 2006.

A. Müller. Stop-loss order for portfolios of dependent risks. *Insur. Math. Econ.*, 21:219–223, 1997.

V. Panchenko. Goodness-of-fit test for copulas. *Phys. A*, 355:176–182, 2005.

N. Papageorgiou, B. Rémillard, and A. Hocquard. Replicating the properties of hedge fund returns. *Journal of Alternative Invesments*, 11:8–38, 2008.

A. J. Patton. Modelling asymmetric exchange rate dependence. *International Economic Review*, 47(2):527–556, 2006.

A. J. Patton. Copula-based models for financial time series. In Thomas Mikosch, Jens-Peter Krei, Richard A. Davis, and Torben Gustav Andersen, editors, *Handbook of Financial Time Series*, pages 767–785. Springer Berlin Heidelberg, 2009.

A. J. Patton. A review of copula models for economic time series. *J. Multivariate Anal.*, 110:4–18, 2012.

B. Rémillard. Goodness-of-fit tests for copulas of multivariate time series. Technical report, SSRN Working Paper Series No. 1729982, 2010.

B. Rémillard. Non-parametric change point problems using multipliers. Technical report, SSRN Working Paper Series No. 2043632, 2012a.

B. Rémillard. Specification tests for dynamic models using multipliers. Technical report, SSRN Working Paper Series No. 2028558, 2012b.

B. Rémillard and O. Scaillet. Testing for equality between two copulas. *J. Multivariate Anal.*, 100:377–386, 2009.

B. Rémillard, N. Papageorgiou, and F. Soustra. Copula-based semiparametric models for multivariate time series. *J. Multivariate Anal.*, 110:30–42, 2012.

T. Roncalli. *La Gestion des Risques Financiers*. Gestion. Economica, 2004.

J. H. Shih and T. A. Louis. Inferences on the association parameter in copula models for bivariate survival data. *Biometrics*, 51:1384–1399, 1995.

M. Sklar. Fonctions de répartition à n dimensions et leurs marges. *Publ. Inst. Statist. Univ. Paris*, 8:229–231, 1959.

R. W. J. van den Goorbergh, C. Genest, and B. J. M. Werker. Bivariate option pricing using dynamic copula models. *Insurance: Mathematics and Economics*, 37:101–114, 2005.

J. J. Xu. *Statistical Modelling and Inference for Multivariate and Longitudinal Discrete Response Data*. PhD thesis, University of British Columbia, 1996.

Chapter 9

Filtering

In this chapter we will study the Kalman filter, a very efficient adaptive method to predict the values of a non-observable Gaussian process, called the signal, given observations obtained from a linear transformation of the signal. Then we will define the general problem of filtering, to finish with a class of Monte Carlo filters. We will also tackle the problem of the estimation of the parameters.

9.1 Description of the Filtering Problem

The filtering problem is about predicting non observable random vectors $Z_1, Z_2, \ldots$, called the signal, by using observations $Y_1, \ldots, Y_n$ linked in a certain way to the signal. In fact, what we really want to do is to find the conditional distribution of Z_{i+k} given $Y_1, \ldots, Y_i$, for $k \geq 0$. We also want to rapidly incorporate new observations $Y_{i+1}, Y_{i+2}, \ldots$. Here are some examples of signals in financial applications: stochastic volatility, detention yield, betas of dynamic portfolios, regimes. Observations could be, for example, the returns of an asset, values of futures contracts, etc.

To be able to do computations, we assume that the signal is a Markov process. We start by giving an example of filtering problem in the financial literature.

Example 9.1.1 *In Schwartz [1997], the author proposed to model the relationship between a commodity spot price S, its convenience yield δ and a futures contract F. For the joint dynamics of the spot price and the convenience yield, he assumed that*

$$\begin{aligned} d\ln\{S(t)\} &= \{\mu - \delta(t) - \sigma_1^2/2\}\, dt + \sigma_1 dW_1(t) \\ d\delta(t) &= \kappa\{\alpha - \delta(t)\}dt + \sigma_2 dW_2(t). \end{aligned}$$

where W_1 and W_2 are correlated Brownian motions, with correlation ρ. Assuming a constant short-term interest r, the value at time t of the futures prices of the commodity with time to maturity τ, is given by

$$F(t) = S(t)e^{A(\tau)-\delta(t)B(\tau)},$$

where $B(\tau) = \frac{1-e^{-\kappa\tau}}{\kappa}$*. The exact form of* $A(\tau)$ *is given in Appendix 9.A. The role of the signal is played here by* $Z_i = (\ln\{S(ih)\}, \delta(ih))^\top$*, for some fixed time interval* h*. Then, one can write*

$$Z_i = c + GZ_{i-1} + \eta_i,$$

for some vector c *and some matrix* G*, where the error terms* η_i *are independent centered Gaussian random vectors. See Appendix 9.A for more details. Here, the observations are* $Y_i = \ln\{F(ih)\}$*, so they are related in a linear way to the signal* Z_i *viz.*

$$Y_i = A(\tau_i) + H_i Z_i + \varepsilon_i,$$

with $H_i = \left(1, -\frac{1-e^{-\kappa\tau_i}}{\kappa}\right)$*, and* $R_i = 0$*.*

9.2 Kalman Filter

This ingenious algorithm has been proposed in a seminal paper by Kalman [1960]. It gives a way to update recursively the conditional distribution of the signal given the information obtained from the observations $Y_1, \ldots, Y_i$, when the combined system (Y, Z) is Gaussian.

9.2.1 Model

The dynamical evolution of the signal is modeled by

$$Z_i = \mu_i + F_i Z_{i-1} + G_i w_i, \tag{9.1}$$

where $F_i \in \mathbb{R}^{m\times m}$, $G_i \in \mathbb{R}^{m\times q}$, and $w_i \sim N_q(0, Q_i)$. It is assumed that $(w_i)_{i\geq 1}$ is an independent sequence. The relationship between the observations and the signal is

$$Y_i = d_i + H_i Z_i + \varepsilon_i, \tag{9.2}$$

where $Y_i, d_i \in \mathbb{R}^r$, $H_i \in \mathbb{R}^{r\times m}$, and $\varepsilon_i \sim N_r(0, R_i)$. It is assumed that $(\varepsilon_i)_{i\geq 1}$ is an independent sequence. In addition, Z_{i-1} is independent of w_i, and the errors terms w_i and ε_i are independent.

For $i \geq 1$, set

$$\begin{array}{lcl} Y_{i|i-1} & = & E(Y_i|Y_{i-1}, \ldots, Y_1), \\ Z_{i|i-1} & = & E(Z_i|Y_{i-1}, \ldots, Y_1), \\ Z_{i|i} & = & E(Z_i|Y_i, \ldots, Y_1), \\ P_{i|i-1} & = & E\left\{(Z_i - Z_{i|i-1})(Z_i - Z_{i|i-1})^\top|Y_{i-1}, \ldots, Y_1\right\}, \\ P_{i|i} & = & E\left\{(Z_i - Z_{i|i})(Z_i - Z_{i|i})^\top|Y_i, \ldots, Y_1\right\}, \\ e_i & = & Y_i - Y_{i|i-1}, \\ V_i & = & E\left(e_i e_i^\top|Y_{i-1}, \ldots, Y_1\right). \end{array} \tag{9.3}$$

The usefulness of the system of equations (9.3) is shown in the next proposition. Its proof is given in Appendix 9.D.1.

Proposition 9.2.1 *Suppose that $Z_0 \sim N_m(Z_{0|0}, P_{0|0})$. Then, all conditional distributions are Gaussian.*

In particular, the conditional distribution of Z_i given $Y_1, \ldots, Y_i$ is Gaussian, with mean $Z_{i|i}$ and covariance matrix $P_{i|i}$, while the conditional distribution of Z_{i+1} given $Y_1, \ldots, Y_i$ is Gaussian, with mean $Z_{i+1|i}$ and covariance matrix $P_{i+1|i}$.

Finally, the conditional distribution of Y_i given $Y_1, \ldots, Y_{i-1}$ is Gaussian, with mean $Y_{i|i-1}$ and covariance matrix V_i. Moreover, the conditional means and covariances can be computed recursively using Algorithm 9.2.1 below.

Algorithm 9.2.1 (Kalman Filter Algorithm) *Given, $Z_{0|0}$ and $P_{0|0}$, we have, for all $i \geq 1$,*

$$\begin{array}{lcl} Z_{i|i-1} & = & \mu_i + F_i Z_{i-1|i-1}, \\ Y_{i|i-1} & = & d_i + H_i Z_{i|i-1}, \\ e_i & = & Y_i - Y_{i|i-1}, \\ P_{i|i-1} & = & F_i P_{i-1|i-1} {F_i}^\top + G_i Q_i {G_i}^\top, \\ V_i & = & H_i P_{i|i-1} {H_i}^\top + R_i, \\ Z_{i|i} & = & Z_{i|i-1} + P_{i|i-1} {H_i}^\top {V_i}^{-1} e_i, \\ P_{i|i} & = & P_{i|i-1} - P_{i|i-1} {H_i}^\top {V_i}^{-1} H_i P_{i|i-1}. \end{array} \tag{9.4}$$

$K_i = P_{i|i-1} {H_i}^\top {V_i}^{-1}$ is often called the "Kalman gain" matrix. It follows that

$$Z_{i|i} = Z_{i|i-1} + K_i e_i = (I - K_i H_i) Z_{i|i-1} + K_i (Y_i - d_i),$$

$$P_{i|i} = (I - K_i H_i) P_{i|i-1}.$$

MATLAB vs C

When working with MATLAB, the matrices H_i, F_i should be defined as 3-dimensional arrays, with the last index being the period i. This way, $H(:,:,i)$ will be recognized as a matrix. Note that $H(i,:,:)$ is not considered a matrix in MATLAB.

On the other hand, using C, the time period i should be the first component, because in that case, $H[i]$ is interpreted as a matrix in C.

9.2.2 Filter Initialization

Kalman equations depend on the initial values $Z_{0|0}$ and $P_{0|0}$ which are the (unknown) mean and covariance matrix of the initial distribution of Z_0. In most cases, taking $Z_{0|0} = 0$ and $P_{0|0} = aI_m$, $a > 0$ large enough will produce good results.

Initialization of $P_{0|0}$

If the diagonal elements of $P_{0|0}$ are too small, it could slow down the convergence of the algorithm or even lead to numerical inconsistencies since there are many matrix inversions involved.

9.2.3 Estimation of Parameters

In most applications, the models will depend on unknown parameters. Using Proposition 9.2.1 and the multiplication formula (A.19), we can compute the joint distribution of $Y_1, \ldots, Y_n$. In fact, the log-likelihood L is given by

$$-L = \frac{1}{2}\sum_{i=1}^{n} \ln(\det V_i) + \frac{1}{2}\sum_{i=1}^{n} e_i^\top V_i^{-1} e_i + \frac{nr}{2}\ln(2\pi). \tag{9.5}$$

One can the use the maximum likelihood principle to estimate the parameters.

Remark 9.2.1 *In theory, one should select a sub-sample consisting of the first n_0 observations and estimate the parameters, then run the filter, without re-estimating the parameters.*

However, it is common practice in the industry to estimate the parameters over a rolling window, using the last n_0 observations, in case there are changes in the evolution and/or measurement equations. However that slows down the algorithm a bit.

9.2.4 Implementation of the Kalman Filter

The system of equations (9.8) is correct in an ideal world (where precision is infinite), but in the real world, there are rounding errors that could affect the computations. For example, the matrix $P_{i|i}$ is supposed to be positive definite. It may happen, due to rounding errors, that $P_{i|i}$ is not positive definite, creating problems if we have to compute $\sqrt{\det(P_{i|i})}$.

9.2.4.1 Solution

Using Proposition 9.B.1 in Appendix 9.B, we have

$$P_{i|i}^{-1} = P_{i|i-1}^{-1} + H_i^\top R_i^{-1} H_i \tag{9.6}$$

and

$$K_i = P_{i|i-1} H_i^\top V_i^{-1} = P_{i|i} H_i^\top R_i^{-1}.$$

As a result, there is no need to compute V_i anymore, since

$$Z_{i|i} = Z_{i|i-1} + P_{i|i} H_i^\top R_i^{-1} e_i. \tag{9.7}$$

It follows from (9.6) that $P_{i|i}^{-1}$ is positive definite, so the same will be true for its inverse $P_{i|i}$. Note that if R_i does not depend on i, the representation of the Kalman gain K_i with R_i^{-1} instead of V_i^{-1}, is much faster to compute.

Summarizing, we have an alternative way to express the Kalman Filter algorithm, not depending on V_i, and such that $P_{i|i}$ is positive definite.

Algorithm 9.2.2 (Kalman Filter Algorithm) *Given, $Z_{0|0}$ and $P_{0|0}$, we have, for all $i \geq 1$,*

$$\begin{array}{lcl} Z_{i|i-1} & = & \mu_i + F_i Z_{i-1|i-1}, \\ Y_{i|i-1} & = & d_i + H_i Z_{i|i-1}, \\ e_i & = & Y_i - Y_{i|i-1}, \\ P_{i|i-1} & = & F_i P_{i-1|i-1} F_i^\top + G_i Q_i G_i^\top, \\ P_{i|i}^{-1} & = & P_{i|i-1}^{-1} + H_i^\top R_i^{-1} H_i, \\ Z_{i|i} & = & Z_{i|i-1} + P_{i|i} H_i^\top R_i^{-1} e_i. \end{array} \tag{9.8}$$

Example 9.2.1 *Suppose that in the Kalman model, the transition equation is $Z_i = \alpha + Z_{i-1} + w_i$, with $w_i \sim N(0, Q)$, and that the measurement equation is $Y_i = Z_i + \varepsilon_i$, with $\varepsilon_i \sim N(0, R)$.*

Note that this model has the following financial interpretation: The signal Z_i corresponds to $\ln\{S(ih)\}$, where S is the "fundamental price" of an asset under the Black-Scholes model, and Y_i is the logarithm of the adjusted closing price of the asset at time $t = ih$, if one believes that the closing price is not the fundamental value of the asset. Then $\alpha = h\left(\mu - \frac{\sigma}{2}\right)$, $Q = \sigma^2 h$ and the illiquidity of the asset could be measured by σ_{obs}, where $R = \sigma_{obs}^2 h$.

Applying the Kalman Filter Algorithm 9.2.1 to this simple model, we obtain

$$\begin{aligned} Z_{i|i-1} &= \alpha + Z_{i-1|i-1}, \\ Y_{i|i-1} &= Z_{i|i-1} = \alpha + Z_{i-1|i-1}, \\ e_i &= Y_i - Z_{i|i-1} = Y_i - \alpha - Z_{i-1|i-1}, \\ P_{i|i-1} &= P_{i-1|i-1} + Q, \\ V_i &= P_{i|i-1} + R = P_{i-1|i-1} + Q + R, \\ Z_{i|i} &= Z_{i|i-1} + P_{i|i-1} V_i^{-1} e_i \\ &= \alpha + Z_{i-1|i-1} + \left(\frac{P_{i-1|i-1} + Q}{P_{i-1|i-1} + Q + R} \right) e_i \\ &= \alpha + Z_{i-1|i-1} + K_i e_i, \\ P_{i|i} &= P_{i|i-1} - P_{i|i-1} V_i^{-1} P_{i|i-1} \\ &= R \left(\frac{P_{i-1|i-1} + Q}{P_{i-1|i-1} + Q + R} \right). \end{aligned}$$

As a result, whenever $P_{0|0} > 0$, *we have*

$$P_{i|i} \to P_{\infty|\infty} = -\frac{Q}{2} + \sqrt{\frac{Q^2}{4} + QR}.$$

Setting $U_i = \frac{R}{V_i}$, *we can rewrite some equations to obtain*

$$U_{i+1} = \frac{1}{\frac{Q}{R} + 2 - U_i}, \qquad Z_{i|i} = U_i(\alpha + Z_{i-1|i-1}) + (1 - U_i)Y_i.$$

Finally, note that as $i \to \infty$, $U_i \to \frac{R}{P_{\infty|\infty}+Q+R} < 1$. *Also, if one starts two filters, one from* $(Z_{0|0}, P_{0|0})$ *and one from* $(\tilde{Z}_{0|0}, P_{0|0})$, *then*

$$Z_{i|i} - \tilde{Z}_{i|i} = U_1 \cdots U_i(Z_{0|0} - \tilde{Z}_{0|0}) \to 0.$$

As a result, the filters forget their starting points, producing the same predictions in the long term. Therefore, it is illusory to try to estimate the starting points $Z_{0|0}$ *and* $P_{0|0}$. *However, note that the estimation of* Q *may be sensitive to the values the starting points* $Z_{0|0}$ *and* $P_{0|0}$.

*As an illustration, consider data from Apple (*ApplePrices*) stock on Nasdaq (aapl), from January 13*th*, 2009, to January 14*th*, 2011. The results of Table 9.1 have been obtained with the MATLAB function* EstKalmanBS, *using the proposed model, starting at points* $(0, 1)$ *and* $(0, 10)$. *For sake of comparison, the estimation of* μ *and* σ *from the Black-Scholes model were respectively* 0.7314 *and* 0.2978. *The graphs of the observed log-returns and the predictions are given in Figure 9.1.*

TABLE 9.1: Estimation of the parameters of the signal and observations for Apple, using Kalman filter.

Parameter	Kalman Filter from $(0,1)$	Kalman Filter from $(0,10)$
μ	0.7321 ± 0.4205	0.7314 ± 0.4205
σ	0.2975 ± 0.0255	0.2975 ± 0.0255
σ_{obs}	0.0001 ± 33.4854	0.0001 ± 25.7248

Remark 9.2.2 *Although it is not always true for general Kalman models, we can often show that the Kalman filter is stable in the sense that it will forget its initial values* $Z_{0|0}$ *and* $P_{0|0}$.

Example 9.2.2 *Suppose that in the Kalman model, the transition equation is* $Z_i = \alpha + \phi Z_{i-1} + w_i$, *with* $w_i \sim N(0, Q)$, *and the measurement equation is* $Y_i = Z_i + \varepsilon_i$, *with* $\varepsilon_i \sim N(0, R)$.

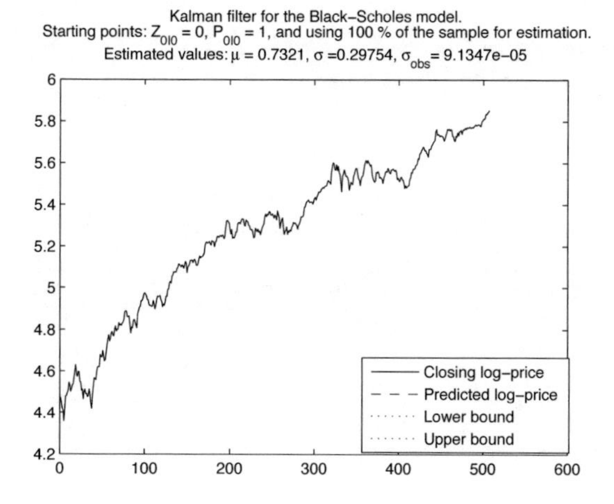

FIGURE 9.1: Observed log-prices and their predictions, using the Kalman filter starting at $(0, 1)$.

Applying the Kalman Algorithm 9.2.1, we obtain

$$\begin{aligned}
Z_{i|i-1} &= \alpha + \phi Z_{i-1|i-1}, \\
Y_{i|i-1} &= \alpha + \phi Z_{i-1|i-1}, \\
e_i &= Y_i - \alpha - \phi Z_{i-1|i-1}, \\
V_i &= \phi^2 P_{i-1|i-1} + Q + R, \\
Z_{i|i} &= \alpha + \phi Z_{i-1|i-1} + \left(\frac{\phi^2 P_{i-1|i-1} + Q}{\phi^2 P_{i-1|i-1} + Q + R} \right) e_i \\
P_{i|i} &= R \left(\frac{\phi^2 P_{i-1|i-1} + Q}{\phi^2 P_{i-1|i-1} + Q + R} \right).
\end{aligned}$$

As a result, whenever $P_{0|0} > 0$, *we have*

$$P_{i|i} \to P_{\infty|\infty} = \frac{-Q - R(1-\phi^2) + \sqrt{Q^2 + 2QR(1+\phi^2) + R^2(1-\phi^2)^2}}{2\phi^2}.$$

As an illustration of the implementation of this Kalman filter, we simulated 250 observations for the model considered here with $\alpha = 0.008$, $\phi = 0.6$, $Q = 0.00001$, *and* $R = 0.001$. *Data can be found in the file* DataKalman. *Using the MATLAB function* EstKalman *with the starting points* $Z_{0|0} = 0$, $P_{0|0} = 1$ *and taking the first* $n_0 = 125$ *observations to estimate the parameters, we obtained the results displayed in Table 9.2. With these parameters, we find that the out-of-sample RMSE (root mean square error) of the signal* Z_i *and the predictions* $Z_{i|i}$, *computed for* $i \in \{n_0, \ldots, n\}$, *is* 0.0045. *Note that the in-sample RMSE using the whole sample is .0043. The graphs in Figure 9.2 illustrate the good quality of the prediction.*

TABLE 9.2: 95% confidence intervals for the estimated parameters for the first 250 observations. The out-of-sample RMSE is 0.0045.

Parameter	True Value	95% Confidence Interval
100α	0.80	0.8069 ± 0.3522
ϕ	0.60	0.6162 ± 0.1281
$100Q$	0.01	0.0083 ± 0.0230
$100R$	0.10	0.1117 ± 0.0422

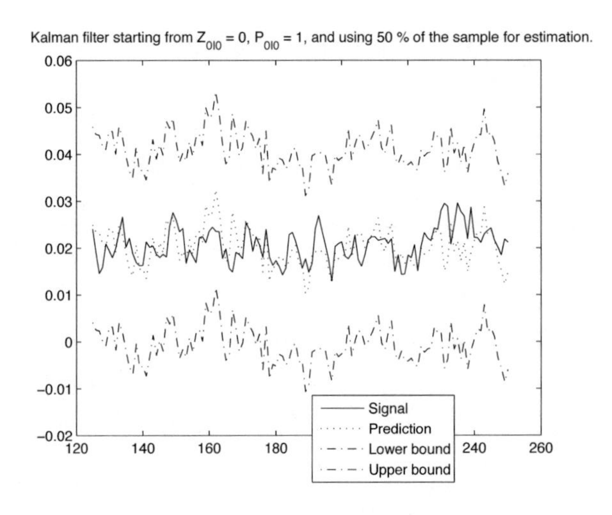

FIGURE 9.2: Graphs of the signal and out-of-sample prediction, with 95% confidence intervals.

Remark 9.2.3 *It might be possible that one cannot estimate all parameters of a model. Taking $\phi = 0$ in Example 9.2.2, one cannot estimate Q and R; one can only estimate $Q + R$. Taking $\phi = 1$ in Example 9.2.2, one recovers Example 9.2.1 and it is possible to get trapped into a local minimum if Q is too small. In fact, one could end up with a null estimation of Q, so the values $Z_{i|i}$ will be linear in i, for i large enough, and $P_{i|i}$ will be almost 0. If one ends up with a zero estimation for Q, one could just reduce the sample size for the estimation.*

9.2.5 The Kalman Filter for General Linear Models

What happens when the Kalman filter is used when the innovations are square integrable but are not Gaussian? The Kalman algorithm will then provide linear predictions $Z_{i|i}$ of Z_i given $Y_1, \ldots, Y_i$. Only in the case of Gaussian innovations are these Kalman predictions the best minimum mean square predictions.

The next question to ask is how can we estimate the parameters? Even if the innovations are not Gaussian, we could use the likelihood method as if they were Gaussian. This methodology is called quasi-maximum likelihood. Conditions of convergence of the quasi-likelihood estimators can be found in Watson [1989]. In general, these conditions are met for the Kalman filter with non-Gaussian innovations. In fact, setting $\ell_i = \ln(\det V_i) + \frac{1}{2}\sum_{i=1}^n e_i^\top V_i^{-1} e_i + \frac{nr}{2}\ln(2\pi)$, and defining

$$J_n = \frac{1}{n}\sum_{i=1}^n \nabla_\theta \ell_i \nabla_\theta \{\ell_i\}^\top, \quad H_n = \frac{1}{n}\sum_{i=1}^n \nabla_\theta^2 \ell_i,$$

we have that $\sqrt{n}(\hat{\theta}_n - \theta) \rightsquigarrow N_p(0, V)$, where V is estimated by $H_n^{-1} J_n H_n^{-1}$. Here, ∇^2 stands for the Hessian matrix. Note that for maximum likelihood estimators, $H_n \approx J_n$, so $V \approx J_n^{-1}$, while in the quasi-maximum likelihood case, $H_n \neq J_n$ in general.

Take for example the following model stochastic volatility studied in Ruiz [1994]: The transition equation is $Z_i = \omega + \phi Z_{i-1} + w_i$, with $w_i \sim N(0, \sigma^2)$, and that the measurement equation is $Y_i = \varepsilon_i e^{Z_i/2}$, with $\varepsilon_i \sim N(0,1)$. The measurement equation can be linearized viz.

$$\ln(Y_i^2) = Z_i + \ln(\varepsilon_i^2), \quad i \geq 1.$$

The independent variables $\ln(\varepsilon_i^2)$ have mean $-\gamma - \ln(2) \approx -1.2704$ and variance $\pi^2/2$, where γ is the Euler constant.[1] In Ruiz [1994], the quasi-maximum likelihood estimators of the previous stochastic volatility model were shown to be asymptotically Gaussian. The methodology is implemented in the MATLAB function *EstKalmanSV*.

Example 9.2.3 *As an illustration, we simulated 2500 observations for the stochastic volatility model considered here with* $Z_0 = \omega/(1-\phi)$, $\omega = -0.1$, $\phi = 0.75$, *and* $\sigma = 0.25$. *These data can be found in the MATLAB file* DataSV. *The results of the estimation, using the first half and the whole sample, are reported in Table 9.3. As one can see from that table, the precision of the estimation is not good. At least, the 95% confidence intervals using the whole sample cover the true values. The out-of-sample RMSE is* 0.4089, *using 1250 observations, while the RMSE, using the whole sample, is* 0.4091.

[1] The Euler constant $\gamma \approx 0.5772$ is defined as the limit of the decreasing sequence $1 + \frac{1}{2} + \cdots + \frac{1}{n} - \ln(n)$.

TABLE 9.3: 95% confidence intervals for the estimated parameters, using quasi-maximum likelihood.

Parameter	True Value	95% Confidence Interval	
		50% of the Sample	100% of the Sample
ω	-0.10	-0.5088 ± 0.4116	-0.2858 ± 0.3785
ϕ	0.75	-0.2680 ± 1.0693	0.3380 ± 0.8041
σ	0.25	0.6641 ± 0.7461	0.6999 ± 0.5801

9.3 IMM Filter

An IMM (interacting multiple model) consists in a dynamical coupling of several linear transition models governed by a Markov chain. The corresponding IMM filter has been popularized by Blom and Bar-Shalom [1988], mainly for radar applications, where the target can switch from a constant speed to a constant acceleration. More precisely, if the non-observable regime is $\lambda_i = j$, the transition and measurement equations are

$$
\begin{aligned}
Z_i &= \mu_i^{(j)} + F_i^{(j)} Z_{i-1} + G_i^{(j)} w_i^{(j)}, \\
Y_i &= d_i^{(j)} + H_i^{(j)} Z_i + \epsilon_i^{(j)},
\end{aligned}
$$

where $w_i^{(j)} \sim N_q\left(0, Q_i^{(j)}\right)$, $G_i^{(j)}$, $d_i^{(j)}$ and $H_i^{(j)}$ are the same as in the Kalman filter model. Moreover, all error terms are assumed to be independent. Finally, $(\lambda_i)_{i\geq 1}$ is a Markov chain with $P(\lambda_i = j|\lambda_{i-1} = k) = p_{kj}$, $k, j \in \{1, \ldots, s\}$. Note that Z is not a Markov process in general, but (Z, λ) is. If fact, (Z, λ) now plays the role of the signal.

9.3.1 IMM Algorithm

Contrary to the Kalman filter in a Gaussian context, the IMM filter is not optimal in the sense that it does not yield the conditional distribution of the signal (Z_i, λ_i) given the observations $Y_1, \ldots, Y_i$. However it is a good approximation in the Gaussian case.

Assume that the initial states $Z_{0|0}^{(j)}$, covariances $P_{0|0}^{(j)}$ and probabilities $\nu_0^{(j)}$ are given for all $j \in \{1, \ldots, s\}$. To go from period $i-1$ to period i, do the following:

- **Interaction step:** For all $j \in \{1, \ldots, s\}$, compute:

- The weighted probabilities

$$\nu_{i-1}^{(kj)} = p_{kj}\nu_{i-1}^{(k)}/\bar{c}_{i-1}^{(j)}, \quad k \in \{1,\ldots,s\}, \tag{9.9}$$

with $\bar{c}_{i-1}^{(j)} = \sum_{k=1}^{s} p_{kj}\nu_{i-1}^{(k)}$.

- The weighted states

$$\hat{Z}_{i-1}^{(0j)} = \sum_{k=1}^{s} Z_{i-1|i-1}^{(k)}\nu_{i-1}^{(kj)}.$$

- The weighted covariances

$$\begin{aligned} P_{i-1}^{(0j)} &= \sum_{k=1}^{s} \nu_{i-1}^{(kj)} P_{i-1|i-1}^{(k)} \\ &\quad + \sum_{k=1}^{s} \nu_{i-1}^{(kj)} \left(Z_{i-1|i-1}^{(k)} - \hat{Z}_{i-1}^{(0j)}\right)\left(Z_{i-1|i-1}^{(k)} - \hat{Z}_{i-1}^{(0j)}\right)^\top. \end{aligned}$$

- **Filtering step:** For any $j \in \{1,\ldots,s\}$, compute the following Kalman equations, where $Z_{i-1|i-1}^{(j)}$ and $P_{i-1|i-1}^{(j)}$ are replaced by $\hat{Z}_{i-1}^{(0j)}$ and $P_{i-1}^{(0j)}$:

$$\begin{aligned} Z_{i|i-1}^{(j)} &= F_i^{(j)}\hat{Z}_{i-1}^{(0j)} \\ P_{i|i-1}^{(j)} &= F_i^{(j)}P_{i-1}^{(0j)}F_i^{(j)\top} + G_i^{(j)}Q_i^{(j)}G_i^{(j)\top} \\ V_i^{(j)} &= H_i^{(j)}P_{i|i-1}^{(j)}H_i^{(j)\top} + R_i^{(j)} \\ K_i^{(j)} &= P_{i|i-1}^{(j)}H_i^{(j)\top}V_i^{(j)-1} \\ e_i^{(j)} &= Y_i - H_i^{(j)}Z_{i|i-1}^{(j)} \\ Z_{i|i}^{(j)} &= Z_{i|i-1}^{(j)} + K_i^{(j)}e_i^{(j)} \\ P_{i|i}^{(j)} &= (I - K_i^{(j)}H_i^{(j)})P_{i|i-1}^{(j)} \\ \Lambda_i^{(j)} &= \exp\left(-\frac{1}{2}e_i^{(j)\top}V_i^{(j)-1}e_i^{(j)}\right)\Big/\sqrt{|V_i^{(j)}|} \\ \nu_i^{(j)} &= \bar{c}_{i-1}^{(j)}\Lambda_i^{(j)}/c_i, \text{ with } c_i = \sum_{k=1}^{s}\Lambda_i^{(k)}\bar{c}_{i-1}^{(k)}. \end{aligned}$$

- **Combination step:** In order to make predictions, set

$$\begin{aligned} Z_{i|i} &= \sum_{j=1}^{s} \nu_i^{(j)} Z_{i|i}^{(j)}, \\ P_{i|i} &= \sum_{j=1}^{s} \nu_i^{(j)} \left\{P_{i|i}^{(j)} + \left(Z_{i|i}^{(j)} - \hat{Z}_i\right)\left(Z_{i|i}^{(j)} - \hat{Z}_i\right)^\top\right\}. \end{aligned}$$

Remark 9.3.1 *The values $Z^{(j)}_{i-1|i-1} - \hat{Z}^{(0j)}_{i-1}$ come from moment matching. In the derivation of the IMM filter, one approximates a convex sum of Gaussian vectors by a Gaussian distribution, making the filter sub-optimal in general. Some justifications of these equations will be given in the next chapter when we will cover regime-switching Markov models.*

9.3.2 Implementation of the IMM Filter

The implantation of the IMM Filter is almost identical to the implementation of the Kalman filter. One can also use the maximum likelihood principle to estimate parameters.

9.4 General Filtering Problem

Set $X_i = (Z_1, \ldots, Z_i)$, $i \geq 1$. We assume that the signal is a Markov process, meaning that the conditional distribution of Z_i given X_{i-1} only depends on Z_{i-1}. The Markov kernel K_i is the expectation operator defined by

$$K_i f(z) = E\{f(Z_i)|Z_{i-1} = z\}. \tag{9.10}$$

It completely defines the conditional distribution of Z_i given X_{i-1}.

The dependence between the observation Y_i and the signal Z_i is not necessarily linear, and the conditional distribution of Y_i given $Z_i = z_i$ has density $\ell_i(y_i, z_i)$. Furthermore, we assume that for any $n \geq 1$, given the signal $X_n = (Z_1, \ldots, Z_n)$, the observations $\mathcal{Y}_n = (Y_1, \ldots, Y_n)$ are independent, and that their joint density is given by

$$\Lambda_n(y, X_n) = \Lambda_n(y_1, \ldots, y_n, X_n) = \prod_{i=1}^{n} \ell_i(y_i, Z_i). \tag{9.11}$$

Remark 9.4.1 *To see that the Kalman model fits into this general model, note that the signal is clearly Markovian and $\ell_i(y_j, z_i)$ is the density at y_j of a Gaussian distribution with mean $d_i + H_i z_i$ and covariance matrix R_i.*

9.4.1 Kallianpur-Striebel Formula

It is easy to check that for any integrable $f(X_{n+m})$, and for any $m \geq 0$, we have

$$\begin{aligned} E(f(X_{n+m})|\mathcal{Y}_n = y) &= E(f(X_{n+m})|Y_1 = y_1, \ldots, Y_n = y_n) \\ &= \frac{E\{f(X_{n+m})\Lambda_n(y, X_n)\}}{E\{\Lambda_n(y, X_n)\}}. \end{aligned} \tag{9.12}$$

In the filtering literature, (9.12) is known as the Kallianpur-Striebel formula. It defines explicitly the optimal filter. However it is not very useful as is because, we cannot use the previous calculations in order to incorporate a new observation Y_{n+1}.

Note that the joint density of $\mathcal{Y}_n = (Y_1, \ldots, Y_n)$ at point $y = (y_1, \ldots, y_n)$ is given by

$$h_n(y) = E\{\Lambda_n(y, X_n)\}. \tag{9.13}$$

Hence, the conditional density $f_{n+1}(y_{n+1}|y)$ of Y_{n+1} given $\mathcal{Y}_n = y$, is

$$\frac{E\{\Lambda_{n+1}(y, y_{n+1}, X_{n+1})\}}{E\{\Lambda_n(y, X_n)\}}. \tag{9.14}$$

By using simulations, we could theoretically estimate the numerator and denominator of (9.12), However, as n becomes large, the product Λ_n will be close to zero, making the Monte Carlo method inapplicable.

9.4.2 Recursivity

To be efficient, a filter must provide an easy way to incorporate the information provided by new observations. We will now state a proposition showing how it can be done in general. To simplify notations, set $\ell_j(z) = \ell_j(y_j, z)$, and $\Lambda_n(x) = \Lambda_n(z_1, \ldots, z_n) = \prod_{k=1}^n \ell_k(z_k)$. According to the Kallianpur-Striebel formula (9.12), the conditional distribution η_n we are interested in satisfies

$$\eta_n(f) = \eta_{n,y}(f) = E\{f(Z_n)|\mathcal{Y}_n = y\} = \frac{\gamma_n(f)}{\gamma_n(1)} \tag{9.15}$$

where

$$\gamma_n(f) = E\{f(Z_n)\Lambda_n(X_n)\}, \tag{9.16}$$

where by convention, $\Lambda_0 \equiv 1$. Using (9.12), we can prove the following proposition, giving the celebrated Zakai recursive equation.

Proposition 9.4.1 (Zakai Equation) *For every $i \geq 1$, we have*

$$\gamma_i(f) = \gamma_{i-1}(K_i[f\ell_i]) \tag{9.17}$$

and

$$\eta_i(f) = \frac{\eta_{i-1}(K_i[f\ell_i])}{\eta_{i-1}(K_i[\ell_i])}. \tag{9.18}$$

In particular, the conditional density of Y_i given $\mathcal{Y}_{i-1}$ is $\eta_{i-1}(K_i[\ell_i])$.

Note that to compute η_i using the Zakai equation, it is not necessary that ℓ_i is a density. Moreover, if we are interested in the joint density of $Y_1, \ldots, Y_i$ in order to estimate parameters, then one can remove from ℓ_i any constant not depending of these parameters.

9.4.3 Implementing the Recursive Zakai Equation

Given the initial distribution $\eta_0 = \gamma_0$ of Z_0, i.e., $\eta_0(f) = \gamma_0(f) = E\{f(Z_0)\}$, the Zakai equation can be used (theoretically) to compute η_n in a recursive way. However, in practice, η_0 is unknown. We already encountered that problem for the Kalman filter, and we assumed η_0 to be Gaussian with mean $Z_{0|0}$ and covariance matrix $P_{0|0}$. As it was the case for the Kalman filter, starting from some initial distribution η_0' is not that important, as long it has a positive density with respect to η_0. Of course the first predictions could be quite bad but after a while, the filter should forget its initial value. In practice we often use a Gaussian distribution with a large variance to start the filter, when no information is available.

9.4.4 Solving the Filtering Problem

There exist several methods for solving the Zakai equation. When the model satisfies the conditions of the Kalman model, then there is an explicit solution. In what follows, we will consider two other methods, namely the computation of the conditional densities, yielding the exact solution, and a Monte Carlo method, producing approximations of the conditional distributions η_i.

9.5 Computation of the Conditional Densities

It happens often that for a continuous time Markov process $\mathcal{Z}$, the conditional distribution of $\mathcal{Z}_t$ given $\mathcal{Z}_s$ has a density $p_{\mathcal{Z}_t|\mathcal{Z}_s=z}(z'|z)$, for all $t > s$. Suppose also that the density $p_0(z)$ of η_0 is given. The aim of the filtering problem is to come up with the conditional density $p_{\mathcal{Z}_t|\mathcal{Y}_i}$. Here we assume that the signal Z is given by $Z_i = \mathcal{Z}_{t_i}$. Using the Zakai equation (9.18), we have the following algorithm:

Algorithm 9.5.1 *To compute the densities of the $\eta_1, \eta_2, \ldots$, do the following steps:*

- *Set $p_{Z_0|\mathcal{Y}_0}(z) = p_0(z)$.*

For each $i \geq 0$, compute

$$
\begin{aligned}
p_{Z_{i+1}|\mathcal{Y}_i}(z) &= \int p_{Z_{i+1}|Z_i}(z|w)p_{Z_i|\mathcal{Y}_i}(w)dw, \\
p_{Y_{i+1}|\mathcal{Y}_i}(y_{i+1}) &= \int p_{Z_{i+1}|\mathcal{Y}_i}(w)p_{Y_{i+1}|Z_{i+1}}(y_{i+1}|w)dw, \qquad (9.19) \\
p_{Z_{i+1}|\mathcal{Y}_{i+1}}(z) &= \frac{p_{Z_{i+1}|\mathcal{Y}_i}(z)p_{Y_{i+1}|Z_{i+1}}(y_{i+1}|z)}{p_{Y_{i+1}|\mathcal{Y}_i}(y_{i+1})}.
\end{aligned}
$$

Equation (9.19) is quite important since it can be used to compute the joint density of $Y_1, \dots, Y_n$, which, according to the multiplication formula A.19, is given by

$$p_{Y_1,\dots,Y_n}(y_1, \dots, y_n) = \prod_{k=1}^{n} p_{Y_k|\mathcal{Y}_{k-1}}(y_k).$$

Note that computing the Algorithm 9.5.1 for the Kalman model leads directly to the Kalman algorithm. See Appendix 9.D.1.

Often, the difficult part in applying Algorithm 9.5.1 is to compute $p_{\mathcal{Z}_t|\mathcal{Z}_s}(z'|z)$. There exist several methods to do so when the process $\mathcal{Z}$ is a diffusion process. For example, one can try to find exact solutions of stochastic differential equations [Kouritzin and Rémillard, 2002] or one can try to solve the associated Kolmogorov equation appearing in Section 9.5.2.

We now look at some ways of computing the conditional densities appearing in Algorithm 9.5.1.

9.5.1 Convolution Method

Suppose that there exists a function H_i such that $H_i(Y_i, Z_i) = \nu_i$, where ν_i has density ϕ. Then one can compute the density $p_{Y_i|Z_i}$. Suppose also that the density $p_{Z_{i+1}|Z_i}(z|w)$ has the form $g(Az - w)$. In this case,

$$p_{Z_{i+1}|\mathcal{Y}_i}(z) = \int p_{Z_{i+1}|Z_i}(z|w) p_{Z_i|\mathcal{Y}_i}(w) dw = \int g(Az - w) p_{Z_i|\mathcal{Y}_i}(w) dw$$

is the convolution $g \star p_{Z_i|\mathcal{Y}_i}$ evaluated at Az and one can use Fourier transforms to speed up the calculations of $p_{Z_{i+1}|Z_i}(z|w)$. See Appendix 9.C. Then, one can compute $p_{Y_{i+1}|\mathcal{Y}_i}(y_{i+1})$ by numerical integration, leading to an expression for $p_{Z_{i+1}|\mathcal{Y}_{i+1}}$.

Example 9.5.1 (Kouritzin et al. [2002]) *In that paper, the authors proposed the following model for electricity prices:*

$$d\mathcal{Z}_t = -\kappa \mathcal{Z}_t dt + dW_t,$$

where W is a Brownian motion. Then

$$Z_t = e^{-\kappa t} \left\{ Z_0 + \int_0^t e^{\kappa s} dW_s \right\}.$$

In the long term, the signal stabilizes to its stationary distribution which is Gaussian with mean 0 and variance $(2\kappa)^{-1}$. So it is natural to assume that $Z_0 \sim N(0, (2\kappa)^{-1})$. It is also assumed that the observations $Y_1, Y_2, \dots$ are taken at t periods $t_1 = \delta, t_2 = 2\delta, \dots$ and

$$Y_i = m + g(Z_i) S_i,$$

where $g(z) = e^{\mu + \sigma z}$ and where $(S_i)_{i \geq 1}$ is a sequence of independent random variables having a symmetric standard stable distribution with parameter $\alpha \in (0, 2]$. The noise S is independent of the signal. It is then easy to check that one can apply Algorithm 9.5.1, using the convolution method.

9.5.2 Kolmogorov Equation

If the signal is related to a process satisfying

$$dZ_t = b(t, Z_t)dt + \sigma(t, Z_t)dW_t,$$

then its density $u(t, z'; t_k, z) = p_{Z_t|Z_{t_{k-1}}}(z'; t, t_{k-1}, z)$, is the solution of the so-called Kolmogorov equation

$$\frac{\partial u}{\partial t} = -\sum_{i=1}^{m} \frac{\partial}{\partial z_i}(b_i u) + \frac{1}{2}\sum_{i=1}^{m}\sum_{j=1}^{m} \frac{\partial^2}{\partial z_i \partial z_j}(a_{ij} u),$$

where $a = \sigma\sigma^\top$, with the condition

$$\lim_{t \downarrow t_k} \int f(z) u(t, z'; t_k, z) dz' = f(z).$$

Then the density can be approximated by computing numerically the solution of Kolmogorov partial differential equation.

9.6 Particle Filters

Since formula (9.18) involves expectations, it is no surprise that one can use particle filters, which are basically Monte Carlo methods to approximate the solution of the Zakai equation. The first particle filter was proposed by Gordon et al. [1993]. Another type of filters, called auxiliary particle filters, were proposed in Pitt and Shephard [1999]. They are both described next. In fact there exists a huge literature on Monte Carlo methods for solving Feynman-Kac type equations, of which the Zakai equation (9.18) is a particular case. See, e.g., Del Moral [2004] and references therein.

9.6.1 Implementation of a Particle Filter

The implementation of the filter is based on the following algorithm, first proposed by Gordon et al. [1993].

Algorithm 9.6.1 *For a large integer N, generate initial values $z_0^{(k)}$, $k \in \{1, \ldots, N\}$ from distribution η_0. Call this population of particles $\mathcal{P}_0$. Then repeat the following steps:*

1. *Prediction: For $i \geq 0$, suppose that*

$$\mathcal{P}_i = \left\{ z_i^{(k)}; k \in \{1, \ldots, N\} \right\}$$

is the population of particles obtained at step i. Then we let these particles evolve according to the Markov kernel, i.e., for each $k \in \{1,\ldots,N\}$, generate $\tilde{z}_{i+1}^{(k)}$ according to the conditional distribution of Z_{i+1} given $Z_i = z_i^{(k)}$. Schematically,

$$z_i^{(k)} \overset{Markov}{\rightsquigarrow} \tilde{z}_{i+1}^{(k)}, \quad k \in \{1,\ldots,N\}.$$

Call this intermediate population $\tilde{\mathcal{P}}_{i+1}$.

2. *Selection: The population $\mathcal{P}_{i+1}$ is obtained by choosing at random N values $z_{i+1}^{(k)}$, $k \in \{1,\ldots,N\}$, amongst the intermediate population $\tilde{\mathcal{P}}_{i+1}$. The probability of choosing $\tilde{z}_{i+1}^{(k)}$ must be as close as possible to the probability*

$$\pi_{i+1}^{(k)} = \frac{\ell_{i+1}\left(y_{i+1}, \tilde{z}_{i+1}^{(k)}\right)}{\sum_{j=1}^{N} \ell_{i+1}\left(y_{i+1}, \tilde{z}_{i+1}^{(j)}\right)}.$$

3. *η_{i+1} is then estimated by the uniform distribution $\hat{\eta}_{i+1}$ on $\mathcal{P}_{i+1}$, i.e., for any function f,*

$$\hat{\eta}_{i+1}(f) = \frac{1}{N}\sum_{k=1}^{N} f\left(z_{i+1}^{(k)}\right).$$

Remark 9.6.1 *There exist certain conditions for the selection of trajectories. It is known that proportional sampling works. In that case, if N_k denotes the number of times $\tilde{Z}_{i+1}^{(k)}$ has been chosen, $k \in \{1,\ldots,N\}$, then the joint distribution of $(N_1,\ldots,N_N)$ is multinomial, with parameters $\pi_{i+1}^{(1)},\ldots,\pi_{i+1}^{(N)}, N$. However, this selection method is slow when N is large. Often, in applications, we take $N = 50000$. A survey of different selection methods in done in Gentil and Rémillard [2008].*

9.6.2 Implementation of an Auxiliary Sampling/Importance Resampling (ASIR) Particle Filter

In the implementation of the previous particle filter, there was no sampling step combined with the prediction step. This is mainly what distinguishes the particle filter from the ASIR particle filter proposed by Pitt and Shephard [1999]. The prediction step is replaced by an "adapted" proposal step, where particles are sampled according to a distribution which may depend on the next observation y_{i+1}. Throughout this section it is assumed that the conditional distribution of Z_{i+1} given Z_i has a density $p_{Z_{i+1}|Z_i}(z|\tilde{z})$ (with respect to some reference measure), and that the observation Y_i has a density $p_{Y_i|Z_i}(y|z)$ with respect to Lebesgue measure.

To describe the algorithm, let $g_{1,i+1}(k,z)$ be a positive function and let $g_{2,i+1}(z|\tilde{z})$ be a conditional density with the same support as $p_{Z_{i+1}|Z_i}(z|\tilde{z})$.

Both functions may depend on y_{i+1}. The implementation of the filter is based on the following algorithm.

Algorithm 9.6.2 *For a large integer N, generate initial values $z_0^{(k)}$, $k \in \{1,\ldots,N\}$ from distribution η_0. Then repeat the following steps:*

1. *For $i \ge 0$, suppose that $z_i^{(k)}$, $k \in \{1,\ldots,N\}$ are the particles obtained in step i with probability distribution $\pi_i^{(k)}$.*

   For $k \in \{1,\ldots,N\}$, compute $\tilde{w}_{i+1}^{(k)} = g_{1,i+1}\left(k, z_i^{(k)}\right)\pi_i^{(k)}$, and set $\tilde{\pi}_{i+1}^{(k)} = \frac{\tilde{w}_{i+1}^{(k)}}{\sum_{j=1}^{M} \tilde{w}_{i+1}^{(j)}}$.

2. *Choose N values $\tilde{z}_{i+1}^{(k)}$, $k \in \{1,\ldots,N\}$, amongst the sample $z_i^{(k)}$, $k \in \{1,\ldots,N\}$, the probability of choosing $z_i^{(k)}$ being $\tilde{\pi}_{i+1}^{(k)}$.*

3. *For $k \in \{1,\ldots,N\}$, generate $z_{i+1}^{(k)}$ according to density $g_{2,i+1}\left(\cdot|\tilde{z}_{i+1}^{(k)}\right)$.*

4. *For $k \in \{1,\ldots,N\}$, compute*

$$w_{i+1}^{(k)} = \frac{p_{Y_{i+1}|Z_{i+1}}\left(y_{i+1}|z_{i+1}^{(k)}\right) p_{Z_{i+1}|Z_i}\left(z_{i+1}^{(k)}|\tilde{z}_{i+1}^{(k)}\right)}{g_{1,i+1}\left(k, \tilde{z}_{i+1}^{(k)}\right) g_{2,i+1}\left(z_{i+1}^{(k)}|\tilde{z}_{i+1}^{(k)}\right)},$$

   and set $\pi_{i+1}^{(k)} = \frac{w_{i+1}^{(k)}}{\sum_{j=1}^{M} w_{i+1}^{(j)}}$.

5. *η_{i+1} is then estimated by the weighted empirical distribution $\hat{\eta}_{i+1}$ of $z_{i+1}^{(k)}$, $k \in \{1,\ldots,N\}$, i.e., for any function f,*

$$\hat{\eta}_{i+1}(f) = \sum_{k=1}^{N} f\left(z_{i+1}^{(k)}\right)\pi_{i+1}^{(k)}.$$

Alternatively, instead of computing the weighted average, one could re-sample from $z_{i+1}^{(k)}$, $k \in \{1,\ldots,N\}$, with probabilities $\pi_{i+1}^{(k)}$, $k \in \{1,\ldots,N\}$. Then, at each step, the weights of the empirical distribution would be $1/N$. The resulting algorithm can then be described as follows:

Algorithm 9.6.3 *For a large integer N, generate initial values $z_0^{(k)}$, $k \in \{1,\ldots,N\}$ from distribution η_0. Then repeat the following steps:*

1. *For $i \ge 0$, suppose that $z_i^{(k)}$, $k \in \{1,\ldots,N\}$ are the particles obtained in step i.*

   For $k \in \{1,\ldots,N\}$, compute $\tilde{w}_{i+1}^{(k)} = g_{1,i+1}\left(k, z_i^{(k)}\right)$, and set $\tilde{\pi}_{i+1}^{(k)} = \frac{\tilde{w}_{i+1}^{(k)}}{\sum_{j=1}^{M} \tilde{w}_{i+1}^{(j)}}$.

2. *Choose* R *values* $\tilde{z}_{i+1}^{(k)}$, $k \in \{1,\ldots,R\}$, *amongst the sample* $z_i^{(k)}$, $k \in \{1,\ldots,N\}$, *the probability of choosing* $z_i^{(k)}$ *being* $\tilde{\pi}_{i+1}^{(k)}$.

3. *For* $k \in \{1,\ldots,R\}$, *generate* $\hat{z}_{i+1}^{(k)}$ *according to density* $g_{2,i+1}\left(\cdot|\tilde{z}_{i+1}^{(k)}\right)$.

4. *For* $k \in \{1,\ldots,R\}$, *compute*

$$w_{i+1}^{(k)} = \frac{p_{Y_{i+1}|Z_{i+1}}\left(y_{i+1}|\hat{z}_{i+1}^{(k)}\right) p_{Z_{i+1}|Z_i}\left(\hat{z}_{i+1}^{(k)}|\tilde{z}_{i+1}^{(k)}\right)}{g_{1,i+1}\left(k,\tilde{z}_{i+1}^{(k)}\right) g_{2,i+1}\left(\hat{z}_{i+1}^{(k)}|\tilde{z}_{i+1}^{(k)}\right)},$$

   and set $\pi_{i+1}^{(k)} = \frac{w_{i+1}^{(k)}}{\sum_{j=1}^{M} w_{i+1}^{(j)}}$.

5. *Resample* N *times amongst* $\hat{z}_{i+1}^{(k)}$, $k \in \{1,\ldots,R\}$ *with probabilities* $\pi_{i+1}^{(k)}$, *to obtain* $z_{i+1}^{(k)}$, $k \in \{1,\ldots,N\}$. η_{i+1} *is then estimated by the empirical distribution* $\hat{\eta}_{i+1}$ *of* $z_{i+1}^{(k)}$, $k \in \{1,\ldots,N\}$, *i.e.*,

$$\hat{\eta}_{i+1}(f) = \frac{1}{N}\sum_{k=1}^{N} f\left(z_{i+1}^{(k)}\right).$$

Remark 9.6.2 *If possible, it is recommended to choose* $g_{1,i+1}$ *and* $g_{2,i+1}$ *such that*

$$g_{1,i+1}(\tilde{z})g_{2,i+1}(z|\tilde{z}) \approx p_{Y_{i+1}|Z_{i+1}}(y_{i+1}|z)\, p_{Z_{i+1}|Z_i}(z|\tilde{z}).$$

In this case, we would get $w_{i+1}^{(k)} \equiv 1$, *so* $\pi_{i+1}^{(k)} = 1/R$, *for each* $k \in \{1,\ldots,R\}$. *When this is the case, it is called a fully adapted algorithm [Pitt and Shephard, 1999].*

We now give some examples of ASIR particle filters. The notation ASIR_j comes from Pitt [2002].

9.6.2.1 ASIR$_0$

By choosing $g_{1,i} \equiv 1$ and $g_{2,i+1}(z|\tilde{z}) = p_{Z_{i+1}|Z_i}(z|\tilde{z})$, we have a particle filter which is similar to the one proposed in Section 9.6.1, and called a sampling/importance resampling (SIR) particle filter or ASIR$_0$.

9.6.2.2 ASIR$_1$

An interesting example of a genuine ASIR particle filter is provided by choosing $g_{2,i+1}(z|\tilde{z}) = p_{Z_{i+1}|Z_i}(z|\tilde{z})$ and $g_{1,i+1}(k,z) = p_{Y_{i+1}|X_{i+1}}\left(y_{i+1}|\mu_{i+1}^{(k)}\right)$, where $\mu_{i+1}^{(k)}$ is associated with the density $p_{Z_{i+1}|Z_i}\left(\cdot\,|\tilde{z}_{i+1}^{(k)}\right)$. For example, $\mu_{i+1}^{(k)}$ could be the expectation of the latter density.

9.6.2.3 ASIR$_2$

Another interesting example of a genuine ASIR particle filter is given in Pitt and Shephard [1999], if one assumes that the density of Z_{i+1} given $Z_i = \tilde{z}$ is Gaussian with mean $\mu_{i+1}(\tilde{z})$ and covariance matrix $A_{i+1}(\tilde{z})$, and the measurement density $p_{Y_{i+1}|X_{i+1}}(y|z)$ is log-concave as a function of z. In this case, $g_{1,i+1}$ and $g_{2,i+1}$ are defined by the equation

$$g_{1,i+1}\left(k, \tilde{z}_{i+1}^{(k)}\right) g_{2,i+1}\left(z|\tilde{z}_{i+1}^{(k)}\right) = C_{i+1} \hat{f}_{i+1}\left(z, \mu_{i+1}^{(k)}\right) p_{Z_{i+1}|Z_i}\left(z|\tilde{z}_{i+1}^{(k)}\right),$$

where C_{i+1} is a constant (independent of parameters and observations), $\hat{f}_{i+1}\left(z, \mu_{i+1}^{(k)}\right)$ is the exponential of the first order Taylor expansion of $\ln\left\{p_{Y_{i+1}|Z_{i+1}}(y_{i+1}|z)\right\}$ as a function of z, about $z = \mu_{i+1}^{(k)}$. In fact,

$$\hat{f}_{i+1}\left(z, \mu_{i+1}^{(k)}\right) = p_{Y_{i+1}|Z_{i+1}}\left(y_{i+1}|\mu_{i+1}^{(k)}\right) e^{a_{i+1,k}^\top\left(z-\mu_{i+1}^{(k)}\right)},$$

where $a_{i+1,k} = \nabla_z p_{Y_{i+1}|Z_{i+1}}\left(y_{i+1}|\mu_{i+1}^{(k)}\right) / p_{Y_{i+1}|Z_{i+1}}\left(y_{i+1}|\mu_{i+1}^{(k)}\right)$. It then follows that $g_{2,i+1}$ is the density of a Gaussian distribution, with mean $\mu_{i+1}(\tilde{z}) + A_{i+1}(\tilde{z}) a_{i+1,k}$ and covariance matrix $A_{i+1}(\tilde{z})$. Moreover,

$$g_{1,i+1}(k, \tilde{z}) = p_{Y_{i+1}|Z_{i+1}}\left(y_{i+1}|\mu_{i+1}^{(k)}\right) e^{\frac{1}{2} a_{i+1,k}^\top A_{i+1}(\tilde{z}) a_{i+1,k} + a_{i+1,k}^\top\left(\mu_{i+1}(\tilde{z}) - \mu_{i+1}^{(k)}\right)}.$$

Finally, note that $p_{Y_{i+1}|Z_{i+1}}(y_{i+1}|z) \le \hat{f}_{i+1}\left(z, \mu_{i+1}^{(k)}\right)$, so

$$w_{i+1}^{(k)} = \frac{p_{Y_{i+1}|Z_{i+1}}\left(y_{i+1}|z_{i+1}^{(k)}\right)}{\hat{f}_{i+1}\left(z_{i+1}^{(k)}, \mu_{i+1}^{(k)}\right)} \le 1, \quad k \in \{1, \ldots, N\}.$$

Example 9.6.1 *In Pitt and Shephard [1999], the authors considered the stochastic volatility model*

$$Y_i = \varepsilon_i e^{Z_i/2}, \qquad Z_{i+1} = \omega + \phi Z_i + \eta_i,$$

where $\varepsilon_i \sim N(0,1)$ *independent of* $\eta_i \sim N(0, \sigma^2)$. *They also took* $\mu_{i+1}^{(k)} = \mu_{i+1}\left(\tilde{z}_{i+1}^{(k)}\right) = \omega + \phi \tilde{z}_{i+1}^{(k)}$, $k \in \{1, \ldots, N\}$. *It follows that*

$$a_{i+1,k} = \frac{y_{i+1}^2}{2} e^{-\mu_{i+1}^{(k)}} - \frac{1}{2}$$

and one can take

$$g_{1,i+1}\left(k, \tilde{z}_{i+1}^{(k)}\right) = e^{\frac{\sigma^2}{2} a_{i+1,k}^2 - a_{i+1,k} - \mu_{i+1}^{(k)}/2}, \quad k \in \{1, \ldots, N\}.$$

9.6.3 Estimation of Parameters

One of the weaknesses of particle filters is the estimation of the parameters. Even if we have an estimation of the joint density of $Y_1, \ldots, Y_n$, it is generally too noisy to be of use for the estimation of parameters.

An ingenious way to estimate parameters for particle filters is to set an a priori distribution for the parameters and consider the parameters as part of the signal, that is (Z_i, Θ_i). This methodology is typical from Bayesian methods in statistics. The a posteriori distribution for Θ is the conditional law of Θ_i given the observations $y_1, \ldots, y_i$. For example, for a variance parameter, one could choose a gamma distribution, while a Gaussian distribution could serve as an a priori distribution for parameters with no constraints. For the Markovian dynamics of Θ_i, one could choose a random walk for components with no constraints, while a multiplicative random walk (meaning that the log is a random walk) would be preferable for positive components.

Another approach was proposed in a technical report [Pitt, 2002]. To try to get rid of the noise, the author suggested to estimate $p_{Y_i|\mathcal{Y}_{i-1}}(y_i)$ by

$$\hat{p}_{Y_i|\mathcal{Y}_{i-1}}(y_i) = \left\{\frac{1}{R}\sum_{k=1}^{R} w_i^{(k)}\right\}\left\{\frac{1}{N}\sum_{k=1}^{N} \tilde{w}_i^{(k)}\right\}, \tag{9.20}$$

using the values computed in Algorithm 9.6.3. It is then shown that the ASIR likelihood

$$\hat{p}(y_1) \times \prod_{i=2}^{n} \hat{p}_{Y_i|\mathcal{Y}_{i-1}}(y_i) \tag{9.21}$$

is an unbiased estimate of the joint density of $Y_1, \ldots, Y_n$. As the likelihood depends on the unknown parameter θ and simulations, it is recommended to use the same seed throughout the steps when the value of θ is changed.

9.6.3.1 Smoothed Likelihood

It is also recommended to change the step 5 in Algorithm 9.6.3 by the following step: Choose a continuous distribution function that approximates the discrete distribution $\left\{\hat{z}_{i+1}^{(k)}, \pi_{i+1}^{(k)}\right\}$, $k \in \{1, \ldots, N\}$ and resample N times from that distribution function. For example, in the one dimensional case, if the discrete distribution is (x_k, π_k), $k \in \{1, \ldots, n\}$, where $x_1 < \cdots < x_n$, one could set $p_1 = \pi_1 + \frac{\pi_2}{2}$, $p_k = \frac{\pi_k + \pi_{k+1}}{2}$, $k \in \{2, \ldots, n-2\}$, and $p_{n-1} = \pi_n + \frac{\pi_{n-1}}{2}$. A good choice for the continuous distribution function would be to define it by its density

$$f(x) = \sum_{k=1}^{n-1} p_k \frac{\mathbb{I}_{(x_k, x_{k+1})}(x)}{x_{k+1} - x_k}.$$

To generate X with density f, it suffices to generate $U \sim \text{Unif}(0,1)$, choose k at random with probability p_k, and then set $X = (1-U)x_k + Ux_{k+1}$. Fast algorithms to sample from f are also proposed in Pitt [2002].

Remark 9.6.3 *Note that the smoothed likelihood can also be computed with the particle filter described by Algorithm 9.6.1. In this case,* $\tilde{w}_i^{(k)} \equiv 1$, *so*

$$\hat{p}_{Y_i|\mathcal{Y}_{i-1}}(y_i) = \frac{1}{N}\sum_{k=1}^{N} w_i^{(k)}.$$

Example 9.6.2 (Stochastic Volatility Model (continued)) *As in Example 9.2.3, we use the 2500 simulated observations for the stochastic volatility model considered by Ruiz [1994]. These data can be found in the MATLAB file* DataSV. *Table 9.4 show the results of the MATLAB function* EstSPFSV *with 1000 particles, implementing a maximum likelihood estimation, and using the smooth particle filter* ASIR_0, *when 1200 and 2500 observations are used. One can see that the estimation results are much better than with the quasi-maximum likelihood method.*

TABLE 9.4: 95% confidence intervals for the estimated parameters, using a smooth particle filter with $N = 1000$ particles. The out-of-sample RMSE is 0.3580.

		95% Confidence Interval	
Parameter	True Value	50% of the Sample	100% of the Sample
ω	−0.10	-0.0337 ± 0.0401	-0.0754 ± 0.0607
ϕ	0.75	0.9073 ± 0.1015	0.8065 ± 0.1424
σ	0.25	0.1579 ± 0.1235	0.2472 ± 0.1291

From Table 9.4, if we use only the first 1250 observations to estimate the parameters, the results are not as good. However, the prediction power, as measured by the out-of-sample RMSE (0.3588) is as good as if we were using the whole sample, which has a 0.3587 *RMSE. One can also check that the estimations are better if we take 60% of the sample for estimation purposes, instead of 50%.*

Remark 9.6.4 *Under the conditions of application of the Kalman filter, the particle filter is not optimal. However it should be close to the Kalman predictions. A smooth particle filter for the model developed in Example 9.2.2 is implemented in the MATLAB function* EstSPF.

9.7 Suggested Reading

For a general reference on Kalman filter, see Harvey [1989]. For applications in finance, see, e.g., Gibson and Schwartz [1990] and Schwartz [1997]. For particle filters and in particular ASIR filters, see Pitt and Shephard [1999] and Pitt [2002].

9.8 Exercises

Exercise 9.1

Consider the following model: $Z_i = \mu + w_i$, $w_i \sim N(0, Q)$, and $Y_i = Z_i + \varepsilon_i$, with $\varepsilon_i \sim N(0, R)$. Write the Kalman equations. Can you estimate all parameters using maximum likelihood?

Exercise 9.2

Consider the Schwartz model for the spot price of a commodity and its convenience yield. Assume that $S(0) = 16$, $\delta(0) = 0.025$, $r = 5\%$, and that the time scale is in years.

(a) Compute the value of futures for 1-month to 12-month maturities, using the following parameters:[2]

μ	κ	α	σ_1	σ_2	ρ	λ
0.142	1.876	0.106	0.393	0.527	0.766	0.198

(b) Plot the graph of the value of the futures as a function of the convenience yield for 1-month and 12 month maturities. For the range of the convenience yield take $[0, 1]$.

Exercise 9.3

Write the Kalman equations corresponding to the Schwartz model where the observations consists in two futures.

Exercise 9.4

Consider the Schwartz model for the spot price of a commodity and its convenience yield.

(a) Simulate a trajectory of $n = 503$ daily values for the spot price and its convenience yield, using the parameters of Table 9.2. Take $S(0) = 16$ and $\delta(0) = 0.025$.

---

[2]In fact these are the estimated parameters for crude oil futures as computed in Schwartz [1997, Table VI].

(b) Use the trajectory simulated in (a) to compute the value, every month (21 days), of five futures contracts with maturities of 2, 3, 6, 9, and 11 months. It means that in January, we have futures expiring in March, April, July, October and December. Next, in February, we have futures expiring in April, May, August, November, and January, and so on. Assume a constant risk-free rate of 5%.

(c) Add independent Gaussian noises ε with variances $0.001h$ to each set of log-prices of the futures and plot the graphs.

Exercise 9.5

Consider the model in Example 9.2.2. Suppose that at some point the minimization algorithm is trapped in a local minimum giving $Q = 0$, which is like starting the minimization with $Q = 0$. In that case, show that you estimate R by the sample variance of the observations and $\frac{\alpha}{1-\phi}$ is estimated by the mean of the observations.

Exercise 9.6

Suppose that for the model $X_i = \alpha + \phi X_{i-1} + w_i$, $Y_i = X_i + \varepsilon_i$, the correlation between the Gaussian error terms ε_i and w_i is ρ. Transform this model into one satisfying the assumptions of the Kalman filter.

Exercise 9.7

Suppose you implement a particle filter. How can you compute a 95% prediction interval for each component of the signal? Can you estimate the (conditional) density of the signal?

9.9 Assignment Questions

1. Construct a MATLAB function for computing the value of a futures under Schwartz model.

2. Construct a MATLAB function to simulate trajectories for Schwartz model.

3. Construct a MATLAB function to estimate the parameters of the Schwartz model when observing d futures. The $n \times d$ matrix of maturities must be an input.

4. Construct a MATLAB function for computing the Kalman equations in the general model.

9.A Schwartz Model

According to Schwartz model,

$$\begin{aligned} d\ln\{S(t)\} &= \{\mu - \delta(t) - \sigma_1^2/2\}\, dt + \sigma_1 dW_1(t) \\ d\delta(t) &= \kappa\{\alpha - \delta(t)\}dt + \sigma_2 dW_2(t). \end{aligned}$$

Hence δ is an Ornstein-Uhlenbeck process, (see Section 5.2.1), given by

$$\delta(u) = \alpha + e^{-\kappa(u-s)}\{\delta(s) - \alpha\} + \sigma_2 \int_s^u e^{-\kappa(u-v)} dW_2(v), \quad u \geq s,$$

according to (5.6). In particular, for a fixed time step h,

$$\delta(ih) = \alpha + e^{-\kappa h}[\delta\{(i-1)h\} - \alpha] + \sigma_2 \int_{(i-1)h}^{ih} e^{-\kappa(ih-u)} dW_2(u),$$

and

$$\begin{aligned} \int_s^t \delta(u)du &= \alpha(t-s) + \{\delta(s) - \alpha\}\left(\frac{1 - e^{-\kappa(t-s)}}{\kappa}\right) \\ &\quad + \sigma_2 \int_s^t \left\{\frac{1 - e^{-\kappa(t-v)}}{\kappa}\right\} dW_2(v). \end{aligned}$$

Set $Z_i = (\ln\{S(ih)\}, \delta(ih))^\top$. Since

$$\ln(S_{ih}) = \ln\{S_{(i-1)h}\} + h(\mu - \sigma_1^2/2) - \int_{(i-1)h}^{ih} \delta(u)du + \sigma_1[W_1(ih) - W_1\{(i-1)h\}],$$

we have $Z_i = c + GZ_{i-1} + \eta_i$, with

$$c = \left(h(\mu - \sigma_1^2/2) + \alpha\left(\frac{1 - e^{-\kappa h}}{\kappa} - h\right), \alpha\left(1 - e^{-\kappa h}\right)\right)^\top,$$

$G = \begin{pmatrix} 1 & \frac{e^{-\kappa h} - 1}{\kappa} \\ 0 & e^{-\kappa h} \end{pmatrix}$, and $\eta_i = (\eta_{1i}, \eta_{2i})^\top$, where

$$\eta_{2i} = \sigma_2 \int_{(i-1)h}^{ih} e^{-\kappa(ih-u)} dW_2(u)$$

has a Gaussian distribution with mean 0 and variance $\sigma_2^2\left(\frac{1-e^{-2\kappa h}}{2\kappa}\right)$, using the properties of stochastic integrals stated in Section 5.A, while

$$\eta_{1i} = \sigma_1[W_1(ih) - W_1\{(i-1)h\}] - \sigma_2 \int_{(i-1)h}^{ih} \left\{\frac{1 - e^{-\kappa(ih-u)}}{\kappa}\right\} dW_2(u)$$

has a Gaussian distribution with mean 0 and variance

$$\sigma_1^2 h + \frac{\sigma_2^2}{\kappa^2}\left\{h - 2\left(\frac{1-e^{-\kappa h}}{\kappa}\right) + \left(\frac{1-e^{-2\kappa h}}{2\kappa}\right)\right\} + 2\rho\sigma_1\sigma_2\left(\frac{1-\kappa h - e^{-\kappa h}}{\kappa^2}\right).$$

Note that η_i is independent of the past $\mathcal{F}_{(i-1)h}$. In addition,

$$\mathrm{Cov}(\eta_{1i},\eta_{2i}) = \rho\sigma_1\sigma_2\left(\frac{1-e^{-\kappa h}}{\kappa}\right) - \frac{\sigma_2^2}{2\kappa^2}\left(1-e^{-\kappa h}\right)^2.$$

Set $a = \alpha - \lambda/\kappa$, where λ is the market price of convenience yield risk, which is assumed to be constant in Schwartz [1997]. Under the equivalent martingale measure Q determined by λ, we have

$$\begin{aligned} d\ln\{S(t)\} &= \left\{r - \delta(t) - \sigma_1^2/2\right\}dt + \sigma_1 dW_1(t) \\ d\delta(t) &= \kappa\{a - \delta(t)\}dt + \sigma_2 dW_2(t). \end{aligned}$$

According to the previous calculations, it follows that

$$\ln\{F(t)\} = \ln\left[E_Q\{S(T)|\mathcal{F}_t\}\right] = \log\{S(t)\} - \delta(t)\left(\frac{1-e^{-\kappa\tau}}{\kappa}\right) + A(\tau),$$

where

$$\begin{aligned} A(\tau) &= \tau\left(r - a + \frac{\sigma_2^2}{2\kappa^2} - \frac{\rho\sigma_1\sigma_2}{\kappa}\right) + \frac{\sigma_2^2}{4}\left(\frac{1-e^{-2\kappa\tau}}{\kappa^3}\right) \\ &\quad + \left(a\kappa + \rho\sigma_1\sigma_2 - \frac{\sigma_2^2}{\kappa}\right)\left(\frac{1-e^{-\kappa\tau}}{\kappa^2}\right) \\ &= r\tau + \left(\frac{\sigma_2^2}{2\kappa} - a\kappa - \rho\sigma_1\sigma_2\right)\left(\frac{e^{-\kappa\tau} - 1 + \tau\kappa}{\kappa^2}\right) \\ &\quad - \frac{\sigma_2^2}{4\kappa}\left(\frac{1-e^{-\kappa\tau}}{\kappa}\right)^2. \end{aligned}$$

9.B Auxiliary Results

Proposition 9.B.1 *If R and Q are positive definite, then $V = R + HQH^\top$ and $P = Q - QH^\top V^{-1}HQ$ are positive definite as well and*

$$P^{-1} = Q^{-1} + H^\top R^{-1}H.$$

Moreover, $QH^\top V^{-1} = PH^\top R^{-1}$.

PROOF. Since $V^{-1}HQH^\top R^{-1} = R^{-1} - V^{-1}$, we have

$$\begin{aligned} P\left(Q^{-1} + H^\top R^{-1}H\right) &= I + QH^\top\left(R^{-1} - V^{-1}\right)H \\ &\quad - QH^\top V^{-1}HQH^\top R^{-1}H \\ &= I. \end{aligned}$$

It then follows that

$$\begin{aligned} QH^\top V^{-1} &= P\left(Q^{-1} + H^\top R^{-1} H\right) QH^\top V^{-1} \\ &= PH^\top V^{-1} + PH^\top R^{-1}(V - R)V^{-1} = PH^\top R^{-1}, \end{aligned}$$

completing the proof.

■

9.C Fourier Transform

The Fourier transform $\hat{f}(\omega)$ of an integrable function $f(x)$, $\omega, x \in \mathbb{R}^m$, is given by

$$T(f)(\omega) = \hat{f}(\omega) = \int e^{ix^\top \omega} f(x)dx.$$

Moreover, if $\hat{f}$ is integrable, i.e., $\int |\hat{f}(\omega)|d\omega < \infty$, then the inverse Fourier transform is given by

$$f(x) = \frac{1}{(2\pi)^m} \int e^{-ix^\top \omega} \hat{f}(\omega)d\omega = \frac{1}{(2\pi)^m} T(\hat{f})(-x),$$

which is also a Fourier transform (up to a constant). There exists very fast algorithms to compute the Fourier transform. The interest of using Fourier transforms in computations is that it transforms a complicated operation called "convolution" into a product. More precisely, if f and g are integrable and $h = f \star g$, called the convolution of f with g, is defined by

$$h(x) = \int f(x - y)g(y)dy,$$

then h is integrable and

$$\hat{h}(\omega) = \hat{f}(\omega)\hat{g}(\omega).$$

In particular, if $\hat{h}$ is integrable, then

$$h(x) = \frac{1}{(2\pi)^m} \int e^{-ix^\top \omega} \hat{f}(\omega)\hat{g}(\omega)d\omega.$$

9.D Proofs of the Results

9.D.1 Proof of Proposition 9.2.1

Since the error terms are Gaussian, and the evolution and measurement equations (9.1) and (9.2) are linear, if the initial (unknown) distribution of

Z_0 is Gaussian, then all conditional distributions are Gaussian, since the joint law is Gaussian.

Now, $p_{Z_0|\mathcal{Y}_0}(z) = p_{Z_0}(z) = \frac{1}{(2\pi)^{p/2}|P_{0|0}|^{\frac{1}{2}}} e^{-\frac{1}{2}(z-Z_{0|0})^\top P_{0|0}{}^{-1}(z-Z_{0|0})}$, and for each $i \geq 1$, set

$$\begin{aligned}
p_{Z_i|\mathcal{Y}_{i-1}}(z) &= \frac{1}{(2\pi)^{p/2}|P_{i|i-1}|^{1/2}} e^{-\frac{1}{2}(z-Z_{i|i-1})^\top P_{i|i-1}{}^{-1}(z-Z_{i|i-1})}, \\
p_{Z_i|\mathcal{Y}_i}(z) &= \frac{1}{(2\pi)^{p/2}|P_{i|i}|^{1/2}} e^{-\frac{1}{2}(z-Z_{i|i})^\top P_{i|i}{}^{-1}(z-Z_{i|i})}, \\
p_{Y_i|\mathcal{Y}_{i-1}}(x_i) &= \frac{1}{(2\pi)^{r/2}|V_i|^{1/2}} e^{-\frac{1}{2}e_i^\top V_i^{-1}e_i}.
\end{aligned}$$

It is easy to check that the Kalman equations are valid for $i = 1$. To prove their validity is general, suppose they are true for $i-1$.

Since the joint law of $Z_0, Z_1, \ldots, Z_i$ and $Y_1, \ldots, Y_i$ is Gaussian, all we have to do is compute the conditional means and expectations. First, $Z_{i|i-1} = E(Z_i|\mathcal{Y}_{i-1}) = \mu_i + F_i E(Z_{i-1}|\mathcal{Y}_{i-1}) = \mu_i + F_i Z_{i-1|i-1}$. Now, $P_{i|i-1} = G_i Q_i G_i^\top + F_i P_{i-1|i-1} F_i^\top$, so the density $p_{Z_i|\mathcal{Y}_{i-1}}$ is correctly defined, as are the formulas for $Z_{i|i-1}$ and $P_{i|i-1}$.

Next, $Y_{i|i-1} = E(Y_i|\mathcal{Y}_{i-1}) = d_i + H_i Z_{i|i-1}$, so $V_i = E(e_i e_i^\top) = R_i + H_i P_{i|i-1} H_i^\top$. Hence the formulas for $Y_{i|i-1}$, e_i and V_i and the density $p_{Y_i|\mathcal{Y}_{i-1}}$ are correct. To conclude, note that

$$p_{Z_i|Z_{i-1}}(z|z_{i-1}) = \frac{1}{(2\pi)^{p/2}|G_i Q_i G_i^\top|^{1/2}} e^{-\frac{1}{2}(z-Z_{i|i})^\top P_{i|i}{}^{-1}(z-Z_{i|i})}.$$

Using Algorithm 9.5.1 or Bayes formula, we have

$$p_{Z_i|\mathcal{Y}_i}(z) = \frac{p_{Z_i|\mathcal{Y}_{i-1}}(z) p_{Y_i|Z_i}(y_i|z)}{p_{Y_i|\mathcal{Y}_{i-1}}(y_i)}.$$

Setting $x = z - z_{i|i-1} = z_{i|i} + P_{i|i-1} H_i^\top V_i^{-1} e_i$, and replacing y_i by $e_i + d_i + H_i z_{i|i-1}$ yields

$$p_{Z_i|\mathcal{Y}_i}(z) = \sqrt{\frac{|V_i||R_i|}{|P_{i|i-1}|}} e^{-\frac{1}{2}W},$$

where

$$\begin{aligned}
W &= x^\top P_{i|i-1}^{-1} x + (e_i + H_i x)^\top R_i^{-1}(e_i + H_i x) - e_i^\top V_i^{-1} e_i \\
&= x^\top P_{i|i}^{-1} x + e_i^\top V_i^{-1} H_i P_{i|i-1} R_i^{-1} e_i + 2 e_i^\top R_i^{-1} H_i x \\
&= (z - z_{i|i})^\top P_{i|i}^{-1}(z - z_{i|i}),
\end{aligned}$$

according to Proposition 9.B.1, if one uses Kalman formulas for $Z_{i|i}$ and $P_{i|i}$. This completes the proof.

■

Bibliography

H. A. P. Blom and Y. Bar-Shalom. The interacting multiple model for systems with Markovian switching coefficients. *IEEE Trans. Auto. Cont.*, 33:780–783, 1988.

P. Del Moral. *Feynman-Kac Formulae.* Probability and its Applications (New York). Springer-Verlag, New York, 2004. Genealogical and interacting particle systems with applications.

I. Gentil and B. Rémillard. Using systematic sampling selection for Monte Carlo solutions of Feynman-Kac equations. *Adv. in Appl. Probab.*, 40(2): 454–472, 2008.

R. Gibson and E. S. Schwartz. Stochastic convenience yield and the pricing of oil contigent claims. *Journal of Finance*, XLV:959–976, 1990.

N. J. Gordon, D. J. Salmond, and A. F. M. Smith. Novel approach to nonlinear/non-gaussian bayesian state estimation. *IEEE Proceedings F, Radar and Signal Processing*, 140(2):107–113, 1993.

A. C. Harvey. *Forecasting, Structural Time Series Models and the Kalman Filter.* Cambridge, 1989.

R. E. Kalman. A new approach to linear filtering and prediction problems. *Transactions of the ASME–Journal of Basic Engineering*, 82(Series D):35–45, 1960.

M. A. Kouritzin and B. Rémillard. Explicit strong solutions of multidimensional stochastic differential equations. Technical report, Laboratory for Research in Probability and Statistics, University of Ottawa-Carleton University, 2002.

M. A. Kouritzin, B. Rémillard, and C. P. Chan. Parameter estimation for filtering problems with stable noise. In *Proceedings of 4th Annual Conference on Information Fusion*, volume I, pages WeB127–WeB130, 2002.

M. K. Pitt. Smooth particle filters for likelihood evaluation and maximisation. Technical Report 651, Department of Economics, University of Warwick, 2002.

M. K. Pitt and N. Shephard. Filtering via simulation: auxiliary particle filters. *J. Amer. Statist. Assoc.*, 94(446):590–599, 1999.

E. Ruiz. Quasi-maximum likelihood estimation of stochastic volatility models. *Journal of Econometrics*, 63(1):289–306, 1994.

E. S. Schwartz. The stochastic behavior of commodity prices: Implications for valuation and hedging. *The Journal of Finance*, LII:923–973, 1997.

M. W. Watson. Recursive solution methods for dynamic linear rational expectations models. *Journal of Econometrics*, 41(1):65–89, 1989.

Chapter 10

Applications of Filtering

In this final chapter, we will see three examples of application of filtering. The first example concerns the estimation of ARMA models, the second example deals with regime-switching Markov models, while the last example is about the so-called beta replication of hedge funds.

10.1 Estimation of ARMA Models

A stochastic process Y_i is called an ARMA(p,q) process if it has the following representation:

$$Y_i = \mu + \sum_{j=1}^{p} \phi_j (Y_{i-j} - \mu) + \varepsilon_i - \sum_{k=1}^{q} \theta_k \varepsilon_{i-k}, \tag{10.1}$$

where the innovations ε_i are independent and identically distributed, with mean 0 and variance σ_ε^2. It is assumed that ε_i is independent of $\mathcal{Y}_{i-1} = (Y_1, \ldots, Y_{i-1})$. Often, the innovations are assumed to be Gaussian. One can rewrite the relationship (10.1) more concisely as

$$\phi(B)(Y_i - \mu) = \theta(B)\varepsilon_i,$$

where $\phi(z) = 1 - \sum_{j=1}^{p} \phi_j z^j$, $\theta(z) = 1 - \sum_{k=1}^{q} \theta_k z^k$ and B is the lag operator, i.e., $BY_i = Y_{i-1}$. In order for the process to admit a stationary distribution, it is necessary and sufficient that the roots of $\phi(z)$ are all outside the unit circle in the complex plane. Also, to obtain a unique representation, one imposes that the roots of $\theta(z)$ are all outside the unit circle in the complex plane.

We will see that in the case of Gaussian innovations, the Kalman filter can be used to predict the process and estimate its parameters. But first we give some examples of ARMA processes.

10.1.1 AR(p) Processes

The simplest example of ARMA apart from the white noise ARMA(0,0) is the AR(1) model, the so-called auto-regressive process of order 1. In this case,

$$Y_i = \mu + \phi(Y_{i-1} - \mu) + \varepsilon_i.$$

It is easy to check that if μ_i is the mean of Y_i, then

$$\mu_i = \mu + \phi^i(\mu_0 - \mu)$$

and if σ_i^2 is the variance of Y_i, then

$$\sigma_i^2 = \phi^{2i}\sigma_0^2 + \sigma_\varepsilon^2\left(\frac{1-\phi^{2i}}{1-\phi^2}\right).$$

Finally, $\text{Cov}(Y_{i+k}, Y_i) = \phi^k\sigma_i^2$. Since $\phi(z) = 1-\phi z$, the root of $\phi(z)$ is $1/\phi$ and $1/|\phi| > 1$ if and only if $|\phi| < 1$. Under this stationarity condition, it follows that $\mu_i \to \mu$, $\sigma_i^2 \to \frac{\sigma_\varepsilon^2}{1-\phi^2}$ and $\text{Cor}(Y_i, Y_{i+k}) \to 0$, all exponentially fast.

One can also prove that for a AR(p) = ARMA(p,0) process, the correlation between Y_i and Y_{i+k} does not depend on i and decreases exponentially fast to 0 as k increases.

Another interesting property is that AR(p) processes forget their initial values. In fact, if one starts two AR(p) process Y and $\tilde{Y}$ from different points $(Y_0, \ldots, Y_{p-1})$ and $(\tilde{Y}_0, \ldots, \tilde{Y}_{p-1})$, and they have the same innovations, then there exists $\rho \in (0,1)$ so that

$$|Y_i - \tilde{Y}_i| \le \rho^{i-p} \max_{j\in\{0,\ldots,p-1\}} |Y_j - \tilde{Y}_j|, \quad i \ge p.$$

The proof is similar to the proof of Proposition 7.1.1.

10.1.1.1 MA(q) Processes

For a MA(q) process, i.e.,

$$Y_i = \mu + \varepsilon_i - \sum_{k=1}^{q} \theta_k \varepsilon_{i-k},$$

Y_{i+j} is clearly independent of $\mathcal{Y}_i$ whenever $j > q$. In particular, the correlation between Y_i and Y_{i+k} does not depend on i and it is zero if $k > q$. One can easily check that for $\theta \neq 0$, the autocorrelation of the MA(1) processes $Y_i = \varepsilon_i - \theta\varepsilon_{i-1}$ and $Z_i = \varepsilon_i - \varepsilon_{i-1}/\theta$ is the same. Indeed, $\text{Cor}(Y_i, Y_{i+1}) = \text{Cor}(Z_i, Z_{i+1}) = -\frac{\theta}{1+\theta^2}$. Using only these autocorrelations, we could not distinguish between the two processes. However, the root of $\theta_Y(z) = 1 - \theta z$ is $1/\theta$, while the root of $\theta_Z(z) = 1 - z/\theta$ is θ. Imposing that the roots are outside the unit circle fixes the representation problem.

10.1.2 MA Representation

From now on, assume that the roots of $\phi(z)$ and $\theta(z)$ are all outside the unit circle. These two conditions imply that $\psi(z) = \theta(z)/\phi(z)$ can be written as an infinite series with a convergence radius greater than 1, i.e., there exists $r_0 > 1$ such that for all $|z| < r_0$,

$$\psi(z) = \sum_{j=0}^{\infty} \psi_j z^j, \tag{10.2}$$

with $\psi_0 = 1$. If the ARMA process Y is stationary, then it can be written with the following MA representation (possibly of infinite order):

$$Y_i = \mu + \psi(B)\varepsilon_i = \mu + \varepsilon_i + \sum_{j=1}^{\infty} \psi_j \varepsilon_{i-j}.$$

Hence

$$\mathrm{Var}(Y_i) = \sigma_\varepsilon^2 \sum_{j=0}^{\infty} \psi_j^2$$

and

$$\mathrm{Cor}(Y_i, Y_{i+k}) = \sum_{j=0}^{\infty} \psi_j \psi_{j+k} \Big/ \sum_{j=0}^{\infty} \psi_j^2.$$

This correlation is of the order of $|z_1|^{-k}$, for k large, where z_1 is the smallest root of $\phi(z)$.

10.1.3 ARMA Processes and Filtering

The following proposition will be useful for implementing a filter and computing the joint density.

Proposition 10.1.1 *Suppose that Y_i is an ARMA(p,q) process with representation $\phi(B)(Y_i - \mu) = \theta(B)\varepsilon_i$, where $r = \max(p, q+1)$. Set $\phi_j = 0$ if $j > p$.*

Further set $Z_i = \begin{pmatrix} \hat{Y}_i(0) - \mu \\ \hat{Y}_i(1) - \mu \\ \vdots \\ \hat{Y}_i(r-1) - \mu \end{pmatrix}$, *with $\hat{Y}_i(j) = E(Y_{i+j}|\mathcal{Y}_i)$. Note that* $E(Z_i) = 0$.

Then Y admits the following Markovian representation:

$$Z_i = F Z_{i-1} + g\varepsilon_i, \qquad Y_i = \mu + a^\top Z_i, \quad i \geq 1,$$

where $a^\top = (1\ 0 \cdots 0)$, $F = \phi_1$ *if* $r = 1$, *and when* $r > 1$,

$$
F = \begin{pmatrix}
0 & 1 & 0 & 0 & \cdots & 0 \\
0 & 0 & 1 & 0 & \cdots & 0 \\
\vdots & & & \ddots & & \\
\vdots & & & & \ddots & \\
0 & 0 & 0 & \cdots & 0 & 1 \\
\phi_r & \phi_{r-1} & \cdots & \cdots & \phi_2 & \phi_1
\end{pmatrix}, \quad g = \begin{pmatrix}
\psi_0 \\ \psi_1 \\ \psi_2 \\ \vdots \\ \vdots \\ \psi_{r-1}
\end{pmatrix},
$$

with $(\psi_j)_{j\geq 1}$ *as defined by* (10.2).

The proof is given in Appendix 10.D.1.

10.1.3.1 Implementation of the Kalman Filter in the Gaussian Case

Before implementing the Kalman filter in the Gaussian case for ARMA models, set $f_i^2 = V_i/\sigma_\varepsilon^2$, $U_{i|i-1} = P_{i|i-1}/\sigma_\varepsilon^2$, and $U_{i-1|i-1} = P_{i-1|i-1}/\sigma_\varepsilon^2$, $i \geq 1$. Since Z is stationary, we can compute $U_{0|0}$. The result appears in the next proposition, proven in Appendix 10.D.2.

Proposition 10.1.2

$$
U_{0|0} = \sum_{k=0}^{\infty} (F^k g)(F^k g)^\top.
$$

Next, using Kalman equations (9.8), we have, for all $i \geq 1$:

$$
\begin{array}{lcl}
Z_{i|i-1} & = & F Z_{i-1|i-1}, \; Z_{0|0} = 0, \\
Y_{i|i-1} & = & \mu + a^\top F Z_{i-1|i-1}, \\
e_i & = & Y_i - \mu - a^\top F Z_{i-1|i-1}, \\
U_{i|i-1} & = & F U_{i-1|i-1} F^\top + g g^\top, \\
f_i^2 & = & a^\top U_{i|i-1} a, \\
Z_{i|i} & = & F Z_{i-1|i-1} + U_{i|i-1} a e_i / f_i^2, \\
U_{i|i} & = & U_{i|i-1} - U_{i|i-1} a a^\top U_{i|i-1} / f_i^2.
\end{array}
\tag{10.3}
$$

Here, $U_{1|0} = U_{0|0}$ since $Z_{1|0} = 0 = Z_{0|0}$. Note that the system of Kalman equations (10.3) does not involve the parameter σ_ε.

Remark 10.1.1 *Iterating the transition equation yields*

$$
Z_{i+k} = F^k Z_i + \sum_{j=0}^{k-1} F^j g \varepsilon_{i+k-j}.
$$

As a result, using the properties of conditional expectations, we obtain

$$\begin{aligned} Z_{i+k|i} &= E\{Z_{i+k}|Y_i,\dots,Y_1\} = F^k Z_{i|i}, \\ P_{i+k|i} &= E\left\{(Z_{i+k} - Z_{i+k|i})(Z_{i+k} - Z_{i+k|i})^\top |Y_i,\dots,Y_1\right\} \\ &= F^k P_{i|i}(F^k)^\top + \sigma_\varepsilon^2 \sum_{j=0}^{k-1} (F^j g)(F^j g)^\top. \end{aligned}$$

It follows from the proof of Proposition 10.1.2 that $F^k \to 0$ as $k \to \infty$, so for a given i, the long term prediction is a Gaussian distribution with mean $Z_{0|0} = 0$ and variance $P_{0|0} = \sigma_\varepsilon^2 U_{0|0}$, i.e., long term prediction is the stationary distribution.

10.1.4 Estimation of Parameters of ARMA Models

Because the joint density of the observations $Y_1, \dots, Y_n$ is an output of the Kalman filter, one can use the latter to estimate the parameters of an ARMA process.[1] Indeed, the likelihood L satisfies

$$\begin{aligned} -\ln(L) &= \frac{1}{2}\sum_{i=1}^n \ln(\det V_i) + \frac{1}{2}\sum_{i=1}^n e_i^\top V_i^{-1} e_i + \frac{n}{2}\ln(2\pi) \\ &= n\ln(\sigma_\varepsilon) + \sum_{i=1}^n \ln(f_i) + \frac{1}{2\sigma_\varepsilon^2}\sum_{i=1}^n \frac{e_i^2}{f_i^2} + \frac{n}{2}\ln(2\pi). \end{aligned}$$

As a result, using the maximum likelihood principle, σ_ε^2 is estimated by the mean of the squared normalized residuals, that is

$$\hat\sigma_\varepsilon^2 = \frac{1}{n}\sum_{i=1}^n \frac{\hat e_i^2}{\hat f_i^2},$$

where $\hat e_i$ and $\hat f_i$ are computed from the Kalman equation 10.3 by plugging in the estimated parameters.

Example 10.1.1 (AR(1) Model) *In this case, $r = p = 1$, $g = a = 1$, $F = \phi$, hence $U_{0|0} = 1/(1-\phi^2)$. Also $e_1 = Y_1 - \mu$, so $f_1^2 = 1/(1-\phi^2)$, $Z_{1|1} = e_1$, and $U_{1|1} = 0$. It is easy to check that $U_{i|i} = 0$ for $i \ge 1$, $U_{i|i-1} = 1$ and $f_i^2 = 1$ for any $i > 1$. In addition, $Z_{i|i} = (Y_i - \mu)$ if $i \ge 1$ and for $i \ge 2$, $e_i = Y_i - \mu - \phi(Y_{i-1} - \mu)$, and $Z_{i|i-1} = \phi(Y_{i-1} - \mu)$. It follows that the maximum likelihood estimators are approximately*

$$\begin{aligned} \hat\mu &= \bar Y, \\ \hat\phi &= \frac{\sum_{i=2}^n (Y_i - \bar Y)(Y_{i-1} - \bar Y)}{\sum_{i=1}^n (Y_i - \bar Y)^2}, \end{aligned}$$

[1]Note that some statistical packages use the Kalman filter for the estimation of ARMA models.

and

$$\hat{\sigma}_\varepsilon^2 = \frac{1}{n-1}\sum_{i=2}^n \left\{Y_{i-1} - \bar{Y} - \hat{\phi}(Y_{i-1} - \bar{Y})\right\}^2.$$

Example 10.1.2 (MA(1) Model) *Here,* $r = 2$, $g = \begin{pmatrix} 1 \\ -\theta \end{pmatrix}$, $a = \begin{pmatrix} 1 \\ 0 \end{pmatrix}$ *and* $F = \begin{pmatrix} 0 & 1 \\ 0 & 0 \end{pmatrix}$. *Also* $U_{0|0} = \begin{pmatrix} 1+\theta^2 & -\theta \\ -\theta & \theta^2 \end{pmatrix}$. *Hence* $e_1 = Y_1 - \mu$ *and* $f_1^2 = 1 + \theta^2$. *By iteration we find* $e_{i+1} = Y_{i+1} - \mu + \theta e_i / f_i^2$, *and*

$$f_i^2 = \sum_{j=0}^{i} \theta^{2j} \Big/ \sum_{j=0}^{i-1} \theta^{2j} = \frac{1 - \theta^{2(i+1)}}{1 - \theta^{2i}}.$$

The expression of f_i^2 *explains why it is much more difficult to estimate the parameters of ARMA models than the parameters of AR models.*

10.2 Regime-Switching Markov Models

Definition 10.2.1 *A stochastic process* $(Y_i, \tau_i)_{i\geq 1}$ *with values in* $\mathbb{R}^d \times E$, $E = \{1, \ldots, r\}$, *is a regime-switching Markov model if there exist densities* $(f_k)_{k=1}^r$ *and a transition matrix* Q *such that*

(i) *The (non-observable) regimes* $(\tau_i)_{i\geq 1}$ *form a Markov chain on* E, *with transition matrix* Q, *and stationary distribution* ν.

(ii) *Given the regimes* $\tau_1 = j_1, \ldots, \tau_n = j_n$, *the observations* $Y_1, \ldots, Y_n$ *are independent, with densities* $f_{j_1}, \ldots, f_{j_n}$ *respectively.*

It follows that if the initial distribution of τ_0 is the stationary distribution ν, then the distribution of Y_k is also stationary, with a density given by the mixture

$$f(y) = \sum_{j=1}^r \nu_j f_j(y). \tag{10.4}$$

Note that by definition, (Y, τ) is a Markov process. In general, Y is not a Markov process, unless the regimes are independent. In the latter case, the observations Y_i are independent with a density given by (10.4).

10.2.1 Serial Dependence

As shown in the next proposition, the sequence of observations Y is serially dependent, mainly due to the dependence induced by the regimes.

Proposition 10.2.1 *Suppose that for every $j \in \{1,\ldots,r\}$, the density f_j has mean μ_j and covariance matrix A_j, Then, if τ_0 has distribution ν, it follows that for any $k \geq 1$,*

$$\mu = E(Y_i) = \sum_{l=1}^{r} \nu_l \mu_l, \tag{10.5}$$

$$A = \mathrm{Cov}(Y_i, Y_i) = \sum_{l=1}^{r} \nu_l A_l + \sum_{l=1}^{r} \nu_l \mu_l \mu_l^\top - \mu\mu^\top, \tag{10.6}$$

$$\Gamma_k = \mathrm{Cov}(Y_i, Y_{i+k}) = \sum_{l=1}^{r}\sum_{j=1}^{r} \nu_l \mu_l \mu_j^\top \left\{(Q^k)_{ij} - \nu_j\right\}. \tag{10.7}$$

Remark 10.2.1 *If Q is ergodic, there exist $C > 0$ and $a \in (0,1)$ such that for all $k \geq 1$, $\max_{j,l\in\{1,\ldots,r\}} \left|(Q^k)_{jl} - \nu_l\right| \leq Ca^k$. It then follows from (10.7) that $\mathrm{Cov}(Y_i, Y_{i+k})$ converges exponentially fast to 0, as $k \to \infty$, i.e., for some $C_1 > 0$, we have $\|\mathrm{Cov}(Y_i, Y_{i+k})\| \leq C_1 a^k$, for all $k \geq 1$. The behavior of the covariance is therefore similar to the one observed for ARMA processes.*

Since the regimes are not observable, we have to rely on filtering techniques to be able to predict them, given the observations Y. This is the content of the next section.

10.2.2 Prediction of the Regimes

Since the process (Y, τ) satisfies the filtering model requirements, with Y being the observation process and τ being the (Markovian) signal process, one can use the recursive Zakai equation to predict the regimes. According to the Kallianpur-Striebel formula 9.12, we have

$$\gamma_i(j) = E\left\{\mathbb{I}(\tau_i = j)\prod_{l=1}^{i} f_{\tau_l}(y_l)\right\},$$

and

$$\mathcal{Z}_i = \sum_{l=1}^{r} \gamma_i(l) = E\left\{\prod_{l=1}^{i} f_{\tau_l}(y_l)\right\} = f_{1:i}(y_1,\ldots,y_i),$$

where $f_{1:i}$ is the joint density of $Y_1,\ldots,Y_i$, $j \in \{1,\ldots,r\}$, $i \geq 1$. Then, according to the Zakai equation (9.17), we have, for any $k \geq 1$ and $j \in \{1,\ldots,r\}$,

$$\gamma_i(j) = f_j(y_i)\sum_{l=1}^{r} \gamma_{i-1}(l) Q_{lj}, \quad i \geq 1. \tag{10.8}$$

Finally, from (9.18), we end up with

$$\eta_i(j) = P(\tau_i = j | Y_1,\ldots,Y_i) = \frac{\gamma_i(j)}{\mathcal{Z}_i}, \tag{10.9}$$

for every $j \in \{1,\ldots,r\}$, and $i \geq 1$. Note that

$$P(\tau_{i+1} = j|Y_1,\ldots,Y_i) = \sum_{l=1}^{r} \eta_i(l) Q_{lj}.$$

In view of applications, it is preferable to rewrite (10.9) only in terms of η, i.e.,

$$\eta_i(j) = \frac{f_j(y_i)}{\mathcal{Z}_{i|i-1}} \sum_{l=1}^{r} \eta_{i-1}(l) Q_{lj}, \quad i \geq 1, j \in \{1,\ldots,r\}, \tag{10.10}$$

where

$$\mathcal{Z}_{i|i-1} = \sum_{j=1}^{l} \sum_{l=1}^{r} f_j(y_i) \eta_{i-1}(l) Q_{lj} = \sum_{j=1}^{l} f_j(y_i) P(\tau_i = j|Y_1,\ldots,Y_{i-1})$$

is the conditional density of Y_i at y_i, given $Y_1,\ldots,Y_{i-1}$.

10.2.3 Conditional Densities and Predictions

For any $i \geq 1$, the conditional density of Y_{i+1} given $Y_1,\ldots,Y_i$, denoted by $f_{i+1:1}$, can be expressed as a mixture, viz.

$$f_{i+1:1}(y_{i+1}|y_1,\ldots,y_i) = \mathcal{Z}_{i+1|i} = \sum_{l=1}^{r} \sum_{j=1}^{r} \eta_i(l) Q_{lj} f_j(y_{i+1}) \tag{10.11}$$

$$= \sum_{j=1}^{r} W_{j,i} f_j(y_{i+1}), \tag{10.12}$$

where

$$W_{j,i} = \sum_{l=1}^{r} \eta_i(l) Q_{lj}, \quad j \in \{1,\ldots,r\}.$$

Note that the last two equations are also valid when $i = 0$. It follows from (10.11) that the prediction of Y_{i+1}, given $Y_1,\ldots,Y_i$ is

$$\sum_{l=1}^{r} \sum_{j=1}^{r} \eta_i(l) Q_{lj} \mu_j.$$

Furthermore, confidence intervals can be constructed for the prediction of Y_{i+1}, using the quantiles of $f_{i+1:i}$. Similarly, using the Markov property of the regimes, we find that for $k \geq 1$, the conditional density of Y_{i+k} given $Y_1,\ldots,Y_i$ is

$$f_{i+k:i}(x) = \sum_{l=1}^{r} \sum_{j=1}^{r} \eta_i(l) \left(Q^k\right)_{lj} f_j(x), \tag{10.13}$$

which is also a mixture of the densities $(f_j)_{j=1}^r$, but with weight $\sum_{l=1}^r \eta_i(l)\left(Q^k\right)_{lj}$ for the regime j, $j \in \{1,\ldots r\}$. In particular, the prediction of Y_{i+k} is

$$\sum_{l=1}^r \sum_{j=1}^r \eta_i(l)\left(Q^k\right)_{lj} \mu_j.$$

Again, one can build confidence intervals for the prediction of Y_{i+k} using the quantiles of $f_{i+k:i}$. Finally, if the Markov chain τ is ergodic, then the conditional density of Y_{i+k} given $Y_1,\ldots,Y_i$, converges to the stationary density

$$f(x) = \sum_{i=1}^r \nu_i f_i(x).$$

Hence, for long term predictions, the behavior of Y becomes independent of its past.

In the next section we will see how the parameters of the model can be estimated, using a powerful methodology called the EM (Expectation-Maximization) algorithm.

10.2.4 Estimation of the Parameters

Since we have incomplete data, the regimes being non-observable, a well-suited estimation methodology is the EM algorithm, proposed in a similar context by Dempster et al. [1977]. For specific applications to more general models called Hidden-Markov models, see Cappé et al. [2005].

To apply the EM algorithm, described in Appendix 10.A, we need to compute the following probabilities:

$$\lambda_i(j) = P(\tau_i = j|Y_1,\ldots,Y_n),$$

and

$$\Lambda_i(j,l) = P(\tau_i = j, \tau_{i+1} = l|Y_1,\ldots,Y_n),$$

for $i \in \{1,\ldots,n\}$ and for all $j,l \in \{1,\ldots,r\}$, since the joint density of $Y_1,\tau_1,\ldots,Y_n,\tau_n$ is given by

$$\left\{\sum_{l=1}^r \eta_0(l) Q_{lj_1}\right\}\left(\prod_{i=2}^n Q_{j_{i-1},j_i}\right)\prod_{i=1}^n f_{j_i}(y_i).$$

We will see next how to implement the E-step in general. However, we will only treat the M-step for Gaussian densities.

10.2.4.1 Implementation of the E-step

The following implementation has been proposed by Baum et al. [1970] and is often called the Baum-Welsh algorithm.

First, define, for $i \in \{1,\ldots,r\}$,

$$\begin{aligned}
\bar{\gamma}_n(j) &= 1, \\
\bar{\gamma}_i(j) &= \sum_{l=1}^{r} \bar{\gamma}_{i+1}(l) Q_{jl} f_l(y_{i+1}), \quad i \in \{1,\ldots,n-1\}.
\end{aligned}$$

Then, for any $i, j \in \{1,\ldots,r\}$, one can verify that

$$\lambda_i(j) = \frac{\gamma_i(j)\bar{\gamma}_i(j)}{\sum_{\alpha=1}^{r} \gamma_i(\alpha)\bar{\gamma}_i(\alpha)}, \quad i \in \{1,\ldots,n\}, \tag{10.14}$$

$$\Lambda_i(j,l) = \frac{Q_{jl}\gamma_i(j)\bar{\gamma}_{i+1}(l) f_l(y_{i+1})}{\sum_{\alpha=1}^{r}\sum_{\beta=1}^{r} Q_{\alpha\beta}\gamma_i(\alpha)\bar{\gamma}_{i+1}(\beta) f_\beta(y_{i+1})}, \quad i \in \{1,\ldots,n-1\}, \tag{10.15}$$

and $\Lambda_n(j,l) = \lambda_n(j) Q_{jl}$. Note that the definitions of λ and Λ are unchanged if γ is replaced by η and if $\bar{\gamma}_i$ is normalized so that $\sum_{j=1}^{r} \bar{\gamma}_i(j) = 1$. These substitutions are more stable numerically.

10.2.5 M-step in the Gaussian Case

From now on, we assume that the densities $f_1,\ldots,f_r$ are Gaussian, with mean $(\mu_j)_{j=1}^r$ and covariance matrices $(A_j)_{j=1}^r$. As it was the case in Example 10.A.1, it is relatively easy to implement the M-Step for the Gaussian mixtures. It only suffices to update the parameters $(\nu_k)_{k=1}^r$, $(\mu_k)_{k=1}^r$, $(A_k)_{k=1}^r$ and Q, by setting, for any $j, l \in \{1,\ldots,r\}$,

$$\begin{aligned}
\nu_j' &= \sum_{i=1}^{n} \lambda_i(j)/n, \\
\mu_j' &= \sum_{i=1}^{n} y_i w_i(j), \\
A_j' &= \sum_{i=1}^{n} (y_i - \mu_j')(y_i - \mu_j')^\top w_i(j), \\
Q_{jl}' &= \sum_{i=1}^{n} \Lambda_i(j,l) \Big/ \sum_{i=1}^{n} \lambda_i(j) = \frac{1}{n}\sum_{i=1}^{n} \Lambda_i(j,l) \Big/ \nu_j',
\end{aligned}$$

where $w_i(j) = \lambda_i(j) \Big/ \sum_{k=1}^{n} \lambda_k(j)$. Note that ν' is not the stationary distribution of Q' since for all $j \in \{1,\ldots,r\}$,

$$\sum_{l=1}^{r} \nu_l' Q_{lj}' = \frac{1}{n}\sum_{i=1}^{n}\sum_{l=1}^{r} \Lambda_i(l,j) = \frac{1}{n}\sum_{i=2}^{n+1} \lambda_i(j) = \nu_j' + \frac{\lambda_{n+1}(j) - \lambda_1(j)}{n} \neq \nu_j'.$$

However,

$$\max_{j \in \{1,\ldots,r\}} \left| \sum_{l=1}^{r} \nu_l' Q_{lj}' - \nu_j' \right| \leq 1/n.$$

Hence, as n becomes large, ν' is very close to the stationary distribution of Q'.

Remark 10.2.2 *In Cappé et al. [2005], it is shown that the estimator EM of* η_0*, when* $\eta_0 \neq \nu$*, is* $\eta_0' = \lambda_1$*.*

It is interesting to note that after each iteration of the EM algorithm, the first two moments of the sample are fitted. See Appendix 10.B for more details.

10.2.6 Tests of Goodness-of-Fit

Using the Rosenblatt transform defined in Appendix 10.C, Rémillard et al. [2010] proposed a test of goodness-of-fit for the regime-switching Gaussian model with r regimes. Its validity is proven in Rémillard [2011]. For the choice of the number of regimes, they suggested to choose the smallest r for which the P-value of the goodness-of-fit test for r regimes is greater than 5%.

The idea of the test is similar to the one used for copulas, the main difference being that the Rosenblatt transforms depend on the index i. For a general dynamic model with parameter θ, let $\Psi_{i,\theta}$ be the Rosenblatt transform associated with the conditional distribution of Y_i given $Y_1, \ldots, Y_{i-1}$. Then a test of goodness-of-fit can be based on the empirical distribution function

$$D_n(u) = \frac{1}{n}\sum_{i=1}^{n} \mathbb{I}(\hat{W}_i \le u), \quad u \in [0,1]^d,$$

where the pseudo-observations $\hat{W}_i$ are defined by $\hat{W}_i = \Psi_{i,\theta_n}(Y_i)$, $i \in \{1,\ldots,n\}$. Under the null hypothesis, $\hat{W}_i \approx W_i \sim C_\perp$, so D_n should be compared to $C_\perp$ and tests of goodness-of fit can be based on statistics of the form $S_n = \psi_n(\mathbb{D}_n)$, where $\mathbb{D}_n = \sqrt{n}\,(D_n - C_\perp)$. For example, one could take the Cramér-von Mises statistics

$$S_n^{(B)} = \int_{[0,1]^d} \mathbb{D}_n^2(u)du$$

or

$$S_n^{(C)} = \sum_{i=1}^{n} \left\{ D_n\left(\hat{W}_i\right) - C_\perp\left(\hat{W}_i\right)\right\}^2.$$

In what follows, all computations were done with $S_n^{(B)}$.

Example 10.2.1 (S&P 500 Index) *As an example, consider the S&P 500 index over the period April 17*th*, 2007 to December 31*st*, 2008. As the series displays periods of high volatility, as illustrated in Figure 10.1, it is natural to try to model it by a Gaussian regime-switching model. Since we are using daily data, we assume that the mean and volatility of each regime are given*

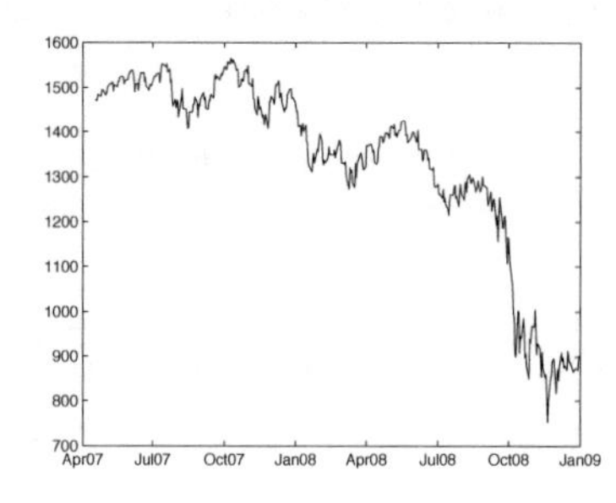

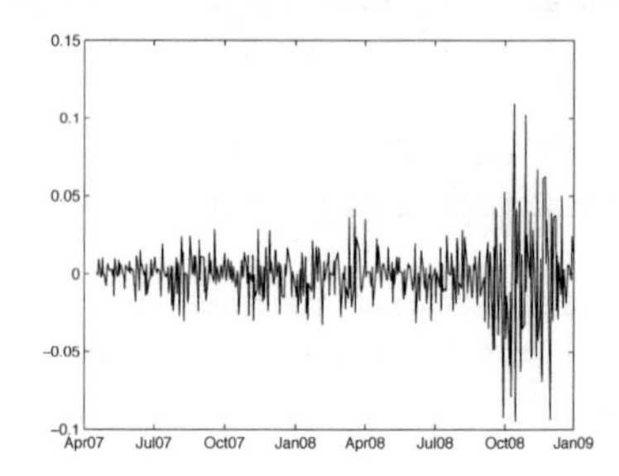

FIGURE 10.1: Values and log-returns of the S&P 500 from April 17th, 2007 to December 31st, 2008.

respectively by $\mu_j h$ *and* $\sigma_j \sqrt{h}$, *where* $h = 1/252$. *This parametrization will be justified in Section 10.2.7. The parameters were estimated by the MATLAB function* EstHMM1d.

Using the Cramér-von Mises test based on the Rosenblatt transform, we find a P-value of 0% for one and two regimes, and a P-value of 52% for three regimes. $N = 10000$ *bootstrap samples were generated to estimate the P-values. Therefore, we could choose a Gaussian model with three regimes. The corresponding parameters are given in Table 10.1 and the transition matrix* Q *is given by* (10.16). *For more details on the estimation and goodness-of-fit tests, see Rémillard et al. [2010]. From these results, we see that if the regime is in state 1, where the mean is positive and the volatility is small, it goes almost always to state 2, where the regime is characterized by a negative mean and large volatility; it can also switch to state 3 which has an even larger negative mean and much larger volatility. Once it is in one of these two states, the regime will stay there with high probability, sometimes going back to state 1, mimicking what was happening in the stock markets at those turbulent times.*

TABLE 10.1: Parameter estimations for 3 regimes.

Regime i	μ_i	σ_i	ν_i	$\eta_n(i)$
1	0.7196	0.0502	0.1890	0.0060
2	-0.3555	0.2197	0.6514	0.9697
3	-1.2032	0.7034	0.1597	0.0243

The corresponding transition matrix is given by

$$Q = \begin{pmatrix} 0.0034 & 0.9837 & 0.0129 \\ 0.2891 & 0.7109 & 0.0000 \\ 0.0000 & 0.0153 & 0.9847 \end{pmatrix}. \qquad (10.16)$$

Finally, using formula (10.13), *one can check the behavior of the condi-*

tional density of the predictions for 1-day, 5-day, together with the long term behavior. The associated densities are displayed in Figure 10.2.

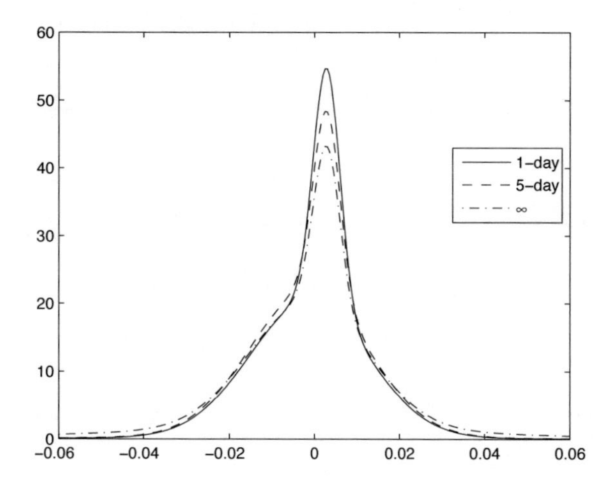

FIGURE 10.2: Forecasted densities for the log-returns of the S&P 500, as of December 31st, 2008.

Example 10.2.2 (Regime-Switching Model for Apple) *As another example, consider the values of Apple, from January 13th, 2009, to January 14th, 2011, included in the MATLAB file* ApplePrices. *So far we tried to fit three models to this data set: The Black-Scholes model, a GARCH with Gaussian innovations, and a GARCH model with GED innovations. Only the latter was not rejected. We will now try to fit a Gaussian regime-switching model. Since we are using daily data, we assume that the mean and volatility of each regime are given respectively by $\mu_j h$ and $\sigma_j\sqrt{h}$, where $h = 1/252$. The P-values for the goodness-of fit test are given in Table 10.2. Therefore we may conclude that 3 regimes suffices, since the P-values for 1 and 2 regimes are below 5%. The estimated parameters for the case of 3 regimes are given in Table 10.3.*

TABLE 10.2: P-values of the goodness-of-fit tests for Apple, using $N = 10000$ replications.

Number of Regimes	1	2	3
P-value (%)	0.0	2.3	82.4

TABLE 10.3: Parameter estimations for 3 regimes.

Regime i	μ_i	σ_i	ν_i	$\eta_n(i)$
1	0.5119	0.4191	0.2947	0.0193
2	1.3326	0.2652	0.3449	0.5412
3	0.2378	0.1363	0.3604	0.4395

The corresponding transition matrix is given by

$$Q = \begin{pmatrix} 0.96436 & 0.03558 & 0.00006 \\ 0.00009 & 0.13843 & 0.86148 \\ 0.02905 & 0.79524 & 0.17571 \end{pmatrix}.$$

The densities of forecasts for 1-day, 5-day, together with the stationary density for Apple are plotted in Figure 10.3.

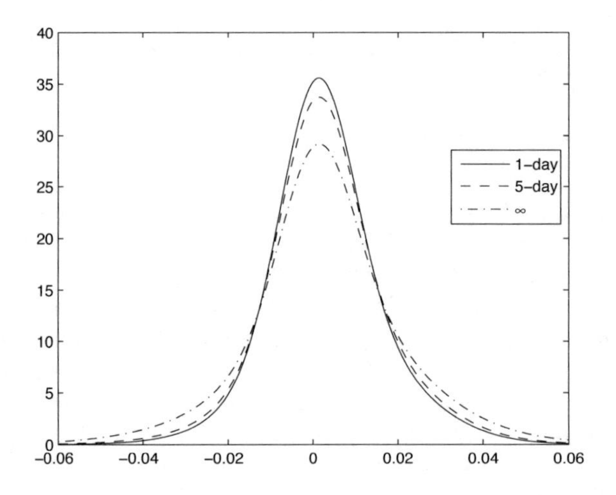

FIGURE 10.3: Forecasted densities for the log-returns of Apple as of January 14th, 2011.

10.2.7 Continuous Time Regime-Switching Markov Processes

One can also see the regime-switching Markov models as approximations of continuous time regime-switching models. Indeed, let Λ be the infinitesimal generator of a continuous time Markov chain τ with state space E, i.e.,

$$P(\tau_t = l|\tau_0 = j) = \left(e^{t\Lambda}\right)_{jl} = \sum_{k=0}^{\infty} \frac{t^k}{k!} \left(\Lambda^k\right)_{jl}, \quad j,l \in \{1,\ldots,r\}.$$

See Appendix 8.A for details. Suppose that $K_1, \ldots K_r$ are Markov kernels. A continuous time regime-switch Markov process (X,τ) is then defined the following way, starting from (x_0, i):

- Generate a random time $T_1 \sim \text{Exp}(\lambda_i)$, with mean $1/\lambda_i = -1/\Lambda_{ii}$. For $t \in [0, T_1)$, set $\tau_t = i$. Further let X_i be a càdlàg Markov process with kernel K_i starting from x_0 at $t = 0$. Set $X_t = X_i(t)$, for $t \in [0, T_1)$;
- For $k \geq 1$, if $\tau_{T_k-} = j$, choose the regime l with probability Λ_{jl}/λ_j, and generate $T_{k+1} - T_k \sim \text{Exp}(\lambda_j)$. For $t \in [T_k, T_{k+1})$, set $\tau_t = l$. Also, generate a càdlàg Markov process X_l with kernel K_l, starting from $X(T_k-)$ at time T_k, and set $X(t) = X_l(t)$, for $t \in [T_k, T_{k+1})$.

Example 10.2.3 *Consider continuous Gaussian processes $X_i(t) = \mu_j t + \sigma_j W_j(t)$, where $W_1, \ldots, W_r$ are independent Brownian motions. Then the associated regime-switching process is called a regime-switching Brownian motion. If $\tau_t = j$ for all $t \in [T_k, T_{k+1})$, then $X(t) = X_i(t) - X_j(T_k) + X(T_k)$ on the same time interval. In the same way, we can define regime-switching Lévy processes. For more details, see Rémillard and Rubenthaler [2009].*

To generate a discrete time regime-switching model observed at times $0, h, 2h, \ldots$, we could define a transition probability matrix Q by setting $Q = I + \Lambda h$, provided h is small enough. Given $\tau_i = j$, we could generate a value $X_j(ih)$ from kernel K_j, starting from X_{i-1} at time $(i-1)h$. Specializing to a regime-switching Brownian motion, it means that at every step, we generate a Gaussian random vector with mean $\mu_j h$ and covariance matrix $A_j h$, which is exactly the model we considered. Note that it also justifies the parametrization we used in Examples 10.2.1 and 10.2.2. In these examples, the parameters μ_j and σ_j can be seen as the parameters of a regime-switching Brownian motion.

10.3 Replication of Hedge Funds

Several investment firms recently launched financial products for the replication of the monthly returns of hedge funds indices, using a portfolio of highly liquid assets. In many cases, these firms used moving-window regression, like Morgan Stanley, while other firms like Innocap and Société Générale, used filters. The latter firm even wrote a technical report on the subject [Roncalli and Teïletche, 2007].

To describe the problem, let $R_i^{(F)}$ be the monthly return of the index to replicate at period i. The goal is to be as close as possible to the target $R_i^{(F)}$ by using a portfolio of $p+1$ liquid assets, traded monthly, whose returns are denoted $R_i^{(1)}, \ldots, R_i^{(p+1)}$, and the weights of the assets in the portfolio are respectively $\beta_i^{(1)}, \ldots, \beta_i^{(p+1)}$, with $\sum_{j=1}^{p+1} \beta_i^{(j)} = 1$. Here of course, we choose assets that can be shorted, and that could be related to hedge funds strategies. Typically, $R_i^{(p+1)}$ corresponds to the return of the 1-month Libor, so the return $R_i^{(P)}$ of the portfolio at period i is given by

$$\begin{aligned} R_i^{(P)} &= \sum_{j=1}^{p} \beta_i^{(j)} R_i^{(j)} + \left(1 - \sum_{j=1}^{p} \beta_i^{(j)}\right) R_i^{(p+1)} \\ &= R_i^{(p+1)} + \sum_{j=1}^{p} \beta_i^{(j)} \left(R_i^{(j)} - R_i^{(p+1)}\right). \end{aligned}$$

Set $\beta_i^\top = \left(\beta_i^{(1)}, \ldots, \beta_i^{(p)}\right)$ and

$$H_i = \left(R_i^{(1)} - R_i^{(p+1)}, \ldots, R_i^{(p)} - R_i^{(p+1)}\right),$$

so that there is no more linear constraint on the weights β_i, other than not depending on the future and respecting the limits imposed by the back-office. Because of the relationship

$$R_i^{(P)} = R_i^{(p+1)} + H_i \beta_i,$$

it is natural to use the excess returns $Y_i = R_i^{(F)} - R_i^{(p+1)}$ for the hedge fund, so that we have to choose β_i so that $H_i \beta_i$ is as close as possible to Y_i.

10.3.0.1 Measurement of Errors

A measure often used in practice to assess the performance of the replication is the RMSE (root mean square error), also called "tracking error" in a replication context. The in-sample RMSE is computed as

$$RMSE = \sqrt{\sum_{i=t_0}^{T-1} (Y_i - H_i \beta_i)^2 / (T - t_0)}.$$

A better measure of the tracking error is the out-of-sample RMSE given by

$$RMSE = \sqrt{\sum_{i=t_0}^{T-1} (Y_{i+1} - H_{i+1} \beta_i)^2 / (T - t_0)},$$

because $Y_{i+1} - H_{i+1}\beta_i$ is the real tracking error.

10.3.1 Replication by Regression

This methodology simply consists in estimating the coefficients β_i using a linear regression over the last n observations, with H as the independent variables. More precisely,

$$\begin{aligned}\beta_i &= \arg\max_{\beta} \sum_{k=i-n+1}^{i} (y_k - H_k\beta)^2 \\ &= \left(\sum_{k=i-n+1}^{i} H_k^\top H_k\right)^{-1} \sum_{k=i-n+1}^{i} H_k^\top y_k.\end{aligned}$$

In practice $n = 24$ is used, corresponding to two years of data.

10.3.2 Replication by Kalman Filter

The idea here is to treat the replication strategy β as a signal. For the Kalman filter, or any filter, we must propose the evolution equation for the signal β. In Roncalli and Teïletche [2007], it is suggested to take a random walk model, i.e.,

$$\beta_i = \beta_{i-1} + w_i, \qquad w_i \sim N_p(0, Q_i),$$

the measurement equation being

$$Y_i = H_i\beta_i + \varepsilon_i, \qquad \varepsilon_i \sim N(0, R_i).$$

10.3.3 Example of Application

In what follows, we try to replicate[2] the returns of a general hedge fund index using a portfolio composed of the following factors: S&P500 Index TR, Russel 2000 Index TR, Russell 1000 Index TR, Eurostoxx Index, Topix, US 10-year Index, 1-month LIBOR. The data, provided by Innocap, are from April 1997 to October 2008. In Table 10.4, we report descriptive statistics of the replication portfolio and the index for the in-sample and out-of-sample performances.

The Kalman filter performs better than the regression method, even if the transition equation is very simple. Note that the parameters for the Kalman filter have been estimated using all observations. Figures 10.4 and 10.5 compare the returns of the index with those obtained from the regression method, computed with a rolling-window of 24 months, and the returns computed with the Kalman filter, for the simple model proposed by Roncalli and Teïletche [2007].

The compounded values of $1 for the returns of the index and the replication portfolios are displayed in Figure 10.6. Finally, the evolution of the

[2]We do it without real efforts of improving the model for confidentiality reasons.

portfolio weights β_i for the regression method and the Kalman filter are displayed in Figures 10.7 and 10.8.

TABLE 10.4: In-sample and out-of-sample statistics for the tracking error (TE), Pearson correlation (ρ), Kendall tau (τ), mean (μ), volatility (σ), skewness (γ_1), and kurtosis (γ_2).

Portfolio	In-Sample Statistics						
	TE (%y)	ρ	τ	μ (%y)	σ (%y)	γ_1	γ_2
Target Ptf	-	1.00	1.00	8.12	7.72	-0.59	5.45
Regression	10.58	0.93	0.75	8.79	8.32	-0.69	5.22
Kalman	8.54	0.95	0.78	9.68	7.75	-0.59	5.53
Portfolio	Out-of-Sample Statistics						
	TE (%y)	ρ	τ	μ (%y)	σ (%y)	γ_1	γ_2
Target Ptf	-	1.00	1.00	8.12	7.72	-0.59	5.45
Regression	19.27	0.83	0.66	9.30	9.86	-0.11	6.34
Kalman	14.71	0.86	0.67	9.97	8.20	-0.40	5.63

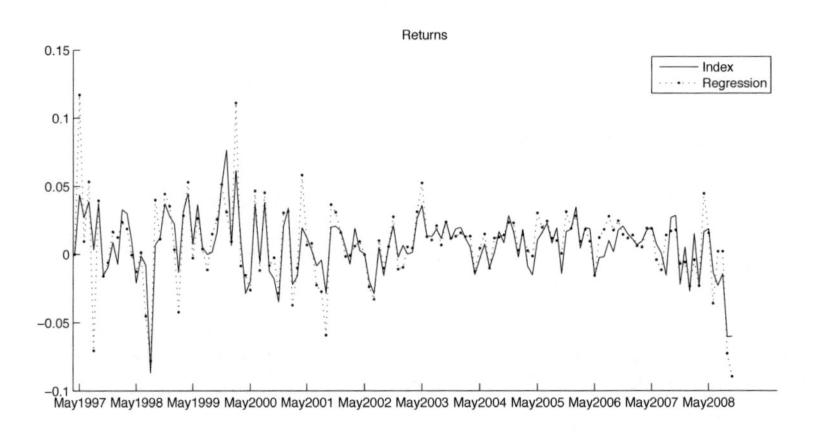

FIGURE 10.4: Returns of the replication portfolio when using the regression method.

As a final remark, note that according to Figures 10.7 and 10.8, the portfolio weights are more stable with the Kalman filter than with the rolling-window regression, meaning less transaction fees.

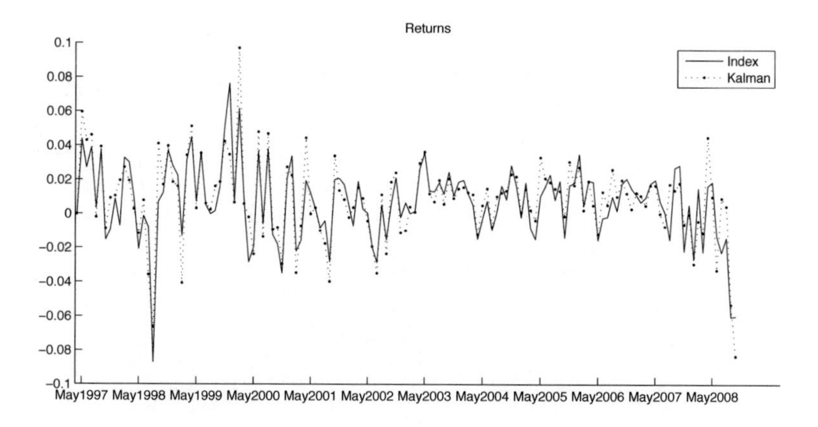

FIGURE 10.5: Returns of the replication portfolio when using the Kalman filter.

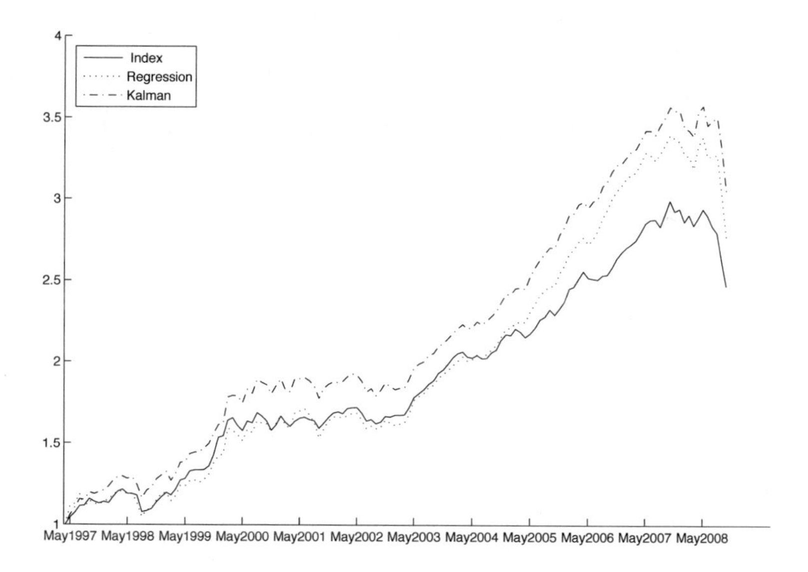

FIGURE 10.6: Compounded values of $1 for the returns of the index and the replication portfolios.

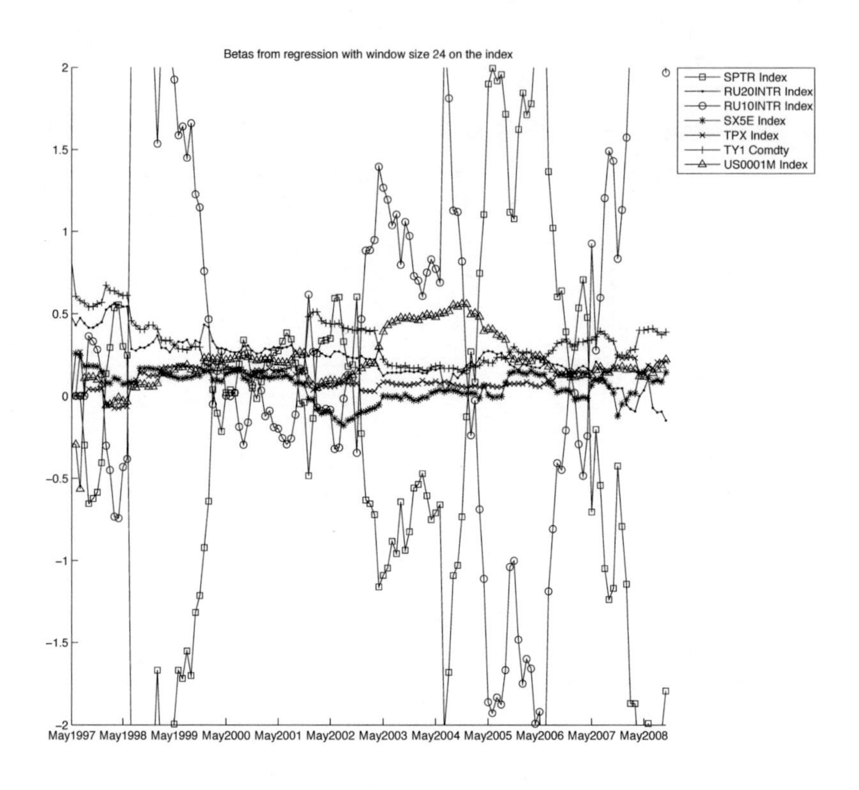

FIGURE 10.7: Evolution of the weights β_i of the replication portfolio, computed with a rolling window of 24 months.

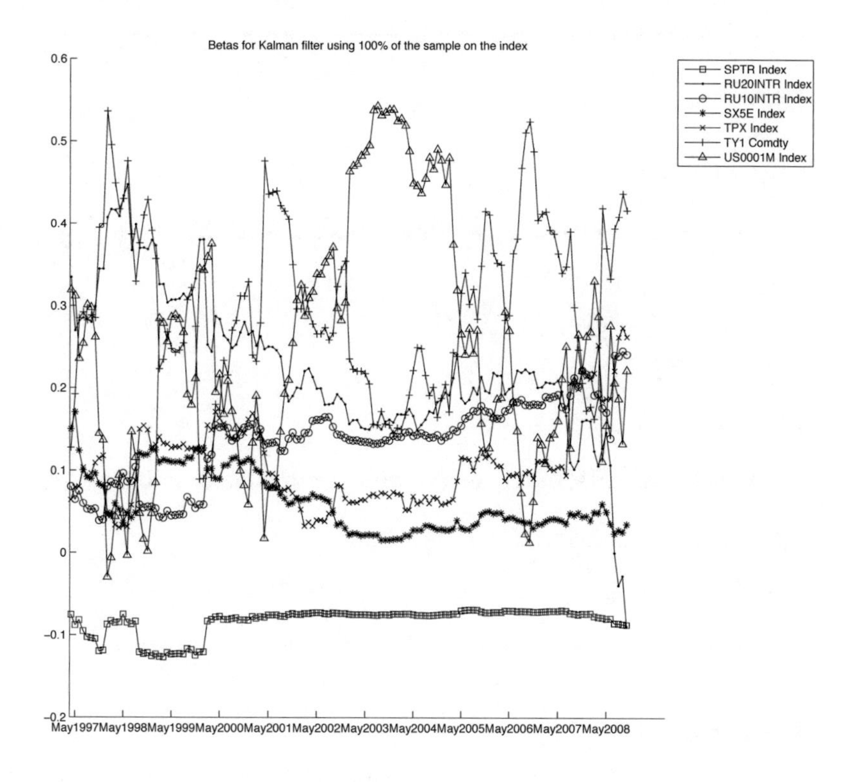

FIGURE 10.8: Evolution of the weights β_i of the replication portfolio, computed with the Kalman filter.

10.4 Suggested Reading

Hamilton [1990] is a good introduction to regime-switching models. For hedge fund replication, one can consult the technical papers Roncalli and Teïletche [2007] and Roncalli and Weisang [2009], who also considered particle filters in a replication context.

10.5 Exercises

Exercise 10.1

For a $MA(1)$ model, show that for all $i \geq 1$, $Z_{i|i} = \begin{pmatrix} Y_i - \mu \\ -\theta \frac{e_i}{f_i^2} \end{pmatrix}$ and $U_{i|i} = \begin{pmatrix} 0 & 0 \\ 0 & \theta^2 \frac{(f_i^2-1)}{f_i^2} \end{pmatrix}$, with $f_{i+1}^2 = 1 + \theta^2 - \frac{\theta^2}{f_i^2}$, $e_1 = Y_1 - \mu$, and $e_{i+1} = Y_{i+1} - \mu + \theta \frac{e_i}{f_i^2}$.

Exercise 10.2

Consider an ARMA(1,1) model with parameters $\phi_1 = 0.4$ and $\theta_1 = -0.5$.

(a) Using Proposition 10.1.1, find F, g, and a.

(b) Compute $U_{0|0}$.

(c) Using (10.3), compute f_i^2 and $U_{i|i}$ for $i \in \{1, \ldots, 10\}$.

Exercise 10.3

For an $ARMA(1,1)$ model, show that for all $i \geq 1$, $Z_{i|i} = \begin{pmatrix} Y_i - \mu \\ \phi(Y_i - \mu) - \theta \frac{e_i}{f_i^2} \end{pmatrix}$ and $U_{i|i} = \begin{pmatrix} 0 & 0 \\ 0 & \theta^2 \frac{(f_i^2-1)}{f_i^2} \end{pmatrix}$, with $f_{i+1}^2 = 1 + \theta^2 - \frac{\theta^2}{f_i^2}$, and $e_{i+1} = Y_{i+1} - \mu - \phi(Y_i - \mu) + \theta \frac{e_i}{f_i^2}$. Also $U_{0|0} = \begin{pmatrix} 1 + \frac{\psi^2}{1-\phi^2} & \psi + \phi \frac{\psi^2}{1-\phi^2} \\ \psi + \phi \frac{\psi^2}{1-\phi^2} & \frac{\psi^2}{1-\phi^2} \end{pmatrix}$, $e_1 = Y_1 - \mu$ and $f_1^2 = 1 + \frac{\psi^2}{1-\phi^2}$.

Exercise 10.4

Consider an ARMA(2,2) model with parameters $\phi_1 = 0.4$, $\phi_2 = -0.1$ and $\theta_1 = -0.3$, $\theta_2 = 0.1$.

(a) Using Proposition 10.1.1, find F, g, and a.

(b) Compute $U_{0|0}$.

(c) Using (10.3), compute f_i^2 and $U_{i|i}$ for $i \in \{1, \ldots, 10\}$.

Exercise 10.5

Consider an ARMA(p,q) and set $r = \max(p, q)$. Show that the coefficients

$\psi_1, \ldots, \psi_r$, from the MA representation of an ARMA(p,q) satisfy the following equations:

$$\psi_j = \phi_j - \theta_j + \sum_{k=1}^{j-1} \psi_k \phi_{j-k}, \quad j \in \{1, \ldots, r\}.$$

Exercise 10.6

Consider the regime-switching model of Example 10.2.2.

(a) What is distribution of the regimes for the next trading day, i.e., January 18th, 2011?

(b) Generate 100000 observations from the 1-day forecast density and compute a 95% confidence interval.

(c) On January 18th, 2011, the closing price of Apple was \$340.65. Does it belong to the 5% confidence interval?

(d) Based on (c), compute the new distribution for the regimes.

Exercise 10.7

Two models for the log-returns of Apple were not rejected: GARCH(1,1) with GED innovations, and a regime-switching model with three Gaussian regimes. Which one would you choose? Motivate your choice.

Exercise 10.8

Using parametric bootstrap to compute P-values, you get an estimated P-value of 3%, using $N = 100$ replications.

(a) Can you safely conclude that the real P-value is less than 5%?

(b) What is the solution?

Exercise 10.9

For the replication of hedge funds indices, suppose that there are constraints on the values of the positions β, for example $\sum_{j=1}^{p} \left|\beta_i^{(j)}\right| \leq 2$. Can you still use the Kalman filter? What should you do in this case?

10.6 Assignment Questions

1. Construct a MATLAB function to compute the coefficients $\psi_1, \ldots, \psi_{r-1}$ arising from a MA representation of an ARMA(p,q) model. Here $r = \max(p, q+1)$.

2. Construct a MATLAB function for the Kalman equations of an ARMA(p,q) model.
3. Construct a MATLAB function to simulate a regime-switching model with Gaussian regimes.
4. Construct a MATLAB function to simulate a regime-switching Brownian motion observed a periods $h, 2h, \ldots, nh$.
5. Construct a MATLAB function to perform the test of goodness-of-fit based on the Rosenblatt transform for one-dimensional regime-switching models. Use the MATLAB function *EstHMM1d*.

10.A EM Algorithm

Here is a description of the EM algorithm, used for finding the maximum likelihood estimate in some difficult settings, as proposed by Dempster et al. [1977].

Suppose that we want to estimate the parameter θ based on a sample $Y_1, \ldots, Y_n$ from a distribution with density $g(y_1, \ldots, y_n; \theta)$. Often, the observations $Y_1, \ldots, Y_n$ are a function of a sample of random vectors $X_1, \ldots, X_n$ from density $f(x_1, \ldots, x_n; \theta)$, which are not completely observable.

Example 10.A.1 (Gaussian Mixtures) *Suppose that the density g of the observations is a mixture of two Gaussian densities, i.e.,*

$$\begin{aligned} g(y;\theta) &= \prod_{i=1}^{n} \left\{ p\phi\left(\frac{y_i-\mu_1}{\sigma_1}\right)/\sigma_1 + (1-p)\phi\left(\frac{y_i-\mu_2}{\sigma_2}\right)/\sigma_2 \right\} \\ &= \prod_{i=1}^{n} \{ p g_1(y_i;\mu_1,\sigma_1) + (1-p) g_1(y_i;\mu_2,\sigma_2) \}, \end{aligned}$$

where ϕ is the density of a standard Gaussian variable and $\theta = (p, \mu_1, \mu_2, \sigma_1, \sigma_2)$. In this case, one could define $X_i = (Y_i, \tau_i)$, where the regimes $\tau_1, \ldots, \tau_n$ are independent, with $P(\tau_i = 1) = p$, $P(\tau_i = 2) = 1 - p$, and the joint density of $X_1, \ldots, X_n$ is

$$\begin{aligned} f(x;\theta) &= f(y_1, t_1, \ldots, y_n, t_n; \theta) \\ &= \prod_{i=1}^{n} p^{2-t_i}(1-p)^{t_i-1} g_1(y_i; \mu_{t_i}, \sigma_{t_i}). \end{aligned}$$

We recover $g(y;\theta)$ by summing over all possible regimes, i.e.,

$$g(y;\theta) = \sum_{t_1=1}^{2} \cdots \sum_{t_n=1}^{2} f(y_1, t_1, \ldots, y_n, t_n; \theta).$$

To implement the EM algorithm, we proceed as follows:

Algorithm 10.A.1 *Define* $Q_y(\theta', \theta) = E_\theta \{\ln f(X;\theta')|Y = y\}$, *and start with an initial guess* $\theta^{(0)}$. *Then repeat the following steps until convergence:*

- *(E-Step) Compute* $Q_y\left(\theta, \theta^{(k)}\right)$.
- *(M-Step) Find* $\theta^{(k+1)} = \arg\max_\theta Q_y\left(\theta, \theta^{(k)}\right)$.

Then, one should have $\theta^{(k)} \to \theta^\star$, *where* $\theta^\star$ *is the maximum likelihood estimator of* $g(y;\theta)$, *i.e.,* $\theta^\star = \arg\max_\theta g(y;\theta)$.

Example 10.A.2 (Gaussian Mixtures (Continued)) *For the Gaussian mixtures in Example 10.A.1, we have*

$$\begin{aligned}
\lambda_i(\theta) &= P_\theta(\tau_i = 1|Y_1 = y_1, \ldots, Y_n = y_n) \\
&= \frac{p g_1(y_i;\mu_1,\sigma_1)}{p g_1(y_i;\mu_1,\sigma_1) + (1-p) g_1(y_i;\mu_2,\sigma_2)}.
\end{aligned}$$

Since $\ln\{f(X;\theta)\}$ *is given, up to a constant, by*

$$\sum_{i=1}^n \left\{(2-\tau_i)\ln(p) + (\tau_i - 1)\ln(1-p) - \frac{1}{2}\left(\frac{Y_i - \mu_{\tau_i}}{\sigma_{\tau_i}}\right)^2 - \ln(\sigma_{\tau_i})\right\},$$

we obtain

$$\begin{aligned}
Q_y(\theta, \theta^{(k)}) &= c_1 + \sum_{i=1}^n \left[\lambda_i\left(\theta^{(k)}\right)\ln(p) + \left\{1 - \lambda_i\left(\theta^{(k)}\right)\right\}\ln(1-p)\right] \\
&\quad -\frac{1}{2}\sum_{i=1}^n \lambda_i\left(\theta^{(k)}\right)\left(\frac{Y_i - \mu_1}{\sigma_1}\right)^2 \\
&\quad -\frac{1}{2}\sum_{i=1}^n \left\{1 - \lambda_i\left(\theta^{(k)}\right)\right\}\left(\frac{y_i - \mu_2}{\sigma_2}\right)^2 \\
&\quad -\sum_{i=1}^n \left[\lambda_i\left(\theta^{(k)}\right)\ln(\sigma_1) + \left\{1 - \lambda_i\left(\theta^{(k)}\right)\right\}\ln(\sigma_2)\right],
\end{aligned}$$

where $c_1 = -\frac{n}{2}\ln(2\pi)$. *Therefore*

$$
\begin{aligned}
p^{(k+1)} &= \frac{1}{n}\sum_{i=1}^{n} \lambda_i\left(\theta^{(k)}\right), \\
\mu_1^{(k+1)} &= \frac{1}{np^{(k+1)}}\sum_{i=1}^{n} \lambda_i\left(\theta^{(k)}\right) y_i, \\
\mu_2^{(k+1)} &= \frac{1}{n\left(1-p^{(k+1)}\right)}\sum_{i=1}^{n} \left\{1-\lambda_i\left(\theta^{(k)}\right)\right\} y_i, \\
\sigma_1^{(k+1)} &= \left[\frac{1}{np^{(k+1)}}\sum_{i=1}^{n} \lambda_i\left(\theta^{(k)}\right)\left(y_i-\mu_1^{(k+1)}\right)^2\right]^{1/2}, \\
\sigma_2^{(k+1)} &= \left[\frac{1}{n\left(1-p^{(k+1)}\right)}\sum_{i=1}^{n} \left\{1-\lambda_i\left(\theta^{(k)}\right)\right\}\left(y_i-\mu_2^{(k+1)}\right)^2\right]^{1/2}.
\end{aligned}
$$

Remark 10.A.1 *In the general case of a mixture*

$$g(y;\theta) = \prod_{i=1}^{n}\left\{\sum_{j=1}^{r} p_j g_1(y_i;\alpha_j)\right\},$$

with $\theta = (p,\alpha) = (p_1,\ldots,p_r,\alpha_1,\ldots,\alpha_r)$, *it is easy to check that*

$$\ln f(x;\theta) = \sum_{i=1}^{n}\sum_{j=1}^{r} \mathbb{I}(\tau_i = j)\left[\ln(p_j) + \ln\{g_1(y_i;\alpha_j\}\right],$$

so for any $i \in \{1,\ldots,n\}$ *and* $j \in \{1,\ldots,r\}$,

$$\lambda_{ij}(\theta) = P_\theta(\tau_i = j|Y_1 = y_1,\ldots,Y_n = y_n) = \frac{p_j g_1(y_i;\alpha_j)}{\sum_{k=1}^{r} p_k g_1(y_i;\alpha_k)}.$$

As a result,

$$Q_y(\theta',\theta) = \sum_{i=1}^{n}\sum_{j=1}^{r} \lambda_{ij}(\theta)\left[\ln(p_j') + \ln\{g_1(y_i;\alpha_j')\}\right].$$

Hence

$$p_j^{(k+1)} = \frac{1}{n}\sum_{i=1}^{n}\lambda_{ij}\left(\theta^{(k)}\right), \quad j \in \{1,\ldots,r\},$$

and

$$\alpha_j^{(k+1)} = \arg\max_{\alpha_j}\sum_{i=1}^{n}\lambda_{ij}\left(\theta^{(k)}\right) g_1(y_i;\alpha_j).$$

10.B Sampling Moments vs Theoretical Moments

The mean and the covariance matrix of the observations are respectively

$$\bar{x} = \frac{1}{n}\sum_{i=1}^{n} y_i \quad \text{and} \quad S = \frac{1}{n}\sum_{i=1}^{n}(y_i - \bar{x})(y_i - \bar{x})^\top.$$

According to formula (10.5), we have

$$\mu' = \sum_{j=1}^{r} \nu_j' \mu_j' = \frac{1}{n}\sum_{j=1}^{r}\sum_{i=1}^{n} \lambda_i(j) y_i = \frac{1}{n}\sum_{i=1}^{n} y_i = \bar{x}.$$

Next,

$$\begin{aligned}
\sum_{j=1}^{r} \nu_j' A_j' &= \frac{1}{n}\sum_{j=1}^{r}\sum_{i=1}^{n} \lambda_i(j)(y_i - \mu_j')(y_i - \mu_j')^\top \\
&= \frac{1}{n}\sum_{j=1}^{r}\sum_{i=1}^{n} \lambda_i(j) y_i y_i^\top + \frac{1}{n}\sum_{j=1}^{r}\sum_{i=1}^{n} \lambda_i(j)\mu_j'(\mu_j')^\top \\
&\qquad - \frac{1}{n}\sum_{j=1}^{r}\sum_{i=1}^{n} \lambda_i(j) y_i (\mu_j')^\top - \frac{1}{n}\sum_{j=1}^{r}\sum_{i=1}^{n} \lambda_i(j)\mu_j' y_i^\top \\
&= \frac{1}{n}\sum_{i=1}^{n} y_i y_i^\top - \sum_{j=1}^{r} \nu_j' \mu_j' (\mu_j')^\top.
\end{aligned}$$

Hence, by formula (10.6), we obtain

$$\begin{aligned}
A' &= \sum_{j=1}^{r} \nu_j' A_j' + \sum_{j=1}^{r} \nu_j' \mu_j' (\mu_j)'^\top - \mu'(\mu')^\top \\
&= \frac{1}{n}\sum_{i=1}^{n} y_i y_i^\top - \sum_{j=1}^{r} \nu_j' \mu_j' (\mu_j')^\top + \sum_{j=1}^{r} \nu_j' \mu_j' (\mu_j)'^\top - \bar{x}\bar{x}^\top \\
&= \frac{1}{n}\sum_{i=1}^{n} (y_i - \bar{x})(y_i - \bar{x})^\top = S.
\end{aligned}$$

10.C Rosenblatt Transform for the Regime-Switching Model

The aim here is to find, for $i \geq 0$, the Rosenblatt transform Ψ_{i+1} corresponding to the conditional distribution of Y_{i+1} given $Y_1 = y_1, \ldots, Y_i = y_i$.

To this end, let $k \in \{1,\ldots,r\}$ be given and let Z_k be a random d-dimensional vector with density f_k. For all $j \in \{1,\ldots,d\}$, denote by $f_{k,1:j}$ the density of $\left(Z_k^{(1)},\ldots,Z_k^{(j)}\right)$, and by $f_{k,j}$ the density of $Z_k^{(j)}$ given $\left(Z_k^{(1)},\ldots,Z_k^{(j-1)}\right)$. Further let $F_{k,j}$ be the distribution function associated with the density $f_{k,j}$, where $F_{k,1}$ is the distribution function of $Z_k^{(1)}$. The Rosenblatt transform

$$z = (z_1,\ldots,z_d) \mapsto T_k(z) = \left(F_{k,1}(z_1), F_{k,2}(z_1,z_2),\ldots,F_{k,d}(z_1,\ldots,z_d)\right)^\top$$

is such that the random vector $T_k(Z_k)$ is uniformly distributed over $[0,1]^d$. For example, in the bivariate Gaussian case, where f_k is the density of $N_2(\mu_k,\Sigma_k)$, with

$$\Sigma_k = \left(\begin{array}{cc} v_k^{(1)} & \rho_k\sqrt{v_k^{(1)}v_k^{(2)}} \\ \rho_k\sqrt{v_k^{(1)}v_k^{(2)}} & v_k^{(2)} \end{array}\right),$$

then $f_{k,2}$ is the density of a Gaussian distribution with mean $\mu_k^{(2)} + \beta_k\left(y_k^{(1)} - \mu_k^{(1)}\right)$ and variance $v_k^{(2)}\left(1-\rho_k^2\right)$, where $\beta_k = \rho_k\sqrt{v_k^{(2)}/v_k^{(1)}}$. Since the density of the conditional distribution is given by (10.12), which is a mixture of the densities $f_1,\ldots,f_r$ with weights $W_{1,i},\ldots,W_{r,i}$, it follows that for all $z_1,\ldots,z_d \in \mathbb{R}$,

$$\Psi_{i+1}^{(1)}(z_1) = \sum_{k=1}^r W_{k,i}F_{k,1}(z_1),$$

and for $j \in \{2,\ldots,d\}$, we have

$$\Psi_{i+1}^{(j)}(z_1,\ldots,z_j) = \frac{\sum_{k=1}^r W_{k,i}f_{k,1:j-1}(z_1,\ldots,z_{j-1})F_{k,j}(z_j)}{\sum_{k=1}^r W_{k,i}f_{k,1:j-1}(z_1,\ldots,z_{j-1})}.$$

The proof that these expressions are indeed correct follows by differentiating $\Psi_{i+1}^{(j)}(z_1,\ldots,z_j)$ with respect to z_j, for all $j \in \{1,\ldots,r\}$. In fact,

$$\begin{aligned} \frac{d}{dz_j}\Psi_{i+1}(z_j) &= \frac{\sum_{k=1}^r W_{k,i}f_{k,1:j-1}(z_1,\ldots,z_{j-1})f_{k,j}(z_j)}{\sum_{k=1}^r W_{k,i}f_{k,1:j-1}(z_1,\ldots,z_{j-1})} \\ &= \frac{\sum_{k=1}^r W_{k,i}f_{k,1:j}(z_1,\ldots,z_j)}{\sum_{k=1}^r W_{k,i}f_{k,1:j-1}(z_1,\ldots,z_{j-1})}, \end{aligned}$$

which, according to (10.12) is the conditional density of $Y_{i+1}^{(j)}$ given $Y_1,\ldots,Y_i,Y_{i+1}^{(1)},\ldots,Y_{i+1}^{(j-1)}$.

10.D Proofs of the Results

10.D.1 Proof of Proposition 10.1.1

Recall that $Y_i - \mu = \sum_{j=1}^{p} \phi_j (Y_{i-j} - \mu) + \varepsilon_i - \sum_{k=1}^{q} \theta_k \varepsilon_{i-k}$. Since $\hat{Y}_i(0) = Y_i$, we have $Y_i = \mu + (1\ 0 \cdots 0) Z_i$. It remains to find the transition equation for the signal Z_i. First, note that Y_i has the MA representation

$$Y_i = \mu + \sum_{k=0}^{\infty} \psi_k \varepsilon_{i-k},$$

where $\psi(z) = \theta(z)/\phi(z)$. Hence, for $j \geq 0$,

$$\begin{aligned} \hat{Y}_i(j) &= \mu + \sum_{k=j}^{\infty} \psi_k \varepsilon_{i+j-k} = \mu + \psi_j \varepsilon_i + \sum_{k=j+1}^{\infty} \psi_k \varepsilon_{i+j-k} \\ &= \psi_j \varepsilon_i + \hat{Y}_{i-1}(j+1), \end{aligned}$$

so we have a representation for the first $r-1$ components of Z_i. Finally, since

$$Y_{i+r-1} - \mu = \sum_{j=1}^{p} \phi_j (Y_{i+r-1-j} - \mu) + \varepsilon_{i+r-1} - \sum_{k=1}^{q} \theta_k \varepsilon_{i+r-1-k},$$

we obtain

$$\hat{Y}_{i-1}(r) - \mu = \sum_{j=1}^{p} \phi_j \left\{ \hat{Y}_{i-1}(r-j) - \mu \right\} = \sum_{j=1}^{r} \phi_j \left\{ \hat{Y}_{i-1}(r-j) - \mu \right\},$$

since $r \geq q+1$ and $r \geq p$. Consequently,

$$\hat{Y}_i(r-1) - \mu = \psi_{r-1}\varepsilon_i + \hat{Y}_{i-1}(r) - \mu = \psi_{r-1}\varepsilon_i + \sum_{j=1}^{r} \phi_j \left\{ \hat{Y}_{i-1}(r-j) - \mu \right\}.$$

Therefore,

$$Z_i = \begin{pmatrix} 0 & 1 & 0 & 0 & \cdots & 0 \\ 0 & 0 & 1 & 0 & \cdots & 0 \\ \vdots & & & \ddots & & \\ \vdots & & & & \ddots & \\ 0 & 0 & 0 & \cdots & 0 & 1 \\ \phi_r & \phi_{r-1} & \cdots & \cdots & \phi_2 & \phi_1 \end{pmatrix} Z_{i-1} + \begin{pmatrix} 1 \\ \psi_1 \\ \psi_2 \\ \vdots \\ \vdots \\ \psi_{r-1} \end{pmatrix} \varepsilon_i,$$

completing the proof.

■

10.D.2 Proof of Proposition 10.1.2

By stationarity

$$U_{0|0} = E(Z_0 Z_0^\top)/\sigma_\varepsilon^2 = F U_{0|0} F^\top + g g^\top.$$

In then follows that for all $k \geq 0$,

$$F^k U_{0|0} (F^k)^\top - F^{k+1} U_{0|0} (F^{k+1})^\top = (F^k g)(F^k g)^\top.$$

Summing over k yields that for any $n \geq 1$,

$$U_{0|0} = F^{n+1} U_{0|0} (F^{n+1})^\top + \sum_{k=0}^{n} (F^k g)(F^k g)^\top.$$

Now, the eigenvalues of F satisfy $\lambda^r - \sum_{j=1}^{r} \phi_j \lambda^{r-j} = 0$, i.e., $\phi(1/\lambda) = 0$. By stationarity of Y, if follows that all eigenvalues $\lambda_1, \ldots, \lambda_r$ are in the unit circle, proving that the spectral radius of F, given by $\max_{j \in \{1,\ldots,r\}} |\lambda_j|$, is strictly smaller than 1. Hence $F^n \to 0$ as $n \to \infty$, so

$$U_{0|0} = \sum_{k=0}^{\infty} (F^k g)(F^k g)^\top.$$

■

Bibliography

L. E. Baum, T. Petrie, G. Soules, and N. Weiss. A maximization technique occurring in the statistical analysis of probabilistic functions of Markov chains. *Ann. Math. Statist.*, 41:164–171, 1970.

O. Cappé, E. Moulines, and T. Rydén. *Inference in Hidden Markov Models.* Springer Series in Statistics. Springer, New York, 2005.

A. P. Dempster, N. M. Laird, and D. B. Rubin. Maximum likelihood from incomplete data via the EM algorithm. *J. Roy. Statist. Soc. Ser. B*, 39: 1–38, 1977.

J. D. Hamilton. Analysis of time series subject to changes in regime. *J. Econometrics*, 45(1-2):39–70, 1990.

B. Rémillard. Validity of the parametric bootstrap for goodness-of-fit testing in dynamic models. Technical report, SSRN Working Paper Series No. 1966476, 2011.

B. Rémillard and S. Rubenthaler. Optimal hedging in discrete and continuous time. Technical Report G-2009-77, Gerad, 2009.

B. Rémillard, A. Hocquard, and N. A. Papageorgiou. Option Pricing and Dynamic Discrete Time Hedging for Regime-Switching Geometric Random Walks Models. Technical report, SSRN Working Paper Series No. 1591146, 2010.

T. Roncalli and J. Teïletche. An alternative approach to alternative beta. Technical report, Société Générale Asset Management, 2007.

T. Roncalli and G. Weisang. Tracking problems, Hedge Fund replication and alternative Beta. Technical report, SSRN Working Paper Series No. 1325190, 2009.

Appendix A

Probability Distributions

In this chapter, we define probability distributions which are used in the book, together with some of their properties and characteristics. We also see how to find the joint density of several random variables. Finally, the Rosenblatt transform is introduced, which is often used for simulations and goodness-of-fit tests.

A.1 Introduction

We begin with some definitions and basic results from probability theory.

Definition A.1.1 *A random variable X is a quantitative function associated with the results of a random experiment. Its random behavior is determined by its distribution function F_X defined for all $x \in \mathbb{R}$ by*

$$F_X(x) = P(X \leq x).$$

A random vector $X = (X_1, \ldots, X_k)$ is a vector for which all components are random variables. Its random behavior is also determined by its distribution function F_X defined for all $x \in \mathbb{R}^k$ by

$$F_X(x) = F_{X_1,\ldots,X_k}(x_1, \ldots, x_k) = P(X_1 \leq x_1, \ldots, X_k \leq x_k).$$

In what follows we consider mainly two types of distribution for random variables: discrete distributions and absolutely continuous distributions. First, the distribution of a random variable X is said to be discrete if its distribution function F is constant at all but a countable set of points. We say that the distribution is continuous if the associated distribution function F is continuous. Finally, the distribution is absolutely continuous if the associated distribution function F has representation

$$F(x) = \int_{-\infty}^{x} f(y)dy,$$

for some non negative function f, called the density of X.

Remark A.1.1 *One can show that a function F is a distribution function if and only if*

1. *$F \geq 0$ and F is non decreasing;*

2. *F is right continuous, i.e., $\lim_{h \downarrow 0} F(x+h) = F(x)$, for all $x \in \mathbb{R}$;*

3. *$\lim_{x \to -\infty} F(x) = 0$ and $\lim_{x \to \infty} F(x) = 1$.*

We may thus conclude that there exists distribution functions which are neither purely discrete nor purely continuous.

Remark A.1.2 *Even if a distribution function is right continuous it does not imply that it is continuous. In fact, $\lim_{h \downarrow 0} F(x-h) = F(x-) = P(X < x)$, so*

$$P(X = x) = F(x) - F(x-),$$

i.e., x is a discontinuity point of F if and only if $P(X = x) > 0$. Note that the set of discontinuity points is at most countable.

A.2 Discrete Distributions and Densities

Recall that the distribution of a random variable X is discrete if there exists a countable set $S = \{x_1, x_2, \ldots\}$ so that $P(X \in S) = 1$. In this case, the distribution of X is uniquely determined by its discrete density $f = f_X$, i.e., $f \geq 0$, $\sum_i f(x_i) = 1$ and $P(X = x_i) = f(x_i)$. As a result,

$$P(X \in B) = \sum_{x \in B} f(x).$$

Remark A.2.1 *Since $P(X \in S) = 1$, one has $f(x) = 0$, if $x \notin S$. For discrete distributions, knowing the discrete density is in general more useful than knowing the distribution function.*

We can now define important characteristics of distributions.

A.2.1 Expected Value and Moments of Discrete Distributions

Definition A.2.1 *Let X be a discrete random variable with discrete density f. Then the pth moment of X, denoted $E(X^p)$, is defined by*

$$E(X^p) = \sum_x x^p f(x), \quad \textit{whenever} \sum_x |x|^p f(x) < \infty.$$

The first moment $E(X)$ is the expected value of X. Moreover, if X is non negative and $\sum_x xf(x) = \infty$, then one sets $E(X) = \infty$. More generally, for a given function g, we define

$$E\{g(X)\} = \sum_x g(x)f(x), \quad \text{whenever} \sum_x |g(x)|f(x) < \infty.$$

Remark A.2.2 *For higher order moments, it is more natural to use central moments defined as $E[\{X - E(x)\}^p]$. In particular, the variance of X is defined as the second central moment*

$$Var(X) = E[\{X - E(X)\}^2] = \sum_x \{x - E(X)\}^2 f(x).$$

Finally, the moment generating function is defined by

$$M_X(u) = E\left(e^{uX}\right) = \sum_x e^{ux} f(x),$$

for all $u \in \mathbb{R}$ for which the right-hand side is finite.

Here is an interesting result for the evaluation of the expectation of random variables taking values in $\mathbb{N} \cup \{0\} = \{0, 1, \ldots\}$.

Proposition A.2.1 *If X is a random variable such that $P(X \in \mathbb{N} \cup \{0\}) = 1$, then*

$$E(X) = \sum_{n=0}^{\infty} P(X > n). \tag{A.1}$$

PROOF.

$$\begin{aligned} E(X) &= \sum_{n=0}^{\infty} nf(n) = \sum_{n=1}^{\infty}\sum_{k=1}^{n} f(n) = \sum_{k=1}^{\infty}\sum_{n=k}^{\infty} f(n) \\ &= \sum_{k=1}^{\infty} P(X \ge k) = \sum_{n=0}^{\infty} P(X > n). \end{aligned}$$

■

Example A.2.1 *Suppose that X is a random variable such that $P(X > n) = (1-p)^{n-1}$, $n \ge 1$. Then*

$$E(X) = \sum_{n \ge 1} (1-p)^{n-1} = \frac{1}{1-(1-p)} = \frac{1}{p}.$$

A.3 Absolutely Continuous Distributions and Densities

Suppose that the random variable X has an absolutely continuous distribution with density f, i.e.,

$$P(X \le x) = F_X(x) = \int_{-\infty}^{x} f(y)dy, \quad \text{for all } x \in \mathbb{R}.$$

It follows that for any Borelian set $A \in \mathcal{B}$[1],

$$P(X \in A) = \int_A f(x)dx.$$

In particular, $P(X = x) = 0$ for any $x \in \mathbb{R}$. Using the fundamental theorem of calculus, we may conclude that the density is the derivative of F at all but a possibly countable set of points. Note that there exists random variables that are neither discrete nor absolutely continuous and such that $P(X \in A) = 0$ for every countable set $A \subset \mathbb{R}$.

A.3.1 Expected Value and Moments of Absolutely Continuous Distributions

Definition A.3.1 *Let X be a random variable with an absolutely continuous distribution with density f. The expected value of X, denoted $E(X)$, is defined by*

$$E(X) = \int_{\mathbb{R}} xf(x)dx, \qquad \textit{whenever} \int_{\mathbb{R}} |x|f(x)dx < \infty.$$

If X is non negative and $\int_{\mathbb{R}} xf(x)dx = \infty$, then one sets $E(X) = \infty$. More generally, for measurable[2] functions g, set

$$E\{g(X)\} = \int_{\mathbb{R}} g(x)f(x)dx, \quad \textit{whenever} \int_{\mathbb{R}} |g(x)|f(x)dx < \infty.$$

The variance of X is defined by

$$Var(X) = E[\{X - E(X)\}^2] = \int_{\mathbb{R}} \{x - E(X)\}^2 f(x)dx.$$

Finally, the moment generating function is defined by

$$M_X(u) = E\left(e^{uX}\right) = \int_{\mathbb{R}} e^{ux} f(x)dx,$$

[1] $\mathcal{B}$ is called the Borel sigma-algebra. It is generated by open intervals. It contains all kind of intervals, countable unions of intervals, ... etc.

[2] Rigorously speaking it means that $\{x \in \mathbb{R};\, g(x) \le y\}$ belongs to $\mathcal{B}$. For example, all continuous functions are measurable and pointwise limit of measurable functions are measurable.

for all $u \in \mathbb{R}$ for which the right-hand side is finite. The moment generating function derives its name from the following important characteristic. If the moment generating function is finite on an open interval containing 0, then the pth moment can be obtained as its pth derivative with respect to u evaluated at $u = 0$, i.e.,

$$E(X^p) = \frac{d^p}{du^p} M_X(u)\bigg|_{u=0}.$$

The following result links expectations and distribution functions.

Proposition A.3.1 *If $E(|X|) < \infty$, then*

$$E(X) = \int_0^\infty \{1 - F(x)\}dx - \int_{-\infty}^0 F(x)dx. \tag{A.2}$$

PROOF. One has $E(X) = \int_0^\infty xf(x)dx + \int_{-\infty}^0 xf(x)dx$. Moreover,

$$\begin{aligned} \int_0^\infty xf(x)dx &= \int_0^\infty \int_0^x f(x)dtdx = \int_0^\infty \int_t^\infty f(x)dxdt \\ &= \int_0^\infty \{1 - F(t)\}dt, \end{aligned}$$

and

$$\begin{aligned} -\int_{-\infty}^0 xf(x)dx &= \int_{-\infty}^0 \int_x^0 f(x)dtdx = \int_{-\infty}^0 \int_{-\infty}^t f(x)dxdt \\ &= \int_{-\infty}^0 F(t)dt. \end{aligned}$$

Hence the result.

■

Remark A.3.1 *Formula* (A.2) *is valid for any random variable X, even if X is discrete or does not have a density. Moreover, if X takes values in $\mathbb{N} \cup \{0\}$, then*

$$\begin{aligned} \int_0^\infty P(X > x)dx &= \sum_{n=0}^\infty \int_n^{n+1} P(X > x)dx = \sum_{n=0}^\infty \int_n^{n+1} P(X > n)dx \\ &= \sum_{n=0}^\infty P(X > n), \end{aligned}$$

which is exactly formula (A.1).

As a by-product of Proposition A.3.1, we have the following result.

Corollary A.3.1 *For any random variable X with distribution function F and any $K \in \mathbb{R}$, we have*

$$E\{\max(X-K,0)\} = \int_K^\infty \{1-F(x)\}dx.$$

Moreover, if one side is infinite, so is the other one.

A.4 Characteristic Functions

To define the properties of a random variable, we can rely on its distribution function or even its density, whenever the latter exists. However, in some cases, the distribution function and the densities cannot be computed explicitly. How can we identify the distribution in such cases? We can answer this question by computing its characteristic function, which is a function that determines the distribution. Before defining it, first recall some properties of complex numbers.

Set $i = \sqrt{-1}$. Then $i^2 = -1$, $i^3 = -i$ and $i^4 = 1$. Moreover, from the de Moivre identity, one has

$$e^{ix} = \cos x + i \sin x.$$

The set $\mathbb{C}$ of complex numbers is the set of all $z = a + ib$, with $a, b \in \mathbb{R}$. Furthermore, the absolute value of z is defined as $|z| = \sqrt{z\bar{z}} = \sqrt{a^2+b^2}$, where $\bar{z} = a - ib$ is called the conjugate. We are now in a position to define the characteristic function.

Definition A.4.1 *Let X be a random variable. Its characteristic function is defined by*

$$\phi_X(u) = E\left(e^{iuX}\right) = E\{\cos(uX)\} + iE\{\sin(uX)\}, \qquad u \in \mathbb{R}.$$

Here are some properties of characteristic functions.

Lemma A.4.1 *Let X be a random variable with characteristic function ϕ_X. Then*

(a) $\phi_X(0) = 1$;

(b) $|\phi_X(u)| \le 1$, *for all* $u \in \mathbb{R}$;

(c) ϕ_X *is uniformly continuous;*

(d) $\phi_{cX+d}(u) = e^{idu}\phi_X(cu)$, *for any* $c, d, u \in \mathbb{R}$;

(e) If $E(|X|^n) < \infty$, then the n-th order derivative of ϕ exists and

$$\frac{d^m}{du^m}\phi_X\bigg|_{u=0} = i^m E(X^m), \qquad m \in \{0, \dots, n\}.$$

Remark A.4.1 *Contrary to the moment generating function, the characteristic function always exists for all $u \in \mathbb{R}$.*

The following results show that one can recover the distribution function from the associated characteristic function.

A.4.1 Inversion Formula

Theorem A.4.1 *Let ϕ_X be the characteristic function of a random variable X with distribution function F. Then*

$$F(b) - F(a) + \frac{1}{2}\left(P(X = a) - P(X = b)\right) = \lim_{T\to\infty} \frac{1}{2\pi}\int_{-T}^{T} \left(\frac{e^{-iua} - e^{-iub}}{iu}\right)\phi_X(u)du,$$

and

$$P(X = x) = \lim_{T\to\infty} \frac{1}{2T}\int_{-T}^{T} e^{-iux}\phi_X(u)du.$$

Furthermore, if $\int_{-\infty}^{\infty} |\phi_X(u)|du < \infty$, then F has a continuous density given by

$$f(x) = \frac{1}{2\pi}\int_{-\infty}^{\infty} e^{-iux}\phi_X(u)du.$$

The following corollary shows why ϕ_X is called a characteristic function.

Corollary A.4.1 *If $\phi_X(u) = \phi_Y(u)$ for all $u \in \mathbb{R}$, then X and Y have the same distribution.*

Instead of the characteristic function, we can sometimes use the moment generating function. Some of its properties are stated next.

A.5 Moments Generating Functions and Laplace Transform

Let X be a random variable. Recall that its moment generating function is defined by

$$M_X(u) = E\left(e^{uX}\right)$$

for all $u \in \mathbb{R}$ for which the expectation is finite. When $M_X(u)$ is only finite at $u = 0$, one usually says that the moment generating function does not exist. For a non negative random variable X, its Laplace transform is defined by

$$L_X(u) = E\left(e^{-uX}\right) = M_X(-u), \qquad \text{for all } u \geq 0.$$

Theorem A.5.1 *If $M_X(u)$ exists in a neighborhood of zero, i.e., for all $u \in (-u_0, u_0)$, for some $u_0 > 0$, then all the moments $E(X^n)$ exist and are given by*

$$E(X^n) = \left.\frac{d^n}{du^n} M_X(u)\right|_{u=0} = M_X^{(n)}(0), \qquad n \geq 1,$$

and

$$M_X(u) = \sum_{n=0}^{\infty} \frac{u^n}{n!} E(X^n), \qquad |u| < u_0.$$

Furthermore, the distribution of X is uniquely determined by its moments, i.e., if Y is another random variable such that $E(Y^n) = E(X^n)$ for all $n \geq 1$, then Y has the same distribution as X.

Instead of using the moments of a random variable, one often uses its cumulants. They are defined next.

A.5.1 Cumulants

If the moment generating function M_X of X exists in a neighborhood of zero, then $K_X = \ln(M_X)$ also exists in a neighborhood of zero, and

$$K_X(u) = \ln\{M_X(u)\} = \sum_{n\geq 1} \kappa_n \frac{u^n}{n!}.$$

The cumulant of order n of X is then defined by κ_n, $n \geq 1$. Note that $\kappa_n = K_X^{(n)}(0) = \frac{d^n}{du^n} K_X(u)\big|_{u=0}$.

Accordingly, K_X is called the cumulant generating function. The following result shows how the cumulants and moments are linked together.

Proposition A.5.1 *Suppose that the moment generating function of X exists in a neighborhood of zero. Then the first six cumulants are given by*

$$\begin{aligned}
\kappa_1 &= \mu = E(X); && \text{(A.3)}\\
\kappa_2 &= \sigma^2 = \mathrm{Var}(X); && \text{(A.4)}\\
\kappa_3 &= E\left\{(X-\mu)^3\right\}; && \text{(A.5)}\\
\kappa_4 &= E\left\{(X-\mu)^4\right\} - 3\sigma^4; && \text{(A.6)}\\
\kappa_5 &= E\left\{(X-\mu)^5\right\} - 10\sigma^2\kappa_3; && \text{(A.7)}\\
\kappa_6 &= E\left\{(X-\mu)^6\right\} - 15\sigma^2\kappa_4 - 10\kappa_3^2 - 15\sigma^6. && \text{(A.8)}
\end{aligned}$$

In particular, the skewness coefficient γ_1 *is given by*

$$\gamma_1 = E\{(X-\mu)^3\}/\sigma^3 = \kappa_3/\sigma^3, \tag{A.9}$$

and the kurtosis coefficient γ_2 *is given by*

$$\gamma_2 = E\{(X-\mu)^4\}/\sigma^4 = 3 + \kappa_4/\sigma^4. \tag{A.10}$$

Remark A.5.1 *If* K_X *exists for all* $s \in (a,b)$*, then* $K_X^{(n)}(s)$ *is the cumulant of order* n *of the distribution* Q_s *with density* $e^{sx}/M_X(s)$ *with respect to the distribution of* X*. In other words, for all* u *such that* $a < u+s < b$*,* Q_s *has cumulant generating function* $K_s(u) = K_X(s+u) - K_X(s)$*. In particular, it follows from* (A.4) *that* $K_X''(s)$ *is the variance of distribution* Q_s*, so it is positive for all* $s \in (a,b)$*.*

A.5.1.1 Extension

If X has moments of order $n \geq 1$, then it makes sense to define its cumulants up to order n by the same formulas as in Proposition A.5.1. For example, if the moment of order 4 of X exists, the cumulants of $\kappa_1, \ldots, \kappa_4$ are defined by equations (A.3)–(A.6). Furthermore, the skewness and kurtosis exist and are given respectively by (A.9) and (A.10).

A.6 Families of Distributions

We will now define some families of discrete and absolutely continuous distributions that are used in the book.

A.6.1 Bernoulli Distribution

A random variable X has a Bernoulli distribution with parameter $p \in (0,1)$ if its discrete density is given by

$$f(x) = \begin{cases} p^x(1-p)^{1-x}, & x \in \{0,1\}; \\ 0, & \text{otherwise.} \end{cases}$$

Such a random variable can be associated with a random experiment for which there are only two possible outcomes: E_1 (success) and $E_2 = E_1^c$ (failure). Then $X = \mathbb{I}_{E_1}$ and $p = P(E_1) = P(X=1)$. Its mean, variance, and moment generating function are given respectively by p, $p(1-p)$, and $(1-p) + pe^u$, $u \in \mathbb{R}$.

A.6.2 Binomial Distribution

A random variable X has a binomial distribution with parameters $n \in \mathbb{N}$ and $p \in (0,1)$, denoted $X \sim \text{Bin}(n,p)$, if its discrete density is given by

$$f(x) = \begin{cases} \binom{n}{x} p^x (1-p)^{n-x}, & x \in \{0,\ldots,n\}; \\ 0, & \text{otherwise.} \end{cases}$$

Here, $\binom{n}{x} = \frac{n!}{x!(n-x)!}$, for $x \in \{0,\ldots,n\}$. X represents the number of successes when a Bernoulli experiment is repeated n times, independently, and in the same conditions. One can write

$$X = Y_1 + \ldots + Y_n,$$

where $Y_1,\ldots,Y_n$ are independent and have a Bernoulli distribution with parameter p.

To prove that X has indeed discrete density f, note that

$$\begin{aligned} P(X=k) &= \sum_{1 \le j_1 < \ldots < j_k \le n} P(Y_{j_i} = 1,\ 1 \le i \le k,\ Y_j = 0 \text{ otherwise}) \\ &= \sum_{1 \le j_1 < \ldots < j_k \le n} p^k (1-p)^{n-k} \\ &= \binom{n}{k} p^k (1-p)^{n-k}, \quad k \in \{0,\ldots,n\}, \end{aligned}$$

where $\{j_i\}_{i=1}^k$ are sequences of indices indicating when successes occur. For example, for $n=5$ and $k=3$ one such sequence is $j_1 = 1$, $j_2 = 3$, and $j_3 = 5$. The mean and variance of X are np and $np(1-p)$ respectively. Finally, the moment generating function of X always exists and is given by

$$M_X(u) = E\left(e^{uX}\right) = (1 - p + pe^u)^n, \quad u \in \mathbb{R}.$$

A.6.3 Poisson Distribution

A random variable X has a Poisson distribution with parameter $\lambda > 0$, denoted $X \sim \text{Poisson}(\lambda)$, if its discrete density is given by

$$f(x) = \begin{cases} e^{-\lambda} \frac{\lambda^x}{x!}, & x \in \mathbb{N} \cup \{0\}; \\ 0, & \text{otherwise.} \end{cases}$$

The Poisson distribution is used to model the number of occurrences of a particular event over a fixed period of time. The mean and the variance of X are λ, while the moment generating function of X always exists and equals

$$M_X(u) = E\left(e^{uX}\right) = e^{\lambda(e^u - 1)}, \quad u \in \mathbb{R}.$$

An interesting property of the Poisson distribution is the following.

Proposition A.6.1 *Suppose that the random variables $X_1,\ldots,X_n$ are independent and $X_i \sim \text{Poisson}(\lambda_i)$, $i \in \{1,\ldots,n\}$. Then $X = \sum_{i=1}^n X_i \sim \text{Poisson}(\lambda)$, where $\lambda = \sum_{i=1}^n \lambda_i$.*

A.6.4 Geometric Distribution

A random variable X has a geometric distribution with parameter $p \in (0,1)$, denoted $X \sim \mathrm{G}(p)$, if its discrete density is given by

$$f(x) = \begin{cases} p\,(1-p)^{x-1}, & x \in \mathbb{N}; \\ 0, & \text{otherwise.} \end{cases}$$

X represents the number of independent Bernoulli experiment realized in the same conditions until a first success is observed. The mean and variance of X are respectively $1/p$ and $(1-p)/p^2$, while the moment generating function of X is given by

$$M_X(u) = E\left(e^{uX}\right) = \frac{pe^u}{1-(1-p)e^u}, \quad u < -\ln(1-p).$$

A.6.5 Negative Binomial Distribution

A random variable X has a negative binomial distribution with parameters $r \in \mathbb{N}$ and $p \in (0,1)$, denoted $X \sim NB(r,p)$, if its discrete density is given by

$$f(x) = \begin{cases} \binom{x-1}{r-1}\, p^r\,(1-p)^{x-r}, & x \in \{r, r+1, \ldots\}; \\ 0, & \text{otherwise.} \end{cases}$$

X represents the number of independent Bernoulli experiment realized in the same conditions in order to observe the first r successes. Therefore one can write $X = Y_1 + \cdots + Y_r$, where $Y_1, \ldots, Y_r$ are independent with $Y_k \sim \mathrm{G}(p)$, $k \in \{1, \ldots, r\}$. The mean and variance of X are respectively r/p and $r(1-p)/p^2$. Finally, the moment generating function of X exists and is given by

$$M_X(u) = E\left(e^{uX}\right) = \frac{p^r e^{ru}}{\{1-(1-p)e^u\}^r}, \quad u < -\ln(1-p).$$

The next list of distributions are absolutely continuous distributions. Most of the time, they are defined by their associated density.

A.6.6 Uniform Distribution

A random variable X has a uniform distribution on (a,b), denoted $U \sim \mathrm{Unif}(a,b)$, if its density is

$$f(x) = \frac{1}{b-a}\mathbb{I}(a < x < b) = \begin{cases} \frac{1}{b-a}, & \text{if } a < x < b; \\ 0, & \text{otherwise .} \end{cases}$$

Uniformly distributed random variables are quite important in simulation, as the following result shows.

Theorem A.6.1 *Suppose that $U \sim \text{Unif}(0,1)$ and let F be a distribution function. Define its inverse by*

$$F^{-1}(u) = \inf\{x; F(x) \geq u\}, \qquad u \in (0,1).$$

Then, setting $X = F^{-1}(U)$, one has $X \sim F$, i.e., X has distribution function F.

A.6.7 Gaussian Distribution

A random variable X has a Gaussian distribution with parameters $\mu \in \mathbb{R}$ and σ^2, $\sigma > 0$, denoted by $X \sim N(\mu, \sigma^2)$, if its density f is given by

$$f(x) = \frac{1}{\sigma\sqrt{2\pi}} e^{-\frac{(x-\mu)^2}{2\sigma^2}}, \qquad x \in \mathbb{R}.$$

Note that if $X \sim N(\mu, \sigma^2)$ then $Z = \frac{X-\mu}{\sigma} \sim N(0,1)$. The latter distribution is called the standard Gaussian distribution. The characteristic function of X is given by $\phi_X(u) = E\left(e^{iuX}\right) = \phi(u) = e^{i\mu - u^2\sigma^2/2}$, while its moment generating function $M_X(u)$ is given by $M_X(u) = E\left(e^{uX}\right) = e^{u\mu + u^2\sigma^2/2}$. Indeed,

$$\begin{aligned} M_X(u) &= \int_{-\infty}^{\infty} \frac{1}{\sigma\sqrt{2\pi}} e^{-\frac{(x-\mu)^2}{2\sigma^2}} e^{ux} dx \\ &= \int_{-\infty}^{\infty} \frac{1}{\sigma\sqrt{2\pi}} e^{-\frac{(x-\mu-u\sigma^2)^2}{2\sigma^2}} e^{\frac{(\mu+u\sigma^2)^2-\mu^2}{2\sigma^2}} dx = e^{u\mu + u^2\sigma^2/2}. \end{aligned}$$

Therefore, its cumulant generating function is $K(u) = \ln\{M(u)\} = u\mu + u^2\sigma^2/2$, $u \in \mathbb{R}$. Consequently, $\kappa_1 = \mu$, $\kappa_2 = \sigma^2$, and $\kappa_n = 0$, $n \geq 3$, which implies that the Gaussian distribution is described entirely by its mean and variance. An important property of the Gaussian distribution is the following one.

Proposition A.6.2 *Suppose that the random variables $X_1, \ldots, X_n$ are independent and $X_i \sim N(\mu_i, \sigma_i^2)$, $i \in \{1, \ldots, n\}$. Then $X = \sum_{i=1}^n X_i \sim N(\mu, \sigma^2)$, where $\mu = \sum_{i=1}^n \mu_i$, and $\sigma^2 = \sum_{i=1}^n \sigma_i^2$.*

As the Gaussian distribution can be used to model the log-returns of an asset, the log-normal is used for modeling prices.

A.6.8 Log-Normal Distribution

A random variable Y has a log-normal distribution with parameters (μ, σ^2) if $Y = e^X$, with $X \sim N(\mu, \sigma^2)$, i.e., $\ln(Y)$ has a Gaussian distribution. From Proposition A.8.1, the density of Y is given by

$$f_y(y) = \frac{e^{-\frac{1}{2}\frac{(\ln(y)-\mu)^2}{\sigma^2}}}{y\sigma\sqrt{2\pi}} \mathbb{I}(y > 0), \quad y \in \mathbb{R},$$

since $\left|\frac{d}{dy}\phi^{-1}(y)\right| = \frac{1}{y}$. The following result is used in the proof of the Black-Scholes formula.

Proposition A.6.3 *Suppose that Y has a log-normal distribution with parameters (μ, σ^2). Then, for any $K > 0$, one has*

$$E\{\max(Y-K,0)\} = e^{\mu+\sigma^2/2}\mathcal{N}\left(\frac{\mu+\sigma^2-\ln(K)}{\sigma}\right) - K\mathcal{N}\left(\frac{\mu-\ln(K)}{\sigma}\right),$$

where $\mathcal{N}$ is the distribution function of the standard Gaussian distribution.

PROOF. One can write $Y = e^{\mu+\sigma Z}$, where $Z \sim N(0,1)$. Since $Y > K$ is equivalent to $Z > \frac{\ln(K)-\mu}{\sigma}$, and using $1-\mathcal{N}(z) = \mathcal{N}(-z)$, it follows that for all z,

$$\begin{aligned} E\{\max(Y-K,0)\} &= \int_{\frac{\ln(K)-\mu}{\sigma}}^{\infty} \left(e^{\mu+\sigma z}-K\right)\frac{e^{-z^2/2}}{\sqrt{2\pi}}dz \\ &= e^{\mu+\sigma^2/2}\int_{\frac{\ln(K)-\mu}{\sigma}}^{\infty} \frac{e^{-(z-\sigma)^2/2}}{\sqrt{2\pi}}dz - K\mathcal{N}\left(\frac{\mu-\ln(K)}{\sigma}\right) \\ &= e^{\mu+\sigma^2/2}\int_{\frac{\ln(K)-\mu-\sigma^2}{\sigma}}^{\infty} \frac{e^{-w^2/2}}{\sqrt{2\pi}}dw - K\mathcal{N}\left(\frac{\mu-\ln(K)}{\sigma}\right) \\ &= e^{\mu+\sigma^2/2}\mathcal{N}\left(\frac{\mu+\sigma^2-\ln(K)}{\sigma}\right) - K\mathcal{N}\left(\frac{\mu-\ln(K)}{\sigma}\right). \end{aligned}$$

■

The next distribution is often used to model waiting times or jumping times.

A.6.9 Exponential Distribution

A random variable X has an exponential distribution with parameter $\lambda > 0$, denoted by $X \sim \text{Exp}(\lambda)$, if its density f is given by

$$f(x) = \lambda e^{-\lambda x}\mathbb{I}(x \geq 0), \qquad x \in \mathbb{R}.$$

Note that if $U \sim \text{Unif}(0,1)$, then $X = -\frac{1}{\lambda}\ln(1-U) \sim \text{Exp}(\lambda)$, since $P(X > t) = e^{-\lambda t}$, for $t \geq 0$. The mean and variance are given respectively by $\frac{1}{\lambda}$ and $\frac{1}{\lambda^2}$. The characteristic function of X is given by $\phi_X(u) = E\left(e^{iuX}\right) = \phi(u) = \frac{\lambda}{\lambda-iu} = \frac{1}{1-iu/\lambda}$, while its moment generating function $M_X(u)$ is given by $M_X(u) = E\left(e^{uX}\right) = \frac{\lambda}{\lambda-u}$, if $u < \lambda$. Therefore, if $|u| < \lambda$, the cumulant generating function is

$$K(u) = \ln\{M(u)\} = \sum_{n=1}^{\infty}\frac{1}{n}\left(\frac{u}{\lambda}\right)^n.$$

Consequently, $\kappa_n = \frac{(n-1)!}{\lambda^n}$, $n \geq 1$.

A.6.10 Gamma Distribution

A random variable X has a gamma distribution with parameters (α, β), denoted $X \sim \text{Gamma}(\alpha, \beta)$, when the density of X is

$$f(x) = \frac{1}{\Gamma(\alpha)\beta^\alpha}\, x^{\alpha-1}\, e^{-x/\beta}\, \mathbb{I}_{(0,\infty)}(x), \quad x \in \mathbb{R},$$

and the gamma function is defined by

$$\Gamma(\alpha) = \int_0^\infty t^{\alpha-1}\, e^{-t} dt, \ \ \alpha > 0.$$

It follows that the exponential distribution with parameter λ is a particular case of a gamma distribution corresponding to $\alpha = 1$ and $\beta = 1/\lambda$. Note that the moment generating function $M(u)$ exists for all $u < 1/\beta$, since

$$\begin{aligned} M(u) &= \int_0^\infty \frac{x^{\alpha-1}}{\beta^\alpha \Gamma(\alpha)} e^{-x/\beta} e^{ux} dx \\ &= \frac{1}{(1-\beta u)^\alpha} \int_0^\infty \frac{x^{\alpha-1}}{\left(\frac{\beta}{1-\beta u}\right)^\alpha \Gamma(\alpha)} e^{-x/\left(\frac{\beta}{1-\beta u}\right)} dx \\ &= \frac{1}{(1-\beta u)^\alpha}. \end{aligned}$$

As a result, the cumulant generating function is

$$K(u) = \ln\{M(u)\} = \alpha \sum_{n \geq 1} \frac{(\beta u)^n}{n}, \quad |u| < 1/\beta,$$

so all the cumulants exist and $\kappa_n = \alpha\beta^n(n-1)!$, $n \geq 1$. The mean, variance, and characteristic function are given respectively by $\alpha\beta$, $\alpha\beta^2$, and $\frac{1}{(1-iu\beta)^\alpha}$.

An interesting property of the gamma distribution is that the sum of independent gamma random variables has also a gamma distribution, provided they all have the same parameter β. More precisely, we have the following result.

Proposition A.6.4 *Suppose that the random variables $X_1, \ldots, X_n$ are independent and $X_i \sim \text{Gamma}(\alpha_i, \beta)$, $i \in \{1, \ldots, n\}$. Then $X = \sum_{i=1}^n X_i \sim \text{Gamma}(\alpha, \beta)$, where $\alpha = \sum_{i=1}^n \alpha_i$.*

A.6.10.1 Properties of the Gamma Function

- $\Gamma(1) = 1$;
- $\Gamma(\frac{1}{2}) = \sqrt{\pi}$;
- $\Gamma(1+\alpha) = \alpha\Gamma(\alpha)$, $\alpha > 0$;
- $\Gamma(n) = (n-1)!$, $n \in \mathbb{N}$.

A.6.11 Chi-Square Distribution

A random variable X has a chi-square distribution with ν degrees of freedom, denoted $X \sim \chi^2(\nu)$, if $X \sim \text{Gamma}(\frac{\nu}{2}, 2)$. Its density function is then given by

$$f(x) = \frac{1}{\Gamma(\frac{\nu}{2})2^{\frac{\nu}{2}}} \, x^{\frac{\nu}{2}-1} \, e^{-x/2} \, \mathbb{I}_{(0,\infty)}(x), \quad x \in \mathbb{R}.$$

One can check that if $Z_1, \ldots, Z_n$ are independent $N(0,1)$, then $Z_1^2 + \cdots + Z_n^2 \sim \chi^2(n)$. The mean and variance are given respectively by ν and 2ν. The moment generating function exists and is given by $M(u) = (1-2u)^{-\nu/2}$, for all $u < 1/2$. In addition, the cumulants are given by $\kappa_n = \nu 2^{n-1}(n-1)!$, $n \geq 1$.

The next family of distribution plays a major role in the modeling of interest rates.

A.6.12 Non-Central Chi-Square Distribution

A random variable X has a non-central chi-square distribution with ν degrees of freedom and non-centrality parameter $\mu \geq 0$, denoted $X \sim \chi^2(\nu, \mu)$, if its density is

$$f(x; \nu, \mu) = e^{-\frac{\mu}{2}} \sum_{n=0}^{\infty} \frac{\left(\frac{\mu}{2}\right)^n}{n!} f_{\nu+2n}(x), \quad x \in \mathbb{R},$$

where $f_\nu(x)$ is the density of a chi-square distribution with ν degrees of freedom. Note that its moment generating function is given by

$$M_X(u) = \frac{e^{\frac{u\mu}{1-2u}}}{(1-2u)^{\nu/2}}, \; u < 1/2,$$

while the cumulants are

$$\kappa_n = 2^{n-1}(n-1)! \, (\nu + n\mu) \, , \quad n \geq 1.$$

Remark A.6.1 *When ν is an integer, say $\nu = n$, we can easily verify that, if $Z_1, \ldots, Z_n$ are independent $N(0,1)$, then*

$$X = (Z_1 + \mu_1)^2 + \cdots + (Z_n + \mu_r)^2 \sim \chi^2(n, \mu), \tag{A.11}$$

with $\mu = \mu_1^2 + \cdots + \mu_r^2$. In the general case, for $\nu \geq 1$, $X = (Z_1 + \sqrt{\mu})^2 + X_1 \sim \chi^2(\nu, \mu)$, if $X_1 \sim \chi^2(\nu - 1)$ is independent of $Z_1 \sim N(0,1)$.

A.6.12.1 Simulation of Non-Central Chi-Square Variables

Following Glasserman [2004], to generate random variables from a non-central chi-square distribution, we consider 2 cases:

- Case $\nu > 1$: In this case, if $Y \sim \chi^2_{\nu-1}$ and $Z \sim N(0,1)$ is independent of Y, then $X = Y + (Z + \sqrt{\mu})^2$ has the desired law.

- Case $\nu \leq 1$: Generate $N \sim \text{Poisson}(\mu/2)$. Given $N = n$, generate $X \sim \chi^2_{\nu+2n}$. Then X has the desired law.

An alternative way of writing the density is:

$$\begin{aligned} f(x) &= \frac{1}{2}\left(\frac{x}{\mu}\right)^{\frac{\nu}{4}-\frac{1}{2}} e^{-(x+\mu)/2} I_{\frac{\nu}{2}-1}\left(\sqrt{\mu x}\right) \mathbb{I}(x>0) \\ &= \frac{1}{2x}\left(\frac{x}{\mu}\right)^{\nu/4} e^{-(x+\mu)/2}\left\{\nu I_{\frac{\nu}{2}}\left(\sqrt{\mu x}\right) + \sqrt{\mu x} I_{\frac{\nu}{2}+1}\left(\sqrt{\mu x}\right)\right\} \mathbb{I}(x>0), \end{aligned}$$

where $I_\nu(x)$ is the modified Bessel function of the first kind defined by

$$I_\nu(x) = \sum_{n=0}^{\infty} \frac{\left(\frac{x}{2}\right)^{\nu+2n}}{n!\Gamma(\nu+n+1)}.$$

If ν is an odd integer, then $I_{\frac{\nu}{2}}(x)$ is a simple function to compute. For example, if $x > 0$, we have

$$f(x) = \begin{cases} \frac{1}{\sqrt{2\pi x}} e^{-\frac{(x+\mu)}{2}} \cosh\left(\sqrt{\mu x}\right), & \text{if } r = 1, \\ \frac{1}{2} e^{-\frac{(x+\mu)}{2}} I_0\left(\sqrt{\mu x}\right), & \text{if } r = 2, \\ \frac{1}{\sqrt{2\pi\mu}} e^{-\frac{(x+\mu)}{2}} \sinh\left(\sqrt{\mu x}\right), & \text{if } r = 3, \\ \frac{1}{2}\sqrt{\frac{x}{\mu}} e^{-\frac{(x+\mu)}{2}} I_1\left(\sqrt{\mu x}\right), & \text{if } r = 4, \\ \frac{1}{\sqrt{2\pi\mu^3}} e^{-\frac{(x+\mu)}{2}} \left\{\sqrt{\mu x}\cosh\left(\sqrt{\mu x}\right) - \sinh\left(\sqrt{\mu x}\right)\right\}, & \text{if } r = 5. \end{cases}$$

The MATLAB functions *PDFNChi2* and *SimNChi2* allow respectively to compute the density and to generate random variables for a non-central chi-square distribution. The MATLAB Statistics Toolbox also has the functions *ncx2pdf ncx2cdf*, and *ncx2inv* which computes respectively the density, the distribution function, and its inverse, while *ncx2rnd* can also be used to generate random numbers.

A.6.13 Student Distribution

The well-known Student distribution is also used in financial applications when one needs to model fat tails. Suppose that $X_1 \sim N(0,1)$, $X_2 \sim \chi^2(\nu)$, and that X_1 and X_2 are independent. Then the law of $Y = X_1 \big/ \sqrt{X_2/\nu}$ is a Student distribution with ν degrees of freedom, denoted $Y \sim T(\nu)$. Its density is

$$f_Y(y) = \frac{1}{\sqrt{\pi\nu}} \frac{\Gamma(\frac{\nu+1}{2})}{\Gamma\left(\frac{\nu}{2}\right)} (1 + y^2/\nu)^{-(\frac{\nu+1}{2})}, \quad y \in \mathbb{R}.$$

Note that Y is centered at 0, and its variance is given $\nu/(\nu-2)$ for $\nu > 2$, ∞ if $1 < \nu \leq 2$ and undefined otherwise. From the properties of the gamma function, it follows that the moment of order $2k$ exists if and only if $\nu > 2k$. In the latter case,

$$E\left(Y^{2k}\right) = \nu^k \frac{\Gamma\left(k+\frac{1}{2}\right)\Gamma\left(\frac{\nu}{2}-k\right)}{\Gamma\left(\frac{1}{2}\right)\Gamma\left(\frac{\nu}{2}\right)} = \nu^k \prod_{j=1}^{k} \frac{(2j-1)}{(\nu-2j)}. \tag{A.12}$$

For more details on the expression of the density, see Example A.8.4.

A.6.14 Johnson SU Type Distributions

Another interesting distribution that is used in financial applications is the Johnson type distributions. A random variable X has distribution Johnson-SU with parameters a, b, c, d, with $b, d > 0$, denoted $X \sim J(a, b, c, d)$, if it has the following representation:

$$X = a + b\left\{\frac{e^{(Z-c)/d} - e^{-(Z-c)/d}}{2}\right\} = a + b\sinh\left(\frac{Z-c}{d}\right) = \psi(Z),$$

where $Z \sim N(0,1)$. Using Proposition A.8.1, we can obtain its density as

$$f(x) = \frac{d}{b}\frac{1}{\sqrt{1+\left(\frac{x-a}{b}\right)^2}}\phi\left\{\psi^{-1}(x)\right\},$$

where $\phi(x) = e^{-x^2/2}/\sqrt{2\pi}$ is the standard normal Gaussian density, and $\psi^{-1}(x) = c + d\,H\left(\frac{x-a}{b}\right)$, with $H(x) = \ln\left(x+\sqrt{1+x^2}\right)$. Its law depends on 4 parameters, and it is used in practice because of its flexibility to match the first four moments for an appropriate choice of parameters.

A.6.15 Beta Distribution

In some problems, one needs to model random variables with values in $(0,1)$. One of the most popular distribution on $(0,1)$ is the so-called Beta distribution. A random variable X is said to have a Beta distribution with parameters α, β, denoted $X \sim Beta(\alpha, \beta)$, if its density is given by

$$f(x) = \frac{1}{B(\alpha,\beta)} x^{\alpha-1}(1-x)^{\beta-1}\,\mathbb{I}_{(0,1)}(x), \quad x \in \mathbb{R},$$

where $\alpha, \beta > 0$ and where the Beta function is defined by

$$B(\alpha,\beta) = \frac{\Gamma(\alpha)\Gamma(\beta)}{\Gamma(\alpha+\beta)}.$$

Note that the uniform distribution on $(0,1)$ is a beta distribution with parameters $\alpha = \beta = 1$.

A.6.16 Cauchy Distribution

When it is time to model a distribution taking very large values, one of the first examples coming to mind, in addition to the Student distribution, is the Cauchy distribution. A random variable X is said to have Cauchy distribution with parameters $\mu \in \mathbb{R}$ and $\sigma > 0$, denoted $X \sim C(\mu, \sigma)$, if its density is

$$f(x) = \frac{\sigma}{\pi(\sigma^2 + (x-\mu)^2)}, \qquad x \in \mathbb{R}.$$

The Cauchy distribution has no mean because $E(|X|) = \infty$. Also, the law of X is symmetric about μ, and for any $x \in \mathbb{R}$,

$$F(x) = P(X \le x) = \frac{1}{2} + \frac{1}{\pi} \arctan\left(\frac{x-\mu}{\sigma}\right).$$

Note that the density can be written as $\frac{1}{\sigma} f\left(\frac{x-\mu}{\sigma}\right)$, where f is the density of a Cauchy distribution with parameters $(0, 1)$.

A.6.17 Generalized Error Distribution

As an alternative to the Gaussian distribution, some authors suggested to use the so-called Generalized Error Distribution (GED for short). A random variable X has a Generalized Error Distribution of parameter $\nu > 0$, denoted $X \sim GED(\nu)$, if its density is given by

$$f_\nu(x) = \frac{1}{b_\nu 2^{1+\frac{1}{\nu}} \Gamma(1+\frac{1}{\nu})} e^{-\frac{1}{2}\left(\frac{|x|}{b_\nu}\right)^\nu}, \quad x \in \mathbb{R},$$

where $b_\nu = 2^{-\frac{1}{\nu}} \sqrt{\frac{\Gamma(\frac{1}{\nu})}{\Gamma(\frac{3}{\nu})}}$. If F_α is the distribution function of a gamma distribution with parameters α and $\beta = 1$, then the distribution function F of $X \sim GED(\nu)$ is given by

$$F(x) = \begin{cases} \frac{1}{2} - \frac{1}{2} F_{1/\nu}\left\{\frac{1}{2}\left(\frac{|x|}{b_\nu}\right)^\nu\right\}, & x \le 0, \\ \frac{1}{2} + \frac{1}{2} F_{1/\nu}\left\{\frac{1}{2}\left(\frac{x}{b_\nu}\right)^\nu\right\}, & x \ge 0. \end{cases}$$

It follows that its inverse function is

$$F^{-1}(u) = \begin{cases} -b_\nu 2^{1/\nu} \left\{F_{1/\nu}^{-1}(1-2u)\right\}^{1/\nu}, & 0 < u \le \frac{1}{2}, \\ b_\nu 2^{1/\nu} \left\{F_{1/\nu}^{-1}(2u-1)\right\}^{1/\nu}, & \frac{1}{2} \le u < 1. \end{cases}$$

As a result, to generate $X \sim GED(\nu)$, first generate a gamma variate Y with parameter $\alpha = 1/\nu$ and a uniform variate U on $(0, 1)$. If $U < 1/2$, set

$X = -b_\nu (2Y)^{1/\nu}$, otherwise set $X = b_\nu (2Y)^{1/\nu}$.

We will now give some examples of multivariate distributions.

A.6.18 Multivariate Gaussian Distribution

Let $\mu \in \mathbb{R}^d$ and $\Sigma \in \mathbb{R}^{d\times d}$, where the matrix Σ is symmetric ($\Sigma = \Sigma^\top$), and positive definite ($x^\top \Sigma x > 0$ for all $x \in \mathbb{R}^d, x \neq 0$). A random vector $X = (X_1, X_2, \ldots, X_d)^\top$ has a multivariate Gaussian distribution with parameters $\mu \in \mathbb{R}^d$ and $\Sigma \in \mathbb{R}^{d\times d}$, denoted by $X \sim N_d(\mu, \Sigma)$, if its density is given by

$$f(x) = \frac{1}{(2\pi)^{d/2}|\Sigma|^{1/2}} e^{-\frac{1}{2}(x-\mu)^\top \Sigma^{-1}(x-\mu)}, \quad x \in \mathbb{R}^d,$$

where $|\Sigma| = \det(\Sigma)$.

A.6.18.1 Representation of a Random Gaussian Vector

Proposition A.6.5 *If B is a $p \times d$ matrix, $\mu \in \mathbb{R}^p$, and Z is a vector of d independent standard Gaussian components, i.e., $Z \sim N_d(0, I)$, then $\mu + BZ \sim N_p\left(\mu, BB^\top\right)$.*

Remark A.6.2 *Proposition A.6.5 is often used to generate a Gaussian vector $X = N_d(\mu, \Sigma)$ from d independent standard normal variables. Indeed, if Σ is a symmetric positive definite $d \times d$ matrix, we can find an upper triangular matrix R such that $R^\top R = \Sigma$ (Cholesky decomposition). We then set $B = R^\top$, and $X = \mu + R^\top Z \sim N_d(\mu, \Sigma)$ if $Z \sim N_d(0, I)$.*

The following proposition allows to identify the parameters of a Gaussian random vector.

Proposition A.6.6 *Suppose that $X \sim N_d(\mu, \Sigma)$. Then $E(X) = \mu$, and for $i, j \in \{1, \ldots, d\}$, we have*

$$\left(E\left\{(X-\mu)(X-\mu)^\top\right\}\right)_{ij} = \mathrm{Cov}(X_i, X_j) = \Sigma_{ij}.$$

Also, if $\lambda \in \mathbb{R}^d$, then

$$\lambda^\top X = \sum_{j=1}^d \lambda_j X_j \sim N(\lambda^\top \mu, \lambda^\top \Sigma \lambda).$$

Remark A.6.3 *Σ is called the covariance matrix of X. It follows that the law of a random Gaussian vector is uniquely determined by its mean μ and its covariance matrix Σ.*

The next multivariate distribution is becoming increasingly popular in applications, especially for defining copulas.

A.6.19 Multivariate Student Distribution

A random vector X has a multivariate Student distribution with parameters ν, Σ, denoted $X \sim T(\nu, \Sigma)$, if X can be expressed as

$$X = Z \Big/ \sqrt{V/\nu}, \tag{A.13}$$

where $Z \sim N_d(0, \Sigma)$ is independent of $V \sim \chi^2(\nu)$. From representation (A.13), X_i/σ_i has also a Student distribution with ν degrees of freedom, where $\sigma_i^2 = \Sigma_{ii}$, $i \in \{1, \ldots, d\}$. The density f is given, for all $x \in \mathbb{R}^d$, by

$$f(x) = \frac{\Gamma\left(\frac{\nu+d}{2}\right)}{(\pi\nu)^{\frac{d}{2}}\Gamma\left(\frac{\nu}{2}\right)|\Sigma|^{\frac{1}{2}}}\left(1 + x^\top \Sigma^{-1} x/\nu\right)^{-(\nu+d)/2}. \tag{A.14}$$

In fact, the two previous examples of multivariate distributions are particular cases of elliptical distributions defined next.

A.6.20 Elliptical Distributions

First, a random vector S is uniformly distributed on the sphere $\mathcal{S}_d = \{x \in \mathbb{R}^d; \|x\| = 1\}$, denoted $S \sim \text{Unif}(\mathcal{S}_d)$ if for every rotation matrix[3] B, $BS \stackrel{\text{Law}}{=} S$. Here, $\|x\|^2 = \sum_{j=1}^d x_j^2$, $x \in \mathbb{R}^d$. Let S_d^+ be the set of all symmetric and positive definite $d \times d$ matrices.

Definition A.6.1 *A random vector Y has an elliptical distribution if there exist $\mu \in \mathbb{R}^d$, $\Sigma \in S_d^+$, and a positive random variable $\mathcal{R}$ such that*

$$Y = \mu + \mathcal{R}^{1/2} A^\top S, \tag{A.15}$$

where $A^\top A = \Sigma$, and $S \sim \text{Unif}(\mathcal{S}_d)$ is independent of $\mathcal{R}$. In such a case, one writes $Y \sim \mathcal{E}(\mathcal{R}, \mu, \Sigma)$. In particular, Y has an elliptical distribution with generator g and parameters μ, Σ, denoted $Y \sim \mathcal{E}(g, \mu, \Sigma)$, if the density h of Y is given by

$$h(y) = \frac{1}{|\Sigma|^{1/2}} g\left\{(y-\mu)^\top \Sigma^{-1}(y-\mu)\right\},$$

where

$$\frac{\pi^{d/2}}{\Gamma(d/2)} r^{(d-2)/2} g(r) \tag{A.16}$$

is a density on $(0, \infty)$. In fact, it is the density of $\mathcal{R} = (Y-\mu)^\top \Sigma^{-1}(Y-\mu)$. In particular, if $d = 2$, the density of $\mathcal{R}$ is πg.

Proposition A.6.7 *If $Y \sim \mathcal{E}(\mathcal{R}, \mu, \Sigma)$, then $Z = \Delta^{-1}(Y - \mu) \sim \mathcal{E}(\mathcal{R}, 0, \rho)$, where Δ is the diagonal matrix such that $\Delta_{ii} = \sqrt{\Sigma_{ii}}$, and $\rho_{ij} = \Sigma_{ij}/\sqrt{\Sigma_{ii}\Sigma_{jj}}$, $1 \le i, j \le d$. In addition, if $E(\mathcal{R}) < \infty$, then*

$$E(Y) = \mu, \quad \text{Cov}(Y) = \frac{E(\mathcal{R})}{d}\Sigma \quad \text{and} \quad \text{Cor}(Y) = \rho,$$

[3]A rotation matrix B is a matrix so that $B^{-1} = B^\top$.

Remark A.6.4 *Parameters μ and ρ can be interpreted has the mean and correlation of Y, if $E(\mathcal{R}) < \infty$.*

PROOF. The first part is obvious. For the moments, it suffices to show that $E(S) = 0$ and $\text{Cov}(S) = I/d$, for if this is true, then $E(Y) = \mu$ and $\text{Cov}(Y) = E(\mathcal{R})A^\top \frac{I}{d} A = \frac{E(\mathcal{R})}{d}\Sigma$. First, since $B = -I$ is a rotation matrix, the laws of $-S$ and S are the same, proving that $E(S) = 0$. Next, $\text{Cov}(S) \in \mathcal{S}_d^+$, so there exist a rotation matrix M and a diagonal matrix D such that $\text{Cov}(S) = M^\top DM$. Then, the law of MS is the same as the law of S so

$$\begin{aligned} M^\top DM &= \text{Cov}(S) = \text{Cov}(MS) = M\,\text{Cov}(S)M^\top \\ &= MM^\top DMM^\top = D. \end{aligned}$$

This proves that $\text{Cov}(S)$ is diagonal. Finally, for any rotation matrix B, we have

$$D = \text{Cov}\, S = \text{Cov}\, BS = B\Delta B^\top.$$

Applying the last equation with permutation matrices, we conclude that $\Delta_{11} = \cdots = \Delta_{dd} = \lambda$, for some $\lambda > 0$. Hence $\Delta = \lambda I$. Finally,

$$\lambda d = \text{Trace}\left\{E\left(SS^\top\right)\right\} = E\left\{\text{Trace}\left(SS^\top\right)\right\} = E\left\{\text{Trace}\left(S^\top S\right)\right\} = 1,$$

since $1 = \|S\|^2 = S^\top S$ by definition.

■

The next proposition is quite useful for finding the marginal distributions in the elliptical case.

Proposition A.6.8 *If $Y \sim \mathcal{E}(g, 0, \rho)$, where ρ is a correlation matrix, then the marginal distributions of $Y_1, \ldots, Y_d$ are the same and their common density f is given, for all $x \in \mathbb{R}$, by*

$$f(x) = \frac{\pi^{(d-1)/2}}{\Gamma((d-1)/2)} \int_0^\infty r^{(d-3)/2} g(r + x^2) dr.$$

In addition, if $Y^{(B)}$ is a sub-vector of Y with components $\{Y_i; i \in B\}$, $B \subset \{1, \ldots, d\}$, then $Y^{(B)} \sim \mathcal{E}(g_B, 0, \rho^{(B)})$ for some generator $g^{(B)}$, where $\rho^{(B)}$ is the correlation matrix with components ρ_{ij}, $i, j \in B$.

The proof is based on the following lemma.

Lemma A.6.1 *For any function $f \geq 0$,*

$$\int_{\mathbb{R}^d} f\left(\|x\|^2\right) dx = \frac{2\pi^{d/2}}{\Gamma(d/2)} \int_0^\infty r^{d-1} f\left(r^2\right) dr = \frac{\pi^{d/2}}{\Gamma(d/2)} \int_0^\infty r^{(d-2)/2} f(r) dr.$$

PROOF. Writing $\rho = \begin{pmatrix} \rho_1 & b \\ b^\top & 1 \end{pmatrix}$, we have $\rho^{-1} = \begin{pmatrix} A & c \\ c^\top & d \end{pmatrix}$, where $d = 1/\left(1 - b^\top \rho_1^{-1} b\right)$, $c = -d\rho_1^{-1}b$, and $A = \rho_1^{-1} + d\rho_1^{-1}bb^\top\rho_1^{-1}$. Then, if $x^\top = (y^\top, s)$, with $y \in \mathbb{R}^{d-1}$ and $s \in \mathbb{R}$, we have

$$x^\top \rho^{-1} x = (y - bs)^\top A(y - bs) + s^2.$$

Since $h(\|x\|^2) = \frac{e^{-x^\top \rho^{-1} x/2}}{(2\pi)^{d/2}|\rho|^{1/2}}$ is a density, we obtain

$$\begin{aligned} 1 &= \int_{\mathbb{R}^d} h((\|x\|^2)dx = \int_{\mathbb{R}}\int_{\mathbb{R}^{d-1}} \frac{e^{-(y-bs)^\top A(y-bs)/2 - s^2/2}}{(2\pi)^{d/2}|\rho|^{1/2}} dyds \\ &= \int_{\mathbb{R}}\int_{\mathbb{R}^{d-1}} \frac{e^{-z^\top Az/2 - s^2/2}}{(2\pi)^{d/2}|\rho|^{1/2}} dzds = |\rho|^{-1/2}|A|^{-1/2}. \end{aligned}$$

Hence $|\rho||A| = 1$. To complete the proof, note that from Lemma A.6.1, $f(s)$ can be expressed as

$$\begin{aligned} f(s) &= \int_{\mathbb{R}^{d-1}} |\rho|^{-1/2} g\left((y - bs)^\top A(y - bs) + s^2\right) dy \\ &= |A|^{-1/2}|\rho|^{-1/2} \int_{\mathbb{R}^{d-1}} g\left(\|z\|^2 + s^2\right) dz \\ &= c_{d-1} \int_0^\infty r^{d-2} g(r^2 + s^2) dr \\ &= \frac{\pi^{(d-1)/2}}{\Gamma((d-1)/2)} \int_{s^2}^\infty (t - s^2)^{(d-3)/2} g(t) dt. \end{aligned}$$

■

Example A.6.1 (Pearson Elliptical Families) *Two popular families of elliptical distributions are the Pearson type II and type VII distributions, whose generators are given in Table A.1.*

TABLE A.1: Generators of Pearson type distributions.

Type	Generator	Parameters
II	$g(r) = \mathbb{I}(0 < r < 1)\frac{\Gamma(\alpha + d/2)}{\pi^{d/2}\Gamma(\alpha)}(1 - r)^{\alpha - 1}$	$\alpha > 0$
VII	$g(r) = \frac{\Gamma(\alpha + d/2)}{(\pi\nu)^{d/2}\Gamma(\alpha)}(1 + r/\nu)^{-\alpha - d/2}$	$\alpha, \nu > 0$

Using Proposition A.6.8, one can show that the margin of a Pearson type

II is still a Pearson of type II, with density

$$f(x) = \frac{\Gamma(\alpha + d/2)}{\pi^{1/2}\Gamma(\alpha + (d-1)/2)}(1 - x^2)^{-\alpha-(d-3)/2}, \qquad |x| < 1.$$

Note that for the Pearson type II, $\mathcal{R} \sim \text{Beta}(d/2, \alpha)$*. Similarly, the margin of a Pearson type VII is a Pearson type VII, with density*

$$f(x) = \frac{\Gamma(\alpha + 1/2)}{(\pi\nu)^{1/2}\Gamma(\alpha)}(1 + x^2/\nu)^{-\alpha-1/2}, \quad x \in \mathbb{R}.$$

For the Pearson type VII, $E(\mathcal{R}^p) < \infty$ *if and only if* $p < \alpha$*. In addition,* $W = \mathcal{R}/(\mathcal{R}+\nu) \sim \text{Beta}(d/2, \alpha)$*. The case* $\alpha = \nu/2$ *corresponds to the Student distribution, while the case* $\alpha = 1/2$ *and* $\nu = 1$ *corresponds to the Cauchy.*

A.6.21 Simulation of an Elliptic Distribution

To generate $Y \sim \mathcal{E}(\mathcal{R}, \mu, \Sigma)$, it suffices to simulate $\mathcal{R}$, and S independently, and then apply formula (A.15). To simulate S, one can set $S = Z/\|Z\|$, where $Z \sim N_d(0, I)$. For $d = 2$ or $d = 3$, one can generate S more efficiently. Indeed, for $d = 2$, it suffices to generate $U \sim \text{Unif}(0, 1)$ and set

$$S = (\cos(2\pi U), \sin(2\pi U)).$$

For $d = 3$, it suffices to generate $U, V \sim \text{Unif}(0, 1)$ independently and set

$$S = (\cos(2\pi U)\cos(\pi V), \sin(2\pi U)\cos(\pi V), \sin(\pi V)).$$

A.7 Conditional Densities and Joint Distributions

If X and Y are random vectors with joint density $f_{X,Y}(x, y)$, then the conditional density of Y given X is

$$f_{Y|X}(y|x) = f_{X,Y}(x, y)/f_X(x), \tag{A.17}$$

for all x such that $f_X(x) > 0$. In many situations, we have to compute the joint density of random vectors $X_1, \ldots, X_n$. When these observations are not independent, the following important formula is often used to compute the joint density.

A.7.1 Multiplication Formula

Equation (A.17) can also be written as

$$f_{X,Y}(x, y) = f_{Y|X}(y|x)f_X(x). \tag{A.18}$$

More generally, one has

$$f_{X_1,\ldots,X_n}(x_1,\ldots,x_n) = f_{X_1}(x_1)\prod_{k=2}^{n} f_{X_k|X_{k-1},\ldots,X_1}(x_k|x_{k-1},\ldots,x_1). \quad (A.19)$$

A.7.2 Conditional Distribution in the Markovian Case

As a particular case of the multiplication formula (A.19), suppose that the random vectors $X_1,\ldots,X_n$ are Markovian, i.e., the law of X_k given $X_1,\ldots,X_{k-1}$ only depends on X_{k-1}. Then the conditional density $f_{X_2,\ldots,X_n|X_1}$ of $X_2,\ldots,X_n$, given $X_1 = x_1$, is

$$f_{X_2,\ldots,X_n|X_1}(x_2,\ldots,x_n|x_1) = \prod_{k=2}^{n} f_{X_k|X_{k-1}}(x_k|x_{k-1}). \quad (A.20)$$

A.7.3 Rosenblatt Transform

Using the multiplication formula (A.19) and the change-of-variable theorem (Theorem A.8.1), one obtains the following well-known result:

Theorem A.7.1 (Rosenblatt) *Suppose that $Y_1,\ldots,Y_n$ are random variables and that*

$$F_t(y_1,\ldots,y_t) = P(Y_t \le y_t|Y_1 = y_1,\ldots,Y_{t-1} = y_{t-1}).$$

If we define $U_1 = F_1(Y_1)$, $U_2 = F_2(Y_1,Y_2)$, ..., $U_n = F_n(Y_1,\ldots,Y_n)$, then $U_1,\ldots,U_d$ are independent and uniformly distributed on $(0,1)$. The mapping $F : Y = (Y_1,\ldots,Y_n) \mapsto U = (U_1,\ldots,U_n)$, i.e., $U = F(Y)$ is an invertible mapping called the Rosenblatt transform.

Remark A.7.1 *The Rosenblatt transform can be used for goodness-of-fit tests as well as for simulating random vectors with a given distribution. In fact, if we generate independent and uniformly distributed random variables $U_1,\ldots,U_d$ over $(0,1)$, then $X = F^{-1}(U)$ has the same distribution as the variables $Y_1,\ldots,Y_n$.*

A.8 Functions of Random Vectors

In many applications, we have to deal with transformations of random variables or random vectors. For example, if X has a distribution function[4]

[4] Called the Pareto distribution.

$F(x) = 1 - (x/\sigma)^{-1/\xi}$, $x \geq \sigma > 0$, $\xi > 0$, we would like to know the moments of the random variable $Y = \ln(X/\sigma)$. The density of Y can be computed by using the following proposition.

Proposition A.8.1 *If X is a random variable with density $f_X(x)$ and if $Y = \phi(X)$, where ϕ is strictly monotonic and $\phi'(x) \neq 0$ for all x in the support S_X*[5] *of X, then the density $f_Y(y)$ of the random variable Y exists for all y in $\phi[S_X]$, and it is given by*

$$f_Y(y) = \frac{f_X\left(\phi^{-1}(y)\right)}{\left|\phi'\left(\phi^{-1}(y)\right)\right|} = f_X\left(\phi^{-1}(y)\right)\left|\frac{d}{dy}\phi^{-1}(y)\right|, \quad y \in \phi\left[S_X\right].$$

If $y \notin \phi[S_X]$, then $f_Y(y) = 0$.

Example A.8.1 (Linear Transformation in the Gaussian Case) *Let $X \sim N(0,1)$ and set $Y = \mu + \sigma X$ with $\sigma > 0$. Then the density of Y is obtained as $f_Y(y) = \frac{e^{-\frac{1}{2}\left(\frac{y-\mu}{\sigma}\right)^2}}{\sqrt{2\pi}\sigma}$, i.e., $Y \sim N(\mu, \sigma^2)$.*

Example A.8.2 (Transformation of a Pareto Distribution) *Suppose X has a Pareto distribution with distribution function $F(x) = 1 - (x/\sigma)^{-1/\xi}$, $x \geq \sigma$, where $\xi, \sigma > 0$. What is the density of $Y = \ln(X/\sigma)$? Since $\phi(x) = \ln(x/\sigma)$ is monotonic and $\phi^{-1}(y) = \sigma e^y$, it follows from Proposition A.8.1 that the density of Y is given by*

$$f_Y(y) = \frac{1}{\xi\sigma}\left(\sigma e^y/\sigma\right)^{-1-1/\xi}\mathbb{I}(y > 0) \times (\sigma e^y) = \frac{1}{\xi}e^{-y/\xi}\mathbb{I}(y > 0),$$

from which we conclude that Y has an exponential distribution with parameter $1/\xi$, or a gamma distribution with parameters $\alpha = 1$ and $\beta = \xi$.

In the case of a mapping of n variables, one has an analogous result. Suppose that G is a continuous mapping from an open set $A \subset \mathbb{R}^n$ to a subset of $\mathbb{R}^n$, defined by

$$y = \begin{pmatrix} y_1 \\ y_2 \\ \vdots \\ y_n \end{pmatrix} = G(x) = \begin{pmatrix} G_1(x_1, \ldots, x_n) \\ G_2(x_1, \ldots, x_n) \\ \vdots \\ G_n(x_1, \ldots, x_n) \end{pmatrix}.$$

If G is differentiable, its Jacobian determinant is defined by

$$J_G(x) = J_G(x_1, \ldots, x_n) = \det\begin{pmatrix} \frac{\partial y_1}{\partial x_1} & \cdots & \frac{\partial y_1}{\partial x_n} \\ \vdots & \cdots & \vdots \\ \frac{\partial y_n}{\partial x_1} & \cdots & \frac{\partial y_n}{\partial x_n} \end{pmatrix}. \quad \text{(A.21)}$$

[5]For our purposes, the support S_X can be defined as the set of $x \in \mathbb{R}$ such that $f_X(x) > 0$.

If the Jacobian determinant J_G does not vanish on A, then there exists an open set B so that G is an invertible mapping from A to B and if w is the inverse of G, i.e.,

$$w(y) = \begin{pmatrix} x_1 \\ x_2 \\ \vdots \\ x_n \end{pmatrix} = \begin{pmatrix} w_1(y_1, \ldots, y_n) \\ w_2(y_1, \ldots, y_n) \\ \vdots \\ w_n(y_1, \ldots, y_n) \end{pmatrix},$$

then the Jacobian determinant J_w of w exists and does not vanish on B, according to the famous inverse function theorem. Moreover, $J_w(y) = \frac{1}{J_G\{w(y)\}}$, for all $y \in B$. The following proposition is an application of this result.

Theorem A.8.1 *Suppose that G is an invertible mapping from an open set A to an open set B, and suppose in addition that G is differentiable, with a non-vanishing Jacobian determinant J_G. Let w denotes the inverse mapping.*

If the law of X has a density f_X and $P(X \in A) = 1$, then $Y = G(X)$ is such that $P(Y \in B) = 1$, and the density of Y is given by

$$f_Y(y) = f_X\{w(y)\}|J_w(y)| = f_X\{w(y)\}/|J_G\{w(y)\}|, \quad y \in B.$$

Example A.8.3 (Box-Müller Transform) *Consider the following mapping:*

$$\begin{aligned} Y_1 &= \{-2\ln(X_1)\}^{1/2}\cos(2\pi X_2), \\ Y_2 &= \{-2\ln(X_1)\}^{1/2}\sin(2\pi X_2), \end{aligned}$$

where X_1 and X_2 are independent and uniformly distributed on $(0,1)$. What is the joint law of $Y = (Y_1, Y_2) = G(X_1, X_2)$? Since $y_1^2 + y_2^2 = -2\ln(x_1)$, one can check easily that the mapping G is a bijection from $(0,1) \times (0,1)$ into $\mathbb{R}^2 \setminus \{0\}$. Moreover, its inverse w is given by

$$\begin{aligned} x_1 &= w_1(y_1, y_2) = e^{-\frac{1}{2}(y_1^2+y_2^2)}, \\ x_2 &= w_2(y_1, y_2) = \frac{1}{2\pi}\arctan(y_2/y_1), \end{aligned}$$

since $y_2/y_1 = \tan(2\pi x_2)$. Its Jacobian determinant is

$$\begin{aligned} J_w(y) &= \det\begin{pmatrix} -y_1 e^{-\frac{1}{2}(y_1^2+y_2^2)} & -y_2 e^{-\frac{1}{2}(y_1^2+y_2^2)} \\ -\frac{y_2}{y_1^2}\frac{1}{2\pi\{1+(y_2/y_1)^2\}} & \frac{1}{y_1}\frac{1}{2\pi\{1+(y_2/y_1)^2\}} \end{pmatrix} \\ &= -\frac{e^{-\frac{1}{2}(y_1^2+y_2^2)}}{2\pi}. \end{aligned}$$

Since $f_X(x) \equiv 1$ for all $x \in (0,1)^2$, it follows that the density of Y is

$$f_{Y_1,Y_2}(y_1, y_2) = 1 \times \frac{e^{-\frac{1}{2}(y_1^2+y_2^2)}}{2\pi} = \left(\frac{e^{-y_1^2/2}}{\sqrt{2\pi}}\right)\left(\frac{e^{-y_2^2/2}}{\sqrt{2\pi}}\right),$$

which is the density of two independent standard Gaussian variables.

Remark A.8.1 *The Box-Müller transformation allows one to generate two independent Gaussian variables from two independent uniform variables, so to simulate Gaussian random variables, there is no need to use an approximation by inverting the distribution function of a standard Gaussian.*

Example A.8.4 (Student Distribution) *Suppose that $X_1 \sim N(0,1)$ is independent of $X_2 \sim \chi^2(r)$. What is the law of $T = X_1 \Big/ \sqrt{X_2/r}$?*

Set $Y_1 = T = X_1 \Big/ \sqrt{X_2/r}$ and $Y_2 = X_2$. Then $X_1 = Y_1\sqrt{Y_2/r}$, $X_2 = Y_2$, and the Jacobian determinant is given by

$$J(y) = \det \begin{pmatrix} \sqrt{y_2/r} & \frac{y_1}{2\sqrt{y_2 r}} \\ 0 & 1 \end{pmatrix} = \sqrt{y_2/r}.$$

The joint density of (Y_1, Y_2) is

$$\begin{aligned} f_{Y_1,Y_2}(y_1,y_2) &= \frac{e^{-\frac{y_1^2 y_2}{2r}}}{\sqrt{2\pi}} \times \frac{y_2^{r/2-1} e^{-y_2/2}}{\Gamma(r/2)2^{r/2}} \times \sqrt{y_2/r} \\ &= \frac{y_2^{\frac{r+1}{2}-1} e^{-\frac{y_2}{2}(1+y_1^2/r)}}{\sqrt{2\pi r}\Gamma(r/2)2^{r/2}}. \end{aligned}$$

To find the density of $T = Y_1$, it suffices to integrate the joint density with respect to y_2. One ends up with

$$\begin{aligned} f_{Y_1}(y_1) &= \int_0^\infty f_{Y_1,Y_2}(y_1,y_2)dy_2 \\ &= \frac{1}{\sqrt{\pi r}} \frac{\Gamma(\frac{r+1}{2})}{\Gamma(r/2)} (1+y_1^2/r)^{-(\frac{r+1}{2})}. \end{aligned}$$

This is the density of a Student distribution, as defined in Section A.6.13.

A.9 Exercises

Exercise A.1

Suppose $N \sim \text{Poisson}(\lambda)$ and that conditionally on $N = n$, X has a binomial distribution with parameters (n,p). Find the (unconditional) distribution of X.

Exercise A.2

Suppose that $\Lambda \sim \text{Gamma}(\alpha, \beta)$ and suppose that conditionally on $\Lambda = \lambda$, X has a Poisson distribution with parameter λ.

(a) Show that for any integer $k \geq 0$,

$$P(X = k) = \frac{1}{\text{Beta}(\alpha, k)} \left(\frac{\beta}{1+\beta}\right)^k \left(\frac{1}{1+\beta}\right)^\alpha .$$

(b) Show that the distribution of X is an extension of the negative binomial distribution by choosing α and β appropriately. This is why the law described in (a) is also called the negative binomial distribution.

Exercise A.3

Prove Lemma A.6.1, i.e., for any function $f \geq 0$,

$$\int_{\mathbb{R}^d} f\left(\|x\|^2\right) dx = \frac{2\pi^{d/2}}{\Gamma(d/2)} \int_0^\infty r^{d-1} f\left(r^2\right) dr = \frac{\pi^{d/2}}{\Gamma(d/2)} \int_0^\infty r^{(d-2)/2} f(r) dr.$$

Bibliography

P. Glasserman. *Monte Carlo Methods in Financial Engineering*, volume 53 of *Applications of Mathematics (New York)*. Springer-Verlag, New York, 2004. Stochastic Modelling and Applied Probability.

Appendix B

Estimation of Parameters

In this chapter, we introduce two fundamental methods for estimating parameters of stochastic models: The maximum likelihood principle and the method of moments. We will also determine the asymptotic behavior of the estimation errors for independent and serially dependent observations.

B.1 Maximum Likelihood Principle

Suppose we have n observations $x_1, \ldots, x_n$ of a random vector X for which the joint distribution is a density f_θ depending on an unknown parameter vector $\theta \in \Omega \in \mathbb{R}^p$. The likelihood function $L(\theta)$ is defined as

$$L(\theta) = f_\theta(x_1, \ldots, x_n).$$

The maximum likelihood principle consists in choosing as the estimation of θ the value $\hat{\theta}_n$ maximizing the likelihood function L, i.e.,

$$\hat{\theta}_n = \arg\max_{\tilde{\theta} \in \Omega} L(\tilde{\theta}). \tag{B.1}$$

In other words, the maximum likelihood estimator consists in choosing $\hat{\theta}_n$ such that the likelihood of having observed $x_1, \ldots, x_n$ is maximized.

Remark B.1.1 *Often the likelihood function can be expressed as a product of positive functions so it is easier to maximize the so-called log-likelihood function, i.e., the logarithm of the likelihood function. It also reduces numerical errors. Then*

$$\hat{\theta}_n = \arg\max_{\tilde{\theta} \in \Omega} \ln\{L(\tilde{\theta})\}. \tag{B.2}$$

Recall that a necessary condition for a point θ to be an optimum point in the interior of Ω is that the gradient at that point vanishes. One can also have optimum points on the boundary on the set; for those points, the gradient does not vanish in general. Therefore, if $\hat{\theta}_n$ is in the interior of Ω, then

$$\left.\frac{\partial}{\partial \theta_j} L(\theta)\right|_{\theta=\hat{\theta}} = 0, \qquad \text{for all } j \in \{1, \ldots, p\},$$

or

$$\frac{\partial}{\partial \theta_j} \ln\{L(\theta)\}\Big|_{\theta=\hat{\theta}} = 0, \qquad \text{for all } j \in \{1,\ldots,p\}.$$

Remark B.1.2 *In financial applications, it is often the case that the data consists of prices $S_0,\ldots,S_n$ or returns $R_1,\ldots,R_n$, where $R_i = \ln(S_i/S_{i-1})$, $i \in \{1,\ldots,n\}$. For estimation purposes, one can use either prices or returns. In fact, if f_θ is the joint density of $R_1,\ldots,R_n$, then according to the change-of-variable formula given in Theorem A.8.1, the density of $S_1,\ldots,S_n$ given S_0 at point $(s_0,\ldots,s_n)$ is*

$$g_\theta(s_0, s_1, \ldots, s_n) = \frac{f_\theta(r_1,\ldots,r_n)}{\prod_{i=1}^n s_i},$$

where $r_i = \ln(s_i/s_{i-1})$, $i \in \{1,\ldots,n\}$. It follows that

$$\hat{\theta}_n = \arg\max_{\tilde{\theta}\in\Omega} g_{\tilde{\theta}}(s_0, s_1, \ldots, s_n) = \arg\max_{\tilde{\theta}\in\Omega} f_{\tilde{\theta}}(r_1,\ldots,r_n).$$

Example B.1.1 (Gaussian Observations) *Consider the following model:*

$$X_i = \mu + \sigma\,\varepsilon_i, \quad i \in \{1,\ldots,n\},$$

where the random variables $(\varepsilon_i)_{i\geq 1}$, are independent, and have a standard Gaussian distribution. The likelihood function of $\theta = (\mu,\sigma)$ is then given by

$$L(\mu,\sigma) = f_{\mu,\sigma}(x_1,\ldots,x_n) = \prod_{i=1}^n \frac{1}{\sqrt{2\pi}\sigma} e^{-\frac{(x_i-\mu)^2}{2\sigma^2}}.$$

Hence $\frac{\partial}{\partial\mu}\ln\{L(\mu,\sigma)\} = \frac{1}{\sigma^2}\sum_{i=1}^n (x_i - \mu)$, *and*

$$\frac{\partial}{\partial\sigma}\ln\{L(\mu,\sigma)\} = \frac{1}{\sigma^3}\sum_{i=1}^n (x_i-\mu)^2 - \frac{n}{\sigma}.$$

As a result, it is easy to check that

$$\hat{\mu} = \bar{x},$$

and

$$\hat{\sigma} = s = \sqrt{\frac{1}{n}\sum_{i=1}^n (x_i - \bar{x})^2}.$$

How precise are those estimations? This question will be answered in the next section.

B.2 Precision of Estimators

Estimators such as the maximum likelihood estimator discussed in the last section provide point values. When estimating parameters or making predictions, one should also be interested in the estimation or prediction error implied by the estimator used. Such uncertainty is usually handled with confidence intervals.

B.2.1 Confidence Intervals and Confidence Regions

A confidence interval of level $100(1-\alpha)\%$ for a one-dimensional parameter $\theta \in \mathbb{R}$ is a closed interval of the form $[\hat{\theta}_{n,1}, \hat{\theta}_{n,2}]$, where the lower and upper bounds are statistics such that

$$P(\hat{\theta}_{n,1} \leq \theta \leq \hat{\theta}_{n,2}) \approx 1 - \alpha,$$

when the number of observations is large enough. The same definition applies for a prediction interval, when the parameter is replaced by a random variable. A confidence region of level $100(1-\alpha)\%$ for a d-dimensional parameter θ is a closed set $R(x_1, \ldots, x_n)$ depending on the sample $x_1, \ldots, x_n$ so that

$$P(R \ni \theta) \approx 1 - \alpha,$$

when n is large enough.

Interpretation of a Confidence Region

Because θ is not random, one cannot conclude that the probability of θ being in one region is $1 - \alpha$. Here is the correct interpretation of a confidence interval or a confidence region: If one conducts many experiments, independently and in the same conditions, the percentage of times the true value of the parameter will belong to these (different) regions is approximately $100(1-\alpha)\%$.

In fact, giving a confidence region is making a prediction. One predicts that θ belong to a given confidence region. Then, $100(1-\alpha)\%$ of the predictions should be correct but you do not know which ones!

B.2.2 Nonparametric Prediction Interval

An interesting example of a prediction interval for a new observation is the following: Suppose that $(X_i)_{i\geq 1}$ are independent and have the same (continuous) distribution function F. Then it is easy to check that for any

$i, j \in \{1, \ldots, n\}$, $i < j$,

$$P(X_{n:i} \leq X_{n+1} \leq X_{n:j}) = \frac{j-i}{n+1},$$

where $X_{n:1} < \cdots < X_{n:n}$ are the so-called order statistics of the sample $X_1, \ldots, X_n$. This follows from the simple fact that for any $i \in \{1, \ldots, n\}$, $P(X_{n+1} \leq X_{n:i}) = \frac{i}{n+1}$. Of course, if one assumes a parametric model for the observations, then the prediction interval could be smaller.

B.3 Properties of Estimators

Before studying some of the properties of estimators, one has to define some types of convergence used in probability and statistics.

B.3.1 Almost Sure Convergence

A sequence $(X_n)_{n\geq 1}$ of random variables converges almost surely to a random variable X, denoted $X_n \stackrel{a.s.}{\to} X$, if

$$P\left(\lim_{n\to\infty} X_n = X\right) = 1. \tag{B.3}$$

B.3.2 Convergence in Probability

A sequence $(X_n)_{n\geq 1}$ of random variables converges in probability to a random variable X, denoted $X_n \stackrel{Pr}{\to} X$, if for all $\epsilon > 0$, one has

$$\lim_{n\to\infty} P(|X_n - X| > \epsilon) = 0. \tag{B.4}$$

Note that the almost sure convergence entails the convergence in probability.

B.3.3 Convergence in Mean Square

A sequence $(X_n)_{n\geq 1}$ of random variables converges in mean square to a random variable X, denoted $X_n \stackrel{L^2}{\to} X$, if

$$\lim_{n\to\infty} E\left\{(X_n - X)^2\right\} = 0. \tag{B.5}$$

Note that the mean square convergence entails the convergence in probability.

B.3.4 Convergence in Law

A sequence $(X_n)_{n\geq 1}$ of random vectors converges in law to a random vector X, denoted $X_n \rightsquigarrow X$, if for every point of continuity x of the distribution function F_X of X, one has

$$\lim_{n\to\infty} F_{X_n}(x) = P(X_n \leq x) = P(X \leq x) = F_X(x). \tag{B.6}$$

One also writes $X_n \rightsquigarrow F_X$, since the convergence does not depend on X but rather on its distribution function. Note that for a given $x \in \mathbb{R}^d$, $X_n \stackrel{Pr}{\to} x$ if and only if $X_n \rightsquigarrow x$.

Remark B.3.1 *Convergence in law can also be characterized in the following way: $X_n \rightsquigarrow X$ if and only if for any continuous and bounded function h, we have $E\{h(X_n)\} \to E\{h(X)\}$, as $n \to \infty$. It then follows that if h is continuous and if for for $p > 1$, we have $\sup_{n\geq 1} E\{|h(X_n)|^p\} < \infty$, then we may also conclude that $E\{h(X_n)\} \to E\{h(X)\}$, as $n \to \infty$. This property is very important in financial engineering where the law of an asset X can be approximated by the law of a sequence X_n. This is the basis of the binomial tree approximation of the Black-Scholes model.*

Convergence in law is also very useful for characterizing estimation errors and constructing confidence regions. Note that the main tool for establishing convergence in law is the famous central limit theorem stated in Theorem B.4.1.

Remark B.3.2 *For random vectors, the almost sure convergence, the convergence in probability and the mean square convergence correspond respectively to the convergence of each of their components.*

Here is a result that can be used to establish convergence in law from the convergence of the moment generating function.

Proposition B.3.1 *Suppose that $(X_n)_{n\geq 1}$ is a sequence of random vectors so that for all $n \geq 1$, the moment generating function $M_{X_n}(u)$ is finite for all $|u| < u_0$, for some $u_0 > 0$. If $M_{X_n}(u) \to M_X(u)$ for all $|u| < u_0$, where M_X is a moment generating function of some random vector X, then $X_n \rightsquigarrow X$.*

Example B.3.1 (Convergence of the Binomial Distribution) *Suppose that $X_n \sim \text{Bin}(n, p_n)$. If $p_n \to p \in (0,1)$, then $Z_n = \frac{X_n - np_n}{\sqrt{np_n(1-p_n)}} \rightsquigarrow Z \sim N(0,1)$, while if $np_n \to \lambda \in (0,\infty)$, then $X_n \rightsquigarrow \text{Poisson}(\lambda)$. In fact, in the first case,*

$$\begin{aligned} M_{Z_n}(u) &= e^{-\frac{unp_n}{\sqrt{np_n(1-p_n)}}} \left\{p_n e^{u/\sqrt{np_n(1-p_n)}} + 1 - p_n\right\}^n \\ &= \left\{1 + \frac{u^2}{2n} + o(1/n)\right\}^n \to e^{u^2/2}, \end{aligned}$$

which is the moment generating function of a standard Gaussian variate, while in the second case, setting $\lambda_n = np_n$, one has

$$M_{X_n}(u) = \left\{1 + \frac{\lambda_n}{n}(e^u - 1)\right\}^n \to e^{\lambda(e^u - 1)}$$

which is the moment generating function of a Poisson variate with parameter λ.

Finally, here is an interesting result for the convergence in law.

B.3.4.1 Delta Method

Theorem B.3.1 (Slutzky Theorem) *Suppose that $H : \mathbb{R}^k \mapsto \mathbb{R}^p$ is such that for all $1 \le i \le p$, $1 \le j \le k$, $J_{ij} = \frac{\partial H_i}{\partial \theta_j}$ exists and is continuous in a neighborhood of θ. The matrix J is called the Jacobian matrix.*

If $\sqrt{n}(\hat{\theta}_n - \theta) \rightsquigarrow N_k(0, V)$, then

$$\sqrt{n}\left(H(\hat{\theta}_n) - H(\theta)\right) \rightsquigarrow N_p\left(0, JVJ^\top\right).$$

In particular, for $k = p = 1$, if $\sqrt{n}(\hat{\theta}_n - \theta) \rightsquigarrow N(0, V)$, then

$$\sqrt{n}\left(H(\hat{\theta}_n) - H(\theta)\right)/H'(\hat{\theta}) \rightsquigarrow N(0, V).$$

In practice, J is estimated by $\hat{J}$, where

$$\hat{J}_{ij} = \left.\frac{\partial H_i}{\partial \theta_j}\right|_{\theta = \hat{\theta}_n}.$$

Remark B.3.3 *An interesting application of the delta method is when one wants to get rid of constraints on parameters. For example, if the parameter of interest is $\alpha = (\mu_1, \mu_2, \sigma_1, \sigma_2, \rho)^\top$, with $\sigma_1, \sigma_2 > 0$ and $\rho \in (-1, 1)$, one could set*

$$\theta = (\theta_1, \theta_2, \theta_3, \theta_4, \theta_5)^\top = \left(\mu_1, \mu_2, \ln(\sigma_1), \ln(\sigma_2), \frac{1}{2}\ln\left(\frac{1+\rho}{1-\rho}\right)\right)^\top,$$

so that $\theta \in \mathbb{R}^5$. In this case, θ is the parameter to be estimated, there are no constraints on θ, and $\alpha = H(\theta)$, where

$$H(\theta) = \begin{pmatrix} \theta_1 & 0 & 0 & 0 & 0 \\ 0 & \theta_1 & 0 & 0 & 0 \\ 0 & 0 & e^{\theta_3} & 0 & 0 \\ 0 & 0 & 0 & e^{\theta_4} & 0 \\ 0 & 0 & 0 & 0 & \tanh(\theta_5) \end{pmatrix},$$

with $\tanh(x) = \frac{e^x - e^{-x}}{e^x + e^{-x}}$. *The associated Jacobian matrix is then given by*

$$J = \begin{pmatrix} 1 & 0 & 0 & 0 & 0 \\ 0 & 1 & 0 & 0 & 0 \\ 0 & 0 & e^{\theta_3} & 0 & 0 \\ 0 & 0 & 0 & e^{\theta_4} & 0 \\ 0 & 0 & 0 & 0 & 1 - \tanh^2(\theta_5) \end{pmatrix} = \begin{pmatrix} 1 & 0 & 0 & 0 & 0 \\ 0 & 1 & 0 & 0 & 0 \\ 0 & 0 & \sigma_1 & 0 & 0 \\ 0 & 0 & 0 & \sigma_2 & 0 \\ 0 & 0 & 0 & 0 & 1 - \rho^2 \end{pmatrix}.$$

B.3.5 Bias and Consistency

Let $\hat{\theta}_n$ be an estimator of θ evaluated from a sample of size n. The bias of this estimator is defined by $E(\hat{\theta}_n) - \theta$. The estimator is said to be unbiased if its bias is zero, i.e., $E(\hat{\theta}_n) = \theta$. It is said to be asymptotically unbiased if $\lim_{n\to\infty} E(\hat{\theta}_n) = \theta$. Finally, the estimator $\hat{\theta}_n$ is said to be consistent if it converges in probability to θ.

Remark B.3.4 *If* $a_n(\hat{\theta}_n - \theta) \rightsquigarrow N(0,1)$, *where* $a_n \to \infty$, *then* $\hat{\theta}_n$ *is a consistent estimator of* θ.

One of the most famous results involving convergence in law is the central limit theorem. It is presented next.

B.4 Central Limit Theorem for Independent Observations

The famous central limit theorem basically gives the asymptotic behavior of the estimation error of the mean of a random vector by its empirical analog.

Theorem B.4.1 (Central Limit Theorem) *Suppose that the random vectors* $X_1, \ldots, X_n$, *with values in* $\mathbb{R}^d$, *are independent and identically distributed, with mean* μ *and covariance matrix* Σ. *Set* $\bar{X}_n = \frac{1}{n}\sum_{i=1}^n X_i$.

Then $\sqrt{n}\,(\bar{X}_n - \mu) \rightsquigarrow N_d(0, \Sigma)$. *In particular, if* V *is not singular, then*

$$\sqrt{n}\Sigma^{-1/2}(\bar{X}_n - \mu) \rightsquigarrow N_d(0, I),$$

and

$$n(\bar{X}_n - \mu)^\top \Sigma^{-1}(\bar{X}_n - \mu) \rightsquigarrow \chi^2(d).$$

Remark B.4.1 *In practice,* Σ *is unknown and it is estimated by* S_n, *the so-called sampling covariance matrix, defined by*

$$S_n = \frac{1}{n-1}\sum_{i=1}^n (X_i - \bar{X}_n)(X_i - \bar{X}_n)^\top.$$

S_n is a consistent estimator of Σ. Also, if Σ is not singular, then

$$\sqrt{n} S_n^{-1/2}(\bar{X} - \mu) \rightsquigarrow N_d(0, I),$$

and

$$n(\bar{X} - \mu)^\top S_n^{-1}(\bar{X} - \mu) \rightsquigarrow \chi^2(d).$$

B.4.1 Consistency of the Empirical Mean

Suppose that the univariate observations $X_1, \ldots, X_n$ are independent and have the same distribution. If the mean μ exists, then $\bar{X}_n = \frac{1}{n}\sum_{i=1}^n X_i$ converges almost surely to μ. As a result, $\bar{X}_n$ is a consistent estimator of μ. In addition, if the second moment is finite, then the error made by estimating μ by $\bar{X}_n$ is asymptotically Gaussian. More precisely, according to the central limit theorem (Theorem B.4.1), $\sqrt{n}(\bar{X}_n - \mu) \rightsquigarrow N(0, \sigma^2)$, where σ^2 is the variance of X_i.

Set $s_n = \sqrt{S_n}$, i.e., s_n is the empirical standard deviation.

B.4.2 Consistency of the Empirical Coefficients of Skewness and Kurtosis

Proposition B.4.1 *Suppose that the univariate observations $X_1, \ldots, X_n$ of X are independent and identically distributed, and assume that the moment of order 4 exists. The estimators $\hat{\gamma}_{n,1}$ and $\hat{\gamma}_{n,2}$ of the skewness and kurtosis are respectively defined by*

$$\hat{\gamma}_{n,1} = \frac{1}{n}\sum_{i=1}^n \left(\frac{X_i - \bar{X}_n}{s_n}\right)^3, \tag{B.7}$$

$$\hat{\gamma}_{n,2} = \frac{1}{n}\sum_{i=1}^n \left(\frac{X_i - \bar{X}_n}{s_n}\right)^4. \tag{B.8}$$

If the moment of order 8 exists, then

$$\left(\sqrt{n}\,(\bar{X}_n - \mu), \sqrt{n}\,(s_n - \sigma), \sqrt{n}\,(\hat{\gamma}_{n,1} - \gamma_1), \sqrt{n}\,(\hat{\gamma}_{n,2} - \gamma_2)\right)^\top \rightsquigarrow N_4(0, V),$$

where the covariance matrix V is defined by

$$
\begin{aligned}
V_{11} &= \sigma^2, \quad V_{12} = \frac{\sigma^2}{2}\gamma_1, \quad V_{13} = \sigma\left(\gamma_2 - 3 - \frac{3}{2}\gamma_1^2\right), \\
V_{14} &= \sigma\left(\mu_5 - 2\gamma_1\gamma_2 - 4\gamma_1\right), \quad V_{22} = \frac{\sigma^2}{4}(\gamma_2 - 1), \\
V_{23} &= \frac{\sigma}{2}\left(\mu_5 - \frac{5}{2}\gamma_1 - \frac{3}{2}\gamma_1\gamma_2\right), \quad V_{24} = \frac{\sigma}{2}\left(\mu_6 + \gamma_2 - 2\gamma_2^2 - 4\gamma_1^2\right), \\
V_{33} &= \mu_6 - 3\gamma_1\mu_5 + \frac{9}{4}\gamma_1^2\gamma_2 - 6\gamma_2 + \frac{35}{4}\gamma_1^2 + 9, \\
V_{34} &= \mu_7 - \frac{3}{2}\gamma_1\mu_6 - \mu_5(2\gamma_2 + 3) + 3\gamma_1\gamma_2^2 + \frac{3}{2}\gamma_1\gamma_2 + 6\gamma_1^3 + 12\gamma_1, \\
V_{44} &= \mu_8 - 8\gamma_1\mu_5 - 4\mu_6\gamma_2 + 16\gamma_1^2(\gamma_2 + 1) + \gamma_2^2(4\gamma_2 - 1),
\end{aligned}
$$

with $\mu_j = E\left\{\left(\frac{X_1-\mu}{\sigma}\right)^j\right\}$, $j \in \{1, \ldots, 8\}$. *V can be estimated by the covariance matrix $\hat{V}_n$ of the pseudo-observations $W_{n,i}$, i.e.,*

$$
\hat{V}_{n,jk} = \frac{1}{n-1}\sum_{i=1}^{n} W_{n,ij}W_{n,ik}, \qquad j, k \in \{1, \ldots, d\},
$$

where $W_{ni} = (W_{n,i1}, W_{n,i2}, W_{n,i3}, W_{n,i4})^\top$, *with*

$$
\begin{aligned}
W_{n,i1} &= s_x e_{n,i}, \qquad W_{n,i2} = \frac{s_x}{2}(e_{n,i}^2 - 1), \\
W_{n,i3} &= e_{n,i}^3 - \hat{\gamma}_{n,1} - \frac{3}{2}\hat{\gamma}_{n,1}(e_{n,i}^2 - 1) - 3e_{n,i}, \\
W_{n,i4} &= e_{n,i}^4 - \hat{\gamma}_{n,2} - 2\hat{\gamma}_{n,2}(e_{n,i}^2 - 1) - 4\hat{\gamma}_{n,1}e_{n,i},
\end{aligned}
$$

and $e_{n,i} = \frac{x_i - \bar{x}}{s_x}$, $i \in \{1, \ldots, n\}$. *In particular,*

$$
\hat{V}_{n,33} = \frac{1}{n-1}\sum_{i=1}^{n}\left\{e_i^3 - \hat{\gamma}_{n,1} - \frac{3}{2}\hat{\gamma}_{n,1}(e_{n,i}^2 - 1) - 3e_{n,i}\right\}^2
$$

and

$$
\hat{V}_{n,44} = \frac{1}{n-1}\sum_{i=1}^{n}\left\{e_{ni}^4 - \hat{\gamma}_2 - 2\hat{\gamma}_{n,2}(e_{n,i}^2 - 1) - 4\hat{\gamma}_{n,1}e_{n,i}\right\}^2.
$$

In the Gaussian case, $\gamma_1 = 0$ and $\gamma_2 = 3$, so the covariance matrix V reduces to the diagonal matrix

$$
V = \begin{pmatrix} \sigma^2 & 0 & 0 & 0 \\ 0 & \sigma^2/2 & 0 & 0 \\ 0 & 0 & 6 & 0 \\ 0 & 0 & 0 & 24 \end{pmatrix}. \tag{B.9}
$$

Remark B.4.2 *The MATLAB functions* skewness *and* kurtosis *can be used to estimate the skewness and the kurtosis.*

PROOF. Set $\varepsilon_i = (X_i - \mu)/\sigma$. It follows from the central limit theorem (Theorem B.4.1) that for all $j \in \{1, \ldots, 8\}$,

$$Z_{j,n} = \sqrt{n}\left(\frac{1}{n}\sum_{i=1}^{n} \varepsilon_i^j - \mu_j\right) \rightsquigarrow Z_j \sim N(0, \mu_{2j} - \mu_j^2),$$

where $\mu_j = E\left(\varepsilon_i^j\right)$. In addition, the covariance between Z_j and Z_k is given by $\mu_{j+k} - \mu_j\mu_k$, $j, k \in \{1, \ldots, 4\}$, $\mu_1 = 0$, $\mu_2 = 1$.

Using the delta method (Theorem B.3.1), we have

$$\begin{aligned}
\sqrt{n}(\bar{X}_n - \mu) &\rightsquigarrow \sigma Z_1, \qquad \sqrt{n}(s_n - \sigma) \rightsquigarrow \frac{\sigma}{2} Z_2, \\
\sqrt{n}(\hat{\gamma}_{n,1} - \gamma_1) &\rightsquigarrow G_1 = Z_3 - \frac{3}{2}\gamma_1 Z_2 - 3Z_1, \\
\sqrt{n}(\hat{\gamma}_{n,2} - \gamma_2) &\rightsquigarrow G_2 = Z_4 - 2\gamma_2 Z_2 - 4\gamma_1 Z_1.
\end{aligned}$$

The estimation of V by the pseudo-observations $W_{n,i}$ is a consequence of the last representation. Next, by hypothesis, $\mathrm{Var}(Z_1) = 1$, $\mathrm{Var}(Z_2) = \gamma_2 - 1$, $\mathrm{Cov}(Z_1, Z_2) = \gamma_1$, and $\mathrm{Cov}(Z_1, Z_3) = \gamma_2$. We can then deduce that

$$\begin{aligned}
\mathrm{Cov}(Z_1, G_1) &= \gamma_2 - 3 - \frac{3}{2}\gamma_1^2, \\
\mathrm{Cov}(Z_2, G_1) &= \mu_5 - \frac{5}{2}\gamma_1 - \frac{3}{2}\gamma_1\gamma_2, \\
\mathrm{Cov}(Z_1, G_2) &= \mu_5 - 2\gamma_1\gamma_2 - 4\gamma_1, \\
\mathrm{Cov}(Z_2, G_2) &= \mu_6 + \gamma_2 - 2\gamma_2^2 - 4\gamma_1^2.
\end{aligned}$$

Furthermore,

$$\begin{aligned}
\mathrm{Var}(G_1) &= \mu_6 - 3\gamma_1\mu_5 + \frac{9}{4}\gamma_1^2\gamma_2 - 6\gamma_2 + \frac{35}{4}\gamma_1^2 + 9, \\
\mathrm{Var}(G_2) &= \mu_8 - 8\gamma_1\mu_5 - 4\mu_6\gamma_2 + 16\gamma_1^2(\gamma_2 + 1) + \gamma_2^2(4\gamma_2 - 1), \\
\mathrm{Cov}(G_1, G_2) &= \mu_7 - \frac{3}{2}\gamma_1\mu_6 - \mu_5(2\gamma_2 + 3) + 3\gamma_1\gamma_2^2 + \frac{3}{2}\gamma_1\gamma_2 + 6\gamma_1^3 + 12\gamma_1.
\end{aligned}$$

In the Gaussian case, $\gamma_1 = \gamma_2 - 3 = \mu_5 = \mu_7 = 0$, $\mu_6 = 15$, and $\mu_8 = 105$. Hence V is diagonal, with $V_{11} = \sigma^2$, $V_{22} = \sigma^2/2$, $V_{33} = \mathrm{Var}(G_1) = 6$, and $V_{44} = \mathrm{Var}(G_2) = 24$.

■

B.4.3 Confidence Intervals I

Using the central limit theorem, one can also build confidence intervals. In fact, since $\frac{\bar{X}_n - \mu}{s_n \sqrt{n}}$ is approximately distributed as a standard Gaussian variate when n is large enough, it follows that

$$P(|\bar{X}_n - \mu| \leq 1.96 s_n/\sqrt{n}) \approx P(|Z| \leq 1.96) = 0.95.$$

The interpretation of the confidence interval is the following: In 95% of case, the interval $[\bar{X}_n - 2s_n/\sqrt{n}, \bar{X}_n + 2s_n/\sqrt{n}]$ will contain the true value μ. However, for a given sample, one does not know if μ belongs to the confidence interval. One just predicts that it does, and 95% of the predictions should be true!

B.4.4 Confidence Ellipsoids

When the parameter is multidimensional, i.e., $\theta \in \mathbb{R}^d$, one can define confidence regions which are ellipsoids. In fact, if $\chi^2_{d,\alpha}$ is the quantile of order $1-\alpha$ of a chi-square distribution with d degree of freedom, then according to Remark B.4.1,

$$P\left(n(\bar{X}_n - \mu)^\top V_n^{-1}(\bar{X}_n - \mu) \leq \chi^2_{d,\alpha}\right) \approx 1 - \alpha.$$

It is called a confidence ellipsoid since $E = \{\mu \in \mathbb{R}^d; (a-\mu)^\top V^{-1}(a-\mu) \leq b\}$ is an ellipsoid. In fact, since V can be written as $V = M\Delta M^\top$, where Δ is a diagonal matrix and $MM^\top = I$, we have

$$E = \left\{ Mz + a; \sum_{j=1}^{d} \frac{z_j^2}{\Delta_{jj}} \leq b \right\}.$$

Since M is a rotation matrix, E is composed of a rotation of an ellipsoid centered at the origin, followed by a translation.

B.4.5 Confidence Intervals II

It sometimes happens that the central limit theorem does not apply, which implies that the estimation error is not Gaussian. Nevertheless, in some situations, one can still define confidence intervals. To this end, suppose that $a_n(\hat{\theta}_n - \theta) \rightsquigarrow \Theta$, where Θ has distribution function H. Then

$$\left[\hat{\theta}_n - \frac{H^{-1}(1-\alpha/2)}{a_n}, \hat{\theta}_n - \frac{H^{-1}(\alpha/2)}{a_n}\right]$$

is a confidence interval of level $100(1-\alpha)\%$, since

$$P\left\{\theta + \frac{H^{-1}(\alpha/2)}{a_n} \leq \hat{\theta}_n \leq \theta + \frac{H^{-1}(1-\alpha/2)}{a_n}\right\}$$
$$\approx H\left\{H^{-1}(1-\alpha/2)\right\} - H\left\{H^{-1}(\alpha/2)\right\} = 1 - \alpha.$$

Remark B.4.3 *When H is the distribution function of a standard Gaussian variate, then one recovers the usual confidence interval, since $-H^{-1}(\alpha/2) = -(-z_{\alpha/2}) = z_{\alpha/2}$ and $H^{-1}(1-\alpha/2) = z_{\alpha/2}$.*

B.5 Precision of Maximum Likelihood Estimator for Serially Independent Observations

Suppose that $X_1, \ldots, X_n$ are independent, with density f_θ. Then the log-likelihood is given by

$$L(\theta) = \sum_{i=1}^{n} \ln\{f_\theta(x_i)\}.$$

Proposition B.5.1 *Under weak conditions, see, e.g., Serfling [1980], the estimator $\hat{\theta}_n$ obtained from the maximum likelihood principle is consistent and*

$$\sqrt{n}\,(\hat{\theta}_n - \theta) \rightsquigarrow N_p(0, V),$$

where V is a non-negative definite matrix satisfying

$$\mathcal{I}_\theta V \mathcal{I}_\theta = \mathcal{I}_\theta,$$

and $\mathcal{I}_\theta$ is the Fisher information matrix given, for $j, k \in \{1, \ldots, p\}$, by

$$(\mathcal{I}_\theta)_{jk} = \int \left[\frac{\partial \ln\{f_\theta(x)\}}{\partial \theta_j}\right] \left[\frac{\partial \ln\{f_\theta(x)\}}{\partial \theta_k}\right] f_\theta(x)dx.$$

Then V is the Moore-Penrose pseudo-inverse of $\mathcal{I}_\theta$. Note that the associated MATLAB function is pinv. Also, if $\mathcal{I}_\theta$ is invertible, then $V = \mathcal{I}_\theta^{-1}$.

B.5.1 Estimation of Fisher Information Matrix

In practice, instead of maximizing the likelihood, one often minimizes the negative of the log-likelihood. Therefore, one could estimate $\mathcal{I}_\theta$ by

$$\hat{\mathcal{I}}_n = \frac{1}{n} \sum_{i=1} (y_i - \bar{y})(y_i - \bar{y})^\top, \tag{B.10}$$

with

$$y_i = -\nabla_\theta \ln\{f_{\hat{\theta}_n}(x_i)\} = -\frac{\nabla_\theta f_{\hat{\theta}_n}(x_i)}{f_{\hat{\theta}_n}(x_i)}, \quad i \in \{1 \ldots, n\},$$

where ∇f_θ is the column vector with components $\partial_{\theta_j} f_\theta(x)$, $j \in \{1, \ldots, p\}$. Note that one should have $\bar{y} \approx 0$. The gradient can be evaluated exactly. It can also be approximated numerically by using, for example, the MATLAB function NumJacobian. Another approach used in practice is to estimate $\mathcal{I}_\theta$ by using the Hessian matrix of the $-$log-likelihood function H_n (matrix formed by the second derivatives), i.e., for $1 \le j,k \le p$, one has

$$(H_n)_{jk} = -\sum_{i=1}^n \frac{\partial_{\theta_j}\partial_{\theta_k} f_{\hat{\theta}_n}(x_i)}{f_{\hat{\theta}_n}(x_i)} + \sum_{i=1}^n \frac{\partial_{\theta_j} f_{\hat{\theta}_n}(x_i)}{f_{\hat{\theta}_n}(x_i)} \frac{\partial_{\theta_k} f_{\hat{\theta}_n}(x_i)}{f_{\hat{\theta}_n}(x_i)},$$

so

$$\frac{H_n}{n} = \bar{y}\bar{y}^\top + \hat{\mathcal{I}}_n - \frac{1}{n}\sum_{i=1}^n \frac{\partial_{\theta_j}\partial_{\theta_k} f_{\hat{\theta}_n}(x_i)}{f_{\hat{\theta}_n}(x_i)}.$$

As a result, H_n/n is close to $\hat{\mathcal{I}}_n$ defined by (B.10), and it converges in probability to $\mathcal{I}_\theta$.

One cannot tell in advance which method of estimating the limiting covariance is more precise. However, in some cases, the Hessian matrix might not be positive definite or even non-negative definite, so (B.10) should be the choice for estimating the Fisher information. Note that in any mathematical packages, the Hessian matrix is one of the possible outputs of the minimizing functions (like *fmincon* and *fminunc* in MATLAB).

Example B.5.1 (Gaussian Data (Continued)) *Recall that in that case,*

$$\nabla_{(\mu,\sigma)} \ln\{f(x;\mu,\sigma)\} = \left(\frac{(x-\mu)}{\sigma^2}, \frac{(x-\mu)^2}{\sigma^3} - \frac{1}{\sigma}\right).$$

As a result, the Fisher information matrix is

$$\mathcal{I}_{(\mu,\sigma)} = \frac{1}{\sigma^2} E \begin{pmatrix} Z^2 & Z(Z^2-1) \\ Z(Z^2-1) & (Z^2-1)^2 \end{pmatrix}$$

where $Z = \frac{X-\mu}{\sigma} \sim N(0,1)$. *Since* $E(Z^3) = 0$ *and* $E(Z^4) = 3$, *one obtains that* $\mathcal{I}_{(\mu,\sigma)} = \frac{1}{\sigma^2}\begin{pmatrix} 1 & 0 \\ 0 & 2 \end{pmatrix}$. *Hence* $V = \mathcal{I}^{-1}_{(\mu,\sigma)} = \begin{pmatrix} \sigma^2 & 0 \\ 0 & \sigma^2/2 \end{pmatrix}$. *In addition,* $E(\hat{\mu}_n) = \mu$ *and* $E(\hat{\sigma}^2_n) = (n-1)\sigma^2/n \approx \sigma^2$, *when* n *is large enough. One may conclude from Proposition B.5.1 that* $\begin{pmatrix} \sqrt{n}(\bar{x}_n - \mu) \\ \sqrt{n}(s_n - \sigma) \end{pmatrix} \rightsquigarrow N_2(0,V)$, *where* $V = \begin{pmatrix} \sigma^2 & 0 \\ 0 & \sigma^2/2 \end{pmatrix}$. *In particular,*

$$\mu = \bar{x}_n \pm s_n \frac{z_{\alpha/2}}{\sqrt{n}}, \quad \sigma = s_n \pm s_n \frac{z_{\alpha/2}}{\sqrt{2n}}$$

are confidence intervals of level $1-\alpha$ *for* μ *and* σ *respectively (but not simultaneously).*

Remark B.5.1 *The convergence result could have been obtained directly using Propositions B.4.1 and B.3.1, i.e., by using the central limit theorem and the delta method.*

Example B.5.2 *Here is an example where the covariance matrix V is singular. For the Pareto distribution function $F(x) = 1 - (x/\sigma)^{-1/\xi}$, $x \geq \sigma$, $\xi > 0$, one has $\hat{\sigma}_n = m = \min(X_1, \ldots, X_n)$ and $\hat{\xi}_n = \frac{1}{n}\sum_{i=1}^n \ln(X_i/m)$. It follows that $\sqrt{n}\,(\hat{\xi}_n - \xi) \rightsquigarrow N(0, \xi^2)$, while $\sqrt{n}\,(\hat{\sigma}_n - \sigma) \rightsquigarrow 0$. In this case, $V = \begin{pmatrix} \xi^2 & 0 \\ 0 & 0 \end{pmatrix}$.*

B.6 Convergence in Probability and the Central Limit Theorem for Serially Dependent Observations

In many applications, instead of having independent observations, the latter are serially dependent, i.e., X_i can depend on $X_1, \ldots, X_{i-1}$. In such a case, one needs another version of the central limit theorem. See Durrett [1996].

Theorem B.6.1 *Suppose that a sequence of random d-dimensional vectors $X_1, \ldots, X_n$ is stationary and ergodic, with mean μ. Then $\bar{X}_n = \frac{1}{n}\sum_{i=1}^n X_i$ converges in probability to μ. Suppose in addition that each X_i is a martingale difference, i.e., $E(X_i|X_1, \ldots, X_{i-1}) = \mu$. Then, if the moment of order 2 exists, $\sqrt{n}\,(\bar{X}_n - \mu) \rightsquigarrow N_d(0, \Sigma)$, with $\Sigma = E\left\{(X_i - \mu)(X_i - \mu)^\top\right\}$.*

Remark B.6.1 *In practice, Σ is unknown and it is estimated by Σ_n, the sampling covariance matrix defined by*

$$\Sigma_n = \frac{1}{n}\sum_{i=1}^n (X_i - \bar{X}_n)(X_i - \bar{X}_n)^\top.$$

Moreover, Σ_n is a consistent estimator of Σ.

B.7 Precision of Maximum Likelihood Estimator for Serially Dependent Observations

Suppose that the data $X_1, \ldots, X_n$ are observation from an ergodic stationary time series. Assume that the conditional densities of X_i given $X_1, \ldots, X_{i-1}$, denoted $f_{i,\theta}$ exist and also depend on an unknown parameter $\theta \in \mathbb{R}^p$. Then, using the multiplication formula (A.19), the joint density of $X_2, \ldots X_n$ given X_1 is

$$f_\theta(x_2, \ldots, x_n | x_1) = \prod_{i=2}^{n} f_{i,\theta}(x_i, \ldots, x_1).$$

It follows that if L is the log-likelihood, to estimate θ one can maximize

$$L(\theta) = \sum_{i=2}^{n} \ln\{f_{i,\theta}(x_i, \ldots, x_1, \theta)\}.$$

Therefore, if $\hat{\theta}_n$ in the interior of Ω, then one must have $\hat{\theta}_n$ such that $\nabla L(\hat{\theta}_n) = 0$, where

$$\nabla L(\hat{\theta}_n) = \sum_{i=2}^{n} \frac{\nabla_\theta f_{i,\hat{\theta}}(x_i, \ldots, x_1)}{f_{i,\hat{\theta}_n}(x_i, \ldots, x_1)} = \sum_{i=1}^{n} \ell_i(\hat{\theta}_n).$$

In many cases, one can then show that the maximum likelihood estimator is consistent; one can also evaluate the error of estimation.

Proposition B.7.1 *Under weak assumptions,*

$$\sqrt{n}\,(\hat{\theta}_n - \theta) \rightsquigarrow N_p(0, V),$$

where V is a non-negative definite matrix satisfying

$$\mathcal{I}_\theta V \mathcal{I}_\theta = \mathcal{I}_\theta,$$

and $\mathcal{I}_\theta$ is the Fisher information matrix defined by $\mathcal{I}_\theta = E\left\{\ell_i(\theta)\ell_i(\theta)^\top\right\}$. The latter can be estimated by

$$\hat{\mathcal{I}}_\theta = \frac{1}{n}\sum_{i=1}^{n}(y_i - \bar{y})(y_i - \bar{y})^\top,$$

where $y_i = \ell_i(\hat{\theta})$, or by H_n/n, where H_n is the Hessian matrix of $-L(\hat{\theta}_n)$.

Example B.7.1 *Assume that the data $(X_i)_{i\geq 1}$ satisfy*

$$X_i - \mu = \phi(X_{i-1} - \mu) + \sigma\varepsilon_i,$$

where $|\phi| < 1$, $\sigma > 0$ and the innovations ε_t are independent, with a standard Gaussian distribution. Then, since the conditional distribution of X_i given $X_1, \ldots, X_{i-1}$ is Gaussian, with mean $\mu + \phi(X_{i-1} - \mu)$ and variance σ^2, setting $\theta = (\mu, \phi, \sigma)$, one has

$$\ln\{f_i(X_i, X_{i-1}, \theta)\} = -\ln\left(\sqrt{2\pi}\sigma\right) - \frac{\{X_i - \mu - \phi(X_{i-1} - \mu)\}^2}{2\sigma^2}.$$

Consequently,

$$
\begin{aligned}
\partial_\mu \ln\{f_i(X_i, X_{i-1}, \theta)\} &= (1-\phi)\left\{X_i - \mu - \phi(X_{i-1}-\mu)\right\}/\sigma^2, \\
\partial_\phi \ln\{f_i(X_i, X_{i-1}, \theta)\} &= (X_{i-1}-\mu)\left\{X_i - \mu - \phi(X_{i-1}-\mu)\right\}/\sigma^2, \\
\partial_\phi \ln\{f_i(X_i, X_{i-1}, \theta)\} &= -\frac{1}{\sigma} + \frac{\{X_i - \mu - \phi(X_{i-1}-\mu)\}^2}{\sigma^3}.
\end{aligned}
$$

Then $\ell_i(\theta) = \left((1-\phi)\varepsilon_i, \varepsilon_i(X_{i-1}-\mu), \varepsilon_i^2 - 1\right)^\top/\sigma$, *leading to*

$$
\mathcal{I}_{\mu,\phi,\sigma} = \frac{1}{\sigma^2}\begin{pmatrix} (1-\phi)^2 & 0 & 0 \\ 0 & \frac{\sigma^2}{1-\phi^2} & 0 \\ 0 & 0 & 2 \end{pmatrix}.
$$

As a result,

$$
\mathcal{I}^{-1}_{\mu,\phi,\sigma} = \begin{pmatrix} \frac{\sigma^2}{(1-\phi)^2} & 0 & 0 \\ 0 & 1-\phi^2 & 0 \\ 0 & 0 & \sigma^2/2 \end{pmatrix}.
$$

In addition, by solving $\sum_{i=2}^n \ell_i(\hat\theta) = 0$, *one obtains*

$$
\begin{aligned}
\hat\mu_n &= \bar x_n + \frac{\phi(X_n - \hat\mu_n) - (X_1 - \hat\mu_n)}{n}, \\
\hat\phi_n &= \frac{\sum_{i=2}^n (X_{i-1}-\hat\mu_n)(X_i - \hat\mu_n)}{\sum_{i=2}^n (X_{i-1}-\hat\mu_n)^2}, \\
\hat\sigma_n^2 &= \frac{1}{n}\sum_{i=2}^n \left\{X_i - \hat\mu_n - \hat\phi_n(X_{i-1}-\hat\mu_n)\right\}^2.
\end{aligned}
$$

Therefore, one could set $\hat\mu_n = \bar x_n$, $\hat\phi_n = \frac{\sum_{i=2}^n (X_{i-1}-\bar x_n)(X_i - \bar x_n)}{\sum_{i=2}^n (X_{i-1}-\bar x)^2}$, *and* $\hat\sigma_n^2 = \frac{1}{n}\sum_{i=2}^n \left\{X_i - \bar x_n - \hat\phi_n(X_{i-1}-\bar x_n)\right\}^2$. *By Theorem B.6.1, these estimators are consistent. Furthermore, one deduces from Theorem B.7.1 that*

$$
\left(\sqrt{n}\,(\bar x_n - \mu), \sqrt{n}\,(\hat\phi_n - \phi), \sqrt{n}\,(\hat\sigma_n - \sigma)\right)^\top \rightsquigarrow N_3\left(0, \mathcal{I}^{-1}_{\mu,\phi,\sigma}\right).
$$

It follows that $\sqrt{n}\,(1-\hat\phi_n)(\bar x_n - \mu)/\hat\sigma_n \rightsquigarrow Z_1$, $\sqrt{n}\,(\hat\phi_n - \phi)/\sqrt{1-\hat\phi_n^2} \rightsquigarrow Z_2$ *and* $\sqrt{n}\,(\hat\sigma_n - \sigma)/(\hat\sigma_n/\sqrt{2}) \rightsquigarrow Z_3$, *where* Z_1, Z_2 *and* Z_3 *are independent standard Gaussian variables.*

B.8 Method of Moments

It is sometimes impossibly difficult to use the maximum likelihood principle to estimate parameters. For example, it can be hard or too computationally intensive to evaluate the log-likelihood function. Another methodology which is often practical and easy to implement is the method of moments, when the unknown parameters can be expressed in terms of the first moments. The idea is simply to estimate the moments by the sampling moments and then recover an estimation of the parameters by inversion. For example, if the expectation μ exists, then $\bar{X}_n = \frac{1}{n}\sum_{i=1}^{n} X_i$ is an unbiased estimator of μ. If the variance σ^2 exists, then $Var(\bar{X}_n) = \sigma^2/n$, so $\bar{X}_n$ is a consistent estimator of μ.

More generally, suppose that the unknown p-dimensional parameter θ can be expressed as $g(\mu_1, \ldots, \mu_p)$, where $\mu_j = E(X^j)$, $j \in \{1, \ldots, p\}$. Then, set $\hat{\theta}_n = g(\hat{\mu}_1, \ldots, \hat{\mu}_p)$, where $\hat{\mu}_{n,j}$ is an estimator of μ_j. For example, one could take

$$\hat{\mu}_{n,j} = \frac{1}{n}\sum_{i=1}^{n} x_i^j, \qquad j \in \{1, \ldots, p\},$$

since these estimators are consistent if $\mu_{2j} < \infty$. Using the delta method (Proposition B.3.1), one may conclude that

$$\sqrt{n}(\hat{\theta}_n - \theta) \rightsquigarrow \nabla g(\mu_1, \ldots, \mu_p) Z,$$

where $Z \sim N_p(0, V)$, with $V_{jk} = \mu_{j+k} - \mu_j \mu_k$, and $(\nabla g(\mu_1, \ldots, \mu_p))_{jk} = \frac{\partial g_j(\mu_1, \ldots, \mu_p)}{\partial \mu_j}$, $j, k \in \{1, \ldots, p\}$. If the random variables $X_1, \ldots, X_n$ are independent and identically distributed, with mean μ and variance σ^2, and if the 4th moment exists, then using the central limit theorem (Theorem B.4.1), one has

$$\begin{pmatrix} \sqrt{n}(\bar{x}_n - \mu) \\ \sqrt{n}(s_n^2 - \sigma^2) \end{pmatrix} \rightsquigarrow N_2(0, \Sigma),$$

where

$$\Sigma = \begin{pmatrix} \sigma^2 & \gamma_1 \sigma^3 \\ \gamma_1 \sigma^3 & \sigma^4(\gamma_2 - 1) \end{pmatrix},$$

and γ_1 and γ_2 are respectively the skewness and the kurtosis coefficients defined in Proposition A.5.1. Note that $\gamma_2 = 3$ in the Gaussian case. In addition, one also has $\begin{pmatrix} \sqrt{n}(\bar{x}_n - \mu) \\ \sqrt{n}(s_n - \sigma) \end{pmatrix} \rightsquigarrow N_2(0, \tilde{\Sigma})$, where

$$\tilde{\Sigma} = \sigma^2 \begin{pmatrix} 1 & \gamma_1/2 \\ \gamma_1/2 & (\gamma_2 - 1)/4 \end{pmatrix}.$$

Example B.8.1 *Suppose that $X_1, \ldots, X_n$ are independent observations of $X \sim Gamma(\alpha, \beta)$, $\alpha, \beta > 0$. It follows that $\bar{x}$ and s^2 are unbiased estimators of the mean and variance of X respectively. Since $E(X) = \alpha\beta$ and $Var(X) = \alpha\beta^2$, one can use the method of moments to obtain $\hat{\beta} = s^2/\bar{x}$ and $\hat{\alpha} = (\bar{x}/s)^2$. It then follows that $\nabla g = \frac{1}{\alpha\beta 2}\begin{pmatrix} 2\alpha\beta & -\alpha \\ -\beta^2 & \beta \end{pmatrix}$. In addition, since $E(X^3) = \alpha(\alpha+1)(\alpha+2)\beta^3$ and $E(X^4) = \alpha(\alpha+1)(\alpha+2)(\alpha+3)\beta^4$, then*

$\gamma_1 = \frac{2}{\sqrt{\alpha}}$ and $\gamma_2 = 3 + \frac{6}{\alpha}$. Hence, $\Sigma = \alpha\beta^2 \begin{pmatrix} 1 & 2\beta \\ 2\beta & 2(\alpha+3)\beta^2 \end{pmatrix}$. As a result, one may conclude that $\sqrt{n}((\hat{\alpha} - \alpha), (\hat{\beta} - \beta))^\top \rightsquigarrow N_2(0, A)$, where

$$A = \nabla g \Sigma (\nabla g)^\top = \begin{pmatrix} 2\alpha(\alpha+1) & -2\beta(\alpha+1) \\ -2\beta(\alpha+1) & (2+3/\alpha)\beta^2 \end{pmatrix}.$$

Example B.8.2 (Student Innovations) *Consider the following model:*

$$X_i = \mu + \sigma\, \varepsilon_i, \quad i \in \{1, \ldots, n\},$$

where the ε_i, $i \in \{1, \ldots, n\}$, are independent and identically distributed, with $\varepsilon_i \sim T(\nu)$. Since the density depends on the gamma function, there will be no explicit solution using the maximum likelihood principle. However, assuming $\nu > 4$ and using the method of moments, one has $E(X_i) = \mu$, $\mu_2 = \mathrm{Var}(X_i) = \sigma^2 \frac{\nu}{\nu-2}$, $\gamma_1 = 0$, and $\gamma_2 = \frac{3\nu-6}{\nu-4}$, where the skewness and kurtosis coefficients γ_1 and γ_2 are defined by (A.9) and (A.10). Hence $\nu = \frac{4\gamma_2-6}{\gamma_2-3}$ and $\sigma^2 = \mu_2 \frac{\gamma_2}{2\gamma_2-3}$. As a result, $\hat{\mu}_n = \bar{x}_n$, $\hat{\mu}_2 = s_n^2$, $\hat{\nu}_n = \frac{4\hat{\gamma}_{n,2}-6}{\hat{\gamma}_{n,2}-3}$ and $\hat{\sigma}_n^2 = s_n^2 \frac{\hat{\gamma}_{n,2}}{2\hat{\gamma}_{n,2}-3}$. It then follows from Proposition B.4.1 and the delta method (Theorem B.3.1) that if $\nu > 8$, $\sqrt{n}\,(\hat{\nu}_n - \nu) \rightsquigarrow N\left(0, \sigma_\nu^2\right)$, where

$$\sigma_\nu^2 = \frac{36}{(\gamma_2 - 3)^4} \left\{\mu_8 - 4\mu_6\gamma_2 + \gamma_2^2(4\gamma_2 - 1)\right\},$$

where μ_6 and μ_8 are the standardized 6-th and 8-th moments of the Student distribution. According to formula (A.12), it follows that $\mu_6 = \frac{15(\nu-2)^2}{(\nu-4)(\nu-6)}$ and $\mu_8 = \frac{105(\nu-2)^3}{(\nu-4)(\nu-6)(\nu-8)}$. Hence,

$$\sigma_\nu^2 = \frac{2}{3} \frac{(\nu-4)(\nu-2)^2(\nu^3 - 4\nu^2 + 23\nu - 20)}{(\nu-8)(\nu-6)}. \tag{B.11}$$

B.9 Combining the Maximum Likelihood Method and the Method of Moments

It might happen that one wants to estimate some parameters with a method of moments, while using the maximum likelihood method to estimate the rest of the parameters. So suppose that $\theta = (\theta_1, \theta_2)^\top$, where $\theta_2 = g(\alpha)$, with $E\{h(X_i)|X_{i-1}, \ldots, X_1\} = \alpha$, for some square integrable vector function h. It follows that one can estimate α by $\hat{\alpha}_n = \frac{1}{n}\sum_{i=1}^n h(X_i)$ and $\hat{\theta}_{n,2} = g(\hat{\alpha}_n)$. Using the central limit theorem (Theorem B.6.1) together with the delta method (Theorem B.3.1), we obtain that $\sqrt{n}\,(\hat{\alpha}_n - \alpha) \rightsquigarrow N(0, \Sigma)$, with $\Sigma = \text{Cov}\{h(X_i)\}$, which can be estimated by

$$\hat{\Sigma} = \frac{1}{n-1} \sum_{i=1}^{n} \{h(X_i) - \hat{\alpha}_n\} \{h(X_i) - \hat{\alpha}_n\}^\top, \tag{B.12}$$

while $\sqrt{n}\left(\hat{\theta}_{n,2} - \theta_2\right) = \dot{g}\left(\hat{\alpha}_n\right) \frac{1}{\sqrt{n}} \sum_{i=1}^{n} h(X_i) + o_P(1)$. Here $\dot{g}$ is the Jacobian matrix of the transformation $\alpha \mapsto \theta_2 = g(\alpha)$. The estimation of θ_1 is then defined by $\hat{\theta}_{n,1} = \arg\min_{\theta_1} L(\theta_1)$, where

$$L(\theta_1) = -\sum_{i=1}^{n} \ln\left\{f_i\left(X_i, \ldots, X_1, \theta_1, \hat{\theta}_{n,2}\right)\right\}. \tag{B.13}$$

Here we use the same notations as in Section B.7. Using the central limit theorem (Theorem B.6.1) and the previous convergence results, we obtain the following proposition.

Proposition B.9.1 *Under weak assumptions,* $\sqrt{n}(\hat{\theta}_n - \theta) \rightsquigarrow N_p(0, V)$, *where* $V = \begin{pmatrix} V_{11} & V_{12} \\ V_{12}^\top & V_{22} \end{pmatrix}$, *with* V_{22} *estimated by* $\hat{V}_{22} = \dot{g}\left(\hat{\alpha}_n\right) \hat{\Sigma} \dot{g}^\top\left(\hat{\alpha}_n\right)$, $\hat{V}_{11} = \hat{\mathcal{I}}^{-1}$, *where* $\hat{\mathcal{I}}$ *is the estimation of the Fisher information (as if* θ_2 *were known), i.e., it can be estimated by*

$$\hat{\mathcal{I}} = \frac{1}{n-1} \sum_{i=1}^{n} \ell_i\left(\hat{\theta}_n\right) \ell_i^\top\left(\hat{\theta}_n\right),$$

with $\ell_i(\theta) = \ell_i(\theta_1, \theta_2) = -\nabla_{\theta_1} f_i\left(X_i, \ldots, X_1, \theta_1, \theta_2\right)$, *or by* $\hat{\mathcal{I}} = H_n/n$, *where* H_n *is the Hessian of the log-likelihood function* (B.13). *Finally,* V_{12} *can be estimated by*

$$\hat{V}_{12} = \hat{V}_{11} \left[\frac{1}{n-1} \sum_{i=1}^{n} \ell_i\left(\hat{\theta}_n\right) \{h(X_i) - \hat{\alpha}_n\}^\top\right] \dot{g}^\top\left(\hat{\alpha}_n\right).$$

B.10 M-estimators

An estimation method which increases in popularity is the method of M-estimators. Given a function $\psi(x, \theta)$ and a sample $X_1, \ldots, X_n$, one sets

$$\hat{\theta}_n = \arg\min_{\tilde{\theta} \in \Omega} \sum_{i=1}^{n} \psi(X_i, \tilde{\theta}).$$

For example, the choice $\psi(x, \theta) = |x - \theta|$ leads to the median, while choosing $\psi(x, \theta) = (x - \theta)^2$ leads to the mean. One can also choose functions which are asymmetric with respect to the origin.

B.11 Suggested Reading

For a general reference on estimation, see Serfling [1980]. See also Berndt et al. [1974] for estimation of parameters in dependent sequences.

B.12 Exercises

Exercise B.1

The MATLAB file *DataCauchy* contains observations from a Cauchy distribution (see Appendix A.6.16) with density

$$f(x) = \frac{1}{\sigma\pi(1 + (x - \mu)^2/\sigma^2)}.$$

Assume that the data are independent. We want to estimate μ and $\sigma > 0$. The log-likelihood function to maximize is

$$L(\mu, \sigma) = -n \ln(\sigma) - \sum_{i=1}^{n} \ln\left(1 + (x_i - \mu)^2)/\sigma^2\right).$$

(a) Using the MATLAB function *EstCauchy*, find $\hat{\mu}$ and $\hat{\sigma}$. Note that the observations have been simulated with $\mu = 10$ and $\sigma = 5$!

(b) Extend the function *EstCauchy* so that the covariance matrix between the estimators is an output. Use the exact gradients to estimate the Fisher information.

Bibliography

E. K. Berndt, B. H. Hall, R. E. Hall, and J. A. Hausman. Estimation and inference in nonlinear structural models. *Annals of Economics and Social Measurement*, pages 653–665, 1974.

R. Durrett. *Probability: Theory and Examples.* Duxbury Press, Belmont, CA, second edition, 1996.

R. J. Serfling. *Approximation Theorems of Mathematical Statistics.* John Wiley & Sons Inc., New York, 1980. Wiley Series in Probability and Mathematical Statistics.

Index

absolutely continuous distribution, 407
expected value, 410
moment generating function, 410
variance, 410
adjusted closing price, 2
affine model, 171
algorithm
auxiliary particle filter, 362
Baum-Welsh, 383
bootstrap, 136
EM, 383, 398
goodness-of-fit test, 96
IMM filter, 354
Kalman filter, 347, 349
P-value
change point test, 335
goodness-of-fit test, 95, 317, 318
test of independence, 282
test of serial independence, 94
particle filter, 360, 362
Ali-Mikhail-Haq copula, 308
almost sure convergence, 438
American call option, 10
approximation
saddlepoint, 122
Archimedean copula, 301
ARMA model, 375
ASIR
see auxiliary particle filter, 361
asymmetric double exponential distribution, 187
auxiliary particle filter, 361
algorithm, 362
Baum-Welsh algorithm, 383
Bernoulli distribution, 415
Bessel function
modified
first kind, 422
second kind, 195, 217
beta
distribution, 423
function, 423
bias, 441
binomial distribution, 416
binomial tree, 98, 439
Black-Scholes formula, 9
dividends, 13
extended, 70
Black-Scholes model, 3
estimation, 6
precision, 7
put-call parity, 10
bootstrap, 136
bootstrap algorithm, 136
Box-Müller transform, 432
Brownian bridge, 134
Brownian motion, 2
geometric, 3

call option, 8, 9
Cauchy distribution, 424
central limit theorem
dependent observations, 448
independent observations, 441
change point test, 334
copula, 335
change-of-variable theorem, 432
characteristic function, 412
exponential distribution, 419
Gaussian distribution, 418

characteristics
  Lévy process, 189
  natural, 192
chi-square distribution, 421
Cholesky decomposition, 34
CIR model, 160
Clayton copula, 304
coherent measure of risk, 105
coherent measure of risk
  Axioms, 105
complete market, 183
compound Poisson process, 185
concave order, 108
concordance, 286
confidence interval, 437
confidence region, 437
consistent estimator, 441
continuous distribution, 407
continuous time Markov chain, 331
continuous time martingale, 183
convergence
  in law, 439
  almost sure, 438
  in distribution, 439
  in probability, 438
  in quadratic mean, 438
copula, 271
  Ali-Mikhail-Haq, 308
  Archimedean, 301
  Clayton, 304
  definition, 271
  elliptical, 298
  Farlie-Gumbel-Morgenstern, 309
  Frank, 306
  Gaussian, 298
  goodness-of-fit, 316
  Gumbel, 305
  independence, 297
  multivariate, 293
  Plackett, 310
  simulation, 274, 294
  Student, 299
  test of independence, 315
Cornish-Fisher expansion, 121, 122
correlated Brownian motions, 33
correlated geometric Brownian motions
  simulation, 34
Cox-Ingersoll-Ross model, 160
Cramér-von Mises test
  goodness-of-fit, 67, 95
  independence, 64, 93
    copula, 332
cumulant, 414
  chi-square distribution, 421
  gamma distribution, 420
  generating function, 414
  Lévy process, 189
  Merton model, 189
cumulant generating function
  gamma process, 193
  jump-diffusion process, 185
  Lévy subordinator, 196
  NIG process, 196
  VG model, 194

Davies approximation, 125
de Moivre identity, 412
delta
  Black-Scholes model, 14
    multivariate, 47
delta hedging, 73
delta method, 440
delta-gamma approximation, 117
  normal, 117
density, 407
  discrete, 408
  estimation, 96
dependogram, 64, 94, 316, 333
diffusion process, 70
discrete density, 408
discrete distribution, 407
  moment, 408
  moment generating function, 409
  variance, 409
distribution
  absolutely continuous, 407
  asymmetric double exponential, 187

Bernoulli, 415
beta, 423
binomial, 416
Cauchy, 424
chi-square, 421
continuous, 407
discrete, 407
double exponential, 188
elliptical, 426
exponential, 419
gamma, 420
Gaussian, 418
GED, 233, 424
geometric, 417
inverse gamma, 245
Johnson-SU, 423
Laplace, 188
Linnik , 188
log-normal, 418
multivariate Gaussian, 425
multivariate Student, 299, 426
negative binomial, 417, 434
non-central chi-square, 161, 421
Poisson, 416
positive stable, 306
Student, 422
uniform, 417
weighted-symmetric, 187
weighted-symmetric Gaussian, 187
distribution function, 407
estimation, 109
double exponential distribution, 188
Duan criterion, 235
Duan methodology, 154

early exercise, 10
Edgeworth expansion, 121
elliptical copula, 298
elliptical distribution, 426
simulation, 429
EM algorithm, 383, 398
empirical distribution function, 109
Epanechnikov kernel, 97
Esscher transform, 197
estimation
bias, 441
combined, 452
consistency, 441
copula
nonparametric approach, 314
parametric approach, 312
semi-parametric approach, 312
density, 96
distribution function, 109
expected shortfall, 114
Fisher information, 446
maximum likelihood, 435
method of moments, 450
precision, 437
dependent observations, 448
independent observations, 446
quasi-maximum likelihood, 353
VaR, 111
European
call option, 8, 9
exchange option, 43
put option, 10
quanto option, 45, 46
exchange option, 43
expansion
Cornish-Fisher, 121
Edgeworth, 121
Gram-Charlier, 243
expected shortfall, 104
expected value, 409, 410
expectile, 128
exponential distribution, 419

Farlie-Gumbel-Morgenstern copula, 309
Feller process, 160
Feynman-Kac formula, 240
filter
IMM, 354
Kalman, 346
particle, 360, 361
filtering problem
general, 356

linear, 345
filtration, 28
natural, 28
finite variation, 191
Fisher information, 446, 449
formula
Black-Scholes , 9
Feynman-Kac, 240
Gil-Pelaez, 139
Kallianpur-Striebel, 357
Lévy-Khinchin, 189
Stirling, 123
Fourier transform, 371
Fréchet-Hoeffding bounds, 276
Frank copula, 306

gamma
Black-Scholes model, 15
multivariate, 47
distribution, 420
function, 420
properties, 420
process, 192
GARCH
goodness-of-fit, 230
likelihood ratio test, 230
option pricing, 235
parameter estimation, 228
GARCH model, 223
augmented, 227
EGARCH, 226
GARCH(1,1), 224
GARCH(p,q), 226
GARCH-M, 235
GJR-GARCH, 227
NGARCH, 227
GARCH type model, 88
Gaussian
distribution, 418
copula, 298
distribution
multivariate, 425
kernel, 97
GED, 233, 424
distribution function, 424
simulation, 424
generalized error distribution
see GED, 424
Generalized Inverse Gaussian, 194
generating function
cumulant, 414
moment, 413
geometric Brownian motion, 3
simulation, 4
geometric distribution, 417
geometric random walk, 77
GIG process, 194
Gil-Pelaez formula, 139
goodness-of-fit
copula, 316
Gaussian distribution, 66, 94
test, 94
Gram-Charlier expansion, 243, 245
greeks, 14
Black-Scholes model
multivariate, 47
delta, 14, 47
estimation
likelihood ratio method, 23, 49
pathwise method, 21, 48
gamma, 15, 47
rho, 14, 47
theta, 14, 47
vega, 15, 47
Gumbel copula, 305

hazard rate order, 108
hedge fund replication, 389
Kalman filter, 391
regression, 391
strong, 257, 389
weak, 257
hedging, 72
delta, 73
optimal, 74
Hessian matrix, 447
Heston model, 246
hidden Markov model, 84
Hoeffding identity, 336

IG process, 193

IMM algorithm, 354
IMM filter, 354
implied
    distribution, 16
    spot rate, 158
    volatility, 19
independence
    dependogram, 333
    test, 64, 332
independence copula, 271
infinitely divisible, 190
innovations, 223
inverse distribution function, 418
inverse gamma distribution, 245
Inverse Gaussian process, 193

Jacobian
    determinant, 431
    matrix, 8, 440
Jarque-Berra test, 66
    revisited, 66
Johnson-SU distribution, 423
jump-diffusion process, 185
    cumulant generating function, 185
    Kou model, 187
    Merton model, 186

Kallianpur-Striebel formula, 357
Kalman filter, 346
    algorithm, 347, 349
Kendall function, 280
    estimation, 281
    multidimensional case, 294
Kendall tau, 286
    estimation, 286
    precision, 287
Khmaladze transform, 250
Kolmogorov equation, 360
Kou model, 187
kurtosis, 415
    precision, 442

Lévy measure, 189
Lévy process, 188
    characteristics, 189
Lévy subordinator, 196
Lévy-Khinchin formula, 189
Laplace distribution, 188
Laplace transform, 414
Legendre transform, 123
Libor, 9
likelihood ratio method, 23, 49
likelihood ratio test, 230
Lilliefors test, 67, 96
Linnik distribution, 188
Lipschitz property, 272
Ljung-Box test, 63, 230
log-normal distribution, 418
LRNVR citerion
    see Duan criterion, 235

market price of risk, 148
Markov chain, 260
    absorbing state, 331
    continuous time, 264, 331
Markov kernel, 356
martingale, 29, 183
martingale measure, 183
martingale problem, 148
maximum likelihood estimator, 435
mean excess function, 135
mean square convergence, 438
Merton model, 186
method of moments, 450
modified Bessel function
    first kind, 422
    second kind, 195, 217
moment, 408
    generating function, 409, 410, 413
        binomial distribution, 416
        chi-square distribution, 421
        exponential distribution, 419
        gamma distribution, 420
        Gaussian distribution, 418
        geometric distribution, 417
        negative binomial distribution, 417
        Poisson distribution, 416
multiplication formula, 429

multivariate
  Black-Scholes model
    definition, 33
    estimation, 36
  Gaussian distribution, 425
  Student distribution, 299, 426

natural characteristics, 192
natural filtration, 28
negative binomial distribution, 417, 434
NIG process, 196
non-central chi-square
  distribution, 161, 421
  simulation, 421
Normal Inverse Gaussian
  see NIG, 196

Omega ratio, 128
optimal quadratic hedging, 74
option
  call, 8, 9
  exchange, 43
  put, 10
  quanto, 44
order
  hazard rate, 108
  PQD, 276
  stochastic, 107
  supermodular, 296
order statistics, 109, 438
Ornstein-Uhlenbeck process, 149

parametric bootstrap, 95, 317, 318
particle filter, 360
  auxiliary, 361
  estimation, 365
  implementation, 360
pathwise method, 21, 48
Pearson type II, 300, 428
Pearson type VII, 300, 428
performance measures
  axioms, 126
Plackett copula, 310
Poisson distribution, 416
Poisson process, 184
  simulation, 185
positive quadrant dependence order, 276
positive stable law, 306
PQD order, 276
  multidimensional, 295
prediction interval, 437
probability integral transform
  see Kendall function, 280
put option, 10
put-call parity
  Black-Scholes model, 10
  CIR model, 166

quantile, 135
quanto option, 44
quasi-maximum likelihood, 353

Radon-Nikodym density, 197
random variable, 407
random vector, 407
ratio
  Omega, 128
  Sharpe, 127
  Sortino, 127
regime-switching
  geometric random walk , 84
  Markov model, 380
    continuous time, 388
rho
  Black-Scholes model, 14
    multivariate, 47
Ricatti differential equation, 171
risk-free rate, 9
risk-neutral measure
  see equivalent martingale measure, 11
Rosenblatt transform, 430
  copula, 294
  regime-switching model, 401

saddlepoint approximation, 122
Schwartz model, 345
serial independence, 292
  dependogram, 94
  test, 64, 93

Sharpe ratio, 127
simple order, 107
simulation
  Ali-Kikhail-Haq copula, 308
  Archimedean copula, 304
  bivariate copula, 274
  Clayton copula, 305
  correlated geometric Brownian motions, 34
  elliptical distribution, 429
  Farlie-Gumbel-Morgenstern copula, 310
  Frank copula, 307
  Gaussian copula, 299
  geometric Brownian motion, 4
  Gumbel copula, 306
  multivariate copula, 294
  non-central chi-square, 421
  Poisson process, 185
  positive stable law, 306
  Student copula, 300
skewness, 415
  precision, 442
Sklar theorem
  bivariate, 272
  multivariate, 293
Slutzky theorem
  see delta method, 440
Sortino ratio, 127
Spearman rho, 287, 288
  estimation, 288
  precision, 288
standard Gaussian distribution, 418
Stirling formula, 123
stochastic integral, 175
stochastic order
  concave, 108
  hazard rate, 108
  simple, 107
stochastic volatility
  Heston model, 246
  Hull-White model, 239
Student copula, 299
Student distribution, 422
  multivariate, 299, 426
subordinator, 191, 196
supermodular order, 296

tail conditional expectation, 104
TailVar, 104
tent map copula family, 271
test
  goodness-of-fit, 66
    Cramér-von Mises, 67
    Gaussian distribution, 94
    Lilliefors, 96
  independence
    copula, 315, 332
  Jarque-Berra, 66
    revisited, 66
  Lilliefors, 67
  Ljung-Box, 63
  serial independence, 64, 93
theta
  Black-Scholes model, 14
    multivariate, 47
trading days, 3
transform
  Box-Müller, 432
  Esscher, 197
  Fourier, 371
  Khmaladze, 250
  Legendre, 123
  Rosenblatt, 430

uniform distribution, 417

Value at Risk
  see VaR, 104
van der Waerden rho, 289
  estimation, 290
  precision, 290
VaR, 104
  estimation, 111
variance, 409, 410
Variance Gamma process
  see VG process, 81
Vasicek model, 148
vega
  Black-Scholes model, 15
    multivariate, 47

VG process, 81
 cumulant, 194
volatility, 4
 implied, 19
volatility vector, 35

weighted-symmetric distribution, 187
weighted-symmetric Gaussian
 distribution, 187
worst conditional expectation, 105

Zakai equation, 357
zero-coupon bond
 CIR model, 163
 Vasicek model, 152